Smart Innovation, Systems and Technologies

Volume 34

About this Series

The Smart Innovation, Systems and Technologies book series encompasses the topics of knowledge, intelligence, innovation and sustainability. The aim of the series is to make available a platform for the publication of books on all aspects of single and multi-disciplinary research on these themes in order to make the latest results available in a readily-accessible form. Volumes on interdisciplinary research combining two or more of these areas is particularly sought.

The series covers systems and paradigms that employ knowledge and intelligence in a broad sense. Its scope is systems having embedded knowledge and intelligence, which may be applied to the solution of world problems in industry, the environment and the community. It also focusses on the knowledge-transfer methodologies and innovation strategies employed to make this happen effectively. The combination of intelligent systems tools and a broad range of applications introduces a need for a synergy of disciplines from science, technology, business and the humanities. The series will include conference proceedings, edited collections, monographs, handbooks, reference books, and other relevant types of book in areas of science and technology where smart systems and technologies can offer innovative solutions.

High quality content is an essential feature for all book proposals accepted for the series. It is expected that editors of all accepted volumes will ensure that contributions are subjected to an appropriate level of reviewing process and adhere to KES quality principles.

More information about this series at http://www.springer.com/series/8767

Amaresh Chakrabarti
Editor

ICoRD'15 – Research into Design Across Boundaries Volume 1

Theory, Research Methodology, Aesthetics, Human Factors and Education

Springer

Editor
Amaresh Chakrabarti
Centre for Product Design
and Manufacturing
Indian Institute of Science
Bangalore
India

ISSN 2190-3018 ISSN 2190-3026 (electronic)
Smart Innovation, Systems and Technologies
ISBN 978-81-322-2231-6 ISBN 978-81-322-2232-3 (eBook)
DOI 10.1007/978-81-322-2232-3

Library of Congress Control Number: 2014957705

Springer New Delhi Heidelberg New York Dordrecht London

Printed on acid-free paper

Springer (India) Pvt. Ltd. is part of Springer Science+Business Media (www.springer.com)

Preface

Design is ubiquitous; it pervades all spheres of life, and has been around ever since life has been engaged in purposefully changing the world around it. While some designs have transcended time, designs are always in the process of being evolved. Research into design and the emergence of a research community in this area have been relatively new, its development influenced by the multiple facets of design (human, artefact, process, organisation, ecology, micro- and macro-economy by which design is shaped and which it shapes in turn) and the associated diversification of the community into those focusing on various aspects of these facets, in various applications. Design is complex, balancing the needs of multiple stakeholders, and requiring a multitude of areas of knowledge to be utilised, with resources spread across space and time.

The collection of papers in these two book volumes constitutes the Proceedings of the Fifth International Conference on Research into Design (ICoRD'15) held at the Indian Institute of Science, Bangalore, India during 7–9 January 2015. ICoRD'15 is the fifth in a series of biennial conferences held in India to bring together the international community from diverse areas of design practice, teaching and research. The goals are to share cutting edge research about design among its stakeholders; aid the ongoing process of developing a collective vision through emerging research challenges and questions; and provide a platform for interaction, collaboration and development of the community in order for it to address the global and local challenges by forming and realising the collective vision. The conference is intended for all stakeholders of design, and in particular for its practitioners, researchers, teachers and students.

Of the 265 abstracts submitted to ICoRD'15, 196 were selected for full paper submission. 151 full papers were submitted, which were reviewed by experts from the ICoRD'15 International Programme Committee comprising 180 members from over 131 institutions or organisations from 34 countries spanning six continents. Finally, 118 full papers, authored by over 275 (275 unique authors, actually 335 author entries in 118 papers) researchers from 86 institutions and organisations from 24 countries spanning six continents, were selected for presentation at the conference and for publication as chapters in this book. ICoRD has steadily grown over the last four editions, from a humble beginning in 2006 with 30 papers and 60 participants, through 75 papers and 100 participants in ICoRD'09, 100 papers and 150 participants in ICoRD'11, to 114 papers and 170 participants in ICoRD'13.

ICoRD'15 had 14 sessions with 70 podium papers, and 48 papers with brief podium presentations followed by poster display and discussion. It had keynotes from prominent researchers and practitioners from around the world such as: Vincent Floderer from CRIMP, France, Kristin Wood from Singapore University of Technology and Design, Singapore, John Gero from George Mason University, USA, Richard Gardner from The Boeing Company, USA, Sudhakar Nadkarni from Welingkar Institute of Management, India, and Rishikesha Krishnan from Indian Institute of Management Indore, India. It had two panel discussions on "Publishing Research Papers" and "Practice of Design", and five workshops, on Design Cognition, Sustainability, Emotional Engineering, Paper Crumpling, and Design Innovation Centre (DIC) Hubs in India. From 2015, ICoRD started giving **ICON^3** awards (acronym for **I**CoRD **O**utstanding **C**ontribution to desig**N** scie**N**ce and educatio**N**) to outstanding contributors to design education and research. Professor Sudhakar Nadkarni and Prof. John Gero were selected as **ICON^3** awardees for 2015, respectively for their outstanding contributions to design education and design research.

The chapters in this book together cover all three major areas of products and processes: functionality, form and human factors. The spectrum of topics range from those focusing on early stages such as creativity and synthesis, through those that are primarily considered in specific stages of the product life cycle, such as safety, reliability or manufacturability, to those that are relevant across the whole product life cycle, such as collaboration, communication, design management, knowledge management, cost, environment and product life cycle management. Issues of delivery of research into design, in terms of its two major arms: design education and practice, are both highlighted in the chapters in this book. Foundational topics such as the nature of design theory and research methodology are also major areas of focus. It is particularly encouraging to see in the chapters the variety of areas of application of research into design—aerospace, healthcare, automotive and white goods are but a few of the sectors explored. The theme of this year's conference and of this book is "Design Across Boundaries", where boundaries are manifold and span many dimensions—economy, culture, age, gender, religion, caste, class, education, family, digitalisation, geography (rural/urban) and so on.

The book has two volumes. Volume I focuses on Design Theory, Research Methodology, Aesthetics, Human Factors and Education. Volume II focuses on Design Creativity, Sustainability, Design for X, Enabling Technologies, Design Management and Applications in Practice. Volume I broadly focuses on form and human factors, while Volume II focuses on functionality.

On behalf of the Steering Committee, Advisory Committee, Local Organising Committee and Co-Chairs, I thank all the authors, delegates, institutions and organisations that participated in the conference and the International Programme Committee for their support in organising ICoRD'15 and putting this book together. I am thankful to the Design Society and Design Research Society for their kind endorsement of ICoRD'15. I thank Indian Institute of Science (IISc), Bangalore and its Centre for Product Design and Manufacturing, for their support of this event. I also wish to place on record and acknowledge the enormous support provided by Mr. Ranjan B.S.C., Ms. Kumari M.C. and Ms. Nishath Salma of IISc in managing the review process, and in preparation of the conference programme and this book, and the large and dedicated group of student-volunteers of Indian Institute of Science, Bangalore in the organisation of the conference. Finally, I thank Springer, especially its Editor Ms. Swati Meherishi, for the wonderful support extended in the publication of this book and for sponsoring books and book coupons for Distinguished Paper Awards and **ICON^3** Awards respectively.

Amaresh Chakrabarti

Conference Organisation

Steering Committee

B. Gurumoorthy	Indian Institute of Science, India
John Gero	George Mason University, USA
Kota Harinarayana	National Aeronautical Laboratories, India (Chair)
Uday Athavankar	Indian Institute of Technology Bombay, India
Udo Lindemann	Technical University of Munich, Germany

Advisory Committee

Ananda Kumar	LPerspective, India
Anjan Das	Confederation of Indian Industry, India
Aparajita Ojha	IIIT D&M Jabalpur, India
Arabinda Mitra	International Division, Department of Science and Technology, India
V.S. Arunachalam	Center for Study of Science, Technology and Policy, India
Arun Jaura	Traktion, India
Ashok Jhunjhunwala	Indian Institute of Technology Madras, Chennai, India
Bala Bharadvaj	Boeing Research and Technology Center, India
R. Gnanamoorthy	IIIT D&M, Kancheepuram, India
Imre Horvath	Delft University of Technology, The Netherlands
Larry Leifer	Stanford University, USA
P.J. Mohanram	Indian Machine Tool Manufacturers Association, India
T.S. Mruthyunjaya (Chairman Emeritus, CPDM)	Indian Institute of Science, India

K. Radhakrishnan	Indian Space Research Organisation, India
Raman Saxena	USID Foundation, India
S. Sainath	LPerspective, India
Sam Pitroda	National Innovation Council, India
Denis Bertin	Airbus India
P.S. Subrahmanyam	Aeronautical Development Agency, India
Lt. Gen. V.J. Sundaram	National Design and Research Forum, India
Udayant Malhoutra	CII National Committee on Design, India

Programme Chair

Amaresh Chakrabarti Indian Institute of Science, India

Co-chairs

Lucienne Blessing	University of Luxembourg, Luxembourg
Gaur Ray	Indian Institute of Technology Bombay, India
Raghu Prakash	Indian Institute of Technology Madras, Chennai, India
Steve Culley	University of Bath, UK
Tim McAloone	Technical University of Denmark, Denmark
Toshiharu Taura	Kobe University, Japan
Umberto Cugini	Politecnico di Milano, Italy

Programme Committee

A.V. Gokula Vijaykumar	University of Strathclyde, UK
Ahmed Kovacevic	City University London, UK
Ajith Kumar	T. A. Pai Management Institute, India
Akin Kazakci	Ecole de Mines, France
Alex Duffy	University of Strathclyde, UK
Alison Mckay	University of Leeds, UK
Amitabha Mukerjee	Indian Institute of Technology Kanpur, India
K. Ananthasuresh	Indian Institute of Science, India
Anindya Deb	Indian Institute of Science, India
Anja Maier	Technical University of Denmark, Denmark
Anshuman Tripathy	Indian Institute of Management Bangalore, India
Ashitava Ghosal	Indian Institute of Science, India
Ashok Goel	Georgia Institute of Technology, USA
Ashok Iyer	Manipal University Dubai Campus, UAE

Ashutosh Tiwari	Cranfield University, UK
Axel Thallemer	Universität für künstlerische und industrielle Gestaltung, Austria
Aylmer Johnson	University of Cambridge, UK
Aziz Bouras	Qatar University, Qatar
Ben Hicks	University of Bath, UK
Benoit Eynard	Universite de Technologie de Compiegne, France
Bernard Yannou	École Centrale Paris, France
Biren Prasad	CERA Institute, USA
Bishakh Bhattacharya	Indian Institute of Technology Kanpur, India
C. Amarnath	Indian Institute of Technology Bombay, India
Cees de Bont	Hong Kong Polytechnic University, Hong Kong
Chalapathi Rao Nori	Indian Institute of Science, India
Christian Redlinghuys	University of Cape Town, South Africa
Christian Weber	Technische Universität Ilmenau, Germany
Christopher Magee	Massachusetts Institute of Technology, USA
Claudia Eckart	Open University, UK
Clement Fortin	École Polytechnique de Montreal, Canada
Craig Vogel	University of Cincinnati, USA
Dan Braha	University of Massachusetts, USA
Dan McAdams	Texas A&M University, USA
Darlie O. Koshy	Apparel Export Promotion Council, India
Dave Brown	Worcester Polytechnic Institute, USA
David Rosen	Georgia Institute of Technology, USA
Debkumar Chakrabarti	Indian Institute of Technology Guwahati, India
Denis Cavallucci	INSA Strasbourg, France
Dibakar Sen	Indian Institute of Science, India
Edwin Koh	National University of Singapore, Singapore
Ehud Kroll	Technion, Israel
Elena Mulet	Universitat Jaume I, Spain
Emmanuel Caillaud	Universite de Strasbourg, France
Eric Blanco	Institut Polytechnique de Grenoble, France
Eswaran Subrahmanian	Carnegie Mellon University, USA
Filippo Salustri	Ryerson University, Canada
Francesco Ferrise	Politecnico di Milano, Italy
Gabriela Goldschmidt	Technion, Israel
Gaetano Cascini	Politecnico di Milano, Italy
Gaur Ray	Indian Institute of Technology Bombay, India
Georges Fadel	Clemson University, USA
Glen Mullineux	University of Bath, UK
Guenther Seliger	Technical University of Berlin, Germany
Gul Kremer	Pennsylvania State University, USA
B. Gurumoorthy	Indian Institute of Science, India
Helen Petrie	University of York, UK
Henry Ming	Shanghai Jiao Tong University, China

Herbert Birkhofer	TU Darmstadt, Germany
I.S. Jawahir	University of Kentucky, USA
Indira Thouvenin	Universite de Technologie Compiegne, France
J.E. Diwakar	Indian Institute of Science, India
Jami Shah	Arizona State University, USA
Janet Allen	University of Oklahoma, USA
Jaywant Arakeri	Indian Institute of Science, India
Joaquim Macia	Polytechnic University of Catalunya, Spain
Johan Malmqvist	Chalmers University of Technology, Sweden
Jonathan Borg	University of Malta, Malta
Joost Duflou	Katholieke Universiteit Leuven, Belgium
Jozef Duhovnik	University of Ljubljana, Slovenia
Jozsef Vancza	MTA SZTAKI, Hungary
K. Sudhakar	Indian Institute of Technology Bombay, India
K. Munshi	Indian Institute of Technology Bombay, India
Kalyanmoy Deb	Michigan State University, USA
Kees Dorst	University of Technology Sydney, Australia
Kemper Lewis	University of Buffalo, The State University of New York, USA
Kikuo Fujita	Osaka University, Japan
Kristina Shea	Technische Universität München, Germany
Lauri Koskela	University of Salford, UK
Li Shu	University of Toronto, Canada
LIN Rungati	National Taiwan University of Arts, Taiwan
Linda Schmidt	University of Maryland, USA
L.S. Ganesh	Indian Institute of Technology Madras, India
Lucienne Blessing	University of Luxembourg, Luxembourg
M. Manivannan	Indian Institute of Technology Madras, India
Maik Maurer	Technische Universität München, Germany
Mandeep Singh	School of Planning and Architecture New Delhi, India
Marco Aurisicchio	Imperial College London, UK
Marco Cantamessa	Politecnico di Torino, Italy
Maria Yang	Massachusetts Institute of Technology, USA
Mario Storga	University of Zagreb, Croatia
Mario Fargnoli	University of Rome, La Sapienza, Italy
Martin Grimhelden	Royal Institute of Technology, Sweden
Martin Steinart	Norwegian University of Science and Technology, Norway
Mary Matthew	Indian Institute of Science, India
Mary Thompson	Technical University of Denmark, Denmark
Ming xi Tang	Hong Kong Polytechnic University, Hong Kong
Mitchell Tseng	The Hong Kong University of Science and Technology, Hong Kong
Monica Bordegoni	Politecnico di Milano, Italy

Monto Mani	Indian Institute of Science, India
Moreno Muffatto	Universita di Padova, Italy
Nilesh Vasa	Indian Institute of Technology Madras, India
Ozgur Eris	Delft University of Technology, The Netherlands
P. Radhakrishnan	PSG Institute of Advanced Studies, India
P. Rodgers	Northumbria University, UK
P.V.M. Rao	Indian Institute of Technology Delhi, India
Panos Papalambros	University of Michigan, USA
Peter Torlind	Lulea University of Technology, Sweden
Peter Childs	Imperial College London, UK
Petra Badke-Schaub	Delft University of Technology, The Netherlands
Phillippe Girard	University of Bordeaux, France
Pieter Vermass	Delft University of Technology, Netherlands
Prabir Sarkar	Indian Institute of Technology Ropar, India
Pradeep Yammiyavar	Indian Institute of Technology Guwahati, India
Puneet Tandon	IIIT D&M Jabalpur, India
Rachuri Sudarsan	National Institute of Standards and Technology, USA
Raghu Echempati	Kettering University, USA
Raghu Prakash	Indian Institute of Technology Madras, India
Rajkumar Roy	Cranfield University, UK
Ralph Bruder	Technical University of Darmstadt, Germany
Ravi Punekar	Indian Institute of Technology Guwahati, India
Ravi Poovaiah	Indian Institute of Technology Bombay, India
Reiner Anderl	Technical University of Darmstadt, Germany
Ricardo Sosa	Singapore University of Technology and Design, Singapore
Richard Liu	Chang Gung University, Taiwan
Rikard Soderberg	Chalmers University of Technology, Sweden
Rina Maiti	Indian Institute of Science, India
Roman Zavbi	University of Ljubljana, Slovenia
Rosario Vidal	Universitat Jaume I, Spain
S. Krishnan	Center for Study of Science, Technology and Policy, India
S. Vinodh	National Institute of Technology Tiruchirappalli, India
Sangarappillai Sivaloganathan	United Arab Emirates University, UAE
Santosh Jagtap	Lund University, Sweden
Satish Kailas	Indian Institute of Science, India
Sean Hanna	University College London, UK
Serge Rohmer	Université de Technologie de Troyes, France
Shayne Gooch	University of Canterbury, New Zealand
Shilpa Ranade	Indian Institute of Technology Bombay, India
Simon Bolton	Birmingham Institute of Art and Design, UK
Somwrita Sarkar	University of Sydney, Australia

Srinivas Kota	Birla Institute of Technology and Science Pilani, India
Srinivasan Venkataraman	Technische Universität München, Germany
Stanislav Hosnedl	University of West Bohemia, Czech Republic
Stephen C.-Y. Lu	University of Southern California Los Angeles, USA
Steve Culley	University of Bath, UK
Steve Evans	University of Cambridge, UK
Steven MacGregor	University of Girona, Spain
Subir Saha	Indian Institute of Technology Delhi, India
Susy Verghese	Indian Institute of Technology Madras, India
Tamotsu Murakami	The University of Tokyo, Japan
Tim McAloone	Technical University of Denmark, Denmark
Tjamme Wiegers	Delft University of Technology, The Netherlands
Tom Vaneker	University of Twente, The Netherlands
Torben Lenau	Technical University of Denmark, Denmark
Tracy Bhamra	Loughborough University, UK
Umberto Cugini	Politecnico di Milano, Italy
Vesna Popovic	Queensland University of Technology, Australia
Vijay Srinivasan	National Institute of Standards and Technology, USA
Vinod Vidwans	FLAME School of Communication, India
Vishal Singh	Aalto University, Finland
William Ion	Strathclyde University, UK
Winifred Ijomah	University of Strathclyde, UK
Wolfgang Wimmer	Vienna University of Technology, Austria
Yasushi Umeda	University of Tokyo, Japan
Ying-Chieh Liu	Chang Gung University, Taiwan
Yong Chen	Jiao Tong University, China
Yong Se Kim	Sungkyunkwan University, Korea
Yong Zeng	Concordia University, Canada
Yoram Reich	Tel Aviv University, Israel
Yosef Oehmann	Technical University of Denmark, Denmark
Yoshiki Shimomura	Tokyo Metropolitan University, Japan
Yoshiyuki Matsuoka	Keio University, Japan
Yrjo Sotamaa	Aalto University, Finland
Yukari Nagai	Japan Advanced Institute of Science and Technology, Japan

Local Organising Committee

G.K. Ananthasuresh	Indian Institute of Science, India
Jaywant Arakeri	Indian Institute of Science, India

Anindya Deb	Indian Institute of Science, India
J.E. Diwakar	Indian Institute of Science, India
Ashitava Ghosal	Indian Institute of Science, India
Satish Vasu Kailas	Indian Institute of Science, India
Rina Maiti	Indian Institute of Science, India
Monto Mani	Indian Institute of Science, India
Mary Matthew	Indian Institute of Science, India
N.V.C. Rao	Indian Institute of Science, India
N.D. Shivakumar	Indian Institute of Science, India
Dibakar Sen	Indian Institute of Science, India
Anumeha Rai	Indian Institute of Science, India
Ashish Verma	Indian Institute of Science, India
Bismita Nayak	Indian Institute of Science, India
Kaushik Choudhury	Indian Institute of Science, India
Dharam Deo Prasad	Indian Institute of Science, India
E. Govindaprasath	Indian Institute of Science, India
Govinda Sharma	Indian Institute of Science, India
Anirudha Bhattacharjee	Indian Institute of Science, India
Murali Krishna	Indian Institute of Science, India
P. Krupakar	Indian Institute of Science, India
Nipun Patil	Indian Institute of Science, India
Salil Sapre	Indian Institute of Science, India
Samrat Sankhya	Indian Institute of Science, India
Shivam Raina	Indian Institute of Science, India
Unnati Jain	Indian Institute of Science, India
Minu Pradeep	Indian Institute of Science, India
Mangaldas Budho Gaonkar	Indian Institute of Science, India
C.S. Susmith	Indian Institute of Science, India
P.S. Suvin	Indian Institute of Science, India
S. Tushar Pawar	Indian Institute of Science, India
A. Vignesh Kumar	Indian Institute of Science, India
S. Chandra Mouli	Indian Institute of Science, India
Chandana Venkatayogi	Indian Institute of Science, India
Dawn Varghese	Indian Institute of Science, India
J.R. Hari Prakash	Indian Institute of Science, India
Hari Narayanan	Indian Institute of Science, India
V.S. Krishna Prasad	Indian Institute of Science, India
M.C. Kumari	Indian Institute of Science, India
Pavan Sridharan	Indian Institute of Science, India
Nilakantha Singh Deo	Indian Institute of Science, India
Biplab Sarkar	Indian Institute of Science, India
B.S.C. Ranjan	Indian Institute of Science, India
B. Santhi	Indian Institute of Science, India
B. Damayanthi Jesudas	Indian Institute of Science, India
Divyanshu Joshi	Indian Institute of Science, India

Gajanan Kulkarni	Indian Institute of Science, India
S. Harivardhini	Indian Institute of Science, India
Kiran Ghadge	Indian Institute of Science, India
N. Madhusudanan	Indian Institute of Science, India
Manoj Kumar Mahala	Indian Institute of Science, India
Nitesh Batia	Indian Institute of Science, India
Praveen T. Uchil	Indian Institute of Science, India
Suman Devadula	Indian Institute of Science, India
Sonal Keshwani	Indian Institute of Science, India
Shakuntala Acharya	Indian Institute of Science, India
V. Prajwal	Indian Institute of Science, India
K. Eazhil Selvan	Indian Institute of Science, India
Vineeth Muralidharan	Indian Institute of Science, India
Anshul Mittal	Indian Institute of Science, India
Rahul Kanyal	Indian Institute of Science, India
G. Ranga Srinivas	Indian Institute of Science, India
Nishath Salma	Indian Institute of Science, India
R. Ravindra	Indian Institute of Science, India
Rohit John Varghese	Indian Institute of Science, India

Contents

About the Conference

Design is ubiquitous; it pervades almost all spheres of life, and has been around as long as life has taken up the task of purposefully changing the world around it. Research in design and the emergence of a research community in this area has been relatively new, its development influenced by the multiple facets of design (human, artefact, process, organization, and the micro- and macro economy by which design is shaped) and the associated diversification of the community into those focusing on various aspects of these individual facets, or various applications. Design is complex, balancing the needs from multiple stakeholders, and requiring a multitude of areas of knowledge to be utilized, from resources spread across space and time.

ICoRD'15 is the fifth in a series of conferences intended to be held every 2 years in India to bring together the international community from diverse areas of design practice, teaching and research, to: showcase cutting edge research about design to the stakeholders; aid the ongoing process of developing and extending the collective vision through emerging research challenges and questions; and provide a platform for interaction, collaboration and development of the community in order for it to take up the challenges to realize the vision. The conference is intended for all stakeholders of design, and in particular for its practitioners, researchers, teachers and students.

The theme of ICoRD'15 is "Design Across Boundaries." Human society is ridden with boundaries: economic, political, religious, educational, cultural, racial, caste and class-based, age-based, ability-based, environmental, oriental-occidental, rural–urban, traditional–modern and so on. Yet designs must transcend boundaries to create value for people from all walks of the global society. Specific forms of design across boundaries are already being encapsulated in for instance "inclusive design" which aims to help melt age- and ability-based boundaries, or "design for the bottom-of-the-pyramid" which extends design to those with extreme economic challenges. The theme of ICoRD'15 is a celebration of the power of design to be pervasive. It is also a call to the design and its research fraternity to acknowledge boundaries, if only to transcend them and reach the benefits of design to everyone.

The conference has:

- Invited presentations from eminent international experts and practitioners;
- Presentations of refereed papers as podium, poster, panel or theme presentations;
- Industrial sessions to present perspectives from industry and studies in practice.

ICoRD'15 is co-located with the 3rd International Conference on Design Creativity (3rd ICDC) held in Bangalore, India, 12–14 January, 2015.

About the Editor

Amaresh Chakrabarti is professor of Engineering Design at Centre for Product Design and Manufacturing, Indian Institute of Science (IISc), Bangalore. He holds a BE in Mechanical Engineering from University of Calcutta (now IIEST), India, an ME in Mechanical Design from IISc, and a Ph.D. in Engineering Design from University of Cambridge, UK. After Ph.D., he led for 10 years the Design Synthesis team at the EPSRC Centre for Excellence Engineering Design Centre at University of Cambridge. His interests are in design synthesis and creativity, eco-design and sustainability, and product informatics. He has authored/edited 10 books, over 250 peer-reviewed articles, and has six patents granted/pending. He co-authored DRM, a methodology used widely as a framework for doing engineering design research. He is an Associate Editor, AI EDAM, Area Editor, Research in Engineering Design (Springer), Regional Editor, Journal for Remanufacturing (Springer), and Advisory Editor for seven international journals including Journal of Engineering Design, Clean Technologies and Environmental Policy (Springer), and International Journal of Design Creativity and Innovation. Professor Chakrabarti has been on the Advisory Board of Design Society, UK, where he is currently a member of the Board of Management. He is a member of the CII National Committee on Design, India, and member of the Jury for India Design Mark, India Design Council. He founded IDeaSLab—the first laboratory in India for research into design creativity, sustainability and innovation. He is Chair for International Conferences on Research into Design (ICoRD), 22nd CIRP Design Conference (CIRP Design 2012), 3rd International Conference on Design Creativity (3rd ICDC 2015), and vice-Chair for AI in Design (AID) and Design Computing and Cognition (DCC) Conferences. He is an Honorary Fellow of the Institution of Engineering Designers, the peer society under the UK Royal Charter in engineering design. Seven of his papers won top paper awards in various international conferences.

Part I
Design Theory and Research Methodology

A Case Study in Participatory Service Design for Rural Healthcare System in India Using a Pattern Language

Pramod Khambete, Uday Athavankar, Pankaj Doke, Ratnendra Shinde, Debjani Roy, Sujit Devkar, Sanjay Kimbahune and Sujata Chaudhary

Abstract Application of a Service and Touch Point design pattern language for a rural healthcare service in India with the participation of a multidisciplinary team comprising designers, IT professionals, business analysts and doctors was studied. We discuss the process followed, and the benefits of using a service design pattern language. Specifically: the participants could start with a few patterns and progressively acquire an understanding of the patterns through mutual assistance; the patterns and associated sharable artefacts facilitated communication and continuity of thought process among the team members; the team could dynamically refine a set of appropriate patterns and were guided to rich, comprehensive and innovative solutions. The real life constraints neither hindered the progress nor the quality of the solutions. The potential of embedding the pattern language in a socio-technical system to mediate the process was demonstrated. The study provides several theoretical and practical insight concerning the use of pattern languages.

Keywords Service design · Participatory design · Multidisciplinary design · Design patterns · Pattern language · Healthcare

P. Khambete (✉) · U. Athavankar · D. Roy · S. Chaudhary
Industrial Design Center, Indian Institute of Technology Bombay, Mumbai, India
e-mail: pramodkhambete@iitb.ac.in; pramod.khambete@gmail.com

P. Doke · S. Devkar · S. Kimbahune
Tata Consultancy Services, Mumbai, India

R. Shinde
Seth G. S. Medical College and KEM Hospital, Mumbai, India

A. Chakrabarti (ed.), *ICoRD'15 – Research into Design Across Boundaries Volume 1*, Smart Innovation, Systems and Technologies 34, DOI 10.1007/978-81-322-2232-3_1

1 Introduction: Service Design, Pattern Language and Multidisciplinary Team Work

The objective was to study the use of a pattern language, its influences on the service design process and outcomes in a multidisciplinary setting. Particular areas of interest were: how the teams acquired a common understanding of the patterns, the team dynamics and the impact of real life constraints in multidisciplinary team work.

The theoretical underpinnings of Service Design, a relatively new design domain are still maturing [1]. The heterogeneous, intangible nature of services, the temporal dimensions and relational components introduce complexity in service design [2]. The value of services is co-created through a customer's experience in varied, contextualised, service encounters in which interactions with several Touch Points constituting a coherent ecosystem take place [3]. Service design calls for multidisciplinary participation that poses challenges due to the diverse backgrounds, skill sets, and experience of the collaborators. Such teams need a shared vocabulary for creative exchange to effectively engage in a process comprising creative conceptualization, realization (or implementation), and evaluation [4]. Boundary Objects, that enable communication and coordination among the participants, can enable the synthesis of their unique, though at times partial knowledge [5]. Design patterns (service design patterns, in our case) could serve as boundary objects in service design endeavours.

Design patterns and pattern languages [6] abstract the invariants in the successful solutions to the recurring problems in a domain and guide towards creative design solutions in a specific problem situation. They provide a "Metadesign environment" [7] for effective collaboration. According to Thomas et al. [8] a pattern languages offers a fertile socio-technical system, a flexible structure to formulate and creatively solve complex problems. They serve as *lingua franca*, offering a trans-disciplinary vocabulary [9].

Designing rural healthcare services in India, a complex, arguably a wicked problem, calls for multidisciplinary collaboration. Pattern languages were found to be relevant for participatory design of healthcare systems [10]. A validated pattern language for designing Services and Touch Point Ecosystems [11] was used in this study. Before proceeding, we briefly touch on few concepts in relation to service design and the pattern language used by us.

Touch Points are the entities with which a customer interacts to commence or progress with a service encounter [12]. They can be human (e.g. a health worker), or non-human (e.g. a portal). A Touch Point Ecosystem is a network of Touch Points that operates coherently to provide desirable service experience. The pattern language used by us comprises 116 design patterns. (Two illustrative pattern summaries: TOUCH POINT AT THE HUB—Customers need a single place where they can go for getting information, or with the desire to commence an encounter, or if they want to find an alternative if an encounter with a Touch Point is abandoned. Create an accessible Touch Point which performs the function of understanding the customer needs, and hands over to the appropriate Touch Point, SMOOTH TOUCH POINT HANDOVER: A customer might commence an encounter with a Touch Point, but other Touch Points could come into play later. The Touch Point Ecosystem design and the back end systems should ensure that the customer does not feel discontinuity or need to spend additional costs or efforts in order to accomplish the encounter goal).

2 Designing Healthcare Services in India

Government of India periodically implements programs, new techniques and mechanisms for improving the efficiency and effectiveness of rural health care, but no specific design approach is used [13]. However, various design approaches, such as Experience Based Design [14] and design thinking [15], as well as active engagement among co-designers and tools have been recommended [16]. Several systemic interventions for healthcare service design have been attempted in India. For instance, financial incentives to poor families coupled with reliable immunisation service in the proximity (e.g. immunisation camps in villages) led to significant increase in full immunisation in those localities [17] and videos on mobile phone shown by Accredited Social Health Activists (ASHA) motivated the beneficiaries and the influential community members to opt for maternal healthcare interventions [18]. Customer focus, intermeshed innovations in technology and business models, and deployment of socio-technical systems comprising technology and people are essential for success [19, 20]. For example, mHealth, a mobile platform, developed by Tata Consultancy Services [21] enables coordinated interventions by ASHAs in the field and medical experts located elsewhere. In this context, mother and child healthcare services in the rural population provided by a Primary Health Centre (PHC) located in Thane district, Maharashtra, India were selected for the study.

3 Methodology

3.1 Domain Familiarization and Data Collection

Initially, the doctors in the team sensitised others to rural healthcare operations and the challenges such as the social contexts, communication barriers and non-compliance. Subsequently, extensive data collection was done using a mix of methods: workshops with the doctors to uncover the service experience goals, contexts and issues; field visits for participant observation, semi structured interviews and critical incident method [22]. All key stakeholders, i.e. doctors, health workers and beneficiaries were targeted.

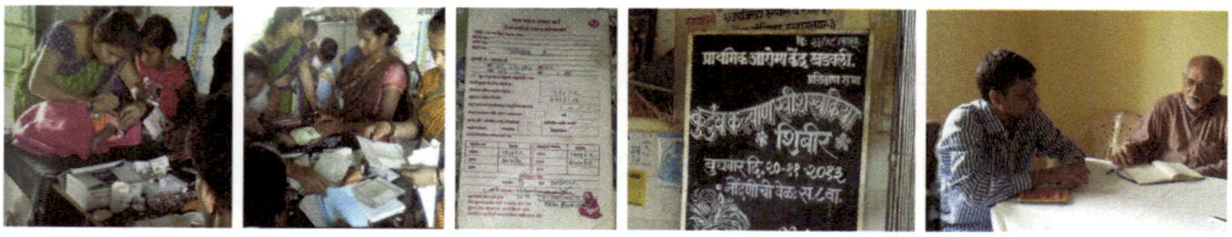

Fig. 1 Illustrative pictures related to immunisation camps, child health records, PHC servicescape and stakeholder interviews

The data comprising audio and video recordings, photographs, field note and artefacts used in the service delivery (illustrative examples in Fig. 1) were analysed using the qualitative data analysis software Atlas.ti. Key service encounters and Touch Points were identified. Thematic Analysis was done to gain understanding of the end user behaviours, service contexts, social dynamics, constraints, facilitating factors, end beneficiary experiential aspects and important issues.

After considering the possible candidates (e.g. antenatal health care, polio immunisation …), routine immunisation service was selected. The reasons: long temporal dimension (5 years); several, varied service encounters with different goals and contextual demands (e.g. immediately on birth, at least five doses over five years at differing intervals, migration of the families …); the need of harmonious, cumulative interaction of the beneficiaries with several Touch Points (e.g. Doctors, PHC staff, ASHAs, technology based Touch Points) and dependence on a strong service infrastructure.

3.2 Study of the Application of the Pattern Language in a Multidisciplinary Design Setting

The multidisciplinary team comprised designers, IT professionals, business analysts, doctors specialised in social medicine and social sciences professionals. Some of the areas of interest in this study were: the differences in the use of the pattern language by participants belonging to different disciplines; ease in understanding the design guidance given by patterns despite varied backgrounds; influence of a participant's background and level of expertise in using a pattern language on the team interactions and the outcomes; and the role of the pattern language as a common, discipline agnostic mode of communication.

The team was introduced to the concepts in design, service design, the pattern language, and user experience design methods and techniques like persona, scenario, service blueprinting and customer experience journey maps in preparatory workshops.

Ten half or full day workshops were conducted over a period of 3 months. No significant intervention was done to overcome the real life constraints, such as full availability of all team members in order to maintain the ecological validity.

The pattern language was made available in the form "Pattern Cards" (approx. 10 cm × 8 cm which gave only the pattern name, summary of the design guidance, examples of implementation and names of connected patterns) and as a navigable MSExcel© workbook, with the freedom to use either or both as needed. The outcome was conceptual design of service encounters, Touch Point design ideas, and identification the operations infrastructure needed for the service. The team worked various modes: mix of group and individual work, full and partial participation, and sequential participation. Video and audio recordings of the workshops was analyzed to identify significant insights.

The stages and the key activities carried out were:

1. Preparatory workshops, explained previously.
2. Creation of Personas, Scenarios, Service Blueprints, and Experience Journeys: The participants collaboratively and iteratively created two personas and two critical service encounter scenarios based on the data. Customer experience journey maps were created to depict service encounter stages (e.g. informing the mother of the date of immunisation, ensuring that she goes to the camp), sub goals at each stage (e.g. convincing the mother to opt for immunisation of her child), the contextual aspects (e.g. past post-immunisation unpleasant experiences, need to take leave from work), and likely service failure causes. The sharable artefacts produced at this stage (e.g. persona posters and journey maps) evolved in the subsequent phases and provided continuity in subsequent stages.
3. Mapping patterns to experience journeys: two teams of 3–4 members each independently worked on the scenarios and assigned patterns to each stage in the experience journey that addressed the contextual service design requirements based on their collective judgment (Fig. 2). The facilitators intervened only to provide minimal clarification when requested, and occasional nudging to maintain the pace. As a result the teams used their own strategies. One team split the pattern cards set among themselves, and individually selected the patterns. Later, they discussed and agreed on the patterns to be used. The other team mainly referred to the MSExcel© workbook, using the pattern cards occasionally, and mapped the patterns using sticky notes. Pace and iterative progress was encouraged (e.g. they did not select a pattern if in doubt, as per Alexander's suggestion [6]).

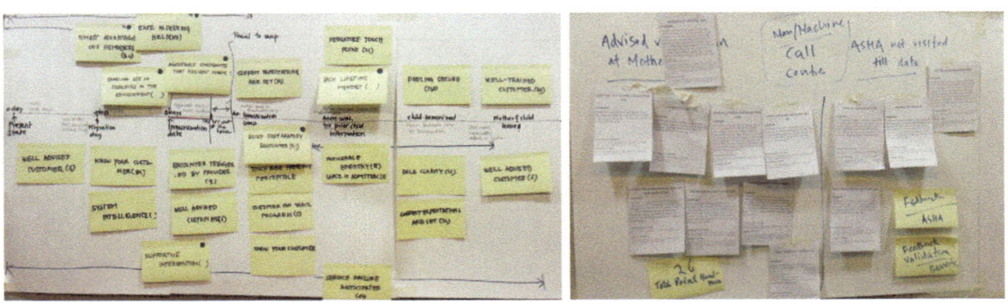

Fig. 2 Different approaches to associating patterns with stages in an experience journey map

4. Detailing pattern and experience journey map, and solution synthesis: The patters set identified in stage 3 was refined and patterns related to the identified patterns were used for design guidance and ideation. At times the teams felt the need of additional information and sought inputs from doctors, healthcare workers and visited the field to fill the gaps in their understanding before refining the pattern selection.
5. Debriefing, analysis and insight consolidation with individual teams: The participants reflected together on their experience and shared views on the pattern language, the process, their perceptions about the advantages and disadvantages of using a pattern language for service design, and experience related to the multidisciplinary team work.
6. Insight consolidation with the full team: Since all members could not participate throughout, the insights were shared with the entire team for validation.

4 Insights from the Study and Discussion

The insights are grouped under different headings below for convenience. However, several of them straddle the categories. Also, they are discussed in situ to retain the context.

4.1 Patterns and Pattern Language

Typically, participants could gain adequate understanding of the design guidance based only on the pattern name, brief summary and examples of implementation. Examples from other service domains were helpful (e.g. example from ecommerce or banking served as analogies), though the participants suggested that more examples from the service domain of their interest would be advantageous. This suggestions seems reasonable, but could lead to loss of cross-fertilisation of creative ideas from other service domains. There is a need to balance the benefits and disadvantages. Progressively enlarging a repository of examples from various service domains could help in this regard.

Looking up the connected patterns helped the participants to understand more about the pattern under consideration, clearly indicating the power of the pattern language as against individual patterns. It was suggested that organising the patterns in different ways would be useful (e.g. grouping based on the nature of the design problem). The need for an organising principle that binds a pattern language together is well-known [23]. The suggestion strengthens the idea of multiple organising principles proposed in [11].

The patterns did act as a common, discipline agnostic means of communication. After using the patterns for a while, the participants used the pattern names to express their ideas. This as well was consistent with [8] and [11].

4.2 Process and Team Dynamics

Most of the team members did not have formal knowledge of pattern languages and service design, except that provided in the preparatory workshops. The doctors had developed healthcare interventions without following formal design methods in the past. The industry participants were familiar with the concept of software design patterns, but not with service design. These differences did not hamper the participants from comfortably commencing with service design and using the patterns. It seems the patterns played a facilitating role ("... working on a scenario and choosing patterns for solutions was helpful. I can use it in the future"). Use of the personas, scenarios and experience journey maps was beneficial and the participants referred to the personas and scenarios throughout the process.

The participants initially selected patterns based on the name, examples and summary. This is consistent with the findings in [11]. Largely, they resorted to discussion within the team, rather than reading the pattern in depth for clarification. These discussions helped clarifying the design ideas behind the patterns, and to decide the appropriateness of a pattern to the stage in the experience journey. Some participants opted to read the pattern details for clarity. It appears perspectives from different disciplines about the same pattern were helpful. The benefits though tapered off if the discussion prolonged. As mentioned previously the facilitator gently discouraged it without forcing. Participants agreed that this constraint was very helpful. It suggests that there should be some mechanism to allow only optimum discussion. This can be a challenge as the optimum duration might vary from team to team. Perhaps a combination of technology and process is needed (e.g. someone in the team acting as a timekeeper).

The teams found it easier to progress smoothly if they initially considered a small set of patterns. This observation is consistent with [11] where it was found that having a large consideration set of patterns initially causes confusion. The relevant patterns missed initially did get included as the design progresses. The connectedness inherent in a pattern language ensured it. However, it seems a combination of the process and a techno-social environment would help in evolving from a small to a comprehensive set of patterns.

As noted before, the teams worked in different modes—synchronous and asynchronous, jointly and individually, as well as sequentially (i.e. joining midway for part of the design work). The freedom seems to have helped. In this context, the patterns along with the artefacts (e.g. the post-it notes bearing the pattern names stuck to the experience journey diagrams) enabled externalisation and a record of the thought processes. This helped in overcoming the limitations of the discontinuities in the team interactions ("the diagrams and patterns helped us move forward"). This is an important finding highlighting the need to combine externalisation mechanisms with the pattern language to facilitate continuity in the team interaction.

Many a times a pattern triggered solution ideas immediately. This is not surprising as the problem space and solution space do co-evolve [24]. However, there was a counterproductive tendency to debate and refine the idea immediately. The participants agreed that recording the idea, the associated patterns and moving on was more effective.

Occasionally the facilitator (an "expert") participated to clarify a pattern when the team was confused and stuck. It suggests that the design team should include at least one such member. Timely access to domain experts as well was essential. Interestingly, it seems the presence of a trained designer could at times be counterproductive ("I thought he (the designer in the team) was an expert and I should go along"). Obviously, the aspects such as the culture, personalities of individual team members and size of the team could play a role in triggering such perceptions. These findings have implications on the team composition and the design process (e.g. mandated mix of individual and team work, inclusion of a pattern for consideration based on voting as against discussion …). The socio-technical environment can play a role in this regard as well.

4.3 Perceived Benefits

The pattern language led to richer and innovative design solutions ("Pattern helped, gave direction and triggered ideas", "a (single) pattern triggered several ideas"). Dorst and Cross [25] found that designers can be more creative if they are able to understand and frame a problem in their own way, and follow a design approach aligned to their own goals and situations. Framing the problem through the scenarios, personas and experience journey maps created by the teams themselves, integration of the pattern language in the artefacts and the freedom to externalise the thought processes through means they were comfortable with seems to have provided such a conducive environment. The participants reported that the pattern language provided a structured way and directed their thought processes but did not constrain them from creative explorations. It helped them make sure that every aspect of a service encounter was addressed ("patterns make you sure that you don't leave this out, patterns guide/prepare the foundation to work further").

The power of analogies in design is well recognised and role of patterns as "productive units of analogical transfer" was noted [26]. We indeed found that patterns, and in particular the examples, triggered analogies which led to richer understanding of the problems, as well as creative ideas ("… can we have something like an hour glass to remind the mother of the vaccination date …").

4.4 Other Insights

Due to the constraints in our study, most of the participants underwent only basic familiarisation of the pattern language. However, participants suggested more intensive familiarisation and time to "play with the patterns" before commencing the design would be more effective. This points to the possibility of gamification of some activities. The socio-technical environment made available to the teams could also incorporate game elements. While the participants were unanimous that they were able to maintain continuity with the help of the artefacts they created, they suggested that formal knowledge handover and periodically refreshing the knowledge of patterns would help.

5 Summary of Conclusions and Directions for Future Work

We confirmed past knowledge regarding the effectiveness of patterns and pattern languages as design tools, in general and specifically for service design. The facilitating role they can play in multidisciplinary team work was underscored. We as well uncovered theoretical and practical implications which have been highlighted in the previous sections. However, we summarise a few. The pattern name, examples and summary seem adequate to acquire working comprehension of a pattern. The interconnections of patterns and realisation of their harmonious relationship furthers the comprehension. We uncovered the advantages of a multidisciplinary team for service design (e.g. clarifying each other's understanding, bringing the perspectives from their domain to enrich the design …). The finding that it is not a perquisite that all team members should have in-depth of knowledge of the pattern language at the beginning, provided they have easy access to "pattern language experts" and domain experts is important. The understanding increases as the patterns are used. Design strategies and processes that motivate the participants to initiate the use of a pattern language and acquire deeper understanding over time seem appropriate. The effectiveness of sharable artefacts built around patterns in transfer of knowledge and supporting continuity of thought process in face of real life constraints was a significant finding. The process followed in the study seems to be suitable for service design practice. These findings have significant implications on practice in terms of team composition and the service design process.

Since in practice service design projects could span a long duration, with the team members dispersed in time and geography, a socio-technical environment (e.g. software and sharable tangible artefacts) seems necessary to support the collaboration. We have pointed out areas where such a tool can play a positive role (e.g. externalisation of thinking, creating a repository of team knowledge, and mediation in the multidisciplinary processes). It can incorporate several features informed by the insights above.

We scoped the study around one important service in rural healthcare. Studies encompassing additional healthcare services as well as other service domains would add to the knowledge. The voluntary participants in our study were not able to devote time continuously. While this limitation did lead to interesting insights mentioned earlier, studies with continuous participation are likely to provide additional insights.

We believe ours was the first study of use of a pattern language in service design in a multidisciplinary setting. These findings are significant new additions to the extant knowledge. Since pattern language used is the first (and at this time only) one in service design domain, we hope the findings would benefit service design practitioners as well. This study was the initial phase of a longer program in which we hope to build on the knowledge. Happily, the medical stakeholders intend to implement some of the 50+ service design ideas generated and make a difference to the lives of the beneficiaries.

Acknowledgments This study was carried out under the aegis of TCS—IIT Bombay Research Cell funded by Tata Consultancy Services. We thank doctors from SGS Medical College and KEM Hospital, Mumbai, notably Dr. Sunita Shanbhag, Dr. Mrudula Solanki and Dr. Pawan Sable for sharing their rich experience and participation, and the staff and doctors associated with the chosen Public Healthcare Centre who helped in the data collection stage.

References

1. Sangiorgi, D.: Building up a framework for service design research. In: Proceedings of 8th European Academy of Design Conference, pp. 415–420, Aberdeen, Scotland (2009)
2. Kimbell, L.: Designing for service as one way of designing services. Int. J. Des. **5**(2), 41–52 (2011)
3. Prahalad, C., Ramaswamy, V.: Co-creating unique value with customers. Strategy Leaders. **32**(3), 4–9 (2004)
4. Mamykina, L., Candy, L., Edmonds, E.: Collaborative creativity. Commun. ACM **45**(10), 96–99 (2002)
5. Fischer, G., Shipman, F.: Collaborative design rationale and social creativity in cultures of participation. Hum. Technol. **7**(2), 164–187 (2011)
6. Alexander, C., Ishikawa, S., Silverstein, M., Jecobson, M., Fiksdahl-King, I., Angel, S.: A pattern language. Oxford University Press, New York (1977)
7. Giaccardi, E., Fischer, G.: Creativity and evolution: a metadesign perspective. Dig. Creativity **19**(1), 19–32 (2008)
8. Thomas, J., Lee, A., Danis, C.: Enhancing creative design via software tools. Commun. ACM **45**(10), 112–115 (2002)
9. Erickson, T.: Lingua Francas for design: sacred places and pattern languages. In: Proceedings of DIS '00, pp. 357–368, Brooklyn, New York (2000)
10. Bowen, S., McSeveny, K., Lockley, E., Wolstenholme, D., Cobb, M., Dearden, A.: How was it for you? Experiences of participatory design in the UK health service. CoDesign: Int. J. CoCreation Des. Arts **9**(4), 230–246 (2013)
11. Khambete, P.: Designing desirable service experience: a pattern language framework for designing touch point ecosystems. PhD Thesis, Indian Institute of Technology Bombay, Mumbai, India (May 2013)
12. Khambete, P.: Pattern language for touch point ecosystem: a potent framework for multidisciplinary design. In: CADMC 2011 Cambridge Academic Design Management Conference, Cambridge, UK (2011)
13. Park, K.: Park's Textbook of Preventive and Social Medicine, 21st edn. Banarasidas Bhanot, Jabalpur (2011)
14. Bate, P., Robert, G.: Experience-based design: from redesigning the system around the patient to co-designing services with the patient. Qual. Saf. Health Care **15**(5), 307–310 (2006)

15. Brown, T., Wyatt, J.: Design thinking for social innovation. Stanford Soc. Innov. Rev., 31–35 (2010)
16. Bessant, J., Maher, L.: Developing radical service innovations in healthcare—the role of design methods. Int. J. Innov. Manag. **13**(4), 555–568 (2009)
17. Banerjee, A., Duflo, E., Glennerster, R., Kothari, D.: Improving immunisation coverage in rural India: clustered randomised controlled evaluation of immunisation campaigns with and without incentives. BMJ: Br. Med. J. **340** (2010)
18. Ramachandran, D., Canny, J., Das, P., Cutrell, E.: Mobile-izing health workers in rural India. In: Proceedings of the SIGCHI Conference on Human Factors in Computing Systems, pp. 1889–1898 (2010)
19. Herzlinger, R.: Why innovation in health care is so hard. Harvard Bus. Rev., 58–67 (2006)
20. Coiera, E.: Four rules for the reinvention of health care. BMJ. Br. Med. J. **328**, 1197–1199 (2004)
21. Bondale, N., Kimbahune, S., Pande, A.: mHEALTH-PHC: an ICT tool for primary healthcare in India. IEEE Technol. Soc. Mag. **32**(3), 31–38 (2013)
22. Gremler, D.: The critical incident technique in service research. J. Serv. Res. **7**(1), 65–89 (2004)
23. Fincher, S., Windsor, P.: Why patterns are not enough: some suggestions concerning an organising principle for patterns of UI design. (2000)
24. Maher, M., Poon, J., Boulanger, S.: Formalising design exploration as co-evolution, pp. 3–30 (1996)
25. Dorst, K., Cross, N.: Creativity in the design process: co-evolution of problem–solution. Des. Stud. **22**(5), 425–437 (2001)
26. Goel, A., Bhatta, S.: Use of design patterns in analogy-based design. Adv. Eng. Inform. **18**, 85–94 (2004)

How Different Models of Value Change Affect Emergent Patterns in Design Practice: Agent Based Simulations

Vishal Singh and John Gero

Abstract This paper reports on a computer simulation based approach to studying longitudinal patterns in social emergence of design practice. Design practice is an emergent and not a well-understood social phenomenon, especially in terms of understanding how values associated with different design disciplines influence their design practice. A society of agents, called design agents, representing designers with different design backgrounds, interact with each other and with the concepts associated with different disciplines. The design agents within each discipline are modelled to be attracted towards concepts, knowledge mode, as well towards the other design agents, knower mode. The force of attraction towards the knower or concepts varies between disciplines. A bottom up simulation approach is used to study how different models of value change affect emergent patterns of behaviour. The findings from these simulations have implications for how we can use computation models to study complex social behaviour in design societies.

Keywords Design practice · Legitimation code · Design values · Agent based simulation

1 Introduction

Design societies and communities, like societies in general, can be described and discussed in terms of their values and practices. In design societies, the underlying design values typically guide design practice, as well as determining what is

V. Singh (✉)
Department of Civil and Structural Engineering, Aalto University, Espoo, Finland
e-mail: Vishal.Singh@aalto.fi

J. Gero
Department of Computer Science and School of Architecture, University of North Carolina, Charlotte, NC, USA

A. Chakrabarti (ed.), *ICoRD'15 – Research into Design Across Boundaries Volume 1*, Smart Innovation, Systems and Technologies 34, DOI 10.1007/978-81-322-2232-3_2

considered as good design and what designers aspire to. The differences in design values across different design disciplines are reflected in the academic literature as well the wider public discourse within each of these disciplinary communities. For example, in engineering design the established value assessment approaches emphasize behavioral characteristics of the designed artefact such as performance, reliability and robustness. These assessments are typically disassociated or distanced from the designer. In contrast, in disciplines such as fashion design and architecture, the value assessment often puts considerable emphasis on aspects such as individual expressions or uniqueness that are considered as much a critic of the designer as that of the designed artefact, leading to comparatively greater pull towards the values associated with iconic designers or designs.

In order to better understand and manage design practices in any community, it is important to better understand the dependencies between design values and design practice. However, changes in design values and design practice occur gradually over an extended period of time, which means data over longitudinal periods is needed to observe these changes and the mutual dependencies between design values and the design practice. Observing such trends is data and resource intensive, and often extremely challenging in terms of obtaining sufficient data within the span of a single research project. Nonetheless, the study is important and critical, because many decisions about managing and improving design practices are often based on our limited understanding of the relationships between design values and design practice. For example, recently there has been a greater push for creating multi-disciplinary design societies, which among other benefits, are expected to facilitate exchange and sharing of design values across different disciplines, in the belief that shared values will enrich design practice. It is often not clear which design values we expect to be exchanged or shared through these multidisciplinary societies, and we do not have an adequate understanding of the long-term implications of these expected exchanges of values. Typically the effects of such multidisciplinary design environments are studied through short-term projects, with significant challenges in avoiding noise around the research variables and parameters. While such empirical studies provide useful insights into differences across disciplinary silos and multi-disciplinary environments, the short duration of these studies and experiments provides little opportunity to study long-term patterns.

The research reported in this paper focuses on the longitudinal patterns of changes in design practices resulting from the changes in design values in a multi-disciplinary design environment. This 'what-if' study focuses on trying to understand the global trends that are likely to emerge as a result of changes in design values over time at individual levels, rather than on the mechanisms of value change at the individual level. Agent-based simulations are used to carry out a longitudinal study. A computational model is created to simulate a society of designers, who interact with each other and the design concepts utilizing the value systems of their respective design disciplines. The design values and concepts are defined at an abstract level such that the simulation results need to be interpreted in a context that can be broadly approximated to these abstractions. These abstractions are based on our current understanding of the relative design values across the three exemplary societies of fashion design, architecture and engineering, studied through the lens of legitimation code theory [1, 2]. For example, based on Carvalho et al. [2], it is assumed that fashion designers have a greater attraction towards leading designers (knowers), compared to engineering designers, who have a greater attraction towards design concepts (knowledge) than towards the leading engineering designers. Assumptions such as the force of attraction become the key parameters of the computational model, which are then varied across different simulation cases. This paper describes this simulation model, the conceptual basis guiding the underlying assumptions in the model, and preliminary results from the different scenarios studied by varying some of these parameters.

2 Background

The simulations are built on an existing computational model developed and previously reported in Singh and Gero [3, 4]. Legitimation code theory (LCT) [1, 2] is adopted as the underlying framework to model the relationships between design values and design practice. A brief background to LCT and use of computational models in social simulations is provided.

2.1 LCT, Design Values and Social Influence

Legitimation describes what is acceptable or normative in a society, typically viewed as some form of unwritten 'rules of the game' [5]. LCT provides the theoretical basis to explain how unwritten rules of normative practice emerge in and guide a knowledge society. Carvalho [6] and Carvalho et al. [2] use LCT to explain how design practice and recognition within a social group are driven through both knowledge and knower modes [7], i.e., the design practices emerge and evolve under the influence of the social structure as well as the knowledge structure. For example, in engineering disciplines design values are mostly associated with the knowledge structure, whereas in fashion design and architecture, design values are equally linked to social structure so that design values are also influenced by knowers and their design values.

2.2 Agent Based Models and Social Simulations

Computational social simulations are an established method to test and generate socially-related hypotheses [8, 9]. These simulations aim to provide a complementary research method and infrastructure that can reduce the time, cost and resource requirements when generating and testing promising theories, especially in scenarios that require longitudinal studies.

3 Description of the Simulation Model and the Experiment Design

Building on Carvalho et al. [2], a society of design agents with different design backgrounds is modelled such that all design agents are attracted towards concepts, i.e., knowledge mode, as well towards the other design agents, i.e., knower mode, which influences their design values. The force of attraction towards the knower or concepts varies across disciplines. The emergent design practice is shown through a plot in a two dimensional space defined by the social and knowledge axes. Design agents higher up the social axis exert higher knower force while the concepts higher up the knowledge axis exert higher knowledge force.

The computational model is implemented in MASON [10], a java based multi-agent system. Following Carvalho [6], the three disciplinary backgrounds considered are architecture, fashion design and engineering. The key assumptions in the model, already described in previous papers [3, 4], are reiterated briefly in Table 1.

The social influence exerted on any agent (A^i) by another agent is described in Table 1. This influence is a function of the distance between them, and their InfluenceRadius, which defines how socially influential they are. Similarly, agents are also attracted towards concepts (C^i), which have an InfluenceRadius. Agents are attracted towards other agents who are higher in their social dimension, and pushed upward with relatively lesser force by agents that are behind them along the social dimension. Similarly, agents are attracted towards concepts that are higher than they are along the knowledge axis. A disciplinary factor, constant K, is used to account for the relative knowledge and knower pulls across the different disciplines. As an initial assumption K is set to be one order different between fashion design and architecture and between architecture and engineering.

Additional assumptions made in these simulations about the change in values are listed in Table 2.

The gap (G) between two agents is the distance along their social axis (agents' position being A_x^i). G is positive if the other agent has a higher position along the

Table 1 Key assumptions in the simulation model

Aspects to model	Assumed relationships	Assumed values
Knower mode Agent (A^1) – agent (A^2) attraction	K **x** (InfluenceRadius A^1 **x** InfluenceRadius A^2)/ (sq. of social distance between A^1 and A^2)	For design agents IF discipline is architecture $K = 100$; IF fashion $K = 1$; IF engineering $K = 1{,}000$
Knowledge mode Agent (A^1) – concept (C^1) attraction	K **x** (InfluenceRadius A^1 **x** InfluenceRadius C^1)/(sq. of distance $A^1 - C^1$)	For design agents IF discipline is: architecture $K = 100$; IF fashion $K = 1$; IF engineering $K = 1{,}000$
Growth of concepts Concept (C^1) – concept (C^2) attraction	$K \times$ (InfluenceRadius $C^1 \times$ InfluenceRadius C^2)/(sq. of distance $C^1 - C^2$)	IF C^1 and C^2 belong to same discipline $K = 100$ ELSE $K = 1$

Table 2 Assumptions in the simulation model for studying the role of design values

Aspects to model	Assumed relationships	Assumed values
A status gap G is required for agent A^1 to influence agent A^2 $G = A^2_x - A^1_x$, i.e., gap in Social positions of $A^1\left(A^1_x\right)$ and $A^2\left(A^2_x\right)$	G has to be greater than the minimum influence threshold T^i such that $G > T^i \times A^1_x$	T^i is the first simulation parameter varied across different cases
Disciplinary influence is mediated by value change. A coefficient of Value Change V^c is introduced. Each agents V^c at any given time is measured across three dimensions V^{cA} (architecture), V^{cE} (Engineering), and V^{cF} (fashion)	Value change ${}^1V^c$ of A^1 while interacting with A^2 (with value change ${}^2V^c$) is $V^c = W_A \times {}^1V^{cA} \times {}^1V^{cA} + W_E \times {}^1V^{cE} \times {}^1V^{cE} + W_F \times {}^1V^{cF} \times {}^1V^{cF}$ where W_A, W_E and W_F are disciplinary weight distribution such that $W_A + W_E + W_F = 1$	Weight distribution, i.e., $W_O{:}W_S{:}W_T$ ratio is the second simulation parameter varied across different cases

Number of interactions between A^1 and A^2 also determines V^c. This factor follows a bell curve. As interactions increase the influence increases exponentially followed by an exponential decay social axis, such that an agent is likely to be influenced by and attracted towards the other agent. However, it is assumed that there needs to be a minimum gap threshold (T^i) between the two agents for an agent to be attractive to the other. For example, if T^i is assumed to be 1.2 in a given simulation, then for an agent at position 100 units along the social axis, the other agent must at least be at the position 120+ units to be attractive and influential for the first agent. Threshold is taken as a parameter to understand how perceived status gaps and thresholds may influence the emergent social patterns.

In addition, a coefficient of value change (V^c) is introduced such that an agent's value change is the weighted average of its values associated with each of the three disciplines (Architecture-V^{cA}, Fashion-V^{cF}, Engineering-V^{cE}). This assumption is conceptually critical, because if we assume that in a multi-disciplinary society agents are likely to be influenced by the design values across the other disciplines, we are implicitly assuming that they have some intrinsic recognition and understanding of the values of the other disciplines. How much weight the agents give to the design values across the other disciplines may vary. Accordingly, the value change coefficient for an agent is assumed to be the weighted mean of the value change coefficients across the three disciplines (own discipline—W_O: second discipline—W_S: third discipline—W_T). For an architecture agent $W_O = W_A$, for engineering agent $W_O = W_E$, and for a fashion design agent $W_O = W_F$. This weight distribution ratio is taken as the second the parameter, whose influence will be studied to determine how design values emerge in a multi-disciplinary society as a function of interdisciplinary and intradisciplinary sensitivity.

In summary, it is assumed that as design agents interact with each other, they exert influence on each other, which has the potential to change their design values. The likelihood for the change to occur is contingent on how influential the interacting agents are, what is the status difference between the interacting agents, and how much contributions these influences have on a design agent's change in value. The emergent design practice including the knowledge and social dimensions of the design agents and concepts are presented graphically. For simulation cases where the agents' values change during the simulations, the forces of influence change over time.

3.1 Experiment Design and Simulation Scenarios

The simulation model and the parameters listed in Table 2 are used to conduct what-if comparisons across two different simulation scenarios, Table 3. The first set of simulations is conducted to assess the effects of T^i, the threshold status gap needed for agents to influence each other. Three different values for T^i are used, which are 1.2, 1.8 and 2.4. These values are used to conduct comparisons, with the expectation that observable effects of the threshold T^i can be found to understand its role in convergence of design values. Conceptually this parameter compares scenarios where design agents of similar social status influence each other's

Table 3 Research questions, simulation matrix and scenarios

Research questions (what if studies)	Comparative simulation scenarios
Q1: If the status difference determines the change in design values, what is its effect on the emergent design practice?	With three different values of T^i
Q2: How does the design practice vary as a function of the relative contributions of social influence?	With two weight distributions i.e., W_O:W_S:W_T ratio

design values to those scenarios where design agents only change their design values looking up to design agents who are much higher than they are in their social status.

The second set of simulations is conducted to assess the effects of the weight distribution ratio, W_O:W_S:W_T, Table 2. Two different values for W_O:W_S:W_T are used, which are 0.50:0.25:0.25 and 0.8:0.1:0.1. Conceptually, in the first case the agents are more receptive to design values of agents from other disciplines than in the second case.

At the commencement of the simulation, i.e., at $t = 0$, all the design agents and concepts in the simulation environment start with a pre-defined position on the two dimensional space, defined by their social and knowledge axes.

4 Simulation Results

A summary of the results from these simulation cases is presented in Fig. 1. For each case the results are based on an average of 50 simulations. Each graph in Fig. 1 presents time series plots of agents' movements along the social axis, averaged over all agents from the same discipline. For example, graph A, presents the results from a simulation case where the parameter values are W_O:W_S:$W_T = 0.50$:0.25:0.25 and $T^i = 1.2$.

In all the simulations agents from each discipline showed bimodal behaviour, such that a majority of agents could be aggregated together into one trend while a minority aggregated to show a different trend in their movement along the social axis. Treating all agents as a single aggregation resulted in a high standard deviation in their behaviour, which was reduced when the agents were divided into two groupings. For each simulation case the results are split and plotted as two graphs, as shown in the label using the mode values, indicated either as M (majority) or NM (non-majority). For example, graphs **A** and **A′** correspond to the same simulation case, but A shows the trend observed with majority mode while **A′** shows the trends observed with the non-majority of the agents from each discipline.

The graphs show a power-like trend-line that approximates the observed patterns. The different starting points of the regression lines is an artifact of their calculation, even though all the simulations have the same starting condition, and with nearly all agents are close to each other at the start of the simulation, i.e., at time $t = 0$.

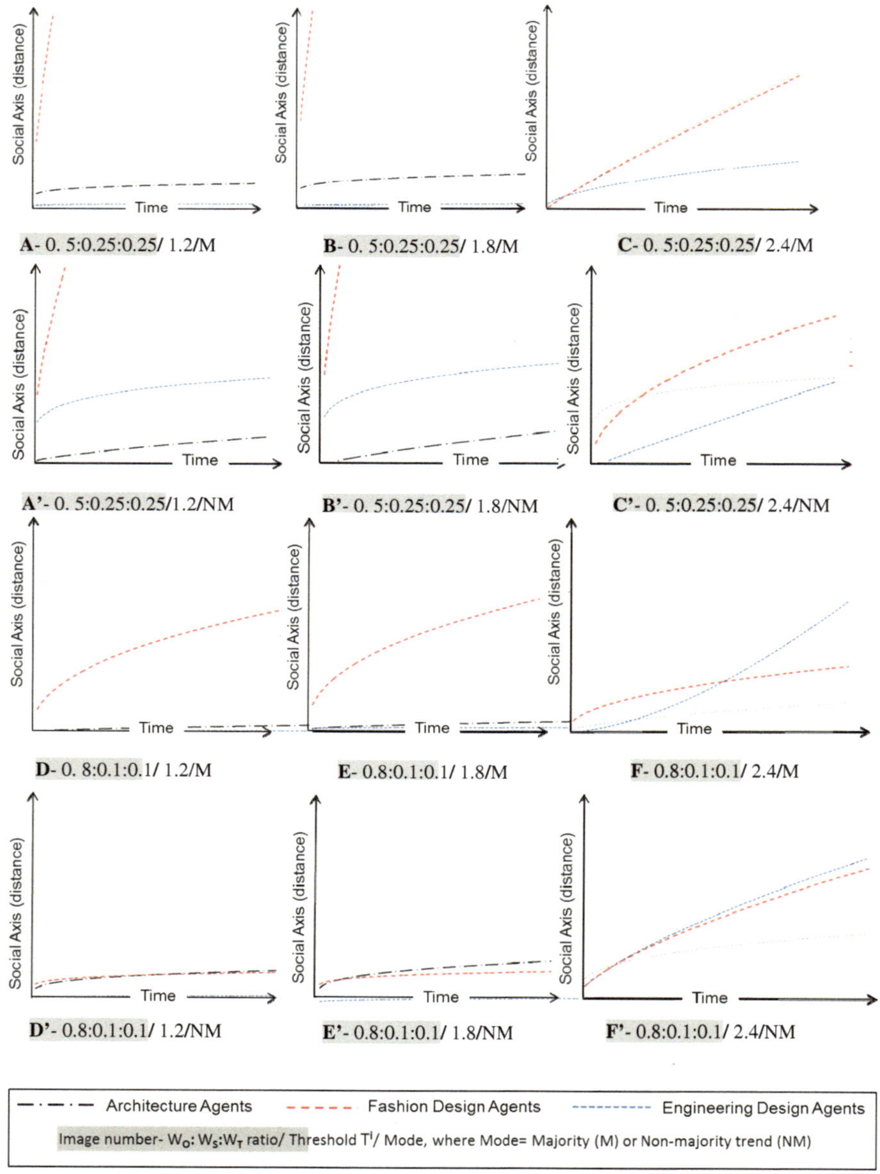

Fig. 1 Average movement of agents along the social axis across different cases

The consistent bimodal trend (comparing any graph X with its counterpart X′) in almost all the cases was an unexpected finding as it was anticipated that typically the differences across the cases would be the slope of the graphs and not necessarily their order. It was expected because of the underlying force assumptions that the graphs corresponding to the fashion agents would have the greatest slope, followed by architecture agents and then the engineering agents. This can be observed in graph **A**. In many cases this expected trend was not observed. A plausible reason for this unexpected result could be the starting positions of the agents at time t = 0. Each agent's starting position was randomly chosen within limits, but once the starting positions were chosen for the first simulation, the positions were retained in all the simulations. Therefore, many of the agents may have been too close to each other (i.e. less than the threshold T^i) in their social dimension, and hence, the social influence was not experienced by them. Even though the initial positions of the agents were not considered as a simulation parameter, it may have been a critical factor in determining the emergent trend.

The effects of gap threshold, T^i (Question Q1, Table 3), can be compared across each row by comparing the trends in A, B and C, with T^i values as 1.2, 1.8 and 2.4 respectively. At lower gap thresholds (Cases A and B), this is where agents are also influenced by their comparable peers, the results are more consistent with the expected order of social growth. When the gap thresholds are much higher (Case C), the convergence is slightly higher, i.e., the slopes are relatively closer than the first two cases. This is likely because of the fact that only fewer agents are able to exert attraction forces on the other agents such that the knower effects are lesser on fashion design and architecture agents, bringing their social growth closer to the engineering agents. This explanation is partly supported by the observed trends in cases D, E and F as well, where the social growth order is skewed towards engineering agents compared to the architecture agents when the gap thresholds are increased. The results also indicate that in the initial starting conditions, there may have been more engineering agents at farther distance to begin with, and hence, when the interdisciplinary exchange (Case F with ratio 0.8:0.1:0.1) were reduced it is the engineering agents that had comparatively higher social growth, unlike the expected trends. These patterns reiterate the conclusion that the initial starting condition of the agents may be an important factor.

The effects of weight distribution ratio, W_O:W_S:W_T (Question Q2, Table 3), can be analyzed by comparing graphs in row 1 (A, B, C) with corresponding graphs in row 3 (D, E, F). While the two sets of results for different weight distribution ratio show that when the interdisciplinary exchange (row 1, with ratio 0.5:0.25:0.25) was greater, the slopes across the three disciplines were more demarcated, and yet the architecture agents moved up the social axis with time. When the gap threshold was increased across the two distributions it showed that weight distributions across disciplinary boundaries had an effect. As noted earlier, the starting conditions may have affected the results.

5 Discussion and Future Work

These preliminary simulation results indicate that more research on the effect of initial conditions of an existing society is needed to determine its significance for how the collective patterns of design trends emerge. In particular, the results indicate that contingent on the initial distribution of agents across their social positions, it is likely that multi-modal patterns will emerge in a society even if all agents have the same mechanisms of value change. Uniformity and collective convergence through a uniform mechanism of exchange may not be supported by the evidence from agent-based simulations. Though the results corresponding to the roles of gap threshold and weight distribution ratio are not conclusive, the results do indicate that gap threshold and weight distribution ratio are interesting parameters to investigate further. For example, as reinforced in the simulation results, lower gap thresholds might increase peer influence along design disciplines with greater knower influence, while higher gap thresholds might lead to skewed trends towards those disciplines that have established leaders or icons that are way ahead of the pack.

Methodologically, these simulations use a simple model with only two parameters, and yet show unexpected results and trends. The results demonstrate the usefulness as well the challenges in using agent-based models to study collective social behaviour. While the observed bi-modal trends appear explicable after post-simulation rationalization, it was not predicted when all agents had similar mechanisms of value change. These results indicate that simpler relationships defined at local levels might have unexpected, emergent outcomes that are difficult to predict. These models can be used to identify such potential behaviour. These models are developed to observe aggregate trends rather than trends at the individual level, and though the actual values of the parameters are not critical, the choice of the values should allow noticeable behavioural change at the global levels. From that viewpoint, the lack of conclusive trends for the two main research questions raises questions such as: What if the gap threshold values that were chosen at 1.2, 1.8 and 2.4 were instead chosen to be 1.2, 2.4 and 4.8? Would that lead to more conclusive comparisons? The experience of developing and using these models indicate that 'calibrating the model' to get useful results is a critical step in social simulations using agent-based models, especially in the context of 'what if' studies of social phenomena that are not well-understood even at the levels of local interactions. The meanings of these values need to be explored.

The next steps in this research are to further calibrate the model and conduct additional simulations. The researchers plan to use initial starting condition as another parameter to investigate how that influences emergent trends. In future, the model will be extended to include other parameters once the roles of the current parameters are better understood.

Acknowledgments This research is partially supported for the US National Science Foundation under grant no. CMMI-1400466. Any opinions, findings, and conclusions or recommendations expressed in this material are those of the authors and do not necessarily reflect the views of the National Science Foundation.

References

1. Maton, K.: Languages of legitimation: the structuring significance for intellectual fields of strategic knowledge claims. Br. J. Sociol. Educ. **21**(2), 147–167 (2000)
2. Carvalho, L., Dong, A., Maton, K.: Legitimating design: a sociology of knowledge account of the field. Des. Stud. **30**, 483–502 (2009)
3. Singh, V., Gero, J.S.: Preliminary simulations of scale and value of legitimation in design practice, DS 75-2. In: Proceedings of the 19th International Conference on Engineering Design (ICED13), Design for Harmonies, vol. 2, Design Theory and Research Methodology, Seoul, Korea, 19–22 Aug 2013 (2013a)
4. Singh, V., Gero, J.S.: Developing a multi-agent model to study the social formation of design practice. In: Chakrabarti, A., Prakash, R.V. (eds.) ICoRD'13. Lecture Notes in Mechanical Engineering, pp. 631–641. Springer, Berlin (2013b)
5. Bourdieu, P.: The field of cultural production, or: the economic world reversed. Poetics **12**(5), 311–356 (1983)
6. Carvalho, L.: A sociology of informal learning in/about design. PhD Thesis, Department of Architecture Planning and Design, The University of Sydney (2010)
7. Maton, K.: On knowledge structures and knower structures. In: Moore, R., Arnot, M., Beck, J., Daniels, H. (eds.) Bernstein: Policy, Knowledge and Educational Research, pp. 44–59. Routledge, London (2006)
8. Carley, K.: Sociology: computational organization theory. Soc. Sci. Comput. Rev. **12**, 611–624 (1994)
9. Sosa, R., Gero, J.S.: A computational study of creativity in design. AIEDAM **19**(4), 229–244 (2005)
10. Luke, S., Cioffi-Revilla, C., Panait, L., Sullivan, K., Balan, G.: MASON: a multiagent simulation environment. Simulation **81**, 517–527 (2005)

Multidisciplinary Design Behaviour Using Sketching and Mental Imagery: A Literature Review and Considerations for Future Research

Mia A. Tedjosaputro, Yi Teng Shih, Patrick Pradel and Chantelle Niblock

Abstract The main behavioural characteristics of the early design stage are that designers move freely between various levels of detail and different cognitive strategies; this is the central tenet of this research. The interplay between the role of sketching and mental imagery during the design process is somewhat mysterious. It is assumed, but not known, how externalisation assists the imagery process and vice versa. In an attempt to demystify this interplay, this paper examines observable behaviour associated with the act of sketching and using mental imagery. It reviews established protocol analytical methods based on two criteria: cognitive strategies and ability to be applied in different design domains. Also, it evaluates the appropriacy of seminal coding schemes for the analysis of mental imagery to better understand moments of insight (idea generation). It concludes that established coding schemes are inadequate methods when used to facilitate analysis of cognitive strategies.

Keywords Sketching · Mental imagery · Protocol analysis · Coding schemes

1 Introduction

The scope of the paper is future needs of research in mental imagery and sketching during the idea generation process. The aim is to review available protocol analysis coding schemes and propose future considerations to facilitate a comparison

M.A. Tedjosaputro (✉)
Faculty of Engineering, University of Nottingham, Ningbo, China
e-mail: mia-ardiati.tedjosaputro@nottingham.edu.cn

Y.T. Shih · P. Pradel
Department of Mechanical, Materials and Manufacturing Engineering, University of Nottingham, Ningbo, China

C. Niblock
Department of Architecture and Built Environment, University of Nottingham, Nottingham, UK

A. Chakrabarti (ed.), *ICoRD'15 – Research into Design Across Boundaries Volume 1*,
Smart Innovation, Systems and Technologies 34, DOI 10.1007/978-81-322-2232-3_3

across design domains (architecture and product design), through the application of sketches and mental imagery.

As far as contribution is concerned, the paper posits the possibility of identifying what is missing from the currently available coding schemes in order to capture the two environments concurrently. The questions to which the authors try to seek the answer is: "How to measure mental imagery?" Key issues are: cognitive aspects in the early design phase; how these aspects can be unfolded; and attention on constructing a more comprehensive coding scheme which is hoped to capture the essence of designers' intentions in both strategies. These aspects delineate the scope of the paper.

2 The Early Design Phase

During this phase, designers tend to intensively generate ideas with various strategies which entails generation of enormous amounts of information (Table 1). It is also comprised of jumps of activities without patterns and across different levels of detail [1, 2]. Apart from the designer him/herself, it also involves the external world, external sources of encoded information and the internal world- previous learning experience [3]. The ability to handle information processing requires cognitive skills.

2.1 Cognitive Strategies in the Early Design Phase

1. Visual analogy
   "The likeness of relations" is notably helpful due to the fact that designers are exposed to rich assortments of visual displays particularly during early stages [4] and it contributes to learning and enhancing creativity concurrently [5].
2. Visual thinking
   The process is a plausible strategy because it aids the process of reasoning by representing and assessing embodiments, in order to achieve a well synthesised final outcome [6].

Table 1 Comparison of cognitive strategies in conceptual phase, sketching and mental imagery

Conceptual phase	Drawing behaviour	Mental imagery behaviour
Visual analogy Visual thinking Shifting attention Multiple representations	Recognition and transformation Inexact strategies Structure of drawings Re-drawing Grouping Ambiguous	Uncertainty Incomplete knowledge Conscious and unconscious processes Recalling long term memory Pre-inventive forms Flash of insight of intuition Prior knowledge

3. Shifting attention
   The shifts are not just in attention, but also different aspects of tasks and modes of activity [7]. This ability has been conjectured as being the critical difference between novice and expert designers.
4. Multiple representation
   The purpose is to aid designers to formulate realities in the problem space, because it assists link creation and is related with figural and conceptual arguments [8].

2.2 The Act of Sketching and Mental Imagery

Due to the nature of visual thinking, sketches are beneficial because of the closeness to visual characteristics and often this is the first step when handling design tasks. It is suggested that sketches can facilitate the transformation of unformed ideas and turn them into inspectable entities for further exploration. The need to sketch is also related to the fact that the design outcome is in the form of drawings to communicate with other involved parties. Some established roles of sketching are to: aid descriptive to depictive translations [9], serve as external memory [10], enable designers to handle different levels of abstraction concurrently [2], suggest special kind of dialectics in design reasoning [11], support lateral transformations [12] and off-load the visual-spatial working memory [13]. The relevant current studies of sketching explore computer supported design, in comparison with free hand pen-to-paper method [14, 15]. Instead of making comparisons between the two kinds (traditional and digital environment); this paper is looking from the perspective of how the proposed exploration of pen-and-paper as dialogue with the mind could enhance development of digital tools.

Mental imagery in the early design phase is co-existent with the use of sketches. The dialogue between them is not a new concept, however research has posited a larger body of knowledge about the latter. In the creative domain, Anderson and Helstup [16] included an external representation (drawing) as additional evidence to inspect the effectiveness of mental imagery while revisiting Finke's experiments. Its role in design problem solving has been posited in literature: simulation of alternatives without the danger of real events, also brings up the idea of fanciful play [17], seeing with the mind's eye and the medium of visual memory [18] and to encode figures about currently non perceived objects [19]. In addition, criticism of mental imagery has drawn attention to the fact that the mental imagery is not sufficient to reduce uncertainty therefore sketches are needed [20] and might impede creativity by reluctance to come out and view design from different angles [3] leading to fixation.

The hypotheses are that the two design environments support the ideation process in unique ways and both are equally substantial; hence an empirical study of how exactly they occur in the design process is necessary. The related studies of research in sketching and mental imagery so far have explored their respective superiorities and supporting roles of each mode, leading to the following findings: externalisation enhances the imagery process by off-loading visuo-spatial working memory [13]; sketching is not an essential tool for expert architects in the early design stages, ascertained by looking at sketching and non-sketching environments [21]; similarly, drawings do not make significant contributions [22]; differences exist between experts and novices in both modes [23]; the significance of drawing lies in its supporting role to mental imagery [24]; the "restructuring" process needs to be enhanced by sketches [20]; and sketching is a special kind of dialectic with the mind [11].

In relation to the previous studies, it is intended to inform the salient issues pertaining to a study of sketching and mental imagery across disciplines (architecture and product design) as a parallel comparison, to better understand current thinking.

3 Protocol Analysis

The urge to understand how designers behave has been the subject of research for decades, with the immediate effect that one is able to the improve design process. The interaction between sketching and mental imagery is always assumed, but not understood; and provides the impetus of this research.

Various ways of investigating designers have been prescribed [2]: interview with designers; observations and case studies; protocol studies; reflection and theorising; and simulation trials. Protocol studies are seen as the most direct method to gain insight of problem-solving activities [25]. Also, the live mode of capturing data means that data will be automatically in chorological order. Accessing designer's internal thoughts, it is hoped to picture the design thinking from the produced drawings and verbalisation. One previously used method of observing the interplaying role between sketches and mental imagery is protocol analysis, which has been used as a main vehicle to unveil designers' activities [13, 17, 21, 26]. Other methods include combining or restructuring tasks using mental synthesis [16, 20, 22, 27]. Protocol analysis is used instead of tasks due to the fact that authors want to reveal the thinking process and to obtain more objective analysis. This in fact is corroborated by the finding of a previous study: participants' comments were contradictory with the analysis, insofar as the respondents felt that the mental imagery exercise was obstructive whereas the experimental results suggested otherwise [21]. This suggests that the empirical study is important from the standpoint of validity.

A number of limitations of this method have been posited in literature. The main argument is about whether protocol analysis is an effective tool to represent internal thoughts, especially non-verbal thoughts. Also, verbal reports are seen as intrusive due to the fact that more than one mechanism (thinking and talking out loud) happen concurrently. However, in the early observation study (by the authors of this report) of two architecture students who participated in the experiment in two different conditions (sketching and mental imagery), the respondents both concluded that the think-aloud method was not intrusive.

The usual processes of protocol analysis are encoding of design episodes, segmentation according to coding schemes and result representations. The selection of the most appropriate coding scheme affects the ability to demystify designers' attention and becomes one of the most fundamental stages in protocol analysis. The remainder of the paper will focus on the consideration related to coding schemes.

3.1 Criteria

There are two criteria suggested, which are: (1) the cognitive aspects, consisting of design activities in the early design phase, drawing behaviour and mental imagery behaviour and also (2) feasibility to be applied in other design disciplines (architecture and product design).

The cognitive aspects have been mentioned in an earlier section, derived from the process of designing which has been identified in literature. The second criterion is concerned with the array of possible applications of the coding scheme in different design domains (architecture and product design) as a means for comparing design activities across disciplines. Differences between architecture or product design processes are not specifically distinguished, thus, the protocol studies are used to reveal the differences.

3.2 Coding Schemes

With no intention of combining the two coding schemes, the authors try to provide a review of what is missing to be able to capture mental imagery concurrently. Suwa et al. [10] is the first scheme which will be explored. It is a development of their previous coding scheme which aimed to explore how external representations are able to crystalise design ideas [28]. The current coding scheme addresses deficiencies appearing in the previous one. The recent one (Table 2) was developed based on how meanings and concepts can be captured by external representations, literature in human cognitive processes (how information is perceived), literature in environmental cognition (where psychological responses are considered) and intensive observation of an architect.

The second coding scheme is the Function-Behaviour-Structure (FBS) framework developed by Gero. It was derived from observation of designers' interaction and models of design reasoning [29]. The underlying concept of this ontology is the idea of the purpose of designing, to transform function (F) to design description (D) through some design activities [30]. In addition, it is characterised with how designers navigate through design problems (the first two categories: Level

Table 2 Categorisations of the Suwa et al. [10] and FBS coding schemes [29]

S&T: major categories	Sub-classes	Description	FBS: dimension	Sub-dimension	Sub-dimension
Physical	D-action L-action M-action	Make depictions Look at previous depictions Other physical actions	Problem domain	Level of abstraction Function behaviour structure	– –
Perceptual	P-action	Attend to visual features of elements Attend to spatial relations among elements Organise or compare elements	Design strategies	Micro strategy	Proposing solution Analysing solution Explicit strategies Top down
Functional	F-action	Explore the issues of interactions Consider psychological reactions		Macro strategy	Bottom up Decomposing the problem Backtracking Opportunistic
Conceptual	E-action G-action K-action	Make preferential and aesthetic evaluations Set up goals Retrieve knowledge			

of Abstraction and Function Behaviour Structure) and strategies which have been used (the latter two dimensions: Micro Strategies and Macro Strategies); refer to Table 2.

Previous protocol analysis studies within the area of sketching and mental imagery have posited the need of constructing a more appropriate coding scheme to be able to make parallel comparison. For instance, Suwa and Tversky's (henceforth S&T) scheme needed to be modified to fit the studies by adding and eliminating categories [13, 21]. This however, makes the comparison rather less parallel between two different environments; also the interaction between environments might not be able to be revealed completely due to different segmentation.

As general comparison, the two coding schemes differ in terms of intended information to be perceived. S&T's scheme is closely related with sketches and paper and contains more spatial aspects than the FBS scheme. Both schemes decompose design actions into the smallest units and study interconnections between units. It is worth mentioning that they share the same features with Linkography developed by Goldschmidt, which notates and analyses design activities by examining links between design moves or design ideas [31]. In addition,

Table 3 Map on how FBS and S&T coding scheme able to capture conceptual phase, drawing and mental imagery behaviour

	Cognitive strategies	S&T	FBS
Conceptual phase behaviour	Visual analogy	–	√
	Visual thinking	√	–
	Shifting attention	√	√
	Multiple representation	–	√
Drawing behaviour	Recognition and transformation	√	√
	Inexact strategies	√	√
	Structure of drawings	√	√
	Re-drawing	√	√
	Grouping	√	√
	Ambiguous	√	√
Mental imagery behaviour	Uncertainty	–	√
	Incomplete knowledge	–	–
	Conscious and unconscious processes	√	√
	Recalling long term memory	√	√
	Pre-inventive forms	√	–
	Flash of insight of intuition	–	–
	Prior knowledge	–	√

to a certain extent, both have been similarly structured by external representation (e.g. the act of sketching).

Referring back to the two criteria, we may then enumerate each. Generally speaking, refer to Table 3, in terms of cognitive aspects; S&T and FBS are both applicable to mapping cognitive aspects in the act of sketching; however, to a certain extent both are less able to perform in the conceptual design phase and mental imagery observed by the mentioned cognitive behaviour in the conceptual design phase and also mental imagery behaviour. Common difficulties include capturing unpredictable events in imagery such as *flash of insight of intuition* and *incomplete knowledge*. This is due to the fact that there is no physical activity on the moment of events. To paraphrase, there is no concrete action.

In S&T, difficulties also occur when dealing with *visual analogy* and *multiple representation*; in that there are no direct actions which can be referred to. The common practice in which designers tend to develop ideas through observation and analysis through analogy cannot be captured as in S&T, as there is a lack of possibility to retrieve long perceived visual images. The K-action class consists of retrieving knowledge by forward and backward reasoning, however it is not detailed enough to decompose *visual analogy*. Dialogue held in *multiple representations* is also less likely to be captured with one single action. It is premature to postulate that the prescribed S&T scheme does not sufficiently capture the visual cognitive process, one of the fundamental activities in the early design phase; however this might suggest that possibly a formula of two or more actions will be able to represent a cognitive strategy. In addition, in terms of *shifting attention*, it can be captured by creating/attending to a new relation between depictions, its associations and re-interpretation. Drawing behaviour is able to be segmented to almost all categories: physical, perceptual and functional; and this is not surprising since it was developed based on sketching activity. Furthermore, ambiguity in mental imagery behaviour (*uncertainty, incomplete knowledge, flash of insight of intuition and prior knowledge*) is less likely to be captured as precisely as ambiguity is captured through drawing behaviour. Similarly in the early design phase, this can be noted, due to the limited actions able to categorise images from long term memory. Similarly with *the process of recalling long term memory* and *pre-inventive forms*; F-actions are able to capture the process of remembering, i.e. a function of depictions but hardly able to code precisely as to what kind of previous memory is involved.

In FBS, the two distinctive main categories are that the first two codes are based on the designer's place in the problem domain, and the last two are based on the designer's strategies [29]. In *visual thinking*, it is deemed to be slightly difficult to capture either positions, problem domain or strategies. The prescribed actions are seen to deal more with moving towards design solutions rather than capturing the thinking process. In a way, it is largely derived from the process of transforming function to design description. In addition, *shifting attention and multiple representations* can be captured by moving in different levels of abstraction. In drawing behaviour, similarly with S&T, it can be categorised well and is laid out in both main categories, "problem domain" and "design strategies". Ambiguity in design behaviour can be categorised in an object's manipulative actions, although under a broad category. As far as mental imagery behaviour is concerned, similarly with S&T, the FBS scheme is less likely to code the behaviour (dealing with *incomplete knowledge, pre-inventive forms* and *flash of insight of intuition*), especially dealing with pre-conceived ideas and uncertainty. Categorised design strategies are more involved with concrete steps of action (for example, proposing solutions or analysing solutions), however, they do not deal with the process of making the decision or adopting certain strategies. For instance, 'consulting external information' does not include how external information is perceived. The reconstructed FBS processes in the newly situated perspective has taken the framework of dealing with the agent's expected world, interpreted world and external world into consideration [32], however, a coding scheme which accommodates this is not available yet.

Furthermore, the second criterion pertaining to the feasibility to make comparison across design disciplines, the S&T scheme is closely related with sketches and paper (physical category-related with physical depictions on the paper) and contains more spatial aspects than the FBS scheme. What makes FBS more applicable in other areas are the models of process design [30] which characterised of transforming *function* (F) into *design description* (D), and the process is applicable to general engineering design. For instance, Kruchten mapped software design into the FBS framework [33].

The aforementioned coding schemes have left the consideration of the suitable scheme to be used to make comparisons between sketching and mental imagery. In terms of cognitive strategies, it is natural to conjecture that neither of the two mentioned schemes are ready for inception. This is similar to the argument of Eastman [3], who posited that there is no traditional protocol study which addresses internal representations used by designers. This remains the status quo up to this point of time, where the area is has seen no real development. However, some issues have been posited which need to be addressed such as: the ability to capture unexpected events, for example dealing with *incomplete knowledge* and *flash of insight of intuition*.

We then revisit similar research of internal and external representations beyond the design thinking area. Referring to Trafton et al. [34] about connecting internal and external representations, internal representation was coded to pure *spatial transformations*-mental operations used by scientists-by segmenting them to *create*, *manipulate* and *comparison*. Trickett and Trafton [35] then continued to explore the use of *conceptual simulation* to resolve *informational uncertainty* using protocol studies; and coded the utterances based on: *conceptual simulation* (visualising the situation, followed by mental operations and then seeing what happens) [34] existence based on mentioned *spatial transformation*, *hypotheses* (for statements which account for a phenomenon) and *reasoning strategies* (*data focus, empirical test, consult a colleague, tie-in with theory* and *domain knowledge and analogical reasoning*). What we can learn from the coding scheme is how mental operations are captured; by identifying references to new representation and an analysis to see the followed utterances to determine the mental operations; other than simply looking at the reasoning strategies. In addition, identification of hypotheses also points out that a hypothesis facilitates conceptual simulation as a source of comparison (alignment process), hence it is fundamental to observe its existence. In terms of the reasoning strategies, patterns are also identified which suggest how hypothesis evaluation takes place.

4 Conclusions

The paper is attempting to posit future needs of methods to capture mental imagery in order to make parallel comparison and identify the interplaying roles between mental imagery and the act of sketching. To answer the question: "How to measure mental imagery?" is achieved by looking at aspects in the early design phase; and posited that the act of sketching and mental imagery have different aspects to be compared. Two criteria have been established to construct a more appropriate coding scheme in the future: the cognitive aspects and feasibility to make comparison across design disciplines (architecture and product design).

It has been noted and observed that capturing mental imagery is not straightforward and therefore the appropriate selection of coding scheme in protocol analysis is indispensable. So far the two most established coding schemes have been compared; namely Suwa and Tversky's and Function-Behaviour-Structure. It is noteworthy that each coding scheme has struggled to measure the use of mental imagery, which has highlighted the need for a comprehensive scheme to address this point. Salient points pertaining to the ability to capture unexpected events such as *flash of insight of intuition* and *incomplete knowledge*; need to be considered for future use; similarly, the possibility to look at patterns of cognitive strategies. It is also suggested that identifying cognitive strategies are not sufficient to unfold mental operations in mental imagery environment. Finding mental operations by looking at the references of new representations and capturing phenomenon statements need to be explored further.

In the light of future development, the considerations posited in this paper will lead to a better understanding on how to construct a more comprehensive coding scheme and will be tested in the next study.

References

1. Dorst, K., Dijkhuis, J.: Comparing paradigms for describing design activity. Des. Stud. **16**, 261–274 (1995)
2. Cross, N.: Natural intelligence in design. Des. Stud. **20**, 25–39 (1999)
3. Eastman, C.: New Directions in design cognition: studies of representation and recall (Chap. 8). In: Eastman, C.M., McCracken, W.M., Newstetter, W.C. (eds.) Design Knowing and Learning: Cognition in Design Education. Elsevier Science, Oxford (2001)
4. Casakin, H., Goldschmidt, G.: Expertise and the use of visual analogy: implications for design education. Des. Stud. **20**, 153–175 (1999)
5. Ozkan, O., Dogan, F.: Cognitive strategies of analogical reasoning in design: differences between expert and novice designers. Des. Stud. **34**, 161–192 (2013)
6. Goldschmidt, G.: How I became a design researcher. In: Chakrabarti, A., Prakash, R.V. (eds.) ICoRD'13. Springer, India (2013)
7. Cross, N.: Design studies special issue: expertise in design. Des. Stud. **25**, 425–426 (2004)
8. Goldschmidt, G.: Capturing indeterminism: representation in the design problem space. Des. Stud. **18**, 441–455 (1997)
9. Fish, J., Scrivener, S.: Amplifying the mind's eye: sketching and visual cognition. Leonardo **23**, 117–126 (1990)
10. Suwa, M., Purcell, T., Gero, J.: Macroscopic analysis of design processes based on a scheme for coding designers' cognitive actions. Des. Stud. **19**, 455–483 (1998)
11. Goldschmidt, G.: The dialectics of sketching. Creativity Res. J. **4**(2), 123–143 (1991)
12. Goel, V.: Sketches of thought. MIT Press, Cambridge (1995)
13. Bilda, Z., Gero, J.S.: The impact of working memory limitations on the design process during conceptualization. Des. Stud. **28**, 343–367 (2007)
14. Tang, H.H., Lee, Y.Y., Gero, J.S.: Comparing collaborative co-located and distributed design processes in digital and traditional sketching environments: a protocol study using the function–behaviour–structure coding scheme. Des. Stud. **32**, 1–29 (2011)
15. Huang, Y.: Investigating the cognitive behavior of generating idea sketches through neural network systems. Des. Stud. **29**, 70–92 (2008)
16. Anderson, R.E., Helstrup, T.: Multiple perspectives on discovery and creativity in mind and on paper. In: Roskos-Ewoldsen, B., Intons-Peterson, M.J., Anderson, R.E. (eds.) Imagery, creativity and discovery: a cognitive perspective. Elsevier Science, Amsterdam (1993)
17. Athavankar, U.A.: Mental imagery as a design tool. Cybern. Syst. **28**, 25–41 (1997)
18. Oxman, R.: The thinking eye: visual re-cognition in design emergence. Des. Stud. **23**, 135–164 (2002)
19. Denis, M.: Mental imagery, psychology of. In: Smelser, N.J., Baltes, P.B. (eds.) International Encyclopedia of the Social and Behavioral Sciences. Pergamon, Oxford (2001)

20. Verstijnen, I.M., van Leeuwen, C., Goldschmidt, G., Hamel, R., Hennessey, J.M.: Sketching and creative discovery. Des. Stud. **19**, 519–546 (1998)
21. Bilda, Z., Gero, J.S., Purcell, T.: To sketch or not to sketch? That is the question. Des. Stud. **27**, 587–613 (2006)
22. Kokotovich, V., Purcell, T.: Mental synthesis and creativity in design: an experimental examination. Des. Stud. **21**, 437–449 (2000)
23. Kavakli, M., Gero, J.S.: Sketching as mental imagery processing. Des. Stud. **22**, 347–364 (2001)
24. Purcell, A.T., Gero, J.S.: Drawings and the design process: a review of protocol studies in design and other disciplines and related research in cognitive psychology. Des. Stud. **19**, 389–430 (1998)
25. Solomon, P.: The think aloud method: a practical guide to modelling cognitive processes. Inf. Process. Manage. **31**, 906–907 (1995)
26. Athavankar, U., Arnab M.: Blindfolded classroom: getting design students to use mental imagery. In: Human Behaviour in Design, pp. 111–120. Springer, Berlin (2003)
27. Finke, R., Slayton, K.: Explorations of creative visual synthesis in mental imagery. Memory Cogn. **16**, 252–257 (1988)
28. Suwa, M., Tversky, B.: What do architects and students perceive in their design sketches? A protocol analysis. Des. Stud. **18**, 385–403 (1997)
29. Gero, J.S., Mc Neill, T.: An approach to the analysis of design protocols. Des. Stud. **19**, 21–61 (1998)
30. Gero, J.: Design prototypes: a knowledge representation schema for design. AI Mag. **11**, 26 (1990)
31. Goldschmidt, G., Tatsa, D.: How good are good ideas? Correlates of design creativity. Des. Stud. **26**, 593–611 (2005)
32. Gero, J.S., Kannengiesser, U.: The situated function–behaviour–structure framework. Des. Stud. **25**, 373–391 (2004)
33. Kruchten, P.: Casting software design in the function-behavior-structure framework. IEEE Softw. **22**, 52–58 (2005)
34. Trafton, J.G., Trickett, S., Mintz, F.: Connecting internal and external representations: spatial transformations of scientific visualizations. Found. Sci. **10**, 89–106 (2005)
35. Trickett, S.B., Trafton, J.G.: What if…: the use of conceptual simulations in scientific reasoning. Cogn. Sci. **31**, 843–875 (2007)

Ways of Drifting—Five Methods of Experimentation in Research Through Design

Peter Gall Krogh, Thomas Markussen and Anne Louise Bang

Abstract Design experiments are claimed to be a core means of inquiry in the research tradition of research-through-design. However, it is rarely articulated how the experiments were carried out in order to test a hypothesis, to begin a fruitful journey into unexplored design terrain or just gradually build knowledge. On the basis of the analysis of ten PhD theses we provide a typology comprised of five forms of design experiments in research-through-design. This provides a general outline of the characteristics which point to the methodological roles that design experiments and design work may acquire in research-through-design. Our typology of design experiments in research-through-design accounts both for relations between major cases and iterations embodied in detailed sketches and prototypes. The purpose of the typology is to provide an overview that respects and account for the less-than-ideal way design research actually happens: process-loops where hypothesis, experiments, and insights concurrently affect one another and result in a drift of research focus and continued adjustment of experiments to stabilize the research endeavour.

Keywords Research-through-design · Methods · Experiments

1 Introduction

Research-through-Design (RtD) is increasingly practiced across engineering-informed and artistic-based design research. Classical processes of research regard "drifting" as a failure since measures and grounds of evaluation can be said to be

P.G. Krogh (✉)
Platform Reform, Aarhus School of Architecture, Aarhus, Denmark
e-mail: peter.krogh@aarch.dk

T. Markussen · A.L. Bang
Design School Kolding, Kolding, Denmark

A. Chakrabarti (ed.), *ICoRD'15 – Research into Design Across Boundaries Volume 1*,
Smart Innovation, Systems and Technologies 34, DOI 10.1007/978-81-322-2232-3_4

in flux. In design, however, "drifting" is a quality measure as it tells the story of a designer capable of continuous learning from findings and of adjusting causes of action. Design Research that does not account for this professional hallmark will fail to gain respect from or build a better basis for design practice. Tolerance of "drifting", however, points to a built-in dilemma of Design Research and, in particular, RtD when sharing knowledge across research disciplines: to what degree can one trust the results of RtD?

Based on 10 exemplary and well-cited PhD theses developed in environments emphasizing artistic quality we describe continued design experimentation in processes of research, and how such chains of experimentation and sketching follow a methodological rigor and how it can be modelled. The selected theses have been developed and written in the Dutch-Scandinavian-Anglo-Saxon tradition and submitted for evaluation in art schools, universities and academies exemplifying the heterogeneity of design research in this tradition. The theses represent the full spectrum of the "Lab, Field, Showroom" taxonomy described by Koskinen et al. Thus the selected theses also exhibit the full spectrum of classical research traditions though they are committed to aesthetic and artistic assessments in design and design research.

Maturation of the theoretical foundation of RtD provide grounds for better-designed research projects and programs. Such work will also help declare the qualities, strengths and weaknesses of RtD in the general landscape of research approaches facilitating interfaces to other research disciplines without adopting a whole set of theories developed in classical research areas and enable design researchers' participation in the language game of research across other disciplines.

This paper is intended in particular to help PhD candidates and applicants, assessors of applications and PhD supervisors in RtD. Although not complete, the typology we present offers a perspective on how case design, actual design work and exploratory sketching as knowledge-building activities can be better substantiated, declared and delimited in the early phases of research work and descriptions. The typology we present is comprised of five forms of design methodologies that we have labelled *accumulative*, *comparative*, *serial*, *expansive* and *probing*.

In this paper we are concerned with the actual internal work activities of RtD processes—designing stuff. We are concerned with how the work itself can be said to exhibit the essential transparency of research, rather than how described knowledge flows in communities and is used to substantiate a research contribution. Thus we are dealing with building and sustaining the trustworthiness of results developed by the single design researcher adopting the RtD methodology, and not how others build upon the generated knowledge.

2 Related Work

RtD is coming of age and numerous researchers have made great and valuable efforts in describing and establishing the foundational theories, methods and approaches of the research area [e.g. 1–8].

Although many of the fundamentals of the research area are well described they also impact different research communities with limited overlap, and yet lack the pervasiveness that research fundamentals need in order to work on equal grounds with other research disciplines and traditions. In recent times there have also been suggestions of using endless and inconsistent prepositions in order to distinguish the methodological role of design experiments from other more established research methodologies [9]. Thus, research *through* design has been defined as different from "research *on* design", "research *into* design", "research *in* design", "research *for* design" and "research *by* design" [10]. As the expanded use of prepositions has only led to obscuring the understanding of RtD, Koskinen and Krogh [11] has recently argued for settling on the term "constructive design research" as the fundamental epistemology. However we will, for the sake of clarity in tradition, stick to the use of "Research-through-Design" (RtD) as conceptualizing research done by means of the skilful practice of design activity revealing research insights.

2.1 On the Details of Working in RtD

In RtD and the definition of its foundations there appear to be two major strands: one that defends the specificity and, compared to other research disciplines, capacity to deliver results that one does not find in other fields e.g. Cross [2], Gaver [12], Seago and Dunne [13] and, to some extent, Zimmerman and Forlizzi [7]. The second strand models the foundations of design research upon already identified research traditions, notably the natural sciences, social sciences and art e.g. Koskinen et al. [4] and Steffen [14].

By the work we present here, we argue that a proper account of the fundamentals of RtD must begin with a close scrutiny of design experiments. In so doing, we agree on the one hand with Gaver and others that this is a prerequisite for acknowledging the specificity of RtD. But, on the other hand, we agree with Koskinen et al., insofar as we argue that we should be careful not to put so much emphasis on its specificity that its possible cross-disciplinary connections and exchanges with other forms of research are lost. This can be avoided if we use the classical vocabulary of research (hypothesis, motivation, research question, experiment), in the development of a methodological explanation of design experiments.

When examining the existing research literature, one is left empty handed when searching for detailed accounts of the process and basic constituents of design experiments. Zimmerman and Forlizzi [7] attempt to develop a formal account of methods used in RtD and suggest that a foundational distinction should be made between two different methodological approaches: (i) a philosophical approach, where researchers wish to "investigate a previously articulated theory through a process of making" (e.g. 'ludic interaction', 'rich interaction', 'aesthetics of interaction', etc.); and (ii) a grounded approach, where researchers focus "on real-world problems by making things that force a concrete framing of the problem".

However, such a meta-level classification is too abstract, and it does not provide insight into how design practice and experiments are used in either of these approaches.

In order to align the methodological foundation of RtD with the practices of professional design Brandt and Binder [15] suggested that design experiments in design research can be better understood as being framed by a "program" and a "research question". In so doing Brandt and Binder assume that RtD can be modelled on the same conditions as a design project where a program or brief is typically used by a client to formulate an assignment for a professional designer. While Brandt and Binder expend much effort in defining the notion of the program and research question, they end up saying surprisingly little about the design experiments themselves.

In the 2012 paper by Bang et al. [16] it is argued that it is counterproductive to the development of RtD to use concepts from outside research to define its foundations and it is more beneficial to provide an account of the knowledge production of the field based on established concepts e.g. hypothesis, experiment, evaluation while declaring how and to what extent design work is similar to and different from more classical scientific disciplines. Such differences are also dealt with in Redström's paper "Some notes on program/experiment dialectics" [17] in which he discusses how design research has a nature of "drifting" that in other research disciplines would be regarded as watering down the research contribution. Traditionally, in science literature, drifting is regarded as bearing the touch of randomness, the uncontrolled, illogical and inconsistent. However, in design research and in particular the professional practice of design, drifting or pursuing alternative opportunities in the vicinity of one's work is an embedded way of arriving at relevant and high quality work. The work we present in this paper is very much in line with these concerns. However, we argue for maintaining the use of classical terms, as they are sufficiently spacious to accommodate the way RtD develops knowledge.

Furthermore there is a slight tendency in design research to also be mostly interested in the final product of a design case. However, in line with Bowers and Gaver [18] we believe it can be of even more importance to declare how one got there—how the design project drifted through and gained insights unintended by its original pursuit—and what knowledge one developed doing so. In line with [4, 12] it is our point of departure that knowledge production in design and RtD can be characterized as fallibilistic. As such any sketch is a question examining parts or the whole of a provisional hypothesis, not in a Popperian sense where a hypothesis is substantiated or falsified, but in a Piercean manner qualitatively informed or questioned through "abductive reasoning" [19] or "bold suggestions" which design experiments and sketches can be characterized as. According to Brandt and Binder [15] experiments in research-through-design are examinations of questions residing in research programs. As such Brandt and Binder do not relate their framework to hypothesizing and accept design research is also a way of describing the urge to explore an interesting concept without a well-defined hypothesis. Bang et al. [16] formulate the act of hypothesizing as a fruitful and "direction-providing" activity in design. Whereas the work of Zimmerman and Forlizzi [7] articulates design research as being theoretically, technically or empirically inspired, Bang et al. [16] follows the line of Brandt and Binder [15] and Zimmerman et al. [8] in voicing the need to include aesthetically and artistically inclined research interests.

3 Typology

As stated in the beginning of the paper, the purpose here is to provide design researchers an extended footing when participating in the language games of research in general. At the same time the intention is to avoid losing the specificity and unique quality they design research has to offer in research in general and society as a whole.

Building on the work of Bang et al. [16] we present a typology depicting how discreet but linked experimental activities in RtD can be described in terms of how they facilitate building knowledge by acting in combination. The classical foundational activity of design is "sketching". It is the dominant means by which ideas are described and evaluated for their quality and appropriateness in responses to a design challenge. Sketches are a (materialized) means for dialogue between the designer and her design challenge. Sketches also act as boundary objects for the designer and stakeholders in a design process [20–22]. Sketches can be temporal materializations of ideas subject to rapid changes, incremental as well as radical changes; sketches can also be materializations of ideas of parts of a whole. As this indicates, experimentation in design research is often intimately linked to evaluation—not necessarily formal and thorough evaluation but at the level of sanity checking. There is no second sketch without even the most rudimentary, theoretical, experiential or aesthetically inclined evaluation of the first. It is, however, beyond the scope of this paper to go deeper into discussing evaluation in RtD.

Processes of development through design and its constituent parts can be likened to babushka dolls, onion layers, ontologies of ideas, free streams of associations, arguments of logic etc. Although such descriptions carry a lot of experiential power when teaching design students the practice of design, it provides little help in enabling research peers from different scientific backgrounds understand the benefits and possibilities for co-research. Along these lines, Gaver [12] argues for annotated portfolios as tool for connecting design artefacts to theoretical concerns and design values across the domain and use of artefacts.

Bang et al. [16] describe a model of how experimentation is the cogwheel of RtD in dialogue with research activities such as hypothesizing, theorizing etc. We build on this model and provide a typology of experimentation in RtD. The model is derived from the analysis of ten theses (Dindler [23], Busch [24], Niedderer [25], Worbin [6], Trotto [26], Ross [27], Fogtmann [28], Frens [29], Borup, Bang [16]) and in particular their reporting on design experimentation. The selected theses have a cultural bias towards the Dutch-Anglo-Saxon-Scandinavian approach, thus other traditions might challenge and extend the typology.

The typology describes five distinct methods of knowledge production through design experimentation: *Accumulative*, *Comparative*, *Serial*, *Expansive* and *Probing* (Table 1). Furthermore, Table 1 includes a graphical representation of each of the methods by characteristic keywords, a graphical model and author names of PhD theses that exemplify the methodology. All the presented methods allow for "drifting"—although to a varying degree. The first category "accumulative" is the least forgiving and "probing" allows for the largest degree of "drifting".

Table 1 Table of typology

Method	Graphic model	Keywords	Exponent(s)
Accumulative		Depth, stacking	Frens
Comparative	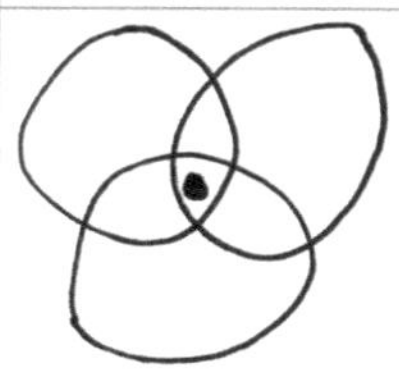	Acknowledging complexity	Ross, Fogtmann
Serial	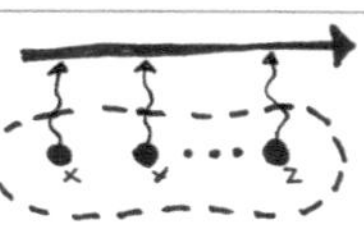	Systematising local knowledge	Lynggaard, Bang
Expansive		Broadening, extending	Dindler, Trotto
Probing	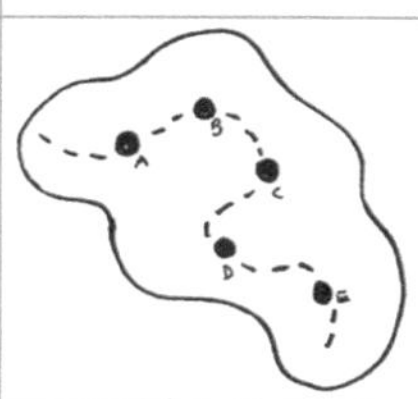	Illogical, artistic, impact oriented	Busch, Worbin

3.1 Accumulative

This method of experimentation can be found in the work of Frens [29]. His design experiments study how tangible interaction might enhance the experience of using a camera.

The design sketches and models are focused on testing specific parts and wholes and are carried out in closed lab settings where the design experiments are evaluated for their cognitive qualities, rather than contextual appropriateness. The work shares many learning and experimental similarities with what happens in technical lab settings where one particular thing is studied, and potentially disturbing elements are excluded for the sake of clarity and rigor in the study. What the study loses in relevance it gains in depth of knowledge on the particular. We use characterizing keywords "depth" and "stacking" to describe the design and experimental interest; the increasing depth of knowing derived from every experiment is iteratively build (layered, stacked) into the next generation of the same version of the camera, a way of stacking knowledge where the final artefact embodies the total knowledge accumulated through the RtD process.

3.2 Comparative

The method of comparative experimentation can be found in the work of Fogtmann [28] and Ross [27]. They both explore their subject by means of a number of design cases—working from or towards a shared platform of comparison. Whereas Ross is interested in ethical and aesthetic aspects of interactive products (in particular lamps) Fogtmann describes the concept Kinesthetic Empathy Interaction (KEI) through a series of design cases each highlighting distinct and overlapping qualities of KEI. The reason for using the experimental method is in both theses to do case-relevant explorations, which cover areas and aspects not yet dealt with in other experiments and to incorporate knowledge from previous experiments. The method may comprise one central design case tried out in a range of contexts or a set of different design cases tried on both identical and different contexts. It may also comprise iterative versions of the same concept changed according to context. The basic approach is that each design experiment should reveal as-yet undocumented additional qualities of a concept and confirm some previously found qualities. In totality, the comparative experiments ideally describe a novel concept, qualify phenomena or add a theoretical distinction to known theory. A characterizing keyword in this model is "acknowledging complexity", which expresses the idea that the design experiments explore the concept by pointing to how it is embedded in a multiplicity of situations. Furthermore the method reveals that a lot of experimental design work done will not necessarily find its way into knowledge production.

3.3 Serial

The method of serial design experimentation denotes how design experiments are being carried out in a certain order or logic of locality determined by how neighbouring experiments in a sequence influence one another.

Complementing the comparative method, knowledge production in the serial method is achieved on the basis of insights gained into the relationships between design experiments that proceed chronologically. In the work of Lynggaard [30] we find each successive experiment is framed on the basis of its predecessor. Each stage generates insights or raises questions that lead the work onward. These pointers provide large and small contributions to the overarching interest in "homing tactics". More specifically, on the basis of ethno-methodological studies, Lynggaard identifies a set of tactics for "making home"—or "homing" as she denotes it. Rather than following a strategic approach, Lynggaard adopts an opportunistic and pragmatic approach where the identified tactics are further explored in concrete design experiments based on equal measures of pragmatic concerns (time, technical request, budget, company interests etc.) and the experiments' capacity to yield additional contributions to the overall research interest. Likewise in the work of Bang [31] we find a series of experiments, where each experiment continually builds on the previous one. In this case, the main interest is an exploration of emotional value of applied textiles from various perspectives, i.e. 'textiles as material', 'textiles as part of an object', and 'textiles as part of an object in an environment'. Inviting stakeholders to participate in this exploration the objective with the experiments was twofold. On the one hand they resulted in the development of an in-depth knowledge of emotional aspects of textile design and, on the other hand, they were stepping stones in developing a structured approach to inviting stakeholders to participate in the industrial textile design process. A key characteristic of this method is: "systematizing local knowledge".

3.4 Expansive

This method articulates the identification of an area as-yet uncovered with the ambition to reveal its qualities, a mode of investigation resembling the work of geographers or biologists mapping new areas. The work of Dindler [23] and Trotto [26] are exemplars of this. Unlike serial experimentation there are no strict successive or linear orders or directions to follow. Experiments and learning from this will contribute new knowledge, as the area is explored. The characterizing keyword for this method is "broadening" and "extending". Rather than deepening our knowledge of a domain, this method widens our perspective and extends the concerns we, as designers, should include in our praxis. Trotto does this through a set of experiments, primarily workshops, that continuously explore new aspects, approaches and techniques for teaching design students to acknowledge that human rights can be enhanced or suppressed in acts of designing and making.

Dindler expands our idea of what "engagement" might mean in interaction design through three diverse experiments including designing engaging artefacts and developing methods for participatory design centred on engagement.

3.5 *Probing*

Exploiting opportunities and exploring design ideas as they emerge through design work is also what characterizes the final method described here: probing.

The approach is widely used in design research and well documented in Worbin [6] and Busch [24], for example. Yet, it is only when we examine probing in relation to the other four methods that its methodological value for design research can be fully grasped. What often characterizes this methodological approach is a personal motivation and engagement in the research pursuit, where the research activities are points of impact in a research field larger than what a single research project can be expected to cover. The choice of experiments in [6, 7, 24] can be characterized as "illogical", "artistic" and "impact oriented". Worbin is interested in the merging of IT and textiles in the very broad sense. Through a number of experiments she highlights recurring and important aspects of this mirage of material and experiential properties and qualities. Busch is interested in hacktivism as part of democratizing fashion production. Both the theses and, presumably, their doctoral studies are logically structured endeavours exploring the qualities of a field. However, theses two are characterized by selecting in an almost eclectic manner wicked, ir-reductive and self-contradictive design settings derived from pursuing opportunities in the environment (as a professional designer would do). From both a practice and research point of view this strongly test their subjects. On the basis of such experiments they make contributions valuable to design research and foster curiosity for the field itself and its neighbouring areas.

4 Reflections on the Typology—and Future Work

Compared to previous discussions and theorizing on RtD our typology contributes to the existing body of knowledge in at least three respects.

First, the five diagrams of the models can serve as an explanatory visual tool for clarifying what form(s) of experimentation will be most relevant for the research question one wants to address. We deliberately allow for the plurality of **forms**. Even though we have managed here to identify a number of PhD theses that are representative of one form of experimentation, it is possible in one design research project to switch between different experimental *modus operandi*. Thus, the five forms are not mutually exclusive as can be seen in Kinch [32]. But their use demands careful consideration of what kind of knowledge interest one has and which form is deemed most appropriate for its exploration.

Secondly, our typology allows for a concise description of different knowledge outcomes that may result from design experimentation: *depth* or *stacking* of knowledge, *acknowledging complexity*, *extending knowledge* of a certain area, and so on. Typically, in the research literature we have consulted such descriptions are not given. Rather knowledge outcomes are classified generally in terms of, for instance, "nascent theory", "conceptual frameworks", "guiding philosophies" or "design implications" [8]. However, it is not irrelevant to ask whether a conceptual framework is "broad" or "systematic". For instance, Forlizzi and Battarbee's [33] framework for understanding user experience in interactive systems is broad as it conceptualizes user experience in terms of basic psychological categories (fluent, cognitive and expressive), while Desmet's [34] framework for emotional design is meticulously worked out as a fine-grained system of emotions defined according to a varied set of distinctive traits and thus systematic. Our typology can help to clarify the nature and generality of knowledge outcomes and eventually to set up valid evaluation criteria for assessing this knowledge.

Third, beyond providing a means to distinguish methods of experimentation in design research, the above typology also to a varying degree depicts a spectrum of methods that have a heavy or light foothold in classical research and science. Accumulative experimentation and serial experimentation have a strong foothold within natural sciences, where controlled experiments are carried out in order to gain deep knowledge of a phenomenon, while expansive experimentation and comparative experimentation are typically represented within social sciences. However, a more elaborated account of how our typology aligns with methods from other research traditions cannot be dealt with here, but is postponed for future work. This future work is aligned with previous work on hypothesizing in RtD [16] and includes a paper in preparation dealing with evaluation in RtD. The three papers are all intended for conferences and will form the basis of a book including a wider study of theses expanding and adjusting the grounds of the claims in the papers.

5 Conclusion

In this paper, we have demonstrated that methodological foundations of RtD can be derived from a careful analysis of how design experiments are used during an inquiry. More specifically we offer a typology for understanding the way in which an experiment may drift throughout a research process. Obviously, since our typology is made up inductively from the analysis of only 10 PhD theses it is not in any way meant to be exhaustive, but merely indicative. It makes visible a potential route for developing a firmer epistemological ground for research practices, which exploit artistically inclined activities and aesthetic practices as their primary vehicle for knowledge production. Our hope is that this can be of help to doctoral students, their supervisors, evaluation committees and research peers who share an interest in grasping the specificity of RtD while also wanting to know how it relates to research traditions outside design research.

Acknowledgments We would like to thank all the international PhD students who participated in the series of three doctoral courses focusing on research-through-design that we organized from 2012–2014 and who helped us critically access the various forms of experimentation in design research. Furthermore we'd like to thank the anonymous reviewers for their valuable and insightful comments and Richard Herriott for helping with editing.

References

1. Archer, L.B.: Whatever became of design methodology? Des. Stud. **1**(1), 17–20 (1979). ISSN 0142-694X
2. Cross, N.: Designerly ways of knowing. Springer, London (2006). ISBN 978-1-84628-300-0, OCLC 63186849
3. Frankel, L. et al.: The complex field of research: for design, through design, and about design. DRS Montreal—Design and Complexity (2010)
4. Koskinen, I., Zimmerman, J., Binder, T., Redström, J., Wensveen, S.: Design Research Through Practice: From the Lab, Field, and Showroom. Morgan Kaufmann, Waltham (2011)
5. Markussen, T., et al.: Dynamic research sketching—a new explanatory tool for understanding theory construction in design research. In: Proceedings of the Design Research Society Conference 2012, Chulalongkorn University Bangkok, Thailand (2012)
6. Worbin, L.: Designing dynamic textile patterns. PhD-Dissertation (2010)
7. Zimmerman, J., Forlizzi, J.: The role of design artifacts in design theory construction. Artifact 2(1), 41–45 (2008)
8. Zimmerman, J., Stolterman, E., Forlizzi, J.: An analysis and critique of research through design: towards a formalization of a research approach. In: Proceedings of the Conference on Designing Interactive Systems, pp. 310–319. ACM Press (2010)
9. Steinö, N., Markussen, T.: Design research between design and research, "when architects and designers write, draw, build? a PhD". In: The 2011 Symposium of the Nordic Association of Architectural Research, Aarhus School of Architecture, 4–6 May 2011
10. Sevaldson, B.: Discussions and movements in design research: a systems approach to practice research in design. FORMakademisk **3**(1), 8–35 (2010)
11. Koskinen, I., Krogh, P.G.: Design Accountability (under submission)
12. Gaver, W.: What should we expect from research through design. In: Proceeding of CHI'12, ACM Austin, Texas (2012)
13. Seago, A., Dunne, A.: New methodologies in art and design research: the object as discourse. Des. Issues **15**, 11–17 (1999)
14. Steffen, D.: Characteristics and interferences of experiments in science, the arts and in design research. In: Proceedings of the NORDES'13 Conference, Copenhagen & Malmö (2013)
15. Brandt, E., Binder, T.: Experimental design research: genealogy–intervention–argument. In: International Association of Societies of Design Research 2007: Emerging Trends in Design, Hong Kong, China, 12–15 Nov 2007
16. Bang, A.L., Krogh, P., Markussen, T., Ludvigsen, M.: The role of hypothesis in constructive design research. Art Res 2012 Mak. Reflecting Underst. (2012)
17. Redström, J.: Some notes on program/experiment dialectics. In Proceeding of Nordes (2011)
18. Bowers, J., Gaver, W.: Annotated portfolios. Interactions **19**(4), 40–49 (2012)
19. Peirce, C.S.: The collected papers of charles sanders peirce, vols. VII–VIII. In: Burks, A.W. (ed.). Harvard University Press, Cambridge (1958)
20. Cross, N.: Discovering design ability. In: Buchanan, R., Margolin, V. (eds.) Discovering Design. Explorations in Design Studies, pp. 105–120. The University of Chicago Press, Chicago
21. Lawson, B., Dorst, K.: Design Expertise. Routledge, Taylor & Francis Group, New York (2009)
22. Schön, D.: The Reflective Practitioner. How Professionals Think in Action. Basic Books, New York (1983)

23. Dindler, C.: Fictional space in participatory design of engaging interactive environments. PhD Dissertation, Aarhus University, Denmark (2010)
24. Busch, O.V.: Fasionable, hacktivism and engaged fashion design. PhD Dissertation, University of Gothenburg, Sweden (2008)
25. Niedderer, K.: Designing the performative object: a study in designing mindful interaction through artefacts. PhD-Dissertation, Falmouth College of Arts, United Kingdom (2004)
26. Trotto, A.: Rights through making. PhD-Dissertation, Eindhoven University of Technology, The Netherlands (2011)
27. Ross, P.: Ethics and aesthetics in intelligent product and system design. PhD-Dissertation, Eindhoven University of Technology, The Netherlands (2008)
28. Fogtmann, M.H.: Designing with the body in mind. PhD Dissertation, Aarhus School of Architecture (2012)
29. Frens, J.W.: Designing for Rich Interaction: Integrating Form, Interaction, and Function (2006). ISBN-10: 90-9020538-1
30. Lynggaard, A.B.: Homing interactions: tactics and concepts for highly mobile people. PhD Dissertation, Aarhus School of Architecture (2012)
31. Bang, A.: Emotional value of applied textiles—dialogue-oriented and participatory approaches to textile design. PhD Dissertation, Kolding School of Design, Denmark (2010)
32. Kinch, S.: Designing for atmospheric experiences—taking an architectural approach to interaction design. PhD Dissertation, Aarhus School of Architecture (2014)
33. Forlizzi, J., Battarbee, K.: Understanding experience in interactive systems. In: Proceedings of DIS2004, 1–4 Aug, pp. 261–268, Cambridge, Massachusetts, USA (2004)
34. Desmet, P.: Designing Emotion. Delft University of Technology, Delft (2002)
35. Sanders, E.B., Stappers, P.J.: Co-creation and the new landscapes of design. CoDesign Int. J. CoCreation Des. Arts **4**(1), 5–18 (2008)

Design Space Configuration for Minimizing Design Information Entropy

Petter Krus

Abstract In this paper concepts for quantitative theory of design are developed. The design process is viewed as a process of increasing the information of the product/system using concept from information theory. The creation and evolution of design space and information entropy are central concepts that are studied in this paper. Furthermore, axiomatic design is discussed in relation to this, and it is shown that the two fundamental axioms are likely to be dependent. The design information entropy is used as a state that reflects both complexity and refinement, and it is argued that it can be useful as some measure of design effort and design quality. It provides a tool to describe the relation between a design, and the design space from which it is derived as well as for quantitative description for various aspects of design, both structural information regarding architecture and connectivity, as well as for parameter values, both discrete and continuous. In this paper it is also shown to be useful as a measure of quality when configuring design spaces, such as in product platforms, and when setting up parameterization for design.

Keywords Design entropy · Information theory · Design space

1 Introduction

During the process of design, information is gradually increased as the design progress, and the uncertainty of the design is reduced. Every design decision makes it more defined and thus reduces the uncertainty of the design, as well as parameter

P. Krus (✉)
Department of Management and Engineering, Division of Fluid and Mechatronic Systems, Linköping University, Linköping, Sweden
e-mail: petter.krus@liu.se

A. Chakrabarti (ed.), *ICoRD'15 – Research into Design Across Boundaries Volume 1*, Smart Innovation, Systems and Technologies 34, DOI 10.1007/978-81-322-2232-3_5

refinement do. It could be argued that this is a central aspect of design. This can be viewed as a learning process, as described in Ullman [1]. Therefore design theory should really be a theory of design information. Although there is a rich literature regarding design process, there has been little or no effort to describe the generation of information in quantitative terms.

A central concept in this study is the *design space*. In fact, design information have no meaning unless it is related to some kind of design space. The design space is here defined as the space within which the design can vary through parameter variations. The parameters are here any coding used to define the design, e.g. design parameter values, information about architecture, components, connectivity etc. A product platform also defines a design space through the different combinations of components that can be chosen. Corresponding to this is also a *functional space*, which is the space with the functional characteristics corresponding to designs within the design space.

Typically, first a system level design space is first refined into a single product/system level specification. In the next step the design space is expanded by adding more details and releasing more parameters into the design space, that is then further refined.

1.1 Information Theory and Design

The classical information theory of communication was founded in 1948 by C.E. Shannon with his paper "A Mathematical Theory of Communication", Shannon [2]. Subsequently it has been recognized that information is a key property of design, and for describing and analyzing the design process. The notion of information theory in design has been introduced by several authors. Notables are Suh [3], Kahn and Angeles [4] and Frey and Jahingir [5]. The two first are discussed later in the text. Frey and Jahingir deals with the transformation of information content in design parameters to information content in the functional characteristics, so is Bras and Mistree [6] where robustness is defined from maximizing the signal to noise ratio, which also deals with the relation between design parameters and functional characteristics. Information theory was also used to analyze design optimization in Krus and Ölvander (Andersson) [7] and [8]. Information theory has also been used to define complexity in software, and notably by Bansiya et al. [9] to describe complexity in object oriented systems, which is very close to general design.

The notion of information is used in the second axiom of N.P. Suh's Axiomatic design, which states that the information content in a design should be minimized, Suh [10].

In Krus [11] design information entropy was introduced as a measure of information needed to define a particular design relative to a design space. In this study this is extended to modular design and to parameterization.

1.2 Design Information Entropy

The definition of Information entropy for the discrete case is defined by Shannon [2]

$$H_d = \sum_{i=1}^{n} p_i \log_2 p_i \tag{1}$$

Here the system can be in n different states with probabilities p_i for each of them.

A more general definition than the information entropy for the discrete case is the differential information entropy for continuous signals, defined by Shannon [2] as:

$$H_c = -\int_{-\infty}^{\infty} p(x)\log_2(p(x))dx \tag{2}$$

The differential entropy of the probability density function $p(x)$ can also be related to another distribution $m(x)$. The result is called the Kullback-Leibler divergence [12] from the distribution $m(x)$. This is the relative entropy, and it is defined as (multi variable case):

$$H_{rel} = \int_{-\infty}^{\infty} \cdots \int_{-\infty}^{\infty} p(x_1 \ldots x_n)\log_2(\frac{p(x_1 \ldots x_n)}{m(x_1 \ldots x_n)})\, dx_1 \cdots dx_n \tag{3}$$

A rectangular distribution of $m(x)$ in the bounded interval $x \in [x_{0,min}, x_{0,max}]$, with $x_R = x_{max} - x_{min}$, would mean that the distribution $m(x)$ of the *design space* is a space of equal possibilities, where no particular region can be considered more likely than another a priori. This ca be can then be rewritten as:

$$H_x = H_{rel}(x) = \int_{x_{min}}^{x_{max}} p(x)\log_2\,(p(x)x_R)dx \tag{4}$$

The H_x is here used here to indicate relative information entropy related to a rectangular distribution, and it has the unit *bits*. It is here defined as the *design information entropy*. For the multidimensional case it can also be written in a more compact form as:

$$H_x = \int_D p(\mathbf{x})\log_2(p(\mathbf{x})S)d\mathbf{x} \tag{5}$$

where D is the design space within where the design $\mathbf{x}$ is defined. S is the size of the design space and is defined as:

$$S = \int_D \mathbf{x}d\mathbf{x} \tag{6}$$

If the range of one variable is divided into equal parts Δx that have the same probability, the probability density distribution will be:

$$p(x) = \frac{x_R}{\Delta x}: \quad x \in [x_{min}, x_{max}]$$
$$p(x) = 0: \quad x \notin [x_{min}, x_{max}] \tag{7}$$

This yields the information content (in bits) for that variable as:

$$H_x = \int_{-x_{min}}^{x_{max}} \frac{x_R}{\Delta x} \log_2(\frac{x_R}{\Delta x})dx = \log_2 \frac{x_R}{\Delta x} = \log_2 \frac{1}{\delta_x} \tag{8}$$

where Δx is the uncertainty of the variable, and x_R its design range. δ_x is introduced as the relative uncertainty in parameters. For the general multivariable case it case this can be expressed simply as:

$$H_x = \log_2 \frac{S}{s} \tag{9}$$

Here S is the size of the design space and s the region of uncertainty for the final design.

2 Design Space

2.1 Design Space for Modular Design

Modular design is an effective approach for defining design spaces that can be used to efficiently produce products with different functional characteristics with a with a low design information content, and low design effort. In fact, the introduction of standardized components means that a much smaller design space is needed to create useful designs, since large design spaces for the individual component are effectively removed.

Modular design is a popular approach to create product families. In modular design, common subsystem or subassemblies are defined, further reducing the need for a large design space. In order to assess how efficient a product platform/ product family is, there is a range of different definitions for commonality indices. One overview is found in Thevenot and Simpson [13]. Interestingly, efficiency index based on design information entropy can also be derived. The components that makes up a product platform, defines a design space. Each particular design/ product then represents design entropy relative to the original design space. Small design information entropy is desirable, since that would imply a low design effort to create a particular variant from the product platform, which would constitute of a limited number of modules/components to achieve a wide variety of products with different functional characteristics. In addition to design information, it

Fig. 1 Modular light utility electric vehicle (*Photo* Peter Hallberg)

is here useful also to introduce information entropy to describe the diversity in the functional space. The simplest would be a simple function of the number of different product variants n_v (with different functional characteristics).

$$H_f = \log_2 n_v \tag{10}$$

Example: The modular light vehicle in Fig. 1, from Hallberg et al. [14], can have two alternative front ends, one and two wheels respectively, and also two alternative rear ends, one with a single wheel and one with the two wheels and accommodating a flatbed for carrying goods. This results in four different possible vehicles: The four wheeler, the two wheeler and two three wheelers, with four different functional characteristics.

To define the design information entropy of each design, the design space has to be defined. A vehicle must have one front section, one centre section and one rear section. There are two front ends, one centre, and two rear ends. This means that a particular design has a design information entropy (with respect to that design space) of:

$$H_x = -\log_2 \frac{s}{S} = \log_2 n_s = \log_2(2 \times 1 \times 2) = 2 \tag{11}$$

Here n_s is the number of unique design defined by the design space.

In reality however, only three of these variants, where considered practical, since the variant with two wheels in front and one in the rear, was impractical for various reasons. This means that the product platform can produce three different viable sets of functional characteristics. The information entropy of the *constrained design space* of viable designs can then be defined as:

$$H_c = \log_2 n_v = \log_2 3 = 1.58 \tag{12}$$

One useful measure for the quality of a modular design would be the relative size of the part of the design space that is outside the constrained design space. That can be formulated as wasted design information entropy as:

$$H_w = H_x - H_c = -\log_2 \frac{S_x/s}{S_c/s} = -\log_2 \frac{S_x}{S_c} \tag{13}$$

then H_w represents waste of design information entropy, and should thus be minimized. It is regarded as waste, since it means that more information is needed to select a concept, without offering more viable concepts, than would be needed without the "waste". In the case of the example, the waste would be 0.42 bits. I.e. it would have been better if that part of the design space was not included in the design space in the first place. If all four had been viable then the wasted entropy would be zero, since the whole design space would represent viable designs. The concept of design entropy waste is similar to the definition in Suh [10] i.e. what is there meant by minimizing "information content", although it is there used in the context of robust design, not product families. Note that this is also very relevant for parameterization of models. I.e. if a parameterized simulation model, or CAD model is used for optimization, the parameterization should be such that as large portion as possible of the design space is within the constraint design space of viable designs, in order to have an efficient optimization. It could also, in the author's opinion, be argued that more "knowledge" is built into a product platform that has little waste of design space.

2.2 Design Space for Parameterization

To consider the quality is very appropriate when defining design parameters in a design. One existing well documented case is the parameterization of aircraft wing plan form. The wing plan form for a wing can be defined with three parameters (disregarding wing sweep). These are wing span, b, root cord, c_r, and tip cord, c_t as defined in Fig. 2. These are here referred to as the primary parameter set. In general, however, it is common to use wing area, S, aspect ratio, AR, and tapering, λ, as parameters instead. See e.g. Torenbeek [15]. These are here referred to as the secondary parameter set. The relationships between these are:

$$\begin{aligned} S &= \frac{b}{2}(c_r + c_t) \\ AR &= \frac{b^2}{S} \\ c_t &= c_r \lambda \end{aligned} \tag{14}$$

To study the design space a data set of existing aircraft are shown in the Table 1.

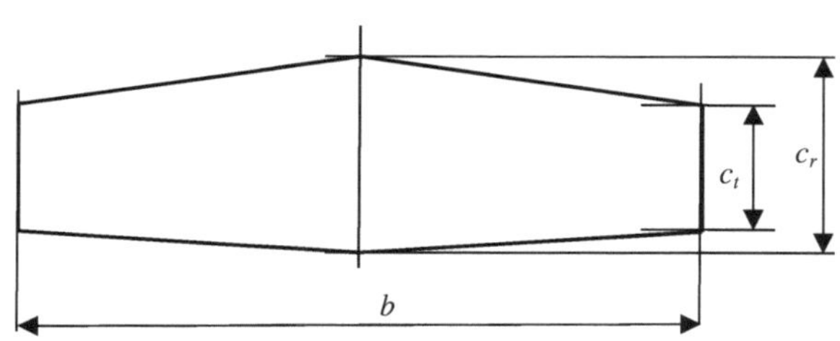

Fig. 2 Wing parameters

Table 1 Parameter values for existing aircraft

	b	cr	ct	S	AR	λ
AN225	88.4	16.5122	3.96293	905	8.63487	0.24
A380	79.75	17.6594	3.53187	845	7.5267	0.2
A320	34.09	5.99394	1.19879	122.6	9.47902	0.2
Gulfstream IV	23.7	5.73191	1.71957	88.3	6.36116	0.3
U2	32	4.83333	0.966667	92.8	11.0345	0.2
F-16	9.96	4.47711	1.11928	27.87	3.55944	0.25
Mirage 2000	9.13	8.5537	0.427685	41	2.0331	0.05
Cessna 172	11	1.59214	1.35332	16.2	7.46914	0.85
Max value	88.4	17.6594	3.96293	905	11.0345	0.85
Min value	9.13	1.59214	0.427685	16.2	2.0331	0.05

Data is from open literature such as Wikipedia and might be inexact, but are just used as an illustrative example

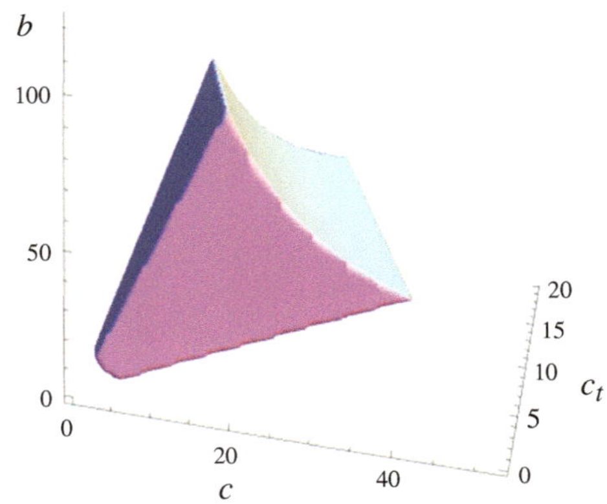

Fig. 3 Design space using the secondary parameter set

Plotting the region spanned by the extreme values in the secondary parameter set, in the coordinates of the primary parameter set gives Fig. 3.

In order to be able to deal with large variation of parameter values it is convenient to use the logarithm of the parameters instead. This can also be seen as reallocation of the probability density function of the design space.

This is also justified by the fact that it is more relevant to define parameter tolerance in relative terms of the nominal parameter value, due to Eq. (9).

In the left part of Fig. 4 the design space defined by the logarithm of the secondary parameters is shown as the slanted block. The parameters of all the studied aircraft are fitted inside this box. To the right, the outer box represents the design space defined by the primary parameters. There is a small part of the design space of the secondary parameters that is outside. This represents an empty part of the design space, which can be regarded as a sort of "waste" of design space. On the other hand

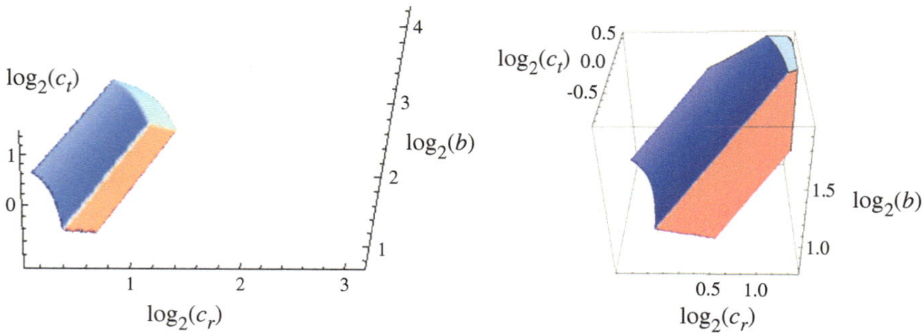

Fig. 4 Design space defined by the logarithms of the secondary parameter set, displayed in the logarithms of the primary parameter set. To the *right* the outer box represents limitations on primary parameters

all the space inside the outer box that is not occupied by the secondary parameter design space is wasted if the primary parameters is used. Doing volume integration over the design space of the second parameter sets reveals that the ratio between its size S_2 and the size of the design space S_2 of the primary parameter set is

$$\frac{S_2}{S_1} = 0.413 \tag{15}$$

This means that the extra amount of information needed to specify a design using the primary parameter set compared to the secondary parameter set to within the same tolerances is:

$$H_{diff} = H_{x,1} - H_{x,2} = \log_2\left(\frac{S_1/s}{S_2/s}\right) = \log_2\left(\frac{S_1}{S_2}\right) = 1.24 \tag{16}$$

Hence this is a measure of the relative quality of the two design spaces. In this way it can be argued that the design information entropy is a way to assess the quality of a parameterization and consequently a design space. An interesting consequence is that this provides a mechanism to optimize the design space itself provided there is a sample sets of designs that can be used to span the design space. There is of course also other consideration regarding parameterization. One is to have a parameterization such as the functional characteristics are as uncoupled as possible with respect to the design parameters. However, this is likely to be coherent with a minimization of waste of design information entropy, since an uncoupled design would mean that the design space can be fit around the design space of the functional characteristics more tightly. Example: For a linear design the functional characteristics can be calculated as:

$$\mathbf{y} = \mathbf{A}\mathbf{x} \tag{17}$$

where **y** is the functional characteristics an **x** is the design parameters. **A** is the design matrix as in Suh [10]. If there are constraints on **y** such that

$$\mathbf{y} \in [\mathbf{y}_{\min}, \mathbf{y}_{\mathbf{max}}] \tag{18}$$

The only way this can coincide with a function design space resulting from design parameters with the limitations

$$x \in [x_{\min}, x_{\mathbf{max}}] \tag{19}$$

is if the matrix **A** is diagonal and hence uncoupled. A consequence of this is that the independence axiom and the information axiom in axiomatic design Suh [10], are not really independent from each other, but that an uncoupled design also tend to minimize the information needed to specify a design. The overriding axiom should be that the best design space formulation is the one that minimize waste of design space.

3 Discussion

The notion of design space is relevant both in design automation, where it is very concrete, and also in manual design, although it is perhaps less obvious there. In this paper design space has been defined by a rectangular probability distribution of the design space. In a manual design it is usually implicit, and can be seen as a more irregular distribution. For design to be effective, it is desirable to have a small design space, but that still contain sufficiently good designs. It could be argued that an experience designer knows early on how to limit the design space to quickly arrive at a design. In a good design space, in the form of a product platform, it is easy to assemble viable designs from a limited set of design elements that can be combined into new products. Design information entropy can also be applied to parameterization of a design. In a good parameterization, all parameter combinations should yield viable designs. In fact the fraction of viable parameter combinations can be seen as a quality of a parameterization, see Amadori et al. [16]. Ideally a good design space should be well aligned with the functional design space of possible requirements. This is consistent with axiomatic design [3] that a design should be as uncoupled as possible.

The concept of design information entropy also provides a sound base for defining creativity as the process of selecting areas for expanding the design space "to think outside the box" which here takes on a more literal meaning. Creativity can also be the process of navigation in highly dimensional multimodal design spaces for concept generation, which on the other hand is simplified by a limited design space, like an efficient product platform, with few unviable designs. One path of future work is to use the result to automate formation of the optimal design spaces in cases where known samples of designs exist, like in aircraft. Another path is to extend the theory into dealing also with unbounded design spaces.

4 Conclusions

In this paper it is demonstrated that introducing design information entropy as a state, can be used for quantitative description for various aspects of design, both regarding structural information regarding architecture, as well as for parameter variables, both discrete and continuous. The design information entropy is a measure of the precision a design is defined within a design space. Since the information entropy increase with a large design space, it is desirable to have a small design space that still contains the good designs. In particular it has been shown that design information entropy can be used as a measure of quality of product platforms for modular design. It has also been shown that it can be used as a measure of the quality of parameterization in design.

References

1. Ullman, D.G.: The Mechanical Design Process. McGraw-Hill Book Co, Singapore (1992). ISBN 0-07-065739-4
2. Shannon, D.: A mathematical theory of communication. Bell Syst. Tech. J. **27**, 379 (1948)
3. Suh, N.P.: The Principles of Design. Oxford University Press, Oxford (1990)
4. Khan, W.A., Angeles, J.: The role of entropy in design theory and methodology. In: Proceedings of CDEN/C2E2 2007 Conference, Winnipeg, Alberta, Canada (2007)
5. Frey, D., Jahangir, E.: Differential entropy as a measure of information content in axiomatic design. In: Proceedings of the 1999 ASME Design Engineering Technical Conference, Las Vegas, USA (1999)
6. Bras, B., Misstree, F.: Compromise design decision support problem for axiomatic and robust design. J. Mech. Des., Trans. ASME. V117 (1995)
7. Krus, P., Andersson, J. (Ölvander): An Information Theoretical Perspective on Design Optimization. ASME, DETC, Salt Lake City, USA (2004)
8. Krus, P., Ölvander, J.: Performance index and meta-optimization of a direct search optimization method. Eng. Optim. **45**(10), 1167–1185 (2013). doi:10.1080/0305215X.2012.725052
9. Bansiya, J., Davies, C., Etzkorn, L.: An entropy-based complexity measure for object oriented design. Theor. Pract Object Syst. **5**(2), 111–118 (1999)
10. Suh. N.P.: Axiomatic Design: Advances and Applications. Oxford University Press, Oxford, USA (2001). ISBN-0-19-513466-5
11. Krus, P.: Information entropy in the design process. In: ICoRD'13: Global Product Development, pp. 101–112, Springer (2013)
12. Kullback, S., Leibler, R.A.: On information and sufficiency. Ann. Math. Stat. **22**, 79–86 (1951)
13. Thevenot, H.J., Simpson, T.W.: Commonality indices for product family design: a detailed comparison. J. Eng. Des. **17**(2), 99–119 (2006)
14. Hallberg, P., Nåbo, M., Krus, P.: Modular sustainable light multi-purpose vehicles, 3rd European Ele-Drive Transportation Conference EET-2008, 11–13, Geneva (2008)
15. Torenbeek, E.: Synthesis of Subsonic Airplane Design. Kluwer Academic Publishers, Dordrecht (1982). ISBN 90-247-2724-3
16. Amadori, K., Lundström, D., Krus, P.: Automated design and fabrication of micro-air vehicles, accepted for publication on Journal of Aerospace Engineering. In: Proceedings of the Institution of Mechanical Engineers Part G [PIG] (2011)

Design Research Model: Establishing a Link Between Design Education, Practice and Theory

Gayatri Menon

Abstract The first part of the paper elucidates the challenges of design research as well as some of the approaches towards design research. The challenges of design are primarily due to the nature of design and the understanding of 'knowledge' and 'truth' in design. Design is a practice based profession and is considered to be dynamic, contextual and creative. This has resulted in a multiplicity of approaches to design as well as a need for new methods and approaches to design. Research to arrive at new methods and approaches in design could seek knowledge from design practice to build design theory. Design education is unique for its pedagogy based on practice. The paper proposes a research model to link design practice, education and research. It is argued that establishing links between educational institutions, industries and researchers could enable the study of design practice and help develop models and frameworks for design.

Keywords Design research · Education · Practice · Research model

1 Introduction

The practice oriented nature of design as a discipline has probably resulted in lack of sufficient interest in the development of design theory and research. We need to understand the various approaches to design research and the challenges faced to develop design theory through research. It is expected that advancing an understanding and developing design theory would also advance the discipline of design.

G. Menon (✉)
Faculty of Industrial Design, National Institute of Design, Ahmedabad, India
e-mail: gayatri@nid.edu; menon.gayatri@gmail.com

A. Chakrabarti (ed.), *ICoRD'15 – Research into Design Across Boundaries Volume 1*,
Smart Innovation, Systems and Technologies 34, DOI 10.1007/978-81-322-2232-3_6

The paper studies various approaches to design research and challenges faced in establishing design theory. Some of the approaches towards research in design include:

1. Research as a part of design process. This is the most common type of research and is carried out by designers to conduct research and collect information while working on design projects. The subject matter of research may vary based on the project undertaken and could deal with studying user, material, technology, cultural practices, market, existing product and similar other facets pertaining to the project.
2. Research in allied areas of design such as human factors, technology, ethnography etc. Designers require knowledge from various fields in order to integrate the same in their practice. Although they are disciplines in their own right, due to the close relation with design profession, research in allied areas has become an important component of design research.
3. Research about design discipline such as the history of design, design criticism etc. The research tries to bring forth a better understanding of the evolution of design as a discipline and brings forth a critical perspective to the happenings in design.
4. Researching design methods. Although the word 'design' is often used as a noun, the primary understanding of design is that of a verb, an activity or a process. An evolution in the understanding and development of design methods could help further both design education and practice. The challenges faced in conducting research in design methods is primarily due to the nature of design as well the understanding of 'knowledge' and 'truth' in design. Design is a practice based profession and is dynamic, contextual and creative. This has resulted in a multiplicity of approaches to design as well as a need for new methods and approaches to design based on the changing context of design.

An understanding of design also leads to further the linkages between design pedagogy, practice and research. Research to arrive at new methods and approaches in design would need to seek knowledge from design practice to build design theory. Design education is also unique for its pedagogy based on practice. Industries and organisations often approach design educational institutions for fresh perspectives and ideas. It is argued that establishing links between educational institutions, industries/organisations and researchers could enable the study of design practice and its various facets.

The paper proposes a research model to link design practice and theory as well as design education and practice. This research model would help in critical reflection of various aspects of design practice and allow the study of design practice within the complexity of real life situation and context. The proposed model elaborates the framework to conduct multiple case study method in an educational setting and further use the same as a research method to link design practice and theory. Researchers could study complex patterns across case studies to build an understanding of methods/approaches in design and create guidelines and frameworks based on the same. It is envisaged that such guidelines/framework would be useful in design education as well as for practicing design professionals.

2 Challenges to Design Research

The challenges to design research are primarily related to the complexities of design discipline itself. Research is primarily related to establishing a body of knowledge. It is also related to what constitutes 'truth'. In order to make sense of the same we need to look into both the nature of knowledge and truth in design as well as what constitutes design discipline.

2.1 The Nature of Design

The meaning and interpretations of 'design' has varied greatly over the period of time and in different cultures. The word 'design' is used both as a noun and as a verb. As a noun it denotes a completed object whereas as a verb it denotes the process of 'becoming'. If technology is 'knowing how' then design is 'envisaging what' [1]. Archer further emphasizes the importance of design as the third great defining characteristic of humankind along with tool making and language use for its capacity for envisaging a non present reality, analyzing it and modeling it externally. The other characteristics of design are its ability to meet human needs through practical results (products, process, systems) embodying values determined by its functional, commercial and social context. Design is described as intentional to distinguish it from serendipity or discovery by chance [1].

Design is contextual, dynamic and creative in nature. Understanding is derived from the whole, since it is only in the light of the whole can we understand the parts. Therefore to investigate the process of design, it is imperative to choose a framework which would encompass the study of design as a 'whole', within a context. Design requires to get inputs from diverse fields of knowledge and bring about a holistic understanding and perspective. The challenge is to translate this understanding into a tangible reality—a new design. This involves lateral thinking, creative processes and execution. Together they constitute the core of design—purposeful creativity, holistic understanding and approach and lateral thinking. Design learning involves learning by thinking and reflecting as well as learning by practice. The methods and cognitive models introduced to designers in a teaching learning context also need to be learnt and internalized through using the same in specific design contexts.

Simon [2] proposed that the science of design could form a fundamental, common ground of intellectual endeavor and communication across the arts, science and technology. The study of design could be an interdisciplinary study accessible to all those involved in the creative activity of making the artificial world. This brings forth the dynamic and multidisciplinary nature of design. Design profession works more and more in multidisciplinary collaborative teams. The idea of multiple stakeholders and the inter connectedness of various parameters pertaining to design has taken root. Design is primarily acting as a subsystem of economy or rather marketing though the sociological view is more towards assuming a network of widely autonomous partial systems that are permanently negotiating their exchange and interaction. Economy is only one subsystem in its network [3]. Thus, design today considers multiple realities: economic, sociological, cultural with equal competence and ease.

The relationship between design and other allied areas such as art, technology etc. and the rapidly changing context in which design operates needs to be looked into. Technological changes, socio-cultural changes, changes in product supply chain and environmental issues are a few contextual changes elucidated [4]. As the inter-relationships and networking among components in the world increases, there is more effect of macro systems on design. Having a goal and vision on how things should change seems to be an integral part of the design process. Design is basic to all activities—the placing and patterning of any act towards a desired goal constitutes a design process. Everyone designs who devises courses of action aimed at changing existing situations into preferred ones [2]. As Bruce Nussbaum writes recently in Business week, these are fundamental changes. "The design profession shifted its core competencies from drawing to thinking, from styling to innovating, from shaping things to visualizing new business paradigms" (Nussbaum 2004).

2.2 Nature of Knowledge and Truth in Design

The nature of 'knowledge' in design may not be fixed or certain. Design involves the knowledge of different physical discipline which may be fixed but design also involves an understanding of human perceptions which is often uncertain. Further design deals not only with materials but primarily with human phenomena and tacit knowledge which is neither fixed nor certain. Design deals with both tangible and intangible aspects—the nature of knowledge in design is fairly broad covering the science of materials, the science of meaning, the science of thinking and doing. Knowledge in design is often embodied, explicit or tacit. 'Knowledge' in design is not based on 'facts' but is rather based on the experience of 'inter-relating various elements' based on the context. This brings us to the idea that if the nature of 'knowledge' in design is dynamic and evolving, the notion of 'truth' in design is also therefore that of being incremental in nature and not absolute.

The notion of truth, as Heidegger advocates is a process of revealing; a progress towards something which is in sight but never fully seen; instead of truth as correctness that can in principle be fully known. The process of investigation is often such that the subject knowledge of the investigator also plays an important role in the same.

3 Design Education and Practise

Design is a practice based profession and design education is expected to equip students to deal with challenges and requirements of professional practice. Due to the emphasis on tacit knowledge embedded in the practice of design, the apprenticeship model of design pedagogy was prevalent in many design institutions. Even in institutions wherein apprenticeship mode of learning is not prevalent, there is a emphasis on 'learning by designing' which forms the core of design education.

Due to the changes happening in design and design thinking, educational institutions are conceptualizing new modes of bringing complex designing experiences in classroom situations. It needs to be considered that in today's times, 'conventional' design is losing its place and the competitive edge is going to those who can innovate. Patel, narrates the experience of starting 'innovation space' program at Arizona university wherein the projects given to students had intentional ambiguity. "Accustomed to more linear projects and course work, these students were suddenly faced with the sort of loosely defined design problem that has become common in the real world" (Patel 2005).

This need for bringing real life experiences in class room situations has also led to closer collaboration between educational institutions and industries. To break away from the set pattern of in house design departments, industries sometimes tend to bring real life projects to classrooms to get fresh perspectives and ideas from students. This linkage between design educational institutions and industries is often a mutually beneficial collaboration which enriches the students with complex design experiences and the industry with fresh new perspective.

4 Proposed Research Model

Figure 1 depicts the proposed research model. The proposed model is based on the collaboration between design educational institutions and industry. Emerging areas in design often have less corporate support and require new approaches. The paper proposes a research model to link design practice and theory as well as design education and practice. This research model would help in critical reflection of various aspects of design practice and allow the study of design practice within the complexity of real life situation and context. The primary stance taken is that reflecting over design practice could lead to a theoretical framework which in turn would help in furthering the practicing profession [5].

4.1 Research Model

Since the objective is to study a current phenomena i.e. design practice in detail and try to understand the process i.e. how and the rationale i.e. why, case study is proposed to be an appropriate method for inquiry. The framework to conduct multiple case study method in an educational setting and further use the same as a research method to link design practice and theory is elaborated.

This model as given in Fig. 1, is particularly helpful in exploratory studies which involve exploring new areas of design interventions wherein organizations do not use a particular design process.

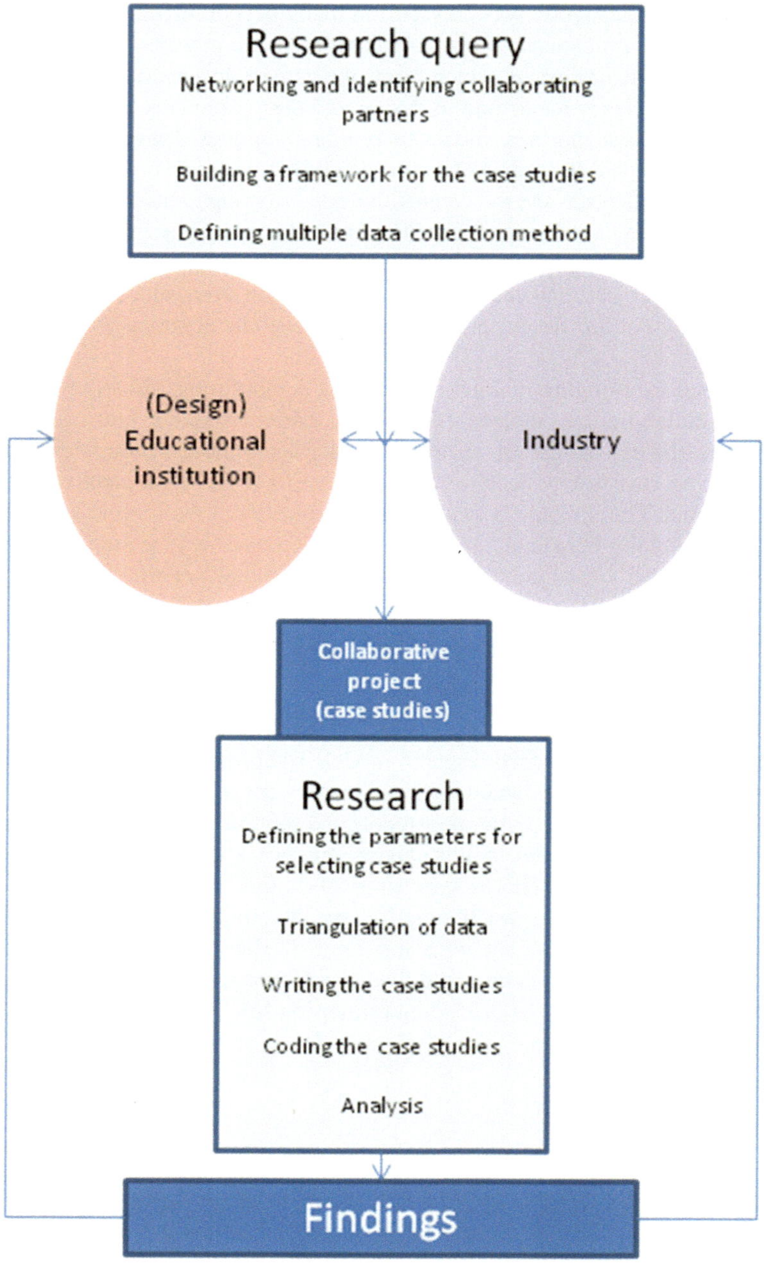

Fig. 1 Design research model linking design education, practice and theory

(a) Networking and identifying collaborating partners
Design projects as part of educational curriculum could become a good possibility for collaboration. After identifying the research objective, industries could be contacted and the project brief outlined together. The internal team consisting of faculty and students, the industry team involving personnel from the industry and external experts if required can be identified.

(b) Building a framework for the case studies
The framework for conducting case studies to investigate various factors pertaining to design practice could include defining the common purpose or objective of the design projects, the specific requirements from the industry including access to its personnel, infrastructure and briefing sessions. It would also include the approximate duration of the project and the resources required for the same.

(c) Multiple data collection method
The information collected for the case studies needs to include multiple data collection methods in order to help in the triangulation of data [6]. The methods may include project reports/project presentation documents, observation notes, interviews with key sources.

(d) Defining the parameters for selecting case studies
All case studies may not be suitable for the study purpose. Parameters for selecting the case study can be identified which may include criteria such as project data that provides a lot of elaboration and detailing in terms of documentation, sketches etc., projects which represented various sectors, projects which extended the understanding of the research area etc. [6]. Based on the criteria, the case studies for the study purpose may be selected for further analysis.

(e) Writing the case studies
The write up of the case studies includes the project context and background, series of action/events that took place and the rationale for the same in terms of a series of thoughts pertaining to the same [6].

(f) Coding the case study write-up
The initial coding may be done based on the research queries. Secondary coding would help in arriving at broad categories [7].

(g) Analysis
Within case analysis to understand unique features related to the research area within each case study may be analyzed. through single case study analysis [8]. This may be followed by a multiple case study analysis approach to replicate the findings in order to further strengthen the same or to extend the theory.

(h) Findings
The findings of the study may be in the form of heuristics, algorithm or conceptual framework. The intention of the study is primarily to make the tacit, often intuitive process of designing more explicit to further the understanding of the domain.

5 Example

The research model as given in Fig. 1 was used to conduct an exploratory research study for opportunity identification in design innovation. The intention of the study was to explore design approaches during opportunity identification phase in (product) innovation.

Case studies involved collaborative projects between academia at NID and industry/organisations. Only projects which entailed the opportunity identification process were entertained. Development of original case studies was a major component of the research method. This included process of developing a framework for generating original case studies, networking and collaborating with industries/organizations and conducting the same.

Triangulation method for case study data collection was carried out and the case study was written based on the same. Triangulation of data collected through interviews, sketches, models, presentation notes and observation notes further helped in consolidating the data. A criteria of selection was developed through which 11 case studies were selected from 36 case studies which were conducted. Multiple case study analysis resulted in a conceptual framework to enable a better understanding of opportunity identification for innovation using a design approach.

While using the research model as given in Fig. 1, it was found that while the industry benefitted from new design directions and concepts, the students benefited through interaction with industry and the support provided by the industry in terms of making their resources available to them for the project.

6 Conclusion

The proposed model as given in Fig. 1 elaborates the framework to conduct multiple case study method in an educational setting and further use the same as a research method to link design practice and theory. Researchers could study complex patterns across case studies to build an understanding of methods/approaches in design and create guidelines and frameworks based on the same. It is envisaged that such guidelines/framework would further help in design education as well as practicing designers.

The model as given in Fig. 1 can be tested and further improvised based on the research query and the possibility of collaborating partners. A model which encompasses collaboration between a variety of educational schools related to design and its allied areas such as art, technology, management with industries/organizations in new areas of design thinking and practice can be further developed.

Acknowledgment I would like to thank Prof. Chakravarthy, Prof. Athavankar and Prof. Malshe from IIT-B for their critical insights and inputs. I would also like to thank students and faculty from NID as well as the industry representatives who took part in the research study based on the proposed research model.

References

1. Archer, B.: Systematic Method for Designers. Council of Industrial Design, London (1965)
2. Simon, H.A.: The sciences of the artificial. Karl Taylor Compton lectures. M.I.T Press, Cambridge (1980)
3. Jonas, W.: A scenario for Design. Des. Issues **17**(2), 64–80 (2001)
4. Chiapponi, M.: A new design curricula for a changing artifact world. In: Scholastic Papers from the International Conference, DETM 2005, pp. 38–42. Ahmedabad: NID (2005)
5. Creswell, J.W., Vicki, L., Clark, P.: Designing and Conducting Mixed Methods research. Sage, Thousand Oaks (2006)
6. Yin, R.K.: Case Study Research: Design and Methods. Sage, Thousand Oaks (2009)
7. Saldana, J.: The Coding Manual for Qualitative Researchers. Sage, London (2009)
8. Yin, R.K.: Applications of Case Study Research. Sage, New Delhi (2012)

Conduplicated Symmetries: Renegotiating the Material Basis of Prototype Research

Arne Berger, Michael Heidt and Maximilian Eibl

Abstract The role of materiality within prototyping recently received a high degree of attention from HCI and design communities. Existing approaches have solely focussed on the materiality of artifacts produced during prototyping. This focus largely has left the materiality of designers and users unaccounted for. The text illustrates how the Somatic-Marker-Hypothesis and Actor-Network-Theory can be employed in order to illustrate these forgotten dimensions of materiality during prototyping: Not only is material shaped during construction of artefacts. Material modifications also comprise alterations within the material of designers' and users' bodies themselves. In order to facilitate these descriptions, a novel approach towards symmetry had to be developed.

Keywords Prototyping · Somatic-marker-theory · Design thinking · Reasoning of designers · Actor-network-theory · Interdisciplinary concept

1 Introduction

Most current research positions on prototypes in interaction design are based on a variety of hypotheses that are all symmetrical in construction. While individual arguments are quite distinct we believe there is an underlying theme that we strive to explore. The following article frames a theoretical backdrop which is inspired by Somatic-Marker-Hypothesis (SMH) from Neuroscience and Actor-Network-Theory (ANT) from Philosophy of Technology.

A. Berger (✉) · M. Eibl
Media Informatics, Technische Universität Chemnitz, 09107 Chemnitz, Germany
e-mail: arne.berger@informatik.tu-chemnitz.de

M. Heidt
Visual Computing, Technische Universität Chemnitz, 09107 Chemnitz, Germany

A. Chakrabarti (ed.), *ICoRD'15 – Research into Design Across Boundaries Volume 1*,
Smart Innovation, Systems and Technologies 34, DOI 10.1007/978-81-322-2232-3_7

Some of the early key publications on prototyping in interaction design, like [5, 7] attempted to form comprehensive lists of what designers and developers can do with which kind of prototype and what users reflectingly see in those artifacts. Ever since, more elaborate propositions have been made, mainly with the goal to establish languages to describe the material factors of prototypes. Beginning with the fidelity-debate [14] this sometimes led to the bold endeavour to "establish a fundamental definition of prototypes" [12].

However, more sophisticated prototyping techniques are constantly needed for broader application and a wider audience [1]. Most theories to describe prototypes in interaction design are symmetric in style. The following chapter describes three schools of prototyping theories; the subsequent chapter introduces SMH and ANT while the third chapter convolutes those for a more radical theoretical approach.

2 Prototypes in Interaction Design

2.1 The Fidelity Debate

One direction of describing prototypes is simple in nature and twisted in reality. The fidelity concept focuses on a variety of continua: low fidelity prototypes are cheap to sketch and help to evolve ideas while high fidelity prototypes are more costly to construct and are used in later process stages to evaluate more elaborated concepts [2]. However, it has been stated quite extensively that the concept of fidelity is broken by design. What if some variables are of low fidelity and others are of high fidelity? Thus the term mixed fidelity has been proposed but is not fit to clearly distinguish different dimensions. Quite similar and in many ways a logic continuation of the fidelity concept is the differentiation of single dimensions as a specification of dimensions in regard to mixed fidelity. Lim et al. [11] and Diefenbach et al. [4] have developed interaction attributes to specify interactions, once again as attribute pairs. The relevance of those attributes have been user tested, implying that those features of prototypes are objectively existing.

2.2 The Inscription Theory

A seminal, yet bold landmark seeks to provide a "fundamental definition of what prototypes are" [12]. The proposed vocabulary is once again symmetric in its architecture and proposes two dimensions: filter and manifestation. A filter is one of the five dimensions appearance, data, functionality, interactivity, spatial structure and is exclusively used by the designer who "screens out unnecessary aspects of the design" [12]. The antagonist to filter is manifestation, which forms the perceived embodiment of the designer's goal. Putting aside the positivistic notion that only designers design prototypes, this proposal also suggests that users dependently recognize the very aspects designers are focusing on. We call proposals of similar nature *inscription theories*, because they suggest a clear role allocation where designers deliberately choose a meaning that they inscribe into material. Two fundamental symmetrical propositions are made. First, designers incorporate a meaning into an artifact for users to explore. Second, prototypes have a clear distinction of form and material.

2.3 The Form-Material Dichotomy

In line with this argument e.g. Jung and Stolterman [8] define form as intention in design and material as evocation in use. They subsequently elaborate this very dichotomy regarding current evaluation methods. They argue that form and materiality are abstract and always need interpretation while failing in prospect of quantitative oriented methodologies. Form and material are blending, their boundaries are blurring in digital artifacts, thus Jung and Stolterman argue very carefully for a third way: an artifact approach. Accordingly, it is not only form that shapes material it is also the material that influences the form. This symmetric proposition of intention and use—is in line with inscription theories but calls for a consideration of "the complex relations that a certain technology makes with other artifacts" [8] thus suggesting the necessity for considering the artifact as an entity with independent meanings and rich interdependencies.

Those observations are by no means complete, but shall count as a blueprint for a variety of typical views in prototyping research. Before we move to our research position, it is essential to explain the already mentioned bordering theories.

2.4 The Figure of Symmetry

During the course of our discussion, we encountered several instances of the notion of *symmetry*. The present paragraph is designed to elucidate both their commonalities as well as respective differences. On the most general level, symmetry denotes a specific strategy of dealing with difference. Symmetry organises elements into a configuration that guarantees invariance regarding one or many transformations. It thus designates a property as remaining *invariant* under a given transformation.

Within the context of inscription theories, it is the meaning/idea/intention that is preserved within the processes of design and use. Additionally, within the discussed example, the processes of filtering and manifestation give rise to a dynamics that can be conceived of as being symmetrical in a literal (geometric) sense: There are two processes, one on each side of the filtering/manifestation division. In a downwards movement, ideas are filtered according to the qualities of the respective prototype. At the same time, an upwards movement is responsible for manifesting filtered ideas. Whereas within form-material theories, a direct relational symmetry can be observed, analogous to symmetry within set-theory: form influences material and vice versa.

3 Metatheoretical Framing

3.1 *Somatic Marker Hypothesis (SMH)*

Human sensory perception is prone to error, intrinsically incomplete and needs to be approximated to be plausibly interpreted. This intuitive, unconscious transcendence of sensory data is key to successful human performance in regard to a complex and ever-changing environment [6]. According to SMH humans embody a cognitive scheme that bonds a certain sensory perception with corresponding somatic markers, thus allowing association of different environmental conditions with specific bodily conditions. Those markers are constantly refined through experience, education and socialisation and are connectors of embodied preferences with exterior factors. They thereby empower better interaction decisions [3]. A somatic marker reliably binds a bodily condition to recurring external situations without the need for logically informed decision making, while improving conscious processes. The cognition of an artifact activates somatic representations, appropriate responses and potential courses of action. A certain system, person or artifact will be rejected or appreciated based on the specific combination of somatic markers it brings about. All those reactions and bondings are inevitably based on the individual's body and perception, constituting the body's individual maps of somatic markers as the main reference point for marking and appraising artifacts [3].

3.2 *Actor Network Theory (ANT)*

Whereas the key elements in SMH are the human body and perception, the role of the artifact hardly crosses that of a passive object. Herein lies one of the main contrasts to ANT. The structure of ANT rises above a general symmetry, which equally includes human and non-human entities. It grasps the social as a circulating entity, as a network of heterogeneous entities connected through a series of transformations. According to this approach agency is not reserved for humans, artifacts can act on their own as well. The main criterion for agency is the capability of modifying a state of affairs by making a difference [9].

Note that within ANT, the notion of symmetry is employed in a distinctive manner. Agency is the property that is preserved while traversing the studied translational network. Whether studying actants that encompass humans or those that do not, the fundamental status of an observed entity does not change. It has to be noted however, that within the context of Latourian theory, the signifier 'symmetry' is employed in a much broader and much more emphatic sense. Not only does it designate the retraction of the allegedly 'pseudo-modern' object-subject dichotomy, it is also employed in order to claim commensurability between science and technology themselves.

In consequence ANT suggests a rethinking of the role of technical artifacts concerning the emergence and maintenance of social connections. ANT brings materiality into focus, as well as relationality—stressing the notion that all entities are produced in relations [10].

4 A Radical Twist

Inscription theories depict prototyping as a "platonic" operation: designers shape prototypes from mute material where they solidify ideas for future reference. This leads to the consistent assumption in inscription theories and to a lesser extent in form-material theories, that the idea can be interpreted reciprocally by users. We argue in line with SMH: material is simply too rich and sensory perception bound to individual bodies and also too limited, that the intention may fail to be transported through the artifact. That does not mean inscription theories are wrong, they are just not adequate for describing the rich effects of prototyping. Inscription can succeed—that is what makes great products great: a designer's intention is well executed and users somatically resonate with it. But more often than not inscription fails, because artifacts have such an enormous sensory complexity beyond the designer's control or inscription of intention. Form-material theories advocate a specific role of the material—e.g. its ability to conduplicate other material and its capability to independently bond relationships with other material. However, up until now this is just another consideration for designers to take into account when forming their intentions. Either way suggests that prototypes can be exhaustively described and to a lesser extent imply that those features exist objectively. We believe both are wrong. In contradiction to prototyping as a platonic operation, and in contradiction to form-material arguments we believe the artifact is neither a mere container, nor does it act on its own. Both theories are by all means useful but not useful to form a persistent meta-theory for subsuming existing theories of prototyping. We rather argue in line with SMH: an artifact is not objectively perceivable, because the only thing that is triggered while perceiving an artifact is the individual map of somatic markers in the individual interpreter. Accordingly there is no objectively existing prototype. There are only compounds of somatic marker configurations that are individually different and that are changing while perceiving.

Whitfield [15] defined aesthetic knowledge as the capability to categorise sensory data and subsequent somatic markers that operates subconsciously and guides our decisions for choosing fitting artifacts. Whitfield's categorical-motivation model is an explanation of whichever somatic marker combination has to be triggered by an artifact to be preferred over other combinations. A slight derivation from a known positive combination is preferred, while confirmed anticipations are equally favoured. Both parameters form the boundaries for somatic conditions. An artifact's configuration has to be recognised as something similar. At the same time this configuration has to be a significant, albeit not immeasurable, derivation from a similar configuration to be considered as sufficiently—but not too much—arousing.

Thus, a prototype is a mere reflection of the designer's somatic marker configuration and that of the interpreting future user. A prototype is more likely to be accepted when its capability of reflecting positive somatic markers is equally symmetric from the designer to the user. This is no simple endeavour. The artifact must infringe current conventions to trigger aesthetic somatic resonance, but must reflect familiar features, habits and traditions to be recognised at all. Great designers seem to inherit somatic marker configurations or at least the capability to simulate such a condition in order to create an artifact that resonates a huge intersomatic overlap, while at the same time triggering enough new marker configurations. Thus prototyping is the art of generating and reflecting as many intersomatic overlaps as possible.

The consequential opposition to platonic prototyping has been proposed as hedonic prototyping [13] constituting the act of collaboratively reflecting somatic markers in the making of a prototype. Hence the artifact is not solely an object to reflect somatic conditions after its creation. Rather the very act of designing-interpreting an artifact is the interwoven reflective practice of individual somatic conditions and as such the necessary condition for designing and interpreting artifacts. In choosing a new mobile phone we may sift through all products on display that look like phones and choosing the one that immediately resonates the most. Depending on previous experience this might be a golden one, the one our work mate uses or the most advertised one. However, deeper somatic reflection will start once we prototype our daily habits with and around the phone. Does it hold on to our aquatic hobbies, the long throw habits of our little ones or the diameter of our forefinger?

5 Hedonistic Prototyping and ANT

Possibly, we employ it differently than its makers intended, in a more decontextualised fashion, more as technological artefact than as ontological commitment. We use ANT as an inspiration to propose a framework for describing prototyping practices in regard to SMH. ANT is in many ways a fitting framework for generalizing the multifaceted relations between designers, users and artifacts. It radically assumes that human and nonhuman actors can be treated the same. ANT is deployed as conceptual agent allowing un/blackboxing to be performed, delegation to be described, essentialism to be dissolved, symmetries to be accounted for.

Prototypes reinscribe different configurations of markers into the bodily material of designers. Together with requirements documents and user studies, designers are aggregated into filters. Within this aggregate designers' and users' markers are activated as part of a process that selects ideas from the respectively constituted design space.

We consider a prototype an amalgamation of different material features. The phone might be golden, waterproof and shaped for large fingers. It is as well a black box of black boxes consisting of potentially hazardous materials mined in third world countries, patents granted on another continent and so on. Depending on whether patents, fingers or environmental damage is in the center of someone's attention, the sum of all of the artifact's materials forms an intention that in turn forms a very specific network of material properties that are in turn the specific intention of the artifact. A description of the likability of the specific artifact depends on individual somatic reflections: is concern for the third world sufficiently weaved into my somatic map or does a golden phone somatically reflect wealth or pretentiousness?

Thus, an individual actor-network is a specifically triggered map of reflecting somatic markers. As much as this argument is prone to linguistic fuzziness, an individual actor-network is embedded within an individual human-artifact relation and thus cannot be exhaustively described through language. Somatic mapping is only possible in reflection in prototyping the experience with an artifact, thus forming an actor-network between humans and non-humans in somatic negotiation of user and artifact.

6 Reflecting Symmetry

Adopting a stance as outlined above allows for construals of prototyping that demystify both the notion of designer's intuition when creating artefact-forms as well as that of user acceptance of the forms created. Material is given form during conscious discussions and poietic actions. At the same time material helps to produce forms out of itself by virtue of redistributing somatic markers within its designerly components. Designerly components can redescribe themselves as material within prototyping processes without incurring narcissistic loss. They are not 'mere' material but are those parts without which material-semiotic networks would not produce.

ANT and SMH as agents for changing perspectives in prototyping are by no means new concepts. However, the plethora of fragments to describe prototyping practices have often remained isolated from each other and as thus eminently delimited from an overarching concept of human-artifact relations. Our proposition serves as a traversing scheme to subsume the symmetric concepts in prototyping research. It has been shown that those are not necessarily axisymmetrical to each other, but can be described as such with the help of ANT and SMH. This does not mean, that current research has to be condemned in the light of an overarching explication. While concepts like the inscription theories may need to be refuted, esp. form-material dichotomies are helpful assumptions for shaping research in design. Hence, the proposed concept is intended to help shaping design research.

Acknowledgments This publication was prepared as a part of the research initiative ValidAX, which is funded by the German Federal Ministry of Education and Research and as a part of the Research Training Group Crossworlds, which is funded by DFG (German Research Foundation). The authors take sole responsibility for the contents of this publication.

References

1. Berger, A.: Design thinking for search user interface design. In: Proceedings of euroHCIR2011, pp. 1–4 (2011)
2. Buxton, B.: Sketching User Experiences. Morgan Kaufmann, Massachusetts (2007)
3. Damásio, A.R.: Descartes' Error: Emotion, Reason, and the Human Brain. Quill (1994)
4. Diefenbach, S., et al.: Ein Interaktionsvokabular: Dimensionen zur Beschreibung der Ästhetik von Interaktion. Usability Professionals **2010**, 27–32 (2010)
5. Floyd, C.: A Systematic Look at Prototyping vol. 1, pp. 1–18. Springer, Berlin (1984)
6. Gigerenzer, G., et al.: Heuristics. Oxford University Press, Oxford (2011)
7. Houde, S., Hill, C.: What do prototypes prototype. Handbook Human-Comput. Interact. **2**, 367–380 (1997)
8. Jung, H., Stolterman, E.: Digital form and materiality: propositions for a new approach to interaction design research. In: Proceedings of the 7th Nordic Conference on Human-Computer Interaction: Making Sense Through Design, pp. 645–654 (2012)
9. Latour, B.: Reassembling the Social. Oxford University Press, Oxford (2005)
10. Law, J.: After ANT: complexity, naming and topology. Soc. Rev. **46**, 1–14 (1998)
11. Lim, Y., et al.: Interactivity attributes: a new way of thinking and describing interactivity. In: Proceedings of the 27th International Conference on Human Factors in Computing Systems, pp. 105–108 (2009)
12. Lim, Y.-K., et al.: The anatomy of prototypes. ACM Trans. Comput.-Human Interact. **15**(2), 1–27 (2008)
13. Petruschat, J.: Tische, Tennisbaelle, kurze Schreie. Prototype! In: Petruschat, J., Adenauer, J. (eds.) Form und Zweck, pp. 287–317 (2012)
14. Virzi, R.A.: What can you learn from a low-fidelity prototype? In: Proceedings of the Human Factors and Ergonomics Society Annual Meeting, pp. 224–228 (1989)
15. Whitfield, T.W.A.: Aesthetics as pre-linguistic knowledge: a psychological perspective. Des. Issues. **21**, 3–17 (2005)

Notes for Rule-Based Design from Cognitive and Visual-Computational Models, Especially Shape Algebras

Paul Varghese

Abstract Some background notes for rule-based design and its models along cognitive and computational lines are discussed; additions to Stiny's shape grammars include shape algebras for multi-dimensional representations; extensions to Knight's visual-computational algebras are proposed for development for application in design and construction, which largely allows intuitive visual conditions.

Keywords Shape grammar · Models of design · Rule-based design · Nonstandard algebras

1 Introduction

Design is a field that encompasses much of human activity. The concept of design can be defined in multiple ways, of which what happens in architectural settings is but a small portion when looked at universally.

Besides the conventional definitions in the design disciplines, such as of *process* as a verb, or *product* as a noun, Stiny's [1] algebraic definition is—"a design is an element is an n-ary relation among drawings, other kinds of descriptions, and correlative devices as needed". Simon [2] states that designers devise "courses of action aimed at changing existing situations into preferred ones"; taken further computationally in terms of *states* and *rules*, design could also be defined as a process which "moves a composition of elements from an existing state A to a desired state B" [3].

A *rule* is a replacement structure, also called a production rule, which allows manipulation of a configuration from an existing state to a preferred state. It could be simply stated in the shape grammatical form that shape $A \rightarrow$ shape B, where A is the left hand side (*LHS*) of the rule-form which would be replaced by the right

P. Varghese (✉)
IntelliARCH, Irinjalakuda, India
e-mail: p.varghese@yahoo.com

A. Chakrabarti (ed.), *ICoRD'15 – Research into Design Across Boundaries Volume 1*,
Smart Innovation, Systems and Technologies 34, DOI 10.1007/978-81-322-2232-3_8

hand side *(RHS), B*. Computationally, rule-based structures and shape grammars had their beginnings in the work of Chomsky in linguistics [4]; later developments included information processing theories of Newell and Simon [5], principally IF-THEN clauses.

At a computational level, the idea of rule-based design has advantages that the shape-change state could be translated into an algorithm/heuristic that allows the optimisation of the design according to a rule form. Parametric sizing could utilise a system of computational optimisation, which as an engineering question within design has relevance in an architectural context too.

Cognitive science typically explains such mechanistic behaviour of representations operated on by procedures to enable changes in mental states, to explicate parts of human behavioural patterns. The use of rules as a part of design is at a subconscious level, in terms of the human design processes; as a formal design process, shape grammars use such methods as part of its formalism [6]. Rule-based processes are an inherent part of the shape grammar methodology; a shape grammar applies rules either as a hand-drawn process or as automated in computer-based programs, such as in interpreters. Such processes have not been sufficiently explored to give definite proof of human design capabilities; it is nevertheless powerful that it can be applied consciously to good effect by designers in the development of design.

2 Computational and Cognitive Studies/Models

Computationally, parts of *if-then* rules can be chained together both backward and forward, using a sequence of rules to answer questions, in human reasoning or to solve problems [5]. The other current cognitive theory is *connectionism*, similar to neural network models; this representation views knowledge as being encoded via connections between simple neuron-like processors (or nodes), similar to neurons connected by synapses, that either excite or inhibit another [7]. Minsky [8] and Schank and Abelson [9] proposed structures of mental representations termed frames, schemas, scripts and prototypes, which attempted to comprehend and model perception, memory, explanation, typically scientific reasoning and discovery.

Explanations of human behaviour also seem to have similarities to computational processes; In a similar manner, designing would happen at informal levels—a novice initially needs to reason out the various processes from scratch, but for an experienced designer, the intermediate states are rapid, some in-between states are dispensed with, and design development happens in quantum leaps. Much of the implicit designing happens in the mind of the designer, in-between states are not, or need not be formally outlined on paper.

At an experimental level, the work of Goel and Grafman [10] involving designers with brain lesions led them to conclude that differing parts of the brain are required in the overall process of design. The right dorsolateral *prefrontal cortex* (PFC) is necessary for open-ended or ill-structured representations and computations, in design; confirmation of this from other sources is also mentioned there. Goel [11], based on his years of experimentation, makes the claim that the neural structures that support precise and unambiguous representations are lateralised in the left PFC, while the those that support the imprecise, abstract and indeterminate representations are similarly so in the right PFC; and that design requires a judicious process of using both of the hemispheres.

The early training of the designer at the conscious level is slow and systematic; as one progresses, such processes become subconscious, reflexive and intuitive, which need not use the conscious brain [11]. The training of architects is intricate; the computations using the visual process are gradual transformations from basic ideas through implicit rule application.

Cross [12], after Rosenman and Gero [13], had stated that at a physical level, novel designs or creativity could belong to four models, i.e., *combination*, *mutation*, *analogy* or *from first principles*. Ideas of expertise and creativity in design has been looked at by several researchers from Akin [14] to Cross and Lawson [15], where some of the comparisons from other areas, such as chess-playing, as well as evaluating methods used by differing designers were reviewed. Gabora [16] identifies cognitive mechanisms with creativity, which "involves the capacity to spontaneously shift back and forth between analytic and associative modes of thought according to the situation"; another prerequisite was heightened sensitivity. Goldschmidt [17] transposes this into design contexts as (a) the brain's architecture controlling memory process, and (b) the designer's preparedness, which would include a large inventory of stored images, either from experience, or from memory. There is currently much interest in creativity studies, or innovation, which has become well-received in these environments, from engineering to product design to business applications.

3 Formal Logical Models and Design

Design, as practiced in architecture, is substantially dependent on the medium of graphical and visual communication to express ideas and concepts. While there is no universally accepted standard for graphical communication, one could see developments for its use as a representational medium. One is aware that various visual and graphical information systems are being worked on as methods for representation.

At a formal/systems level, architects and logicians have debated over the nature and models of design for some time. It was understood that design is not purely *deductive* or analytical, but tends to be mostly *inductive* or of synthetic reasoning. The roots of logic go back to the Greeks and the development of Aristotelian syllogistic reasoning. The early architectural system theorists of the '50s and '60s looked at design as being cycles of analysis and synthesis, and to the two-part model was later added a unit for evaluation, which fed back into the analytical loop. Later additions took on a spiral form with the process advancing over time [18].

The American pragmatist philosopher Peirce [19] had earlier introduced a differing form of synthetic reasoning, which he called *abduction*; based on the information available—the results and the rule, which could involve external assumptions and possibilities to support the results, to then infer the case; it is the inference of a case from a rule and a result. Peirce, it seems, interpreted abduction essentially as an "inferential" creative process of generating a new "explanatory" hypothesis, is mentioned by Magnani [20]; he further stated that it "is the only logical operation which introduces any new ideas" [19]. It has a classical logical form, distinct from deduction and induction. Abduction is sometimes referred to as *reduction*, and also as *retroduction*; March [21] calls it *production* in the context of design.

It would seem that much of design is more akin to an abductive form of reasoning, where not only are the design elements to follow a certain set of conditions, but also involve variations in the arrangements which go beyond the immediate set of stated end-conditions. It would seem that design explores variations into 'what-if' positions, which then form the basis for fresher hypotheses. March [21] opines that this mode is the part that uses *creativity*, which makes *design* different from *science*. Further, Magnani [20] follows Peirce's [19] opinion in stating that the generation of a scientific discovery itself requires abductive reasoning.

Design goes beyond the deduced solution, and opens up the potential to allow other possibilities. In design, it would seem that introducing variations in the arrangements of elements would move the process from single solutions to offering multiple options, without having to let go of the basic analytically derived result. March indicates a model, which cycles through the three forms of reasoning—productive (P), deductive (D) and inductive (I), and proposes that design is composed of PDIPDI… iterations; this fruitfully encompasses all modes of reasoning. March states that inductive reasoning is used to generalise, while design needs productive reasoning to particularise. He states that while science only uses deductive reasoning, it is the abductive component, which makes design a creative exercise. The thinking in such instances, is also to pose a hypothesis (in an original Peircian sense), which would be tested out as the design.

3.1 Shape Grammars

Shape grammars are similar to Lindenmayer (or *L*-) systems [22] and graph grammars in its operation. While having its foundations from the technical fields, much of its development happened in the area of architecture and design; its uses have been summarised from the Palladian plans [23], to the prairie house designs of Frank Lloyd Wright [24]. In the area of product design, it has depicted the development of coffee-makers [25], of motorcycles of the Harley-Davidson brands [26], and of the styling of cars [27]; aiding advances in engineering include the development of Shea's EifForm [28] to robotic applications [29]. Shape grammars have traced the logical design processes used in these instances; its capabilities includes enumeration and alternative generation [23–27], and its uses include reasoning, development of style and recording style change.

3.2 Rule Application in Shape Grammars

A rule application in a shape grammar is when a shape B in a shape C replaces an occurrence of a shape A … a rule is applied to a shape A to transform it into shape B, and is succinctly represented as $A \rightarrow B$ [30]. The application of the rule transforms according to the formula

$$C' = [C - t(A) + t(B)] \quad (1)$$

If the shape A is partitioned, for the elements which make up the shape, where $\{a_n \mid n = 0 \text{ to } 3\}$, where a_n is an independent element of algebra n. The partitioning would be the Cartesian product

$$A = [a_0 \times a_1 \times a_2 \times a_3] \quad (2)$$

Similarly for shapes B, C, and C', such that

$$C' = [c'_0 \times c'_1 \times c'_2 \times c'_3]$$

where C'_i is the shape or subshape in the ith dimension in the algebra (Fig. 1).

Architecturally, Alexander et al. [31] consider the subsistence of *patterns* and their inherent, everyday necessity for design. Computationally, the methodology used in the human brain to recognise patterns in structures are not yet fully understood, but one does recognise these in extremely short time periods [32]. Computers have not yet developed such agile algorithms or heuristics, and hence are subject to *brute-force search* procedures at each stage of the shape recognition process.

The reason for a representation to be more complex is the structure of the shape grammar elements and its environment. In a shape grammar environment, any element of a shape is apparently without structure so that it could be interpreted in multiple manners for shape rules. This apparent lack of structure allows the shape element to be interpreted by differing shape rules as they occur. Technically, this structure or the lack of it, would allow not just the shape, or at a lower level, a graphical editor to respond to the introduction of new rules, but to the occurrence

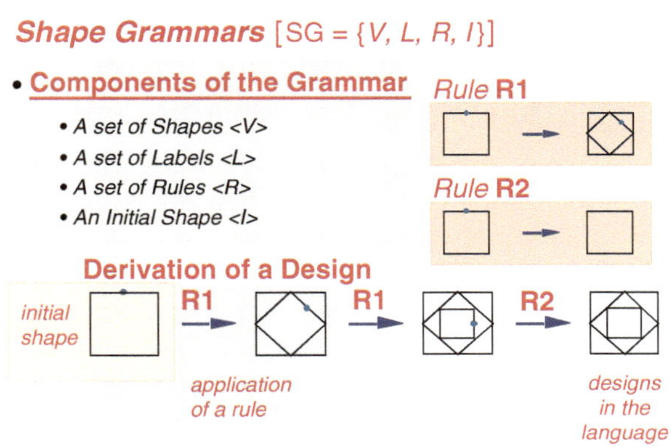

Fig. 1 Typical 2D shape rules and its derivation. *Source* adapted from the original by Stiny [30]

of all possible applicable rules. This is a little harder to do. In this sense, shape grammars are also open to the possibilities of emergent shapes, as well as for the development of preprocessing.

4 Designing Subconsciously Using Rules: Design-Based Rule Behaviour

The designer uses rules, subconsciously and consciously, which aids in the design process as one sketches with rapid mental calculations on forms on which the designer is working. Manipulation of form happens in the mind, eye and with the hand, on tracing paper and with speedy mental calculations. Individuals explore options and alternatives, of which they choose a few for development. The visual representation of the model on paper helps to mentally make computations on forms by implicit or explicit rules of composition, which guide the development. Processes of configuration using the visual medium are rapid; a well-trained designer does these implicitly, so that much of these become subconscious procedures. Cognitively [33], forward-chaining of rules helps combine multiple rules, sometimes weeding out unnecessary intermediate ones. At the same time, backward-chaining of rules helps develop a design from an identified end-result back to the beginning; this could sometimes prove useful in analysing known designs, or to reverse-engineer recognised or approved models or conditions.

Architects and designers in their initial scribbling make use of conceptual structures that start as squiggles, that later graduates to bubble-diagrams or functional relationships, which connect between functions, spaces, services, and similar. The architect develops a chain of possibilities out of which a few are considered as suitable candidates. The generation of alternatives is necessary so that a better range is developed, and the better design candidates chosen. The generation of choices is for options; these follow a set of rules, termed a grammar, to produce a range in a particular *style*; the designer could favour a set of rules following a language, which results in generating designs within a style. The generation of alternatives would be a conscious process of developing variations on a *theme*, made by a collection of rules. The rule set would have individual shape elements, but that could not predict the variety of alternatives generated. Using rule sets could generate diverse and surprising emergent designs, and with some unanticipated by the designer.

Decision-making sometimes happens within known constraints of site conditions, structural or services configurations, ergonomic standards or building byelaws. These constraints and conditions would need to be reinterpreted or converted into grammar rules. The advantage of these conversions is their ease of application, when or where similar conditions exist. A range of rules could also be developed which respond to existing site conditions, or even to differing styles chosen. Rapid prototyping for alternatives development, for client approvals, or for climatic considerations.

A designer in specifying a colour to paint a building or in indicating a particular kind of light switch subconsciously provides a rule for implementation. However, they do not have to explicate, or consciously derive these at each instance. These rules then become implicit, and recede into the background to become part of the accepted design process.

While Stiny's algebra of shapes proposes the use of multidimensional shapes in the same space, Knight [34] proposes different manners of organising shapes in several modes—hierarchically, visually or non-visually, giving the freedom to the designer to set their own priorities. For convenience and to distinguish the differences, Varghese [3] has called them the *standard/normal* and the *nonstandard* forms of the shape grammar or algebra. Knight [34] hierarchically defines them as *composite, hybrid, nonmanifold* and similar, where the definitions dictate the manner in which the shapes are ordered and organised. While these manners of organisation are diverse, it is understood that the *hybrid shape algebra* actually comes closest to the manner that the human mind normally perceives and manipulates shapes.

The author's conviction is that the use of the nonstandard shape algebras such as the *composite*, *nonmanifold* and particularly the *hybrid* shape algebras are areas which have not been sufficiently explored by researchers. Since Knight [34] first mentioned these concepts, Varghese and Merchant [35], and Jowers [36], among few have proposed manners in which they can be put to use in computation. It is understood that this was an area that is closest to intuitive and visual human design methods with respect to rule-based methods; hence, these need to be examined in more depth. This, as an area of research, could be further explored for methods, algorithms or interfaces by design researchers. It presents scope for computational and human-computer interfaces (HCI) for inputting, manipulating, and exploring designs; it has potential for exploration in developing design interfaces, or for specialised applications.

5 Rule-Based Design

Rule-based design could be interpreted as a synonym for design, which explicitly uses rules of construction. Implicitly, it can be argued that partially much of design is rule-based. One could interpret that design as a series of transformations, which takes the design through sequential intermediate states; one could anticipate that rules use this series of intermediate states, where rules change the design between states.

Computational rule-based design is formal in nature; this would mean constructing explicit rules of design for the design process. Such a possibility could extend to use in design development. Typically, such editing works are related to changes and corrections applied throughout a project, especially in relation to changing details and specifications. These happen usually as design progresses, and as decisions are finalised.

It is explicit when the designer uses stages documented on paper as sketches or as computations to go from one state to the next. A novice designer might go through the motions of laying out multiple possibilities before choosing a layout.

A normal shape grammar would have all the elements and their regular operations. The proposed interactions between the elements are allowed by using the *nonmanifold* and *hybrid* algebraic shapes [34, 35]. This increases the number of operations possible for the designer to work with. This is doable in two ways, namely through a design interface, or through rule-based processes. Parts of the design would need reinterpretation as being composed of elements other than purely of geometry, into such as colour, weight, or similar.

The development of rule-based design is inevitable, at least at the conscious level. Rule-based reasoning helps in ways related to preliminary drawings, and in the rapid development of designs. Speedy alterations in design drawings could be effected to designs that need fast decision-making. The development of rule-based design entails that many of the working methods in design could need to change for the future. Rule-based design methods necessitate that design be done by following rules of reasoning, rather than by the compositions of its individual elements; changes of one part might affect at others or the overall design. Rapid decision-making needs swift follow-up, and this could be achieved by substitutions, by using conditionals, or with restricted situations. The development of applications and algorithms that would execute such needs to be investigated further.

Automation within the design and construction process is ongoing, not just in the manufacture of materials, but also in processes such as the assembly at site. Robotics could be widely used, and become a natural extension of construction. Building automation and rule-based systems also need the development of tools that span the spectrum of dimensions. Many possibilities lie in the hands of designers and explorers.

While these might seem to follow the normal rules of construction, the rules in design are different; design rules are intended to give form to the building, to lay out the spaces according to the functions, or for norms of behaviour of the users, which the designer decides in accordance with the design requirements; both design rules and construction rules could be numerous, the numbers depending on the level of detail intended. Rules of construction are normally sequential, and works from the ground up. Design rules would need to be reinterpreted or converted into construction rules, through a thorough analysis of the design, including the construction details, materials and finishes. These could also be coupled with construction scheduling methods like PERT/CPM, and could be charted into tangible schedules. Adoption of rule-based construction could also be adapted to automation, especially in processes such as prefabrication or modular construction.

The examination and workings of shape grammar interpreters [37, 38] demonstrate that design has machine-implementable possibilities, within some stated limitations; some forms of rule-based mechanical processes could be developable; and the idea of rule-based systems could be made practical for design.

Symbol-shape hybrid systems have been implemented. The underlying representational methods of all current systems including the shape interpreters, are dependent on symbolic computation; it is reasoned that this dependence might in the long term put limitations on computational methods, instead of pure shape computations, also referred to by Stiny [4]; an escape from this situation cannot be envisaged at this point; such explorations could contravene Newell and Simon's [39] argument regarding the universality of symbols. Symbol-shape models if used only for shape generation as shown in the research could be developed for current computational applications too.

6 Conclusion

The paper has introduced in a minor way, the environment, scope and direction for the future development of *rule-based systems*. It can be seen that while the process of rules and rule application could be used productively in the computational aspects within systems, the process of visualisation and decision-making could need human input with interpretation. The machine today does a commendable job of the generation of designs, yet into the near future, the evaluation of designs could need human interpretation. The need for looking to cognitive models is to examine the closeness of the human and computational models of design, from which several tentative innovative paths for developments can be laid.

Working from several fronts in the scheme, it can be understood that:

- *much of design happens at the cognitive stage, and that rule-based design can be developed from these levels,*
- *formal models of design need deduction, induction and abductive modes of logic,*
- *the development of rules capture the transformative process in design,*
- *types of nonstandard algebras can be used for better visual decision-making,*
- *rule-based construction and its scheduling is another extension besides design,*
- *the gap between explanations provides scope for further exploration.*

It can therefore be concluded that cognitive and computational rule-based design could have significant relevance in today's design and construction field.

References

1. Stiny, G.: What is a design? Environ. Plan. **17**(1), 97–103 (1990)
2. Simon, H.: The Sciences of the Artificial, 3rd edn. MIT Press, Cambridge (1996)
3. Varghese, P.: Shape algebras and rules in design: bridging the gap between formal and intuitive thinking. Architecture and Regional Planning, Indian Institute of Technology, Kharagpur (2013)
4. Chomsky, N.: Syntactic Structures. Mouton, The Hague (1957)

5. Newell, A., Simon, H.A.: Human Problem-Solving. Prentice-Hall, Englewood Cliffs (1972)
6. Stiny, G.: What rule(s) should I use? Nexus Netw. J. **13**(1), 15–47 (2011)
7. Rumelhart, D.E., McClelland, J.L. (eds.): Parallel Distributed Processing: Explorations in the Microstructure of Cognition. MIT Press/Bradford Books, Cambridge (1986)
8. Minsky, M.: A framework for representing knowledge. In: Winston, P.K. (ed.) The Psychology of Computer Vision, pp. 211–277. McGraw-Hill, New York (1975)
9. Schank, R.C., Abelson, R.P.: Scripts, Plans, Goals, and Understanding: An Inquiry into Human Knowledge Structures. Lawrence Erlbaum, Hillsdale (1977)
10. Goel, V., Grafman, J.: Role of the right prefrontal cortex in ill-structured planning. Cogn. Neuropsychol. **17**(5), 415–436 (2000)
11. Goel, V.: Creating artifacts: integrating cognitive processes and functional anatomy. In: Gero, J. (ed.) Unpublished manuscript, SDC-10. GWU, Washington D.C (2010)
12. Cross, N.: Descriptive models of creative design: application to an example. Des. Stud. **18**, 427–455 (1997)
13. Rosenman, M., Gero, J.: Creativity in design using a design prototype approach. In: Gero, J., Maher, M.L. (eds.) Modelling Creativity and Knowledge-Based Creative Design. Lawrence Erlbaum, New Jersey (1993)
14. Akin, O.: Necessary conditions for design expertise and creativity. Des. Stud., 107–113 (1990)
15. Cross, N., Lawson, B.: Studying outstanding designers. In: Gero, J., Bonnardel (eds.) Studying Designers, pp. 283–287. Sydney, Australia (2005)
16. Gabora, L.: Revenge of the 'neurds': Characterizing creative thought in terms of the structure and dynamics of memory. Creativity Res. J. **22**(1), 1–13 (2010)
17. Goldschmidt, G.: Ubiquitous serendipity: potential visual design stimuli are everywhere. In: Gero, J. (ed.), SDC '10, Washington DC. www.gwu.edu (2010)
18. Lawson, B.: How Designers Think. The Architectural Press, London (1980)
19. Peirce, C.S.: Chance, Love and Logic. Kegan Paul, Trench, Trubnor, London (1923)
20. Magnani, L.: Abductive Cognition: The Epistemological and Eco-cognitive Dimensions of Hypothetical Reasoning. Springer, Berlin (2009)
21. March, L.: Introduction: the logic of design and the question of value. In: March, L. (ed.) The Architecture of Form. Cambridge University Press, Cambridge (1976)
22. Lindenmayer, A.: Mathematical models for cellular interaction in development I—filaments with one-sided inputs. J. Theor. Biol. **18**, 280–289 (1968)
23. Stiny, G., Mitchell, W.J.: The Palladian grammar. Environ. Plan. B **5**(1), 5–18 (1978)
24. Koning, H., Eizenberg, J.: The language of the prairie: Frank Lloyd Wright's prairie houses. Environ. Plan. B **8**, 295–323 (1981)
25. Agarwal, M., Cagan, J.: A blend of different tastes: the language of coffee makers. Environ. Plan. B **25**(2), 205–226 (1998)
26. Pugliese, M., Cagan, J.: Capturing a rebel: modeling the Harley-Davidson brand through a motorcycle shape grammar. Res. Eng. Des. **13**(3), 139–156 (2002)
27. McCormack, J.P., Cagan, J., Vogel, C.M.: Speaking the Buick language: capturing, understanding, and exploring brand identity with shape grammars. Des. Stud. **25**(1), 1–29 (2004)
28. Shea, K., Cagan, J., Fenves, S.J.: A shape annealing approach to optimal truss design with dynamic grouping of members. J. Mech. Des. **119**(3), 388–394 (1997)
29. Antonsson, E.K., Cagan, J. (eds.): Formal Engineering Design Synthesis. Cambridge University Press, Cambridge (2001)
30. Stiny, G.: Introduction to shape and shape grammars. Environ. Plan. B **7**(3), 343–351 (1980)
31. Alexander, C., Ishikawa, S., Silverstein, M.: A Pattern Language: Towns Buildings Construction. Oxford University Press, Oxford (1977)
32. Sinha, P.: Recognizing complex patterns: review. Nat. Neurosci. Suppl. **5**, 1093–1097 (2002)
33. Thagard, P.: Cognitive architectures. In: Frankish, K., Ramsay, W. (eds.) The Cambridge Handbook of Cognitive Science, pp. 50–70. Cambridge University Press, Cambridge (2012)
34. Knight, T.W.: Computing with ambiguity. Environ. Plan. B **30**(2), 165–180 (2003)

35. Varghese, P., Merchant, A.: Modeling shapes for design and computation: hybrid algebras for the designer. In: Proceedings of the 10th International Conference, CAADRIA '05, vol. 2, pp. 79–88. New Delhi (2005)
36. Jowers, I.: Computation with curved shapes: towards freeform shape generation in design. PhD thesis, Department of Design and Innovation, The Open University, UK (2006)
37. Gips, J.: Computer implementation of shape grammars. Retrieved from www.shapegrammar.org (1999)
38. Trešcák, T.: Shape grammar interpreter. Retrieved from http://www.sourceforge.net/ (2009)
39. Newell, A., Simon, H.A.: Computer science as empirical inquiry: symbols and search. Commun. ACM **19**(3), 113–126 (1976)

The Roles of Engineering and Spirit in Product Design

S. Saleem Ahmed

Abstract This is an exploratory study on the nature of product design in which the role of engineering is seen analogous to role of spirit (life). Within the domain of product design if Product Experience (Desmet and Hekkert in Int. J. Des. 1(1):57–66, 2007 [1]) is recognized as one part, then Product Entity is present as the other part. The physical, emotional, and mental aspects that make up three distinct experiences, namely aesthetic experience, emotional experience, and experience of meaning, within the Product Experience, are also present in the Product Entity as embodiments. Any experience, including Product Experience, is something only those who have spirit (life) can realize in the sense that we human beings understand. Any product entity becomes a reality only due to engineering. In other words, engineering enables something to be a product entity and spirit (life) enables something to be a product experience. Therefore, the role of engineering is that of an enabler of product embodiment much like the role of spirit (life) as an enabler in product experience. Understanding of this nature could possibly help tackling the issues in development of product designs better.

Keywords Product entity · Product experience · Engineering · Spirit

1 Introduction

This study has been done with an objective of exploring the very nature of product design when holistic order, a synonym of beauty, is achieved. Initial part of this paper deals with what is holistic order in product design as we understand today and the phenomenology involved in creating such an order. It is being done by defining product at the basic level first and then building on from the key elements

S.S. Ahmed (✉)
CPDM, Indian Institute of Science, Bangalore, India
e-mail: saleem@cpdm.iisc.ernet.in

A. Chakrabarti (ed.), *ICoRD'15 – Research into Design Across Boundaries Volume 1*,
Smart Innovation, Systems and Technologies 34, DOI 10.1007/978-81-322-2232-3_9

in the definition. Ever since Desmet and Hekkert [1] published the Framework of Product Experience, it has been of interest to this author to explore on the other side of product design, i.e. Product Entity. This is dealt with in the next part of this paper. Finally the actualization process of holistic order in product design is explained. In the process, how the role of engineering in the product entity is analogous to the role of spirit (life) in the product experience is established.

2 Holistic Order in Product Design

Beauty is the term being used by philosophers, aestheticians, poets, painters, and thinkers for describing the quality that gives pleasure to the human senses. Other terms like visual order, discursive order, implicate order, and aesthetic order are also being used in the literature. This author is proposing a new term *Holistic Order* to describe the qualities that give pleasures to not only all the physical senses like visual, acoustical, tactual, olfactorial, and gustatorial, but also emotional and cognitive faculties. Detailed discussion on this is given in Sect. 2.5 of this paper.

2.1 Definition of Product

Based on the definitions given by the leading dictionaries like Oxford and Webster, the definition of *product* can be simplified as follows for our purpose: "A product is a MAN-MADE OBJECT which is USEFUL TO SOMEBODY". From this definition, it is evident that there are two things present in a product which make up the whole. One is the 'object' part of it and the other is the 'human' part of it. What makes a product different from an object is the presence of the *human* part in it. In this connection, we are concerned with user profile and human-product interaction.

2.2 Model of Human-Product Interaction

When usefulness to somebody is part of product definition, the 'user' of the product is implied. Though it is possible that even animals can be users, for our present study, let us assume that the user be only a human being. This user can be a man or a woman; an old or an young; a rich or a poor; an educated or an illiterate; with a Western or an Eastern thoughts; a cultured or a unruly; an introvert or an extrovert; somebody creative or a uncreative; smart or a stupid; a normal or a disabled person. Every one of us as a human being is experiencing the world in our own unique way. Physical, emotional, and intellectual differences may affect

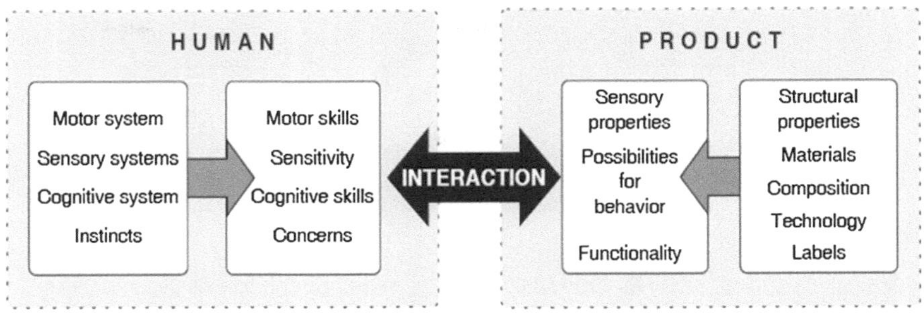

Fig. 1 Model of human-product interaction. *Source* Schifferstein and Hekkert [2]

the way the sensory organs gather data and process the information. This may result in different perception than otherwise what someone may have gotten with normal sensory inputs. However different they may be, all are some kind of human experience any way.

Figure 1 shows the model of human-product interaction as proposed by Schifferstein and Hekkert [2]. This model is exhaustive in nature and not prescriptive. In this model, all the physiological and psychological human systems i.e. motor, sensory, and cognitive, and the corresponding human skills involved have been paired with the sensory and structural properties of the product, functionality, materials, and technology. Also it indicates that there are *human* and *product* sides which are in constant interaction with each other in a typical product experience.

2.3 Product Experience

Desmet and Hekkert [1] proposed a framework of product experience as shown in Fig. 2. In this they have described three kinds of experiences, i.e. aesthetic experience, emotional experience, and experience of meaning, involved in a user-product interaction.

Aesthetic experience is concerned with physical aspects of the product. A product can be beautiful to look at, make a pleasing sound, feeling soft and smooth to touch, or smelling nice. These physical aspects of the product are in turn being detected by the motor, sensory, and cognitive systems of the user. The degree to which a perceptual system manages to detect order or coherence and assess a product's novelty/familiarity typically determines the affect that is generated in the product experience. When the user is pleased by the sensuous shape of a pen, the gentle but firm sound of a car door, or the soft and velvety texture of a couch, these experiences refer to aesthetic experience.

It may be worth noting that the *physical* aspects, like the form, colour, texture, sound, and smell, of the object are responsible for the *physiological* aspects of the user perception, like *sensuous* shape, *gentle* sound, and *velvety* texture.

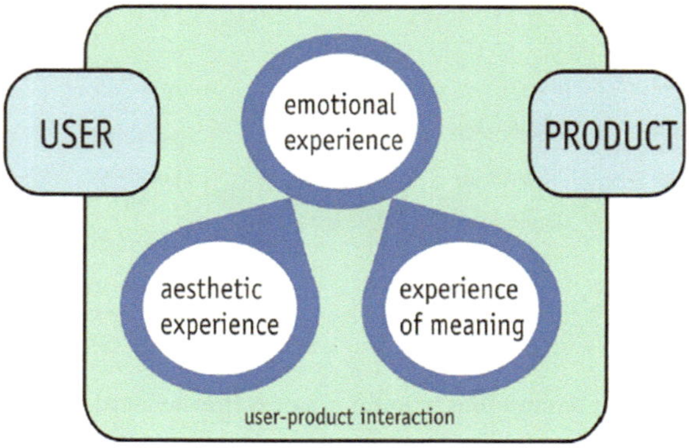

Fig. 2 Framework of product experience. *Source* Desmet and Hekkert [1]

A product that evokes anger will be ignored, one that evokes fascination will be explored, and one that evokes boredom will be put aside. Pleasant emotions pull us towards products that are beneficial, whereas unpleasant emotions will push us away from those that are detrimental for our well-being [3]. When the user is *disappointed* by the limited memory capacity of a digital camera, *delighted* by an innovative solar car for its green design, or *frustrated* by the complexity of a user interface, we can identify these experiences as emotional experiences. It may be noted that the causes for all the emotional experiences, like the *limited memory* of the digital camera, *green design* of the solar car, and *complexity* of the user interface, are all *meta-physical* aspects of the *objects*.

Experience of meaning is concerned with mental (intellectual) aspects. Through cognitive processes, like interpretation, memory retrieval, and associations, we are able to recognize motifs and metaphors, assign personality or other expressive attributes, and assess the personal or symbolic significance of products. Considering a water filter as feminine and very much 'for you', a computer mouse as sleek, and a new table light referring to the seventies, are all examples belonging to the experience of a product's meaning. Again it may be noted that the causes for all the experiences of meaning, like the *feminine attributes* of the water filter, *sleekness* of the computer mouse, *retro looks* of the table light, are all the *meta-physical* aspects of the objects.

Both emotional experience and experience of meaning are psychological in nature for the user of products. As noted earlier these are meta-physical aspects of the objects.

2.4 Product Entity

Going by the doctrine of cause and effect, all aspects present as the *product experience* must be there, in the first place, as some kind of elements in the *product entity*. If so, the physical, emotional, and mental (intellectual) aspects that make up the product experience are present in the product entity as attractiveness, appeal, and meaningfulness. Individually these terms are explained in the following.

Line, Shape, texture, tone, and colour are all physical aspects which are embedded in a product. Conventionally these physical aspects are known as *design elements* [4]. When these elements are arranged in accordance with design principles like proportion, rhythm, balance, and emphasis, aesthetic qualities like harmony and variety are achieved initially. Finally attractiveness (unity) is achieved in a composition of product design. Therefore it can be said that *attractiveness* which embodies all the physical aspects in a product entity is responsible for the aesthetic experience of a product's design.

Helandar and Tham [5] demonstrated the importance of affect for ergonomics, Jordan [6] discussed the role of affect in comfort, and Tractinsky et al. [7], demonstrated a relationship between affect and usability. The *frustration* of carrying a suitcase with an inappropriately small and hard handle leads to low *appeal* due to poor usability. The *delight* of sitting on a comfortable computer chair with a good back support leads to a greater *appeal* towards the product. Therefore it can be said that *appeal* which embodies all emotional aspects in a product entity is responsible for the emotional experience in a product's design.

Meaningfulness is a quality that embodies all the mental aspects in product's design. What meaning a red coloured serving tray may have for a German family could be quite different to what may be for a Chinese family. A photo frame in black colour may never be appreciated well by an Indian as a wedding gift, as it is meant for framing a photo of dead person. The same photo frame may be well received by an American or French due to the meaning attributed in their culture. Piggy banks are never given to Arab children as anything shaped like a pig is not acceptable as a meaningful gift in Arabic culture. Same is not true in Europe or America. Therefore it can be said that *meaningfulness in a* product entity is responsible for the *experience of meaning* in a product's design.

2.5 Holistic Order

A holistic perception of reality—seeing things as whole—requires interdisciplinary focus, and sets into motion the search for a "complete knowledge system" [8]. Oliver [9] proposed a new cultural paradigm, which mandates academic interdisciplinary approach. His model integrates the mind, body, and spirit in knowing, values our connection to nature, and takes into account the evolutionary capacity of the human species. In his view, the achievement of holistic perception can add comprehensiveness and depth to understanding our relationship and expression in the social and universal scheme. Thus holistic approach can integrate as well as interconnect physical, emotional, and mental aspects in the field of inquiry.

Bohm, in his theory of implicate order, discussed quantum physics in terms of positivist knowledge and the interdisciplinary holistic paradigm. He focused on the schism between matter and consciousness [10]. Alexander's four books on The

Nature of Order discussed the connection between architecture and wholeness or life [11]. He had a new vision of architecture to make buildings which have life and profound order. He argued that each thing—regardless of what it is—has some degree of life. This, he felt, furnishes us a single coherent conception of the world.

Systems thinking is the process of understanding how things, regarded as systems, influence one another within a whole. In nature, systems thinking examples include ecosystems in which various elements such as air, water, movement, plants, and animals work together to survive or perish. In organizations, systems consist of people, structures, and processes that work together to make an organization "healthy" or "unhealthy". Systems thinking has its roots in the General System Theory that was advanced by Bertalanffy [12] in the 1940s.

As noted so far in this section, holistic approach to solving problems has been in existence in different fields. In the realm of arts and design, synonyms of *Holistic Order* such as beauty, truth, goodness, order, visual order, and aesthetic order have been in usage for a long time. Beauty, truth, and goodness are subjects of study for philosophers, aestheticians, poets, and thinkers for centuries. Plato (400 BC) discussed beauty in his works on *Theory of Forms* and *Composition in Art* [13]. Aristotle (330 BC) explained the complex nature of physical, metaphysical, physiological, and psychological aspects that contribute to beauty through his works *Metaphysica* and *Poetics* [14].

Baumgarten introduced the term *Aesthetics* in 1735. In his work *Aesthetica* [15], he defined "Aesthetics (the theory of the liberal arts, lower gnoseology, the art of thinking beautifully, the art of the analog of reason) is the science of sensitive cognition". In his other work *Metaphysica* [16], he had discussed various metaphysical and psychological aspects involved in aesthetic appreciation of art.

Kant made great contributions to the aesthetic appreciation of arts. Through his works *Critique of Pure Reason* [17] and *Critique of Judgment* [18], he explained how sensory input, intuition, and intellectual aspects combined to give artistic imagination and aesthetic response. Santayana [19] in his book titled 'The Sense of Beauty' defines: "beauty is a pleasure regarded as the quality of a thing". He further elaborates that the pleasure is a value that we perceive out of the quality of a thing. From this statement it is evident that beauty exists in perception as a value and in things as a quality.

The term Aesthetic Order has been in use only recently. Osborne [20] used it in his paper titled Aesthetic and Other Forms of Order. Haines [21] published a paper titled Aesthetic Order and Lorand [22] authored a book with the same title. Though the broad meaning of the term Aesthetic Order remains the same, the specific meaning varied subtly according to the context of creation of various authors.

In product design, when all the design elements are arranged in some order, it is supposed to be aesthetically pleasing to the users/viewers of the design. For time immemorial, artists/craftsmen/designers have been making best efforts to please the senses of users/viewers through composition of design elements in accordance with the established design principles in all their creations.

According to Ocvirk et al. [4], unity is the highest order attainable in a creation of art/craft/design. In their book titled 'Art Fundamentals', they have given a framework of unity which they claim as a logical and common order of events in the creation of an artwork, though artists often alter the sequence. It may be noted that the entire framework covers only the physical aspects of product design. If so, the other parts of product experience, namely emotional experience and experience of meaning are not accounted for in the framework. The aesthetic experience as described by Desmet and Hekkert [1] is all about what one may perceive through any of the sensory modalities. In other words, aesthetic experience of products is all about the physical aspects only. So, the emotional experience and experience of meaning will have to be accounted for in the framework of holistic order.

Papanek [23] authored a book titled 'Design for the Real World'. In the last three decades, this book remains the most widely read book on design. It has already been translated into more than 20 languages so far. This book is mostly on the *fitness* aspects of design like cultural values, sustainability, ecology, and social change. In this book Papanek stated, "Some furniture, such as a dining table with legs constructed of stainless steel with a marble top, evokes the response to lie down and wait for the doctor to come and extract one's appendix. Nothing about the table says, dine off me". This connotation or communication between the product and the user is about the mental aspect in the holistic order.

Emotional experience and experience of meaning are both psychological in nature. One is about the emotional aspect and the other is about the mental aspect. While the emotional aspect is a sub-conscious experience, the other, namely mental aspect, is a conscious experience. Both together form the psychological experience of a product.

Thus a holistic order in product design is an order of all the physical, emotional, and mental aspects as a whole as well as interconnecting them. These different aspects exist in the dichotomy of product design, i.e. product entity and product experience.

3 Holistic Order in Product Design—Actualization Process

As noted earlier, physical, emotional, and mental aspects are all present in some manner as embodiments in a typical product design. Since product design is a result of some intentional human effort, all these aspects (physical, emotional, and mental) are embodied during the creation of a design. The same are being perceived by the user during the product experience. So, what is being encoded during the creation of a design is being decoded during the experience of the same. This is being termed as the Actualization Process.

3.1 Role of Engineering in Holistic Order of Product Design

Attractiveness, appeal, and meaningfulness are the three constituents of a product's entity. They are all embodying respectively the physical, emotional, and mental aspects of a product's entity. They are all interconnected and being embodied in a product form through the process of engineering. That means, attractiveness, appeal, and meaningfulness are all made into a reality in a product form through engineering. It is important that all these three aspects relate with each other and gel well together. This phenomenon of gelling together of physical, emotional, and mental aspects is the holistic order of the product entity. Role of engineering in this phenomenon is that of an *enabler* without which the product entity would not be a reality.

3.2 Role of Spirit (Life) in Holistic Order of Product Design

As explained in Sect. 2.3, aesthetic experience (physical), emotional experience, and experience of meaning (mental) are the three parts of the product experience. They are all interconnected and being felt during a product experience through spirit or life. Any kind of experience, including a product experience, is a possibility only by somebody with life or spirit. We do not have knowledge of any kind of experience that dead person can feel. Spirit or life is needed in order to have a human experience. That means, product experience is a reality only due to spirit or life. It is important that all the three experiences, i.e. aesthetic experience, emotional experience, and experience of meaning, match well with each other (gel well together). This phenomenon of gelling well together of all the three parts of product experiences is the holistic order of the product experience. Role of spirit or life in this phenomenon is that of an enabler without which the product experience would not be a reality.

3.3 Analogy: Engineering and Spirit

Figure 3 shows the actualization diagram of product design. As noted earlier, product entity and product experience are the two sides of product design. The axis showing the physical, emotional, and mental aspects is common to both sides. Product entity is concerned with the *objects* which are non-living and product experience is concerned with organisms (human in our case) which are living. Physical, emotional, and mental aspects are embodied in a product entity through engineering. Same aspects are felt in a product experience through spirit or life. So, what is encoded on one side of product design is decoded on the other side of the product design. On one side, engineering is the enabler for the reality and on the other side spirit (life) is the enabler for the reality.

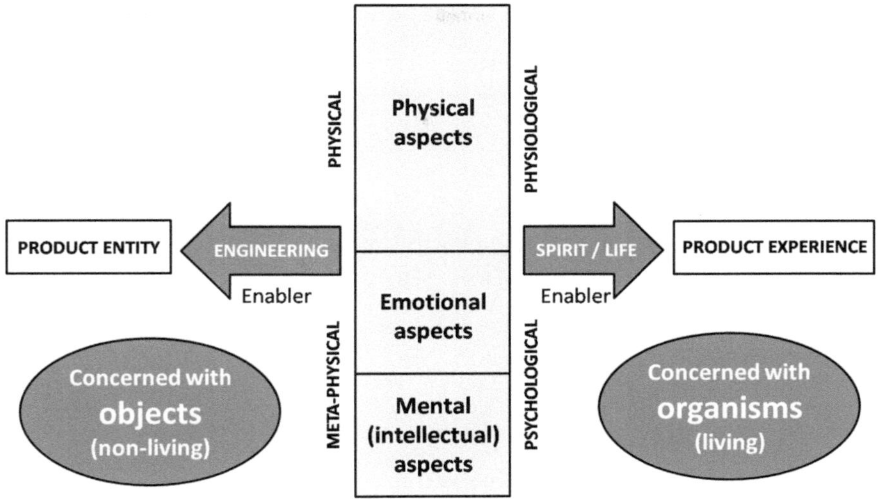

Fig. 3 Actualization diagram of product design

4 Conclusion

Only those who have spirit (life) can experience anything that we normally know of. As product experience is a phenomenon for all with spirit (life) including animals, product entity is a reality due to engineering. In other words, engineering enables something to be embodied in a product entity and spirit (life) enables something to be a product experience. So, we could say that the roles of engineering and spirit (life) in a product design are that of an enabler.

This study has shown how engineering is analogous to spirit (life) in product design. Phenomenology of holistic order and the actualization process in product design have been described in the paper. The analogy between engineering and spirit (life) has been made in the end. Understanding of this analogy and the constituents of product design as physical, emotional, and mental aspects could help tackling the issues of development of product design better.

References

1. Desmet, P., Hekkert, P.: Framework of product experience. Int. J. Des. **1**(1), 57–66 (2007)
2. Schifferstein, H.N.J., Hekkert, P.: Product Experience. Elseviers, UK, Netherlands (2008)
3. Desmet, P.: Designing Emotions. Delft University of Technology, Delft (2002)
4. Ocvirk, O.G., Stinson, R.E., Wigg, P.R., Bone, R.O., Cayton, D.L.: Art Fundamentals: Theory and Practice. McGraw-Hill, USA (2002)
5. Helandar, M.G., Tham: Hedonomics—affective human factors design. Ergonomics **46**(13/14), 1269–1272 (2003)

6. Jordan, P.W.: Pleasure with products: human factors for body, mind, and soul. In: Green, W.S., Jordan, P.W. (eds.) Human Factors in Product Design: Current Practice and Future Trends, pp. 206–217. Taylor & Frasis, London (1999)
7. Tractinsky, N., Katz, A.S., Ikar, D.: What is beautiful is usable. Interact. Comput. **13**(2), 127–145 (2000)
8. Harman, W.: Global Mind Change. Knowledge Systems Inc, Indianapolis (1988)
9. Oliver, D.W.: Education, Modernity, and Fractured Meaning. University of New York Press, Albany (1989)
10. Dabrowski, I.J.: David Bohm's theory of the implicate order: implications for holistic thought processes. Issues Integr. Stud. **13**, 1–23 (1995)
11. Alexander, C.: The Nature of Order, Book One—The Phenomenon of Life. The Center for Environmental Structure, Berkeley (2002)
12. Bertalanffy L.V: General System Theory: Foundations, Development, Applications. George Braziller, New York (1976)
13. Hyland, D.A.: Plato and the Question of Beauty. Indiana University Press, Bloomington (2008)
14. Wood, R.E.: Placing Aesthetics. Ohio University Press, Athens (1999)
15. Baumgarten, A.: Aesthetica (1750) (The German Aesthetic Tradition, Kai Hammermeister. Cambridge University Press, Cambridge 2002)
16. Baumgarten, A.: Metaphysica (1739) (The German Aesthetic Tradition, Kai Hammermeister. Cambridge University Press, Cambridge 2002)
17. Kant, I.: Critique of Pure Reason (1781) (Translation by Guyer, P., Wood, A.: Cambridge University Press, Cambridge 1998)
18. Kant, I.: Critique of Judgment (1790) (Translation by Guyer, P., Mathews, E.: Cambridge University Press, Cambridge 2000)
19. Santayana, G.: The Sense of Beauty. Charles Scribner's Sons, USA (1896)
20. Osborne, H.: Aesthetics and other forms of order. Brit. J. Aesthetics **22**(1), 3–16 (1982)
21. Haines, V.: Aesthetic order. J. Value Inq. **28**, 193–215 (1994)
22. Lorand, R.: Aesthetic Order: A Philosophy of Order, Beauty, and Art. Routledge, London (2002)
23. Papenek, V.: Design for the Real World. Thames & Hudson, UK (1984)

Holistic Approach to Product Design

S. Saleem Ahmed

Abstract Holistic approach to product design is not new. It has been in existence longer than the field of industrial design, which originated only in 1920s. As there had been always attempts to simplify the process of problem-solving through conventional reductionistic and mechanistic approaches, holistic approach to product design possibly never got the patronage that it deserved. In the current scenario, with the increased number of factors involved in solving a product design problem and the field of product design being inter-disciplinary in nature, it may be appropriate to take up holistic approach of wholeness and connectedness. In our context, 'holistic product design' means product design created by emphasizing the importance of the whole and the interdependence of its parts. Holistic is a term used for an entity that is encompassing as well as interconnecting all physical, emotional, and mental aspects. Holistic health and holistic life are some examples for that. When a designer consciously attempts to address all these aspects in his/her creation in a holistic manner, the degree of embodiment of these aspects greatly improve and often a better product design emerges in the process.

Keywords Holistic · Product design · Whole · Parts · Interdependence

1 Introduction

Holistic approach to problem solving has been there already in domains that are purportedly complex like quantum physics, human brain research, landscape ecology, climate change, and pollution control. Dabrowski [1], who investigated the holistic thought processes, reported, "New versions of the sciences as well as other academic disciplines are being formulated in terms of a naturalistic mode

S.S. Ahmed (✉)
CPDM, Indian Institute of Science, Bangalore, India
e-mail: saleem@cpdm.iisc.ernet.in

A. Chakrabarti (ed.), *ICoRD'15 – Research into Design Across Boundaries Volume 1*,
Smart Innovation, Systems and Technologies 34, DOI 10.1007/978-81-322-2232-3_10

of inquiry. The emergent post-positivist paradigm is directed toward a holistic conception of reality, replacing the Cartesian, mechanistic mode of conceptualization, In fact, characteristics of the "whole paradigm" are being identified in a variety of disciplines. New dimensions include a refocusing from simple to complex realities, the movement from the mechanical to the holographic, and the emphasis from linear to mutual causality".

This paper studies how holistic approach to problem solving is helpful to tackling complex issues in different disciplines, how the same can be extended to product design, and what are the implications. In the end, how Visual Order of yester years has become Holistic Order in present day context for product design and also how holistic approach to product design can be an answer to the paradigm shift in the recent times.

2 Holistic Approach in Various Human Endeavours

Bohm discussed theory of quantum physics in terms of positivist knowledge and the interdisciplinary holistic paradigm. He focused on the schism between matter and consciousness [1]. Sperry [2] and Harman [3] propounded holistic approach in their mental health research. An underlying knowledge assumption guiding these premises is that a multi-dimensional nature of illness designates an interdisciplinary "perceptual space", which makes many social and behavioral sciences, and even many nonscientific disciplines relevant to the investigation of health and illness [4]. In the context of this literature, a distinct notion of health and health-seeking behavior from the perspective of the humanities and the social sciences is emerging [5].

Systems thinking is one of the approaches to problem solving, by viewing "problems" as parts of an overall system, rather than reacting to specific part, outcomes or events and potentially contributing to further development of unintended consequences. Systems thinking focuses on cyclical rather than linear cause and effect. Planning and evaluation is one of the many fields where systems thinking is widely popular. However, there is disagreement on what constitutes systems thinking. Some scholars describe it as synonymous with systems sciences (i.e. complexity, chaos, nonlinear dynamics) and others view it as taxonomy—a laundry list of systems approaches [6]. But, basically systems thinking is holistic in nature.

A holistic perception of reality requires interdisciplinary focus, and sets into motion the search for a "complete knowledge system" [3]. Oliver [7] proposed a new cultural paradigm, which mandates academic interdisciplinary approach. His model integrates the mind, body, and spirit in knowing, values our connection to nature, and takes into account the evolutionary capacity of the human species. In his view, the achievement of holistic perception can add comprehensiveness and depth to understanding our relationship and expression in the social and universal scheme. Thus holistic approach can integrate as well as interconnect physical, emotional, and mental aspects in the field of inquiry.

To the holist, fragmentation does not occur only at the level of objective disciplinary content. There is also the subjective factor of consciousness. The numerous versions of holistic and process education are considered to have confluent goals, meaning that cognitive and affective domains merge and seek to integrate the intellect with emotions for responsible action in life. These indicate strong distinction between acts, emotions, and thoughts, which are nothing but physical, emotional, and mental aspects respectively. According to Ferguson [8], Bohm too recognized these prevailing distinctions.

3 Evolution of Knowledge Base of Art and Design Over the Years

Beauty has always been the pursuit of artists and designers in their creations. This has been the case for centuries. Beauty is concerned with the qualities that give pleasure to the senses. The term beauty exists in synonyms like harmony, unity, order, visual order, discursive order, implicate order, and aesthetic order. An overview of all these different yet connected terms is being presented in this section.

Plato discussed on sensual pleasures that result from the compositions of art in his works on theory of forms [9]. Aristotle identified the importance of golden mean (proportion) in his work Poetics [10]. He also explained the principles of rhythm and harmony in his work. His monumental work Metaphysica [11] dealt with the metaphysical aspects like actuality, potentiality, truth, and falsity in the works of art. He placed beauty as the highest attainable quality in a work of art.

Bharata had written 'Natyashastra' in the 6th century. It was about all the aspects that give pleasures in a dance performance. He identified eight different emotional expressions that a performer can bring out in a dance form. In the 11th century Abhinavgupta, an Indian philosopher and aesthetician, wrote commentaries on Natyashastra and also wrote on rasa theory, which is concerned with emotions in art form, and Indian aesthetics [11].

Baumgarten made great contributions through his works Metaphysica [12] and Aesthetica [13]. He first coined the term aesthetics in the year 1735 in his dissertation. Baumgarten described beauty as a result of mixing form with contents and means of expression. He also discussed the trinity of good, truth, and beauty and how they are the highest attainable qualities of social sciences, natural sciences, and arts respectively.

Kant extended the works of Baumgarten and published his works 'Critique of Pure Reason' in 1781 [14] and 'Critique of Judgment' in 1790 [15]. He too propounded beauty as the highest attainable quality in artistic creations. He identified three distinct aspects—sensory input, intuition, and intellectual—that make up beauty in artistic creations. Like Baumgarten, Kant too had a holistic approach to the aesthetic appreciation.

Hogarth [16] in his Analysis of Beauty placed great emphasis on *fitness* by placing it as the very first chapter in his book. Here the term fitness was being used to describe the appropriateness of a design and its elements to the whole. Jones [17] in his Grammar of Ornament proposed 37 general principles of arrangement of form and colour in architecture and decorative arts. He stressed that fitness, proportion, and harmony are a must in all works of arts and architecture. Also his proposition no.4 states that true beauty results from that which the mind feels when the eye, the intellect, and the affections are satisfied from the absence of any want. It may be noted that he connected beauty with the satisfaction of physical (eye), emotional (affections), and mental (intellect) aspects. This shows the holistic approach that Jones had for aesthetic appreciation. His book was primarily written as a guide for designers, artists, and architects. It was hugely influential in art education during the second half of nineteenth century.

Santayana [18] in his 'The sense of beauty' concluded, "Beauty therefore seems to be the clearest manifestation of perfection, and the best evidence of its possibility. If perfection is, as it should be, the ultimate justification of being, we may understand the ground of the moral dignity of beauty. Beauty is a pledge of the possible conformity between the soul and nature, and consequently a ground of faith in the supremacy of the good". His holistic approach is evident from this articulation. This was the last work in the 19th century that had holistic view point on the factors contributing to beauty in a work of art.

From the beginning of the 20th century, the focus of art education changed towards the physical aspects of beauty in America. Dow [19], Ross [20], and Prang [21] had a huge influence in not only American art education but also in the Europe. Dow and Prang propounded *harmony* as the highest attainable quality in a work of art. But, Ross propounded *order* as the highest attainable quality for the same. All of them described on how various design elements when arranged in accordance with the guiding principles result in harmony/order in work of art. It is worth noting that all these are concerned with only physical aspects of beauty and also *fitness* was missed out in the scheme of things then.

4 Origins of Product Design

Industrial design is a term used synonymous with product design. Industrial design is a field concerned with designing of products that are industrially manufactured. The origin of this field started in 1920s after the second industrial revolution. Earlier to this period, products were typically made by craftsmen. They were also involved in designing and selling of the products. With the advent of machinery for industrial production, designing of products became a specialized activity and the field of industrial design was born [22].

When craftsmen were involved in designing and making of products, there were deeper understandings of the users, materials, processes, as well as cultural, ethical, moral, social, and traditional values that the products they made were embodied with. As a result, the creations of the traditional craftsmen were usually holistic in the outcome. In other words, the product designers in their earlier incarnation, i.e. craftsmen, were holistic in their design approach. Their creations were not only concerned with aesthetics, but also emotional and cognitive aspects. So, the results were usually wholesome and interconnected. This holistic approach to design changed when the design activity moved from the hands of the traditional craftsmen to industrial designers. Over years, connect with the society, culture, and users were either lost or weakened. The focus of the designers was more on the aesthetics and the emotional and cognitive aspects were relegated to the background.

Art education during the formative years of industrial design needs to be looked into at this point. Initially, industrial designers were taught through the same art fundamentals as the artists [23]. In addition, they were educated on the materials and processes. Books on art education in the early part of 20th century focused on the *physical* aspects of aesthetics. Dow [19], Ross [20], and Prang [21] dominated the art education till the mid 20th century and all never gave importance to the emotional and cognitive aspects in their books. Art educators such as Arthur Wesley Dow and others who followed him like Denman Ross and Louis Prang emphasized drawing and the elements and principles method as the foundation in American art education. Accordingly, the assigned projects in their textbooks emphasized the same. Even today many foundation courses are still being taught by this elements and principles method. This is primarily a two dimensional orientation; student isolates each element and principle and theoretically combines to create compositions of aesthetically pleasing objects. Several bestselling textbooks in the field of art education are based on this approach. It should be noted that design, in these textbooks, refers to ornament and repeat patterns [24]. So, it is evident that only *physical* aspects of design were in focus and not the *emotional* and *cognitive* aspects. This is a clear departure from the earlier holistic approach of the erstwhile designers of the product, the craftsmen. Designers, who had their foundation courses based on only the physical aspects of the product, only expressed their creations with focus on physical beauty; emotional and cognitive contents were not important. This trend has been continuing for nearly 100 years now. However, inputs on emotional and cognitive aspects were given to the design students through different other books and methods. Methodology of these inputs varied from one school to the other. But, the fact remained that there was no one unified approach to designing for a long time.

5 Current Scenario in Product Design

Young [24] reported in his book on Visual Studies that there was an effort to upgrade the basic art instruction throughout America by setting up a national committee on arts. This committee attempted to solve the problems of visual vocabulary and came up with an Elements and Principles Chart. In this chart the elements of design mentioned are line, form, tone, textures, and colour. Repetition, rhythm, proportion, balance, and emphasis are mentioned as the major principles of design in that chart. Also, harmony and fitness are mentioned as the resulting attributes and beauty as the supreme attainment. It is notable that all the elements and principles mentioned in that chart are concerned with the physical factors only. The metaphysical factors that are responsible for emotional and cognitive aspects are not part of the chart.

Ocvirk et al. [25] authored a book titled "Art Fundamentals" in 2002. According to them artists normally use a logical and common order of events in the creation of an artwork as given in page 33 of their book. That is a framework using elements and principles for achieving unity in an artistic creation. Also their framework is an extension of the elements and principles chart compiled by the American National Committee on Arts. It should be noted that over the years the focus on the physical aspects of creation never changed. The emotional and intellectual aspects were never considered central to creation of designs at the fundamental level. The approach to design has never been holistic.

However, the *emotional* and *intellectual* aspects of design were dealt with separately by many authors like Coomaraswamy [26], Osborne [27], Papanek [23], Haines [28], Mitias [29], Lorand [30], Nagamachi [31], Norman [32], Desmet and Hekkert [33], Schifferstein and Hekkert [34]. Coomaraswamy described the role of emotions in his work on 'Theory of Beauty'. Nagamachi, Norman, and Desmet also detailed the process of emotional involvement in the aesthetic appreciation of products. Papanek emphasized on the role of sustainability and purposefulness in design through his most widely read text book of design today, 'Design for the Real World' [23]. His writings were focused on making a socio-cultural change in the society. Without doubt, Papanek's influential writings were concerned with the intellectual aspects of design in our context.

The term *aesthetic order* has been in usage only since 1980s. Osborne, Haines, and Lorand used it to describe the qualities responsible for giving pleasures to human senses. Lorand's aesthetic order is a lawless order. It does not consist of a priori (external) principles; it is an "internal" order. The principle cannot be abstracted from the particular set and applied to another set. The complex of relationships among the elements determines the aesthetic order of the set; in this sense the set and the principle are one.

6 Holistic Product Design

This is a new term being coined to describe the field of design concerned with the wholeness and interconnectedness of all the elements. The field of product design has been evolving all these years as noted in the earlier sections. It has always been a multi-disciplinary field and the focus kept varying over the years. From the literature that has contributed to the knowledge base of the field of art and design, it has been noted that the focus of interest was about the physical aspects of design for major part of the 20th century and later shifted to the emotional and intellectual aspects.

Noticeably, there has not been an attempt to make a holistic approach to design so far, in spite of fact that the field has long qualified for one. As seen in Sect. 2, all the fields that are multi-disciplinary in nature like brain research, climate change, and landscape ecology, the issues are better addressed by a holistic approach. As seen in Sects. 3, 4 and 5, the field of product design has evolved over the years. In present day context, there are multiple disciplines involved in a typical product design. Conventional reductionistic or mechanistic approach of problem solving is no longer valid for successful tackling of issues concerned with the field. Holistic approach is a natural evolution towards meeting that requirement. Hence, the author proposes to call the field as holistic product design. By doing so, there is a refocus to the field product design on the wholeness and interconnectedness unlike the earlier discrete aspects.

For most part of the history of design, the approach had been holistic. Mechanistic approach, triggered by the industrial revolution, changed the manner design is understood and issues tackled. By calling the field as holistic product design, it is hoped to bring positive changes through quicker and better problem solving. Much like what happened to other multi-disciplinary fields, product design too can benefit from the holistic approach. This change appears to be an evolution of the field in the positive direction as it only improves the scope of accommodating even more complex problems in its pathway. For that matter, it is only a natural way of growth for any prospering field.

7 Conclusion

This study looks at the history of the field of design and suggests an approach that is best suited for the current situation. Design had always been holistic till 1900. But, there was a change in perspective noticed since 1900 when only the *physical* aspects of design were addressed and the *emotional* and *mental* aspects were sidelined. New *holistic product design* will change the perspective of the field in the right direction. The field of design has become more inter/cross/poly/multi-disciplinary in nature and holistic approach is the need of the hour to meet the current scenario. Through *holistic approach to product design* we not only can have more attractive designs, but also more appealing and meaningful ones.

References

1. Dabrowski, I.J.: David Bohm's theory of the implicate order: implications for holistic thought processes. Issues Integr. Stud. **13**, 1–23 (1995)
2. Sperry, R.W.: Structure and significance of the consciousness revolution. J. Mind **8**(1), 37–65 (1987)
3. Harman, W.: Global mind change. Knowledge Systems Inc., Indianapolis, IN (1988)
4. Lyng, S.: Holistic health and biomedical medicine: a counter system analysis. State University of New York Press, Albany, NY (1990)
5. Cassell, E.J.: The place of the humanities in medicine. The Hastings Center, Institute of Society, Ethics, and the Life Sciences, Hastings-on-Hudson, NY (1984)

6. Cabrera, D., et al.: Systems thinking. Evaluation and Program Planning, vol. 31, pp. 299–310 (2008)
7. Oliver, D.W.: Education, modernity, and fractured meaning. University of New York Press, Albany, NY (1989)
8. Ferguson, M.: Physicist Bohm: meaning links mind and matter. Brain/Mind Bull. **10**, 1–2 (1985)
9. Hyland, D.A.: Plato and the Question of Beauty. Indiana University Press, Bloomington (2008)
10. Wood, E.: Placing Aesthetics. Ohio University Press, Athens (1999)
11. Munro, T., Scruton, R.: Eastern Aesthetics, Encyclopedia Britannica (2013)
12. Baumgarten, A.: Metaphysica. The German Aesthetic Tradition, Kai Hammermeister. Cambridge University Press, Cambridge, 1739 (2002)
13. Baumgarten, A.: Aesthetica. The German Aesthetic Tradition, Kai Hammermeister. Cambridge University Press, Cambridge, 1750 (2002)
14. Kant, I.: Critique of Pure Reason. Translation by Paul Guyer and Allen Wood. Cambridge University Press, Cambridge, 1781 (1998)
15. Kant, I.: Critique of Judgment. Translation by Paul Guyer and Eric Mathews. Cambridge University Press, Cambridge, 1790 (2000)
16. Hogarth, W.: Analysis of Beauty. R. Scholey, London (1810)
17. Jones, O.: Grammar of Ornament. Day and Son, London (1856)
18. Santayana, G.: The Sense of Beauty. Charles Scribner's Sons, USA (1896)
19. Dow, A.W.: Composition. Doubleday, Page, and Co, New York (1899)
20. Ross, D.: A Theory of Pure Design. Houghton, Mifflin and Co, Boston and New York (1907)
21. Prang, L.: Art Education. Prang Educational Company, USA (1908)
22. Bürdek, B.E.: DESIGN—History, Theory and Practice of Product Design. Publishers for Architecture, Basel, Switzerland (2005)
23. Papenek, V.: Design for the Real World. Thames & Hudson, UK (1984)
24. Young, F.: Visual Studies. Prentice-Hall/McGraw-Hill, Englewood Cliffs (1985)
25. Ocvirk, O.G., Stinson, R.E., Wigg, P.R., Bone, R.O., Cayton, D.L.: Art Fundamentals: Theory and Practice. McGraw-Hill, USA (2002)
26. Wignesan, T.: Ananda K. Coomaraswamy's aesthetics: Hindu view of art II: theory of beauty. J. Inst. Asian Stud. **14** (1996)
27. Osborne, H.: Aesthetic and other forms of order. Br. J. Aesthetics **22**(1), 3–16 (1982)
28. Haines, V.: Aesthetic order. J. Value Inq. **28**, 193–215 (1994)
29. Mitias, M.: Aesthetic Experience of Architectural Work. JAE (1999)
30. Lorand, R.: Aesthetic Order: A Philosophy of Order, Beauty, and Art. Routledge, London, New York (2002)
31. Nagamachi, M.: Kansei engineering as a powerful consumer-oriented technology for product development. Appl. Ergon. **33**, 289–294 (2002)
32. Norman, D.: Emotional Design. Basic Books, London (2004)
33. Desmet, P., Hekkert, P.: Framework of product experience. Int. J. Des. **1**(1), 57–66 (2007)
34. Schifferstein, H.N.J., Hekkert, P.: Product Experience. Elseviers, UK and Netherlands (2008)

Expanding DRM Framework to Formulate Supreme Causal Models from Research Articles in the Area of Product Disassembly

S. Harivardhini and Amaresh Chakrabarti

Abstract This study focuses on the implementation of a Causal Model for extracting causal relationships from research articles in the area of product disassembly. A Causal Model is a Model with a network of influencing factors to describe situations, as proposed by the DRM framework. In this work, DRM framework is further aided by formulating Supreme Causal Models. Individual Causal Models were generated using causal relationships extracted from research articles; Supreme Causal Models were formulated by collating these individual Models. The methodology adopted includes the following steps: data was collected from forty research articles in the area of product disassembly, which were analyzed using the concept of Causal Model from the DRM framework. Results include individual Causal Models, extracted from research articles, and Supreme Causal Models. The individual Causal Models elaborate some of the existing situations within the area of product disassembly; the Supreme Causal Models give a more complete picture of these existing situation fragments. Also, they represent some of the potential desired situations, such as maximizing profit or minimizing environmental impacts, and provide insights on how to achieve these desired situations.

Keywords DRM's causal model · Causal relationship · Reference model · Impact model

1 Introduction

The Causal Models within the DRM framework [1] is intended to help achieve (i) a comprehensive understanding of a research area through identifying causal relationships from research articles; (ii) identifying major research gaps

S. Harivardhini (✉) · A. Chakrabarti
IDeaS Lab, CPDM, Indian Institute of Science, Bangalore 560012, India
e-mail: vardhini@cpdm.iisc.ernet.in

A. Chakrabarti (ed.), *ICoRD'15 – Research into Design Across Boundaries Volume 1*,
Smart Innovation, Systems and Technologies 34, DOI 10.1007/978-81-322-2232-3_11

in the area by revealing the weaker or non-existent links among the factors in the causal network; and (iii) determining the important parameters of the area, i.e. those which, if strengthened with the help of a support, would enhance the performance of other important parameters. All of these should, as proposed in DRM, lead to the betterment of some aspects in that particular research. In this work, DRM framework is not only implemented but also further strengthened by formulating Supreme Causal Models as a collation of individual Causal Models. Formulation of Supreme Causal Models has helped us in developing significant insights into the way in which various important parameters of product disassembly processes and outcomes influenced one another, as to how strong these influences have been, and whether the influences are positive or negative. Also, the Supreme Causal Models helped us identify the potential parameters within the area of product disassembly that require intervention in the form of support for improvement.

In the current study, DRM's Causal Models have been implemented to extract causal relationships from research articles in the area of Product disassembly. Section 2 discusses the details of Causal Models within the DRM framework, and the important concepts within the framework that were used in this study. Section 3 details on generation of individual Reference Models from causal relationships and elaborates on some of the individual Reference Models developed in our research. Section 4 discusses the formulation of Supreme Causal Models from these individual Reference Models. Discussions and conclusions are provided in Sect. 5.

2 Causal Models Within DRM

DRM Framework [1] distinguishes between two types of networks of influencing factors, Reference Model and Impact Model, to describe respectively two types of situations, existing situation and desired situation. These Causal Models are made up of **nodes**. Each node represents an **Influencing Factor**. These nodes are connected by **links**. Each link represents the causal relationship between two nodes i.e., the causal relationship between two Influencing factors, one factor being the cause and the other being the consequence or effect. Main concepts of these causal models are provided below [1]:

Influencing Factor or Factor: It is an aspect of a situation (existing situation in case of Reference Models, and desired in the case of Impact Models). It is formulated as an **attribute** of an **element** that can be observed, measured or assessed, i.e., for which a so-called **Operational definition** can be formulated.

Operational definition: It defines "what to do" and "what to measure" in a concept.

Links: These are edges connecting the nodes, which show how the factors influence or are desired to influence each other.

Key Factors: These are influencing factors that seem to be the most useful ones to address in order to improve an existing situation.

Success Factors: These are the factors at the 'top' of the network, i.e., at the end of the cause-effect chains that provide the justification of the research.

Success Criteria: These are the desired values of the Success Factors.

Measurable Success Criteria: These are criteria that are linked to the chosen Success Criteria and can be applied to judge the outcomes of the research, given the resources available within the project or programme.

Measurable Success Factors: These are the factors whose desired values are taken as Measurable Success Criteria.

Proxies: When it is not possible to use Success Criteria as Measurable Success Criteria, Measurable Success Criteria should be chosen such that they can serve as reliable indicators (or proxies) for the Success Criteria.

Support: DRM framework proposes that in a research project that involves developing a support, the envisaged support, i.e. the **Intended Support** is defined first, and is realized to such an extent that its core concepts can be demonstrated and core effects evaluated.

Reference Model: This model represents an existing situation in design, and is the reference against which the intended improvements are to be benchmarked.

Impact Model: This model represents a desired situation in design, and represents the assumed impacts of the support to be developed.

3 Extraction of Causal Relationships from Research Articles

In this research, causal relationships were extracted from forty research articles in the area of product disassembly. However, for demonstrating its research approach in this paper, causal relationships extracted from only ten of these articles have been used to generate illustrative Supreme Causal Models used in this paper. These articles were published between 1996 and 2008, covering the following aspects of the research area: integration and optimization of product design for ease of disassembly; Design for Disassembly (DfD) to increase service life of buildings; disassembly process planning; integrated development of assembly and disassembly; enhancing disassembly and recycling planning; evaluation of disassemblability; and development of Design for Remanufacturability guidelines.

The causal relationships are represented in Reference Models, which represent fragments of the existing situation within the research area as reflected in these articles only, since the articles used represent only a section of the state of the art of research in this area (see Appendix 1 for the list of research articles used). "It is important to base the Reference Model on the original statements found in the literature, even if this implies a non-continuous line of argumentation. The Reference Model represents the current understanding as-is" [1]. As per this statement, Reference Models were generated based on the original statements found in the research articles, as shown in Fig. 1.

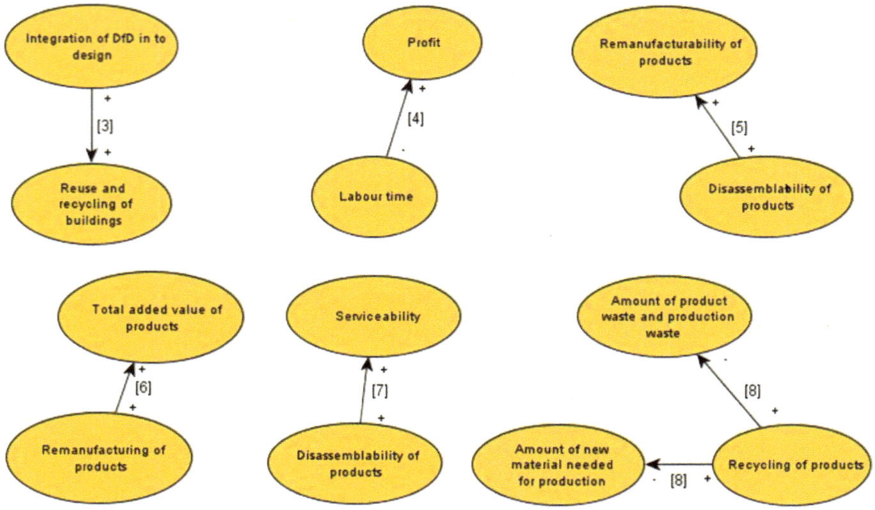

Fig. 1 Some of the causal relationships in the form of reference models extracted from research articles in the product disassembly field (as per DRM's framework, research articles [3–8] are quoted on the links)

4 Formulation of Supreme Causal Models

4.1 Supreme Reference Model

The individual Reference Models as shown in Fig. 1 were collated to form a Supreme Reference Model. The potential individual Reference Models that were to be collated were selected based on a Success Criterion (see Sect. 3). Each Supreme Reference Model was formulated based on this Criterion. The concept of formulation of Supreme Reference Model from the individual islands of Reference Models based on the Success Criterion is new, and introduced for the first time in this study. This concept is based on the following reasoning: Rather than selecting Reference Models at random, there should be a rationale which should drive the selection of individual Reference Models that are to be collated to form a Supreme Reference Model. Thus, a Success Criterion is chosen such that it is closely equivalent to an objective of a research project. This helps construct the complete Supreme Reference Model based on lines of argumentation that are aligned with the goals of the project. Since desired situations (or goals) could vary based on different perspectives with which the area is approached, Success Criteria should also vary. This should be used proactively to build the Supreme Reference Models. It should also be noted that, in some cases, the Supreme Reference Model contains links on assumptions (see Fig. 3). These links represent causal relationships that are hypotheses, since they are neither validated empirically nor established

Fig. 2 Supreme reference Model 1

as original statements in the research articles. In the current study, two Supreme Reference Models (Figs. 2 and 3) were formulated based on two Success criteria: (i) maximizing profit obtained from the End of life phase of a product life cycle; and (ii) minimizing the environmental impact in the Production and End of life phases of the product life cycle.

In Supreme Reference Model 1, most of the Influencing factors were extracted from research article [1], which is quoted on the link as well as shown in Fig. 2. From the figure, it can be seen that Factor "Integration of DfD in to design" has a positive influence on factors such as "Accessibility of parts", "Efficiency of disassembly operations" and "Probability to meet environmental standards". Also, the Factor "Integration of DfD in to design" has a negative effect on Factors namely "Complexity of product structure", "Number of disassembly steps" and "Potential risk of contamination".

Factor "Efficiency of disassembly operations" in turn has a negative effect on "Disassembly cost" and "Disassembly time". Research articles [4], [2] and [10] helped in identifying causal relationships among Factors such as "Disassembly time", "Degree of automation of disassembly", "Retention of material purity", and the Success Factor "Profit obtained from the End of life phase of a product life cycle".

In Supreme Reference Model 2, causal relationships between Factors such as "Disassemblability of products" and "Remanufacturability of products" were

Fig. 3 Supreme reference Model 2

identified from research article [5]. This article also helped in identifying that the Factor "Remanufacturability of products" has a negative influence on the "Use of landfill", "Production of green house gases" and "Level of virgin material and energy used in production", and have positive effects on Factors "Profit" and "Total added value of products" [6]. The existence of a causal relationship between "Remanufacturability of products" and "Production of green house gases" is strongly confirmed by another research article [2], see Fig. 3. It was also found from this article that, "Preservation of functional value of products" tend to reduce the "Impact on environment". Research articles [7] and [8] show respectively the existence of causal relationships between Factors such as "Disassemblability of products" and "Serviceability" and between "Disassemblability of products" and "Recyclability of products". Factor "Recyclability of products" in turn negatively influences "Amount of new material needed in production" and "Amount of product and production waste" [8].

Some of the causal relationships assumed are also shown in Fig. 3. These assumptions were made following the DRM framework, see Sect. 2.4.1 of [1]. It was identified from research article [3] that, "Integration of DfD in to design" positively influences "Reuse and recycling of buildings". Based on this relationship, an assumption was made and a new causal relationship was established between "Integration of DfD in to design" and "Recyclability of products". Also it was assumed that, improving disassemblability improves the ease of isolation of hazardous components. Based on this assumption, a causal relationship was established between "Disassemblability of products" and "Ease of isolation of hazardous components". Similarly, one more assumption was made and a causal relationship was established between factors "Total added value of products" and "Preservation of functional value of products". In this way, two Supreme Reference Models were built by collating individual Reference Models, each based on a different Success Criterion.

4.2 Supreme Impact Model

Two Supreme Impact Models were then derived, as shown in Figs. 4 and 5, from the two Supreme Reference Models in Figs. 2 and 3. These are Impact Models, describing some of the potential desired situations for product disassembly. Each Supreme Impact Model can be taken as a research project with its corresponding Success Criterion as the research goal. With the help of DRM's Causal Model framework, some of the key Factors and Proxies identified for two Supreme Impact Models are shown in Table 1. Table 1 also shows the Success Factors and Success Criteria identified while generating the Supreme Reference Models.

Fig. 4 Supreme impact Model 1

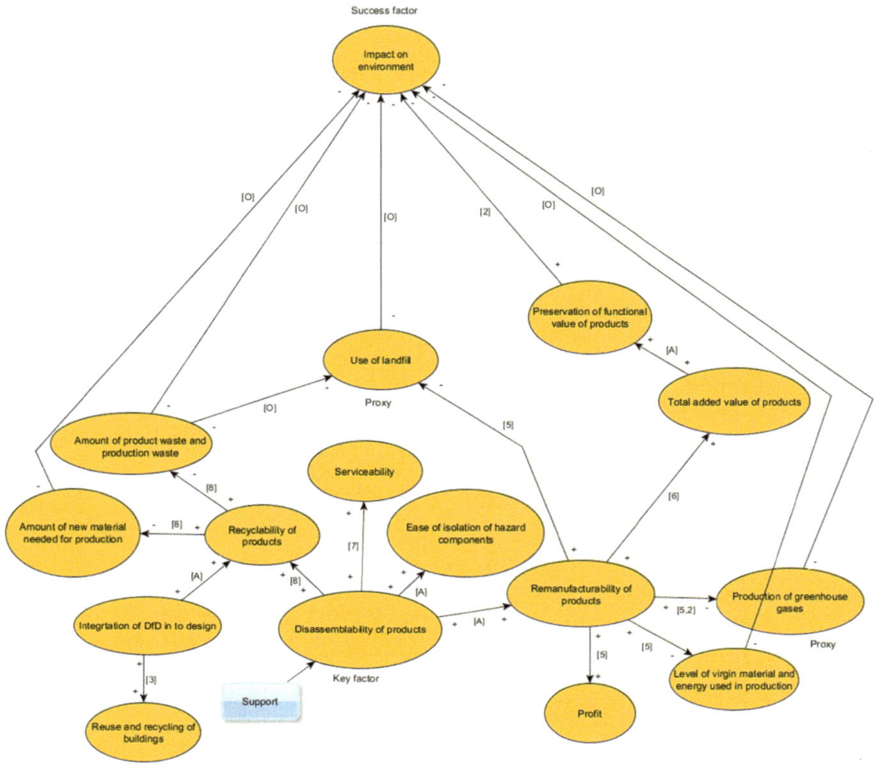

Fig. 5 Supreme impact Model 2

Table 1 Some of the important factors of supreme impact Models 1 and 2

Supreme impact model	Key factor	Proxy	Success factor	Success criterion
1	Integration of DfD in to design	Disassembly time disassembly cost	Profit obtained from end of life phase of the product life cycle	Maximizing the profit obtained from end of life phase of the product life cycle
2	Disassemblability of products	Use of landfill production of green house gases	Environmental impact in the production and end of life phase of the product life cycle	Minimizing the environmental impact in the production and end of life phase of the product life cycle

The Key Factors helped identify some of the most important parameters influencing product disassembly. Two different pieces of support were intended to be introduced; one for improving the performance of key factor associated with each Supreme Impact Model. Supporting both their key factors should, we assumed, help in achieving both the success criteria: maximizing profit and minimizing environmental impact.

In Supreme Impact Model 1, "Integration of DfD in to design" is chosen as the key Factor. It was found from research article [1] (see Appendix 1) that this Factor can influence many other Factors in both positive and negative manners. This reveals the importance of this parameter and thus has been chosen as a Key Factor. Proxies should be chosen as close (i.e. as directly connected) as possible to the Success Factors. The link between Proxy and Success Criteria is assumed to exist either based on existing evidence or based on reasoning (see Sect. 2.5 in [1]). Based on this guideline from the DRM framework, Proxies such as "Disassembly cost" and "Disassembly time" were chosen in order to closely satisfy the research goal.

"Statements that are found in the literature cannot simply be reversed….", if reversed it would be an assumption (see Sect. 2.4.1 in [1]). Based on this guideline, the causal relationship between Factors "Number of components" and "Disassembly cost" is marked as an assumption. Factor "Modularisation" is added to Supreme Impact Model 1, since it is assumed that "Integration of DfD in to design" will improve "Modularisation" and this in turn will minimize the "Complexity of product structure". The relationship 'Reduction of "Disassembly cost" will increase "Profit"' link is marked as [O], which implies that this relationship is established based on one's own investigation (as per DRM guideline given in Sect. 2.4.1 in [1]).

In Supreme Impact Model 2, all the Factors were retained from Supreme Reference Model 2. "Disassemblability of products" was chosen as the Key Factor, for the following reason: (i) this is one of the most important parameters of the product disassembly field, and (ii) enhancing the performance of this factor will improve the performance of many other important aspects of the situations of the research field.

Six new links were added to the Model based on our own investigation (Fig. 5). Among these links, five are causal relationships established between Success Factor "Impact on Environment" and the remaining five factors which include two Proxies: "Use of landfill" and "Production of greenhouse gases".

Of the six Factors that are directly linked to the Success Factor, "Use of landfill" and "Production of greenhouse gases" were chosen as Proxies, since Proxies should be chosen such that they should be measurable quantities, i.e. able to be assessed within the time scale of the project. The sixth link is established between "Amount of product waste and production waste" and "Use of landfill", based on our own investigation.

5 Discussions and Conclusions

We argue that the work reported in this paper enriches the canon of DRM framework by introducing two new supporting elements. One is the concept of "Supreme Causal Model", which integrates multiple, individual Causal Models. Supreme Causal Models give insights into the various important parameters influencing product disassembly, and how they influence one another. The second element is the use of Success Criterion to select individual Reference Models and construct Supreme Causal Models. Using DRM's Causal Models, various aspects of the existing situation and potential desired situations for product disassembly have been identified. These need not be the only desired situations in this area. In our study, two Supreme Causal Models have been formulated based on two separate Success Criteria. More such Models could be formed if Success Criteria were altered around other Success Factors.

Using this approach, the major results obtained in the area of product disassembly are: (i) the individual Reference Models and (ii) the two Supreme Reference Models and two Supreme Impact Models reflecting some aspects of the existing and desired situations in this area. Also, several important parameters within the area were identified: Disassemblability of products and Efficiency of disassembly operations. They influence various other important parameters such as Remanufacturability, Recyclability, Serviceability, Disassembly time, Disassembly cost, and Integration of DfD into design. The last factor influences Accessibility, Complexity of product structure, Disassembly steps, Probability to meet standards, and Material contamination.

Use of the DRM's Causal Model brings two major benefits. The first is that this Model works on the concept of causality. The relationships among important parameters of the field have been understood less ambiguously with the help of causality. Thus, with a higher confidence level, the existence of such Causal relationships could be confirmed. This helped in carrying out research on identifying and resolving current issues associated with the area in a focused manner. The second benefit is that, as per DRM's framework, these Causal relationships in the research articles are represented in a graphical form (See Fig. 1), which made them more readable and easier to understand.

Several difficulties were also faced in using the Causal Models. One was to identify whether a Causal relationship belonged to an existing or a desired situation. Unless the context was thoroughly understood, it was difficult to place them in the right (Reference or Impact) Causal Model. Another difficulty was in selecting individual Reference Models to be included in a Supreme Model. Since, the long and in some cases, the indirect chain of links in the Supreme Model should be known in order to include the appropriate individual Reference Models. Care should be taken not to leave any important Factor while selecting individual Reference Models in formulating Supreme Causal Models. It is better to include all the factors that are linked to the important factors in a Supreme Causal Model while formulating that Model.

Appendix 1 [References as Marked on Links (See Figs. 1, 2, 3, 4 and 5)]

Research articles used in the study	
1	"Integration and optimisation of product design for ease of disassembly", Motevallian, B., Abhary, K., Luong, L. and Marian, R.M., Book: Engineering the Future, ISBN: 978-953-307-210-4, chapter: 16, pp: 317–340, 2010
2	"Efficiency and feasibility of product disassembly:A case-based study", Duflou, J.R., Seliger, G., Kara, S., Umeda, Y., Ometto, A. and Willems, B., CIRP Annals-Manufacturing Technology, vol. 57, no. 2, pp: 583–600, 2008
3	"Designing for disassembly to extend service life and increase sustainability", Crowther, P., 8th International Conference on Durability of Building materials and Components, Vancouver, Canada, 1999
4	"Process planning for product disassembly", Das, S.K. and Naik, S., International Journal of Production Research, vol. 40, no. 6, pp: 1335–1355, 2002
5	"Development of robust design-for remanufacturing guidelines to further the aims of sustainable development", Ijomah W. L., McMahon C.A., Hammond G.P. and S.T. Newman, International Journal of Production Research, vol. 45, no. 18–19, pp: 4513–4536, 2007
6	"Integrated Development of Assembly and Disassembly", Westkamper, E., Feldmann, K., Reinhart, G. and Seliger, G., Annals of the CIRP vol. 48, no. 2, pp: 557–565, 1999
7	"Evaluation of disassemblability to enable design for disassembly in mass production", Desai, A. and Mital, A., International Journal of Industrial Ergonomics, vol. 32, pp: 265–281, 2003
8	"Enhancing disassembly and recycling planning using life-cycle analysis", Kuo, T.C., Robotics and Computer-Integrated Manufacturing, vol. 22, pp: 420–428, 2006
9	"A Graph-Based Approach to Disassembly Model for End-of-Life Product Recycling", Zhang, H.C. and Kuo, T.C., IEEE/CPMT International Electronics Manufacturing Technology Symposium, 1996
10	"Computer Aided Disassembly Planning: State of the Art and Perspectives", Santochi, M., Dini, G. and Failli, F., CIRP Annals-Manufacturing Technology, 51, 2, pp: 507–529, 2002

Reference

1. Blessing, L.T.M., Chakrabarti, A.: DRM, A Design Research Methodology. Springer, New York. Book-ISBN 978-1-84882-586-4 (2009)

The Taguchi Method as a Means to Verify the Satisfaction of the Information Axiom in Axiomatic Design

Sergio Rizzuti

Abstract The paper deals with the crucial phase of identification of the problems that occur during the design process of industrial products. Designers need to identify the nature and the importance of the problems. An interesting approach to this purpose is the Axiomatic Design, which is the basis for the further application of Robust Design techniques. Among the latter the Taguchi method can be integrated with Axiomatic Design in order to discern the best values of the design parameters and also those values least affected by noises. The study of the interaction of the parameters can reveal the presence of problems, when decoupled matrices are analyzed. The discussion will be made considering the different scenarios of uncoupled and decoupled design matrices and the different reasoning the designer must use. All the information collected can then guide the designers to pursue the design of the best products.

Keywords Axiomatic design · Robust design · Design matrix · Information axiom · Taguchi method

1 Introduction

Axiomatic Design methodology [1] has gained a certain attention from designers and academics over the years, in that it gives a formal and analytical data structure that can allow designers to check the validity of the solution under investigation.

The Design Matrix, the core of the method, can validly support the designer only when a solution has been designed and a correlation between functional requirements (FRs) and design parameters (DPs) has been established. Axiomatic

S. Rizzuti (✉)
Department of Mechanical, Energy and Management Engineering, University of Calabria, Ponte Pietro Bucci 46/C, 87036 Rende, CS, Italy
e-mail: sergio.rizzuti@unical.it

A. Chakrabarti (ed.), *ICoRD'15 – Research into Design Across Boundaries Volume 1*, Smart Innovation, Systems and Technologies 34, DOI 10.1007/978-81-322-2232-3_12

Design offers a methodology able to solve conflicts that can emerge when many design parameters together influence several functional requirements. This is the moment in which the designer has to solve the problem, redesigning the solution or searching for modified versions. In any case, the whole matrix must be reviewed and new design architectures must be analyzed. The satisfaction of the independence axiom is a so-called necessary condition that guarantees a certain performance by a product [2, 3].

Generally the design process produces many design alternatives and axiomatic design should be able to suggest the best solution. In any case each design alternative must be correctly dimensioned and the satisfaction of Functional Requirements must be quantified. However, it must be considered that all alternatives satisfy the same functional requirements, even though each alternative is characterized by different design parameters. Also it must be underlined that the relation among DPs and FRs must be identified and quantified, and each relation must be expressed by a set of deterministic equations.

The form of the design matrix influences the way a designer can tackle the design of a reliable solution. If the design matrix is diagonal all (or a group of) design parameters independently influence a macro functional requirement. If the design matrix is triangular a set of design parameters together influence the same functional requirement. In the latter case it is difficult, if not impossible, to investigate the influence of each DP on the FR and verify the satisfaction of the information axiom.

The paper discusses how to use the Taguchi method [4, 5] to verify the influence that each design parameter has on the functional requirements. First of all an Objective Function must be associated with one single functional requirement or to a macro-functional requirement, and the relation to a set design parameters must be identified by investigating the design matrix. Analysing the results of the mean values the most influential parameters can be recognized, by the ANOVA method. This can allow the designer to discern the importance of each parameter, and the Signal to Noise Ratio can suggest the suitable range for each parameter. In the case of a multiple influence of one design parameter on several functional requirements, it is important to verify the mutual interaction among the design parameters and evaluate the kind and level of interaction.

On the basis of this study designers can gain many insights to understand the reliability of a design solution.

2 General Structure of the Design Matrix

The core of Axiomatic Design is the Design Matrix. It establishes the relation between Functional Requirements (FRs) and Design Parameters (DPs). The form it assumes is of type:

$$\{FR_i\} = [A_{ij}]\{DP_j\} \tag{1}$$

(a)

		DP1	DP2		DP3	DP4		DP5	DP6	
			DP21	DP22		DP41	DP42		DP61	DP62
FR1		X								
FR2	FR21		X							
	FR22			X						
FR3					X			X		
FR4	FR41					X				
	FR42						X			
FR5					X			X		
FR6	FR61								X	
	FR62									X

(b)

		DP1	DP2		DP3	DP4		DP5	DP6	
			DP21	DP22		DP41	DP42		DP61	DP62
FR1		X								
FR2	FR21		X							
	FR22			X						
FR3					X					
FR4	FR41					X				
	FR42						X			
FR5								X		
FR6	FR61								X	
	FR62									X

(c)

		DP1	DP2		DP3	DP4		DP5	DP6	
			DP21	DP22		DP41	DP42		DP61	DP62
FR1		X								
FR2	FR21		X							
	FR22		X	X						
FR3					X					
FR4	FR41			X		X				
	FR42			X	X	X	X			
FR5								X		
FR6	FR61	X			X				X	
	FR62		X					X		X

Fig. 1 Form of the design matrix: **a** coupled; **b** uncoupled; **c** decoupled

and the nature of the relation is strictly related to its content. Avoiding demonstration of the need to have a squared form (for which the reader is directed to the discussion made by Suh [1]), in the following we concentrate in discussing the nature of a Design Matrix associated with a design solution that can gain the status of a “valid solution”.

In Fig. 1 the three main types of matrices that can be generated during the conceptual design phase are reported, and they are classified as: (a) coupled; (b) uncoupled and (c) decoupled. The X symbol reported in them represents that a relation is present between one generic Functional Requirement and the Design Parameter associated with it, in the design solution. The nature of this relation must be clarified and it will deepen understanding during the phases of detailed design that follows when the design solution is assessed from many points of views.

The Design Matrix allows the design team to evaluate the macroscopic correctness of the solution and to enhance the intrinsic difficulty of each design alternative. It must be underlined that at this point of the product development, generally, many alternatives have been conceived, and it is really important to know which of them has the greatest chance of success.

The coupled Design Matrix shows an intrinsic flaw in the designing and suggests immediately to the designer team that something must be changed in this design alternative if it is decided to pursue it. In fact two DPs together control two FRs and this occurrence dramatically invalidates the design solution.

The uncoupled and decoupled Design Matrices are two design solutions that can be expanded because their starting points are promising [6]. The nature of the uncoupled DM (see Fig. 1b) is that, at a certain level of investigation and decomposition, the independence of the relation among all Design Parameters and the corresponding Functional Requirements is guaranteed. This condition, also called “ideal case”, is the best starting point from which further investigations can be made, in order to embody an almost robust product, because the best values for each Design Parameter can be identified independently of all others.

Also the decoupled DM of Fig. 1c describes the nature of a design alternative that can be further developed, because it has, at least theoretically, a chance to embody a valid solution. In fact, ordering (and numbering) properly the sequence of the function decomposition, a certain control on the system described by the

DM can be guaranteed acting first by choosing the most promising values for the DPs that are in strict relation to the associated FRs (diagonal terms) and then trying to identify all other DPs that influence each FR.

In any case the independence among DPs and FRs is guaranteed, even if in the second one, the design solution must be subject to a finer investigation.

These considerations are at the basis of the so-called independence axiom, that establishes the main conditions required to a design solution to be further pursued. Axiomatic Design can be employed validly in a wide range of applications, although the discussion in the following paragraphs will be intrinsically oriented to the solution of problems by means of electro-mechanical devices. These kinds of product require more flexibility during all the phases of product development, and the matrix nature of the method allows designers and stakeholders to maintain control of the whole process.

3 Meaning of the Information Axiom

The second step of Axiomatic Design consists in reaching an assessment of the quality of the design alternative. The aim is to quantify the level of success of one solution, and the final intent is to set up a method by which the comparison of several alternatives can be made.

Suh in his 2001 book [1] introduced the following definition of the Axiom of Information: "The information required to a device must be the minimum", where the information is associated with the probability of success. For the case of a single Functional Requirement:

$$I_i = \log_2 1/P_i \tag{2}$$

where P_i is the probability of i-th Functional Requirement to being satisfied. The probability P_i coincides with the area A_{cr} of the solution range that falls inside the design range assigned to the FR.

If the design solution is not good enough, part of its performance being out of the design range, and there is a bias between the mean value of the design solution and the target in the design range. If the probability density function falls completely inside the design range the solution is valid, being without bias.

In order to compare several design alternatives, a consideration like this must be made for all Functional Requirements, and each level of satisfaction must be verified. The comparison among alternative solutions, considering each one as a whole, should be based on a similar index, even though this is just a chimera. Global indices, such as I_{sys} introduced by Suh [1], are not able to enhance pointwise problems that can occur in a single part of the design solution and for decoupled problems introduced the concept of the conditional probability. Frey et al. [6] investigated how to compute the information content in a decoupled design. In any case designers must decide among trade offs and require information about the degree of interaction among Design Parameters. So it is extremely useful to support designer in the decision about the best choice in term of the values that reduce the bias or in term of the values that maximize the probability of success.

4 Integration of Robust Design into the Information Axiom

The concept at the basis of Robust Design [7] is the search for the most suitable solution that experiences the minimal effect of noises during its performance [8]. The Taguchi method is now generally employed in industry and academia and can be validly employed in this design context. It can allow designers to have deeper knowledge about the design solution they are investigating, since the method is extremely adaptable to manage computer simulation [9, 10]. Even though the method has been subjected to a certain level of criticism by several statisticians (for example Box et al. [11]), the argument that Montgomery [12] adopted was encouraging: "In summary, we should support Taguchi's philosophy of quality engineering. However, we must rely on simpler, more efficient methods that are easier to learn and apply to carry this philosophy into practice".

Many researchers have investigated the relation between axiomatic design and Robust Design [13–16] suggesting coherent strategies to support the identification of the best design solution. Bras and Mistree [13] introduced the compromise Decision Support Problem as a method to combine Axiomatic and Robust design by the Taguchi approach. They demonstrated how to determine the most suitable values and tolerances for a given set of parameters, and identify the most suitable principal design parameters. This approach requires the definition of all the relations among Functional Requirements and Design Parameters. They used also reangularity and semangularity to establish the degree of independence of a design solution, even though it must be underlined that these two quantities were no longer employed by Suh after 2001. Gu et al. [14] integrated the analysis of independence of Axiomatic design with Robust Design and used the condition number of the sensitivity matrix, related to the design matrix, as a means to evaluate the degree of independence of a design solution. The design matrix, also in this case, must be fully determined in each component. Xiao and Cheng [15] developed an analytic approach to demonstrate the relation between Axiomatic Design and Robust Design. They used the new insight by Suh [1] and studied the uncoupled and decoupled design matrices. They demonstrated on the basis of some case studies, and with properly probability density functions, why an uncoupled design is more robust than a decoupled design and why this latter is better than a coupled one. More recently Lijuan et al. [16] used the concept of optimization framework to integrate axiomatic design, robust design and reliability-based design, although they needed to use reangularity and semangularity again to configure the optimization framework. Kar underlined the strict relation between axiomatic design and the Taguchi method [17]. Hu et al. [18], in their two part paper that appeared in 2000 in the Triz Journal, investigated the important task of defining the "system output response", which is the necessary step to establish the right relation between a set of Design Parameters and the functional behavior of the device under investigation. There is a great similarity between the Objective Function used in this paper and the concept of system output response. The rules defined in [18] can be really useful to guide designers towards the most suitable simulation of the device.

The main characteristic of the present approach is that it is not necessary to know, during design development, all the relations between Functional Requirements and Design Parameters. Also, the information derived from robust design must orient the choices towards the best design solution. Considering that this activity is carried out during the early phases of embodiment, the designers can modify the aspects that may hide inconsistencies in term of the independence axiom.

The main objective of this process is to identify the solution characterized by the wider range of Design Parameter that control the Functional Requirement. The relation between FRs and DPs can be initially identified from the analysis of the Design Matrix. As said, the DM has a structure that depends mainly on the level of detail that the designer considers interesting for the study. Therefore, instead of concentrating on the DM as a whole, more hints will be derived from the analysis of group of relations FRs-DPs, focusing on single zone or sub-domain.

The example that will be used to illustrate the method is derived from one of the most interesting solutions given to a problem posed by a Firm operating in the sector of automatic drive for roller shutter by the students of the Master level course of Product Design and Development. The Objective Function was associated with the maximum Von Mises stress of a component on which a set of seven Design Parameters were identified. The general schema of the design process can be found in [19].

4.1 Initial Phase Supposing a Sub-domain Structured as an Uncoupled DM

The case is related to a macro-Functional Requirement that has been recognized as dependent on a set of Design Parameters, and the whole influence appears to be uncoupled, that is, each DPi1-DPi7 independently contributes to the performance of the Functional Requirement FRi, as represented in Fig. 2.

From the Robust Design point of view the best range of each DPij (j = 1, 7) must be identified. To this purpose the Design of Experiment, in the sense of the Taguchi method, implemented by means of a set of computer simulations, is the tool that can give designers really useful information: (a) to identify the range of

		...	DPi							...
			DPi1	DPi2	DPi3	DPi4	DPi5	DPi6	DPi7	
...		X								
FRi	FRi1		X							
	FRi2			X						
	FRi3				X					
	FRi4					X				
	FRi5						X			
	FRi6							X		
	FRi7								X	
...										X

Fig. 2 Case of a sub-domain structured as an uncoupled DM

the values for each DP; (b) to have much knowledge about the most influential parameter on the Functional Requirement.

The identification of the proper orthogonal array [5] is the first step for designing the experiment and simulation. Strictly related to this is the identification of the limits (lower and upper) of each parameter to be used in the experiment. During the early phase of design, it is sufficient to use two levels that identify the extremes of the design range of each parameters. Only when the embodiment phase proceeds, the analysis with more factor levels can be justified, the computational and the management effort being higher. In the following the results of an experiment treated with an L_8 (2^7) orthogonal array will be discussed.

The identification of an Objective Function that characterizes the Functional Requirement is the difficult step that attests the level of consciousness of the problem under investigation by designers. This is generally expressed by means of the "loss function", in one of the possible scenarios: target is better, lower is better, higher is better. Also the introduction of noise, in term of source and levels, is the strategic step that can give designers the possibility of discerning not only the better range of value for each parameter, but also which level is less influenced by noise. The latter analysis is supported by studying the Signal to Noise Ratio (SNR), computed by

$$SNR = 10\log_{10}\mu^2/\sigma^2 \tag{3}$$

The graphs reported in Fig. 3a are related to the mean values of each parameters associated with a problem in which the best solution is identified by the minimization of the Objective Function. The graphs of Fig. 3b represent the values of SNR associated with the variation of the scenario with respect three levels of the values of a noise factor.

The analysis of the design solution, in term of Information Axiom must be made evaluating the graphs of Fig. 3b. It shows that the lower level of the Design Parameters: DPi1, DPi4, DPi6 and DPi7, and the upper level of the DPs: DPi2, DPi3 and DPi5 should be selected. All these parameter levels in fact record the values for which there is the greater insensitivity of the solution with respect the variation of a noise.

At the same time it must be checked whether the design ranges, which have been assumed for each Design Parameter, are really valid, or a bias can be

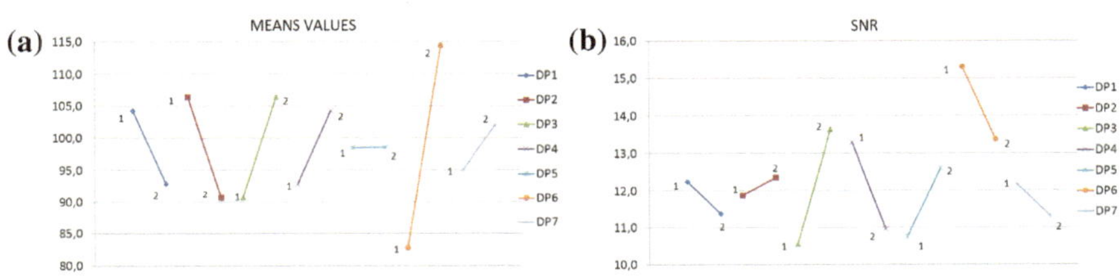

Fig. 3 **a** Behavior of each design parameter, as mean value, on the objective function; **b** graphs of the corresponding SNR associated to each DP

observed. This can be made comparing both graphs of the means values and SNR. Each Design Parameter must present the same level that together satisfies both characteristics. In this example, only DPi2 upper, DPi4 lower, DPi6 lower and DPi7 lower satisfy this rules and could be concluded that these levels are the best ones in that they reduce the bias between design range and system range and that require the lowest level of information since they are those for which there is the minimum influence of noise.

The cases of DPi1 and DPi3 require further investigation since the levels are in contrast with each other and the designer does not have elements to discern the best solution. In fact, new ranges must be tried, and specifically, for DPi1 a new range shifted towards lower values than the lower limit, and for DPi3 a new range shifted towards higher values of the upper limit. The case of DPi5 is almost neutral, in term of nominal dimension, even though the upper value is less influenced by noise.

4.1.1 Identification of the Most Influencing Design Parameters

Also essential for the designers, during the design stages, is to familiarize themselves with the importance of the Design Parameters they have identified as fundamental for one Functional Requirement. Still reasoning on the example of Fig. 3, before trying to solve the remaining problems with new investigations, it is important to know what are the DPs on which the solution mainly depends. The analysis of variance can be a valid aid to discern this aspect. Recalling that the Total Sum of Square is equal to the sum of the Square [5] of all k-th Design Parameters

$$SSTot = SS(DPi1) + \cdots + SS(DPik) \tag{4}$$

The ranking of the most influencing Design Parameters can be listed and further investigation can be carried out. On the basis of the results of Fig. 3a, the most promising parameters are: DPi2 (13.7 %), DPi3 (13.6 %) and DPi6 (55.4 %).

After this selection, a finer investigation can be pursued employing only these three parameters. An experiment with three levels, for example, can introduce new insights into designing.

4.2 Second Phase Supposing a Sub-domain Structured as a Decoupled DM

The discussion of Sect. 4.1, in which the independence among the parameters was assumed, must be further investigated, because at this level of knowledge nothing can be excluded, like a certain degree of interrelation among design parameters, as in the case of Fig. 4.

		...	DPi							...
			DPi1	DPi2	DPi3	DPi4	DPi5	DPi6	DPi7	
...		X								
FRi	FRi1		X							
	FRi2			X						
	FRi3			x	X					
	FRi4					X				
	FRi5						X			
	FRi6			x	x			X		
	FRi7								X	
...										X

Fig. 4 Case of a sub-domain structured as a decoupled DM

The general expression that represents the multiple influence of many Design Parameters on the same Functional Requirement is:

$$FR_i = A_{ii}DP_i + \sum_j A_{ij}DP_j \tag{5}$$

where $A_{ij} = \partial FR_i / \partial DP_j$

This relation, although it represents formally the meaning of this influence, leaves the designer to search for the kind of relation. To this purpose, the data contained in the performed experiments of robust design can offer a first aid to designer. In fact, the interrelation between Design Parameters can now be analyzed.

Concentrating our attention on the relations among DPi2, DPi3 and DPi6, all associated with off diagonal terms not zero in the lower triangular matrix of Fig. 4, we find that they are all collaborative, in that their reciprocal variations do not intersect, as reported in the three graphs of Fig. 5.

By means of this further analysis an answer can reasonably be given to the x placed out of the main diagonal of the Design Matrix. In fact, as can be seen in Fig. 4, some non-zero terms appeared out of the principal diagonal (i.e. A_{23} corresponding to a relation between DPi2 and DPi3). This fact could have appeared since the first phase of construction of the Design Matrix, but, more interestingly, it can emerge after the study of the design solution, after the design of the experiment put into practice by means of the Taguchi method. The combination of the information derived from the analysis of the mean values, the SNR of the Objective Function and the interaction among Design Parameters can guide the designer towards more conscious design choices.

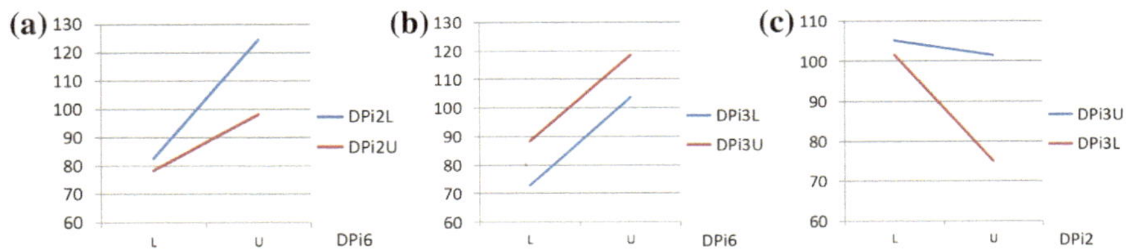

Fig. 5 Interrelation among design parameters: **a** DPi6 versus DPi2; **b** DPi6 versus DPi3; **c** DPi2 versus DPi3

The supposed decoupled Design Matrix that has interrelations A_{ij} different from zero, can be further investigated when these relations are collaborative, as in Fig. 5. If, otherwise, interactions between parameters can emerge, designers have advised that some contradictions are present and the design solution is not related to a decoupled solution, but potentially to coupled one.

Discussing the influence of DPi2 to the Functional Requirement FRi, the meaning of the terms non-zero in the Design Matrix is that:

1. the design range for this design parameter DPi2 must be chosen first as the best suited to the Objective Function associated to the FR. This is related to the X (capital letter) placed on the main diagonal term A_{22};
2. the design range for the value of X, becomes a sort of constraint for the following analyses, where it must be checked if it influences the solution for the other DPs with which it is in relation, checking all the lines in which it appears as an x (lower-case). For this case A_{ii} can be considered as a first order derivative of a multivariable function with constraints:

$$A_{ii} = \partial FR_i / \partial DP_i|_{DP_j} \tag{6}$$

Considering again the design matrix of Fig. 4, the search for the best solution of the design range of DPi6 is subjected to the constraints represented by design parameters DPi2 and DPi3, as:

$$A_{i6} = \partial FR_i / \partial DP_6|_{DP_2, DP_3} \tag{7}$$

5 Conclusions

In the paper a new interpretation of the results obtained by the application of the Taguchi method is presented for the identification of the best design parameters for a problem described in term of the Axiomatic Design approach.

Consulting the variations of an Objective Function, related to the Functional Requirement, in term of the mean values, Signal to Noise ratio and interaction among parameters, the designers have the possibility of reasoning about the design solutions on which they are working.

The paper has underlined how all data can be easily managed by data sheet and do not require complex optimization procedures, which in a certain sense frees the designers from the decision phase about the choices that must be made towards the identification of the best product design.

Acknowledgments The research was supported by a grant of the University of Calabria. The author would like to thank Antonio Rotella and Saverio Parrilla, both MD in Mechanical Engineering, for CAE simulation and data collection related to the design of a device during the course of Product Design and Development at UNICAL.

References

1. Suh, N.P.: Axiomatic Design: Advances and Applications. Oxford University Press, Oxford (2001)
2. Hong, E.-P., Park, G-J.: Modular design method using the independence axiom and design structure matrix in the conceptual and detailed design stage. In: Proceedings of 6th International Conference on Axiomatic Design (ICAD11), pp. 134-141. Daejeon (2011)
3. Cheng X.: Independent Axiom-Based Robust Design for Nonlinear System. Sensors and Transducers, vol. 16(144–151) (2012)
4. Taguchi G.: Taguchi on Robust Technology Development. ASME Press, New York (1993)
5. Phadke, M.S.: Quality Engineering Using Robust Design. Prentice-Hall, Englewood Cliffs (1989)
6. Frey, D.D., Jahangir, E., Engelhardt, F.: Computing the information content of decoupled designs. Res. Eng. Des. **12**, 90–102 (2000)
7. Park, G.J., Lee, T.H., Lee, K.H., Huang, K.H.: Robust design: an overview. AIAA J. **44**(1), 181–191 (2006)
8. Andersson, P.: On robust design in the conceptual design phase: a qualitative approach. J. Eng. Des. **8**(1), 75–89 (1997)
9. Gijo, E.V., Scaria, J.: Product design by application of Taguchi's robust engineering using computer simulation. Int. J. Comput. Integr. Manuf. **25**(8), 761–773 (2012)
10. Hu, Y., Rao, S.S.: Robust design of horizontal axis wind turbines using Taguchi method. J. Mech. Des. **133**, 1–15 (2011)
11. Box, G.E.P., Hunter, J.S., Hunter, W.G.: Statistics for Experimenters. Design, Innovation and Discovery, 2nd edn. Wiley, London (2005)
12. Montgomery, D.C.: Design and Analysis of Experiments, 3rd edn. Wiley, London (1991)
13. Bras, B., Mistree, F.: A compromise decision support problem for axiomatic and robust design. J. Mech. Des. **117**, 10–19 (1995)
14. Gu, P., Lu, S., Spiewak, S.: A new approach for robust design of mechanical systems. Ann. CIRP **53**(1), 129–133 (2004)
15. Xiao, R.B., Cheng, X.F.: An analytic approach to the relationship of axiomatic design and robust design. Int. J. Mater. Prod. Technol. **31**(2–4), 241–258 (2008)
16. Lijuan, S., Jun, Y., Yu, Z.: An integration design optimization framework of robust design, axiomatic design and reliability-based design. Qual. Reliab. Eng. Int. **27**, 959–968 (2011)
17. Kar, A.K.: Linking axiomatic design and Taguchi methods via information content in design. In: Proceedings of 1st International Conference on Axiomatic Design (ICAD 2000), Cambridge, MA, pp. 219–224 (2000)
18. Hu, M., Yang, K., Taguchi, S.: Enhancing robust design with the aid of TRIZ and axiomatic design. TRIZ J. Part I (October 2000), Part II (November 2000) (2000)
19. Rizzuti, S.: Learn to Design by Mapping Information Among Design Methods. ASME IDETC/CIE, Chicago (IL), USA, 12–15 Aug 2012, DETC2012-70834 (2012)

Part II
Design Aesthetics, Semiotics, Semantics

Identification of User Perceptions and Design Parameters of Vehicle Cluster Instruments in Different Cultures

Pratap Kalenahalli Sudarshan, Matthias Wagner, Olesja Marinets and Michaela Kauer

Abstract The study of user perception of products is important in the globalised world where designers increasingly design for international and unfamiliar audiences. Although the semantic differential method (SDM) is a proven method to assess the perceptions of products, previous studies have however neglected the cultural background of users. This study explores the usefulness of SDM in cross-cultural comparisons. This explorative study further examines difference in perceptions of six vehicle cluster instruments across three cultures of the USA, Germany and India through an online survey. From the analysis, nine of the thirteen adjective pairs used were found suitable for intercultural comparison where cultural differences in the evaluation of cluster instruments were observed. Further based on the results, possible implications for design of cluster instruments for the three cultures studied is presented.

Keywords Semantic differential method · Aesthetic perceptions · Cross-cultural comparisons · Cluster instruments

1 Introduction

The role of aesthetics is not to be underestimated even in the design of vehicle cluster instruments (CI) which forms a part of the vehicle dashboard. The CI is one of the most important elements in the interior of the vehicle, as it is always in the view of the driver and supplies the driver with all important information about the state of the car. This aspect is especially important for research of aesthetic perceptions given the global reach of automobiles and the fact that designers are often required to design for audiences in other, unfamiliar cultures [1]. In this

P. Kalenahalli Sudarshan (✉) · M. Wagner · O. Marinets · M. Kauer
Institut Für Arbeitswissenschaft, Technische Universität Darmstadt, Darmstadt, Germany
e-mail: pratapks@yahoo.com

A. Chakrabarti (ed.), *ICoRD'15 – Research into Design Across Boundaries Volume 1*,
Smart Innovation, Systems and Technologies 34, DOI 10.1007/978-81-322-2232-3_13

paper, the user perceptions regarding the design of six vehicle cluster instruments across the cultures of India, Germany and the US are examined with the help of the Semantic Differential Method (SDM). A subsequent objective of this study was to understand the validity of the SDM adjectives in cross-cultural situations.

2 State of the Art

The attractiveness of the product is related to the subjective consumer response to the product properties such as color, shape, texture, etc. which differs from the objective side of the said product properties. This implies that the subjective and objective sides of the product properties need not necessarily coincide with each other. Therefore, the objective importance to a particular product property by the designer may deviate significantly from the actual perceived importance to that product/product property by the user [2]. This is especially possible in cross-cultural situations. There are many studies indicating the influence of culture in design where perceptions have been dealt with in various ways [3, 4]. In this regard, two dimensions of hedonic and pragmatic [5] take an important position in the perception of product characteristics. The hedonic dimension (further divided into stimulation and identification) describes the experiential aspects of product use. In contrast, the pragmatic dimension brings in instrumental properties such as usefulness and functionality [5].

With regard to products and their design, the importance of the vehicle interiors and its direct contact with the consumer during usage has resulted in many studies such as [6–8] examining the perceptions of the car interiors. These studies however have focused on the perception or usability of either the interiors as a whole or as individual elements of the interiors without taking into account cultural backgrounds. In recent times, an assessment regarding the design of the CI was carried out by Herbeth and Blumenthal which focused on the perception of French customers [9]. From the designer's perspective, the design of the CI is important as emotions such as modernity, sporty, futuristic, etc. are expressed through the CI [10]. Jung et al. [11] further postulate that the cluster instrument is one of the essential components with a great influence on effects of the vehicle on the customer.

From the above, a gap in literature is identified as the intercultural backgrounds not being considered in studies thus far regarding perception of vehicle CIs. Such research is especially important in the context of the globalised world where designers are often asked to design for foreign and unfamiliar cultures [1] and therefore need to be well informed about the perceptions of their target (unfamiliar) audience. With this paper, the gap in research is addressed with an exploratory study taking the intercultural aspects in the perception of the CI into consideration. The primary questions answered through this study are; (1) Are the set of adjectives used in the Semantic Differential Method a suitable for a cross-cultural comparison of the CI? (2) Are there differences in how the design parameters are perceived between the cultures?

3 Methodology

The SDM is one of the most popular methods to study perceptions and is characterized by numerous application areas such as psychology, media analysis, market research, etc. The use of the SD method is particularly useful in settings where multiple groups or objects are to be compared.[1] The method of semantic differential was developed by Osgood, Suci and Tannenbaum in 1957 to measure the connotative meaning of concepts [9]. In this study, the SDM was used especially since it has been previously used to study the vehicle interiors in [12] and in particular vehicle CI in [9].

In this paper, the adjectives for the semantic evaluation exercise are taken from the study from Herbeth and Blumenthal [9] as it too was related to the user perceptions of vehicle cluster instruments. The complete set of 13 adjectives from their study was used in this explorative study to examine the difference in perceptions of six vehicle cluster instruments using the semantic differential method across three cultures of the USA, Germany and India (n = 46 for every culture) through an online survey. The six vehicle cluster instruments examined in this study belonged to the 'B'/'Kleinwagen'/'subcompact' segment of vehicles in the three cultures, Fig. 1. The selection of CIs was based on popularity and presence of the designs in the three cultures i.e., 3 designs that were popular in all the three cultures and 3 culture specific designs.[2]

The general structure of the survey involved an introduction about the study, checking the eligibility of the participant, individual evaluation of the six cluster instruments (order appropriately randomized to reduce bias) along the 13 adjective pairs on a 7 point likert scale, a ranking of the CIs based on preference along with the reasoning for rank 1 and rank 6, and finally the collection of general demographic information. The 13 adjectives used are presented below in Table 1.[3] The adjectives and the survey were translated to German with the help of native speakers for the German portion of the survey.

The total of 138 participants across the three cultures belonged to various age groups with all participants in the survey holding a valid car driving license. The data obtained from the completed online questionnaires was analyzed using factor analysis, while the ranking of preferences for the CIs from the participants was collected as qualitative inputs.

[1] Stauche, H. Das semantische Differenzial: Theoretische Grundlage sowie Hilfe für seine PC gestütze Gestaltung und Auswertung. URL:http://www.db-thueringe.de/servlets/DerivateServlet/Derivate-25408/semdiff.pdf.10.08.2013.

[2] Selection was based on sales figures: Source for the sales figures: bestsellingcarsblog.com, last accessed 28.6.2013.

[3] Adjectives extracted in the original paper by 3 stage process using the Repertory Grid Technique, etc. Refer to the paper [9].

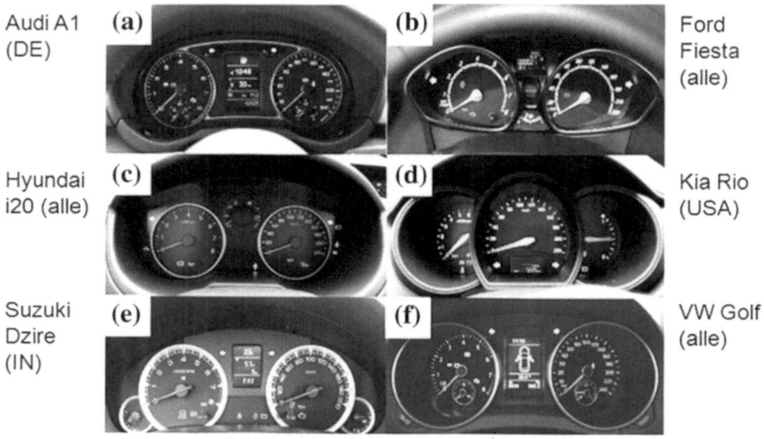

Fig. 1 The six cluster instrument designs used in this study

Table 1 The 13 adjectives used in the semantic differential questionnaire

Unreliable–Reliable	Old-fashioned–Innovative	Down-market–Up-market	Complex–Clear
Dreary–Magical	Sober–Gaudy	Common–Original	Professional–Funny
Coarse–Refined	Unfriendly–Friendly	Unreadable–Readable	Low-tech–High-tech
Ugly–Beautiful			

4 Results

The results of the study are described with regard to the key questions i.e., (1) Are the set of adjectives used in the SDM a suitable for a cross-cultural comparison of the CI? (2) Are there differences in how the design parameters are perceived between the cultures?

4.1 Are the Set of Adjectives Used in the Semantic Differentia Method a Suitable for a Cross-Cultural Comparison of the CI?

For the analysis of this question, a factor analysis was performed to reduce the 13 adjectives used into specific dimensions of pragmatic and hedonic in this study. Based on an expert interview (experts were a Designer and a Psychologist), the 13 adjectives were classified as Hedonic (further divided into Stimulation, Identification) and Pragmatic as per [5], Table 2. The resulting rotated component matrix after factor analysis is shown in Table 3 where the rotation is converted in 3 iterations.

Table 2 Classification of the adjectives based on the expert interviews

Hedonic	Stimulation	Ugly–Beautiful; Common–Original; Dreary–Magical; Sober–Gaudy; Coarse–Refined; Unfriendly–Friendly
	Identification	Old fashioned–Innovative; High tech–Low tech; Down market–Up market; Professional–Funny
Pragmatic	Unreadable–Readable; Complex–Clear; Unreliable–Reliable	

Table 3 Factor analysis with varimax rotation

Rotated component matrix						
	Germany		India		USA	
	Components					
	1	2	1	2	1	2
Clear	**−0.702**	0.256	0.125	0.919	−0.092	0.841
Gaudy	**0.773**	−0.335	**0.865**	0.015	**0.440**	**−0.571**
Friendly	0.294	**0.795**	0.341	**0.830**	0.187	**0.857**
Up market	**0.878**	0.012	**0.751**	0.481	**0.754**	0.114
High tech	**0.875**	0.168	**0.809**	0.554	**0.908**	−0.008
Innovative	**0.879**	0.042	**0.894**	0.251	**0.931**	−0.051
Readable	−0.217	**0.770**	0.009	**0.933**	−0.032	**0.931**
Magical	**0.889**	−0.003	**0.832**	0.396	**0.903**	0.170
Original	**0.890**	−0.120	**0.794**	0.393	**0.931**	−0.150
Refined	**0.815**	0.130	0.494	**0.717**	**0.542**	**0.635**
Beautiful	**0.441**	**0.376**	**0.675**	**0.576**	**0.563**	**0.582**
Funny	0.133	**−0.575**	**0.682**	0.050	0.130	**−0.734**
Reliable	0.022	**0.928**	0.216	**0.707**	0.303	**0.765**

Table 3 shows the amount of loading on the individual factors, thus allowing the assignment of adjectives along one of the two dimensions. From Table 3, for Germany, the adjectives friendly, readable, reliable and funny that load towards the second factor capture the pragmatic attributes. The adjective beautiful loads on both factors with similar high loading. The remaining eight adjectives are associated with the hedonic dimension, except for the adjective clear and funny that have positive loading.

Similarly, for the US (Table 3) the adjectives clear, friendly, readable, funny and reliable were oriented towards the second factor. Gaudy, refined and beautiful exhibit both hedonic and pragmatic characteristics while the remaining five adjectives are regarded as hedonic. The loading of funny and gaudy is negative for the second factor meaning professional and sober designs are regarded to be pragmatic.

Again from Table 3 for India, the adjectives clear, friendly, readable, refined and reliable were considered pragmatic, with the adjective refined having a relatively high loading on the first factor. The adjective beautiful can simultaneously be assigned to both dimensions while the remaining seven adjectives are associated to the hedonic dimension. In the case of up-market and high-tech, relatively high loading was found even on the second factor. The consistently positive charge of all 13 adjectives for India is also noteworthy in this regard.

Overall, the adjective beautiful loads on both factors for all three countries. In four adjectives there are cultural differences in both the assignment and the signs of the loading. For example, the adjective refined loads on the first factor in Germany and on the second factor in India, although the loading on the first is relatively high. In contrast, the adjective refined loads on both factors for the United States with a relatively high loading. Similar differences between the cultures can be seen even in the case of clear, funny and gaudy. These adjectives are highlighted in Table 3. In addition, it is to be noted that the adjective pair friendly–unfriendly associates itself to the pragmatic component in all three countries. This means that the above-described division into hedonic and pragmatic dimensions is applicable only in a limited capacity.

4.2 Are There Differences in How the Design Parameters Are Perceived Between the Cultures?

For this question, the results of the factor analysis and the preferences of the participants expressed in the ranking of the cluster instruments are considered. Since (as seen in Sect. 4.1) the factor analysis of some adjectives do not produce uniform and unanimous results to compare hedonic and pragmatic properties between the different cultures, the number of adjectives is reduced to 8. The factor analysis scores along pragmatic and hedonic dimensions for the three countries are as shown below in Table 4.

Figure 2a–c illustrates visually the participants' responses for the CIs along the two dimensions. The figure also shows the preference and ranking of the cluster instruments by the participants of each country along with some points of reasoning mentioned for the 1st and 6th preferences.

From Fig. 2a, it is seen that the German participants assigned the first three ranks to those designs that performed best on the pragmatic level. The CI of the Ford Fiesta was rated the least in the pragmatic assessment for the German

Table 4 Average scores for the six CIs across the two dimensions for the three cultures

Cluster instrument	Germany		USA		India	
	Pragmatic (avg)	Hedonic (avg)	Pragmatic (avg)	Hedonic (avg)	Pragmatic (avg)	Hedonic (avg)
Audi A1	5.4	4.0	5.1	4.3	5.2	4.2
Ford Fiesta	3.4	4.5	4.1	4.7	4.7	4.7
Hyundai i20	5.4	2.6	5.4	3.2	5.4	3.8
Kia Rio	4.0	4.0	4.5	4.3	5.0	4.7
Suzuki Dzire	4.3	3.8	4.7	3.9	5.1	4.2
VW Golf	5.8	5.2	5.1	5.1	5.2	5.2

(a)

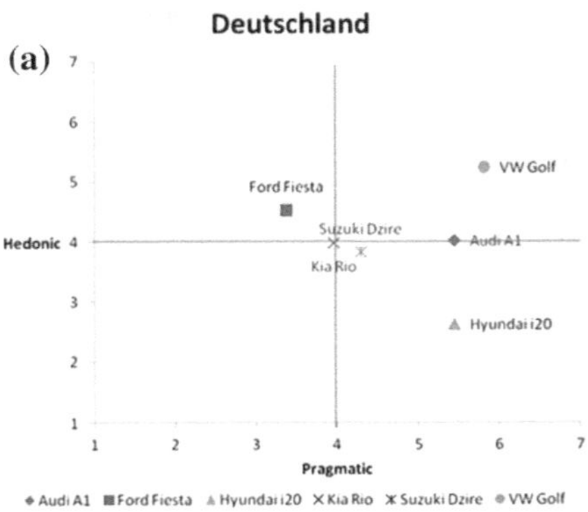

Car	Average Preference Score	Ranking Preference
Audi A1	2.5	2
Ford Fiesta	5.0	6
Hyundai i20	3.8	3
Kia Rio	4.1	5
Suzuki Dzire	4.0	4
VW Golf	1.6	1

Reasons for 1st and 6th preferences	
Clear	Playful
Simple	Too bright
Readable	Cheap
Premium	Distracting

(b)

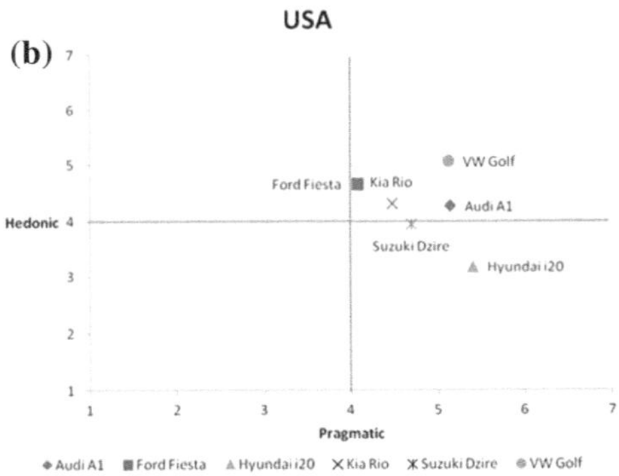

Car	Average Preference Score	Ranking Preference
Audi A1	2.5	2
Ford Fiesta	4.8	6
Hyundai i20	3.7	3
Kia Rio	3.9	5
Suzuki Dzire	3.9	4
VW Golf	2.1	1

Reasons for 1st and 6th preferences	
Clear	Playful
Simple	Too bright
Readable	Cheap
Premium	Distracting

(c)

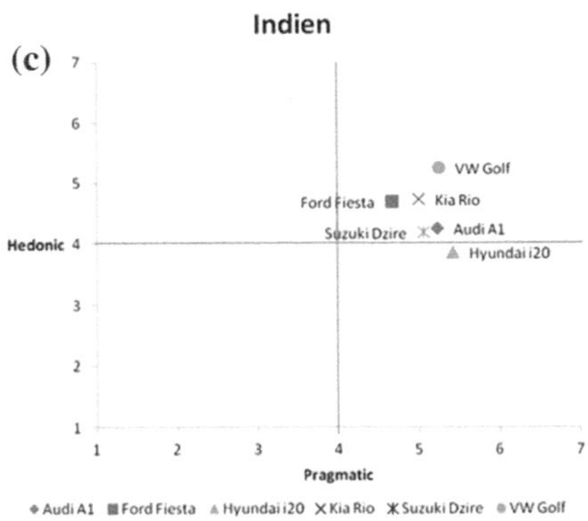

Car	Average Preference Score	Ranking Preference
Audi A1	3.2	2
Ford Fiesta	3.9	5
Hyundai i20	4.3	6
Kia Rio	3.7	4
Suzuki Dzire	3.4	3
VW Golf	2.6	1

Reasons for 1st and 6th preferences	
Technically rich	Looks old
Digitization	Not clear
Exclusive	Low-tech
Stylish	

Fig. 2 **a** The factor analysis for participants' responses from Germany along the two dimensions. **b** The factor analysis for participants' responses from USA along the two dimensions. **c** The factor analysis for participants' responses from India along the two dimensions

participants, thus making it the least preferred. The VW Golf, whose cluster instrument had the highest expression in terms of both pragmatic and hedonic dimension, was thus most preferred.

From Fig. 2b for the US participants, it is seen a similar picture to that of Germany emerges (Fig. 2a), with the designs of VW Golf, Audi A1 and Hyundai i20 being rated the best on the pragmatic scale although with a slightly lesser score than those for Germany. The same can be observed with the scores on the hedonic dimension with the scores by the American participants being slightly lower to that of German participants. This similarity is further reflected in the ranking of preference by the participants and the associated reasoning.

From Fig. 2c, it is seen that the Indian participants considered the pragmatic qualities to be more or less equal for all the designs (a difference of 0.7 between the best and worst, Table 4). However, despite this small difference the perceptions do not seem to be very different to those of their western counterparts. The differences are stronger in the evaluation of the hedonic aspects where the values are higher than those of the western participants. In terms of the rankings, it was seen that the VW Golf and Audi A1 had similar rankings across the three cultures. The least preferred design for the Indian participants was that of the Hyundai i20. It is interesting to note that both the most preferred (VW Golf) and the least preferred (Hyundai i20) scored the most and least with respect to the hedonic aspects.

5 Discussion and Implications for Design

Section 4.1 of the paper examined the suitability of adjectives used in this study for cross-cultural comparison with the help of factor analysis. The adjective pair sober–gaudy is clearly identified in Germany and India as hedonic. In the U.S., however, it has a relatively high loading on both factors, with the loading on the pragmatic factor being higher. According to experts (Table 2), the terms sober–gaudy are hedonic in nature and relate to the visual aesthetics of the CI but, in the U.S., this pair of adjectives seem to appeal to the pragmatic level and bring the functionality or ease of use into question.

The adjective pair professional–funny is regarded in Germany and the U.S. as pragmatic. The negative loading indicates that the adjective professional was considered and not the adjective funny. On the other hand, Indian participants identify the adjective pair professionally–funny as hedonic with a positive loading. Therefore, this pair of terms is considered deviant to the western cultures.

Interestingly, the adjective pair coarse–refined, was associated in Germany as hedonic, in India as pragmatic and in the United States as both the dimensions. Here it seems as if the participants understand coarse–refined in both aesthetic and functional perspectives. Even the adjective pair complex–clear the factor analysis does not provide unanimous results. In Germany it loads with a negative charge on the first factor, in India and the U.S. with a positive charge on the second factor. The adjective pair was however regarded by experts (Table 2) as pragmatic.

With above in mind, it can be concluded that the adjectives generated in France [9] and subsequently used in this study can be used for a cross-cultural comparison with the exception of the four above mentioned pairs. However, for a uniform understanding of the terms in future studies, it is recommended here that synonyms of the adjectives be provided in addition.

With respect to the second objective of this paper—Sect. 4.2, factor analysis of the reduced set (8 adjectives) of adjectives along with a ranking of preferences was considered. For Germany, the results would translate to German participants preferring CI designs that are both pragmatic as well as hedonic features with the pragmatic aspects dominating. This can be explained by the general impressions of Germans, who are seen to be pragmatic, reliable and well structured.[4] This is further supported by the qualitative reasoning given by the participants for their ranking. Similarly, it is seen that the American participants too prefer CI design to take into account both pragmatic and hedonic aspects equally with a preference for higher pragmatic value.

In contrast to the other two countries the hedonic attributes play an important role for Indian participants. This finding is in line with culture-specific features of the Indian culture that is characterized by the influence of Bollywood, religious festivals, etc. leading to openness to new stimuli and a greater enthusiasm towards more playful or 'unusual' features. Further, Indian participants highlighted aspects of style and exclusivity owing to general need of Indians to be exclusive among peers, whereas the least preferred design was described as being old fashioned and low-tech.

Therefore the design implications of this study could mean that for the German and US markets, the CI should be pragmatic and ideally also have hedonic aspects too with clarity and concise structure. In comparison for the Indian population, the CI design should focus on the hedonic aspects to a greater extent. For CI designs targeted for the international market, it is recommended to choose as pragmatic a design as possible and then incorporate culture specific hedonic aspects as required. It can further be noted that for Germany and the US unusual shapes and colors must be avoided with focus on a simple, clear and unambiguous choice of shapes and colors. In contrast, in India, original shapes and colors are generally perceived as positive in comparison.

One of the drawbacks of the present work refers to the narrow selection of the CIs for examination. Therefore, future examination with more number of CIs is recommended to achieve generalizable results. Further, a greater variety in the selection of designs is recommended to better identify culture specific perceptions and preferences. Finally, future studies must focus on a specific group of users which would give deeper insights about a particular target audience.

Finally, based on the results of this study, another way for analysis of the survey results is suggested here through the bi-variate correlation matrix. Using a correlation matrix, one can examine the interdependencies that exist between the

[4] Hofstede's dimension: geert-hofstede.com, last accessed 10.2.2014.

individual adjectives used in the survey, and also investigate whether the adjectives used produce similar positive or negative associations in all three cultures. These aspects are also recommended to be examined in future research in this area.

6 Conclusions

This paper presented an explorative study on design perceptions of vehicle CI across the three cultures of Germany, India and the US using the semantic differential method. In addition, the study examined if the adjectives used in the SDM are suitable to employ in cross cultural situations where 9 out 13 were found to be suitable. The results of the study for both the above objectives are presented and discussed along with possible implications for design. The paper finally, goes on to describe the limitations of this research along with suggested improvements and topics for further research.

References

1. Diehl, J.C., Christiaans, H.H.C.M.: Globalization and cross-cultural product design. In: DS 36: Proceedings DESIGN 2006, the 9th International Design Conference, Dubrovnik, Croatia (2006)
2. Hsu, S.H., Chuang, M.C., Chang, C.C., Hsu, S.H., Chuang, M.C., Chang, C.C.: A semantic differential study of designers' product form perception. Int. J. Ind. Ergon. **25**, 375–391 (2000)
3. Sonderegger, A., Sauer, J.: The influence of socio-cultural background and product value in usability testing. Appl. Ergon. **44**, 341–349 (2013)
4. Tomico, O., Karapanos, E., Lévy, P., Mizutani, N., Yamanaka, T.: The repertory grid technique as a method for the study of cultural differences. Int. J. Des. **3**, 55–63 (2009)
5. Hassenzahl, M.: The interplay of beauty, goodness, and usability in interactive products. Hum. Comput. Interact. **19**, 319–349 (2004)
6. Jindo, T., Hirasago, K.: Application studies to car interior of Kansei engineering. Int. J. Ind. Ergon. **19**, 105–114 (1997)
7. Karlsson, B.S.A., Aronsson, N., Svensson, K.A.: Using semantic environment description as a tool to evaluate car interiors. Ergonomics **46**, 1408–1422 (2003)
8. Liu, Y.C., Wen, M.C.-H.: Comparison of head-up display (HUD) vs. head-down display (HDD): driving performance of commercial vehicle operators in Taiwan. Int. J. Hum. Comput. Stud. **61**, 679–697 (2003)
9. Herbeth, N., Blumenthal, D.: Product appraisal dimensions impact emotional responses and visual acceptability of instrument panels. Food Qual. Prefer. **29**, 53–64 (2013)
10. Faria, A.W.C., Menotti, D., Pappa, G.L., Lara, D.S.D., Araujo, A.de A.: A methodology for photometric validation in vehicles visual interactive systems. Expert Syst. Appl. **39**, 4122–4134 (2012)
11. Jung, G., Kim, S.M., Kim, S.Y., Park, S.: Effects of design factors of the instrument cluster panel on consumers' âffection. In: Proceedings of the International MultiConference of Engineers and Computer Scientists, vol. III, pp. 1–4
12. Buss, S., Chouard, N., Schulte-Fortkamp, B.: Semantic differential tests show intercultural differences and similarities in perception of car-sounds. DAGA **2000**, 1–2 (2000)

Designing Alternative Paradigm for Traditional Visual Storytelling

Saptarshi Kolay and Shatarupa Thakurta Roy

Abstract Indian art is formed in amalgamation of multiple cultural art forms. The visual culture of contemporary India is eclectic in nature with unification of many influences. Apart from the academic Persian and European style, the regional folk or vernacular art has played a significant role in operating its key characteristics. The aim of the paper is to identify the elements that are responsible in the retention of Indian-ness in Indian Art forms by conducting a thorough visual-analysis. It describes an in depth establishment of identity design by extracting the core elements from the age-old tradition of 'Bengal Patachitra'. The process includes a transformation of the conventional paradigm into a new one that is popularly accepted among the urban mass. The paper describes the process of adapting the art style of the narrative scroll paintings by iconic, thematic and semiotic analysis beyond a cultural margin as well as finds its possibilities of adaptation into new media application. The 'Bengal Patachitra' is narrative scroll painting made on paper or mounted cloth using natural colour pigments. The story telling approach of the scroll painting and the corresponding song is an intrinsic part of rural-folk culture of regional Bengal. It acts as a vernacular comic Strip. It has all the potentialities to be translated into an animated film without losing its visual identity. The process of the work involves field research and survey to document the socio-cultural atmosphere. The study is duly meant for the knowledge preservation such as method, material and their use by the artisans (patuas). Instead of superficial adaptation, the identified features as an outcome of methodical visual analysis are readdressed for the alternative version. The attempt is to shift to a different yet appropriate paradigm that suits its present context. The deliverable of the design research culminates into a short animated film. The paper therefore unfolds the design process that is involved in creating a visual identity in reference of a habitual art form.

S. Kolay (✉)
Department of Architecture and Planning, Indian Institute of Technology Roorkee, Roorkee, India
e-mail: saptarshikolay1988@gmail.com

S.T. Roy
Design Programme, Indian Institute of Technology Kanpur, Kanpur, India

A. Chakrabarti (ed.), *ICoRD'15 – Research into Design Across Boundaries Volume 1*, Smart Innovation, Systems and Technologies 34, DOI 10.1007/978-81-322-2232-3_14

Keywords Identity design · Paradigm shift · Knowledge preservation · Visual analysis · Adaptation · Alternative version

1 Aim

The visual language of Indian subcontinent is an amalgamation of varied stylistic identity. Visual language of India changes in every province of India. If we go back in time in different historical eras, we see external cultural influences of many types. These influences are often proven as operational factors for present day visual culture of this country.

The contemporary art practice in India as a result is eclectic [1–3]. Indian artist K.G. Subramanyan (b. 1924), a theoretician and educator, defines contemporary Indian art practice as "the interaction, and maybe the reconciliation, of different cultural forms" [4]. He explains that it not only is indicative of the modern multicultural situation but also can be an important tool in the renewal of culture. For Subramanyan, accepting modern culture as eclectic is one of the first steps of finding what he terms "an internal solution" to the "identity crisis" [1, 5] that marks much of postcolonial Indian art. So there is a need to extract the Indian identity from the contemporary culture of India and establish a visual identity as to be solely Indian and at the same time contemporary in culture.

1.1 Finding Visual Identity of Contemporary India

The contemporary scenario of Indian art is a derivative of two main aspects; firstly the art practice of contemporary painters and artists and secondly the contemporary vernacular art practiced by different vernacular craft-guilds throughout the India. The artisans of these craft-guilds are imbibing the contemporary contexts and events; thus their vernacular art is evolving towards a contemporary vernacular genre, which engages both with age-old traditional vernacular culture and the modern culture. So after assessing these two aspects the key visual characteristics should be extracted to find out the contemporary visual language of India.

1.2 Necessity of Paradigm Shift

Visual identity creates the mental model of people. After identifying the key features of Indian visual language, these features should be transformed into a contemporary paradigm, which is well accessible by mass; as it is important to establish the identity in strong foundation to be acceptable by people. Digitalisation of the paintings that are preserved in the museums is a timely measure. Alongside, converting painting into different mediums such as animated movies and graphic novels, which are already well popular, can be a considerable choice [6] for building awareness among the greater mass. An animated film or graphic novel designed with a regional Indian identity may open the door for spreading the cultural awareness amongst people getting them into an indigenous aesthetic prospect.

2 Methodology

For identifying the Indian visual style of a vernacular art form and adaptation of the style into another paradigm the following three steps are conducted.

2.1 Ethnographic Survey

To analyze the methods and materials as well as the socio-cultural aspects of a vernacular art an ethnographic survey is necessary.

2.2 Analysis and Codifying the Visual Identity

After doing the literature study and ethnographic survey visual analysis of a particular Indian art form is done to derive the key characteristics of the visual form. To identify the features of the art form elements and principals of design of that particular art form is analyzed.

2.3 Adaptation and Final Design

These key characteristics are then finally transformed to create a new visual identity and finally translated into animation. Final design is done using the key features identified to achieve the Indian identity without hampering the visual style of the art form.

3 Ethnographic Survey

Ethnographic survey is conducted to analyse the contextuality, elements and principals of 'Bangal Patachitra' a vernacular graphic narrative of Bengal [7].

Fig. 1 Story of Surpanakha—Jorano Pata: Bengal Patachitra by Moyna Chitrakar. *Source* Author

Bengal Patavhitra or Scroll paintings of Bengal are made by the artisans or 'Patuas' or 'Chitrakars' of craft-guilds. Patuas are also accomplished singers. The mythical narratives and folklores painted on scrolls are carried from village to village, and narrations of the stories are accompanied by folk songs [8, 9]. There are two different kind of Patachitras—'Jorano Pata' (a complete comic-strip with multiple frames: reference Fig. 1) and 'Chouka Pata' (a single framed painting: reference Fig. 2). Bengal has majorly three different styles of Patachitra, i.e. 'Kalighat Patachitra' (reference Fig. 5), 'Midnapur Patachitra' (reference Fig. 4) and 'Birbhum Patachitra'. Each three has distinguishable characteristics.

3.1 Interview of Moyna Chitrakar

Moyna Chitrakar, a painter of Nayagram Midnapur, is famous amongst the Muslim 'Chitrakar' community of Midnapur. She also worked with 'Tara Books' and painted the imageries for 'Sita's Ramayana' [10].

Fig. 2 September 11 attack—Couka Pata by Gurupada Chitrakar. *Source* Indigo Arts Gallery

Besides interview the survey is documented with film and photography. Survey and the documentation is conducted to understand primarily the following aspects:

(a) Tools, techniques, materials, methods and the detailed procedure of the scroll painting.
(b) Structure or visual attributes (elements and principles) of the graphic narrative.
(c) Socio-cultural environment of the craft-guilds of 'Patuas' or scroll painters.

3.2 Contemporary Influences in Bengal Patachitra

Moyna Chitrakar worked with 'Tara books' and developed a graphic narrative called 'Sita's Ramayana' along with designer Samhita Arni. So Patachitra of Nayagram has already experienced a paradigm shift from scroll painting to graphic novel. She is also in the process of making another graphic novel along with 'Tara books', the story of 'Noah's Ark' [11] and the story of 'Mather Teresa'. Artisans of Nayagram also depict several contemporary issues in their paintings, like 'Tsunami' [12], September 11 attack on World Trade Centre of New York (reference Fig. 2), etc.

4 Analysis of Visual Identity and Potentiality of Adaptation into Animation

After the ethnographic survey, a through visual analysis is done on 'Midnapur Patachitra' as well as 'Kalighat Patachitra', which is an urban folk art practiced in Kolkata. Though there is much dissimilarity between these 'Patachitra's, there lies some similitude between them.

4.1 Analysis of Visual Identity of Bengal Patachitra

The elements and principles of visual design of these two types of graphic narratives of Bengal are discussed.

4.1.1 Borders of Panels

Floral border used in 'Bengal Patachitra', as depicted in Fig. 3, is generally consists of a painted creeper with red flowers and green leaves on yellow background. These borders subscribe each frames of 'Jorano Pata' like gutter spaces between each panels of a comic-strip. In case of 'Chouka Pata' these borders are painted in each sides of the panel.

4.1.2 Colour Combinations

Previously the colour used for the painting was organic colours (reference Table 1) on cloths. Now they are using other colours available in markets. They mainly use Poster colour in gouache technique; opaque and transparent both techniques are mixed in several cases.

Fig. 3 Borders used in Midnapur Patachitra. *Source* Painting: Moyna Chitrakar, "Durga Pata". *Photography* Author

Table 1 Organic colours of 'Bengal Patachitra' and their raw materials

Colours	Raw materials
Black	Lamp soot
White	Khari maati, a white soft lime stone
Green	Green seeds
Red/Orange	Palash (*Butea monosperma*)
Crimson red	China-rose and Rose (*Rosa* sp.)
Blue	Aparajita flower (*Clitoria ternatea*)
Yellow	Raw turmeric paste

Source Derived from interview with Moyna Chitrakar

Colour palette of 'Midnapur Patachitra' (reference Fig. 4) is diverse and full with vibrant colours. Flat areas of colours are outlined by black coloured lines. Artisans use to paint the ornaments and jewelleries in fine white lines.

Unlike 'Midnapur Patachitra', 'patua's of Kalighat use less colours for their paintings (reference Fig. 5). Use of black and red is predominant, though there are many paintings painted using different hues. But the main characteristic feature which makes 'Kalighat Patachitra' different from 'Midnapur Patachitra' is the use of gradient shades.

4.1.3 Line Quality

First the flat backgrounds are painted with thicker brushes. Then figures are painted with a comparatively thinner brush. After making all the ornamentations and decorations finally a double-lined border is applied throughout the painting.

Fig. 4 Colour palette of Midnapur Patachitra. *Source* Painting: Moyna Chitrakar, "Sita's Ramayan", Tara Books. *Colour palate* Author

Fig. 5 Colour palette of Kalighat Patachitra. *Source* Painting: website of Peabody Essex Museum. *Colour palate* Author

Artisans of Nayagram, Midnapur employ extremely heavy, bold out-lines. The use of spontaneous black out-lines in their paintings enhances the composition. Whereas in 'Kalighat Patachitra', black outlines are often replaced with gradient of coloured patches.

4.1.4 Human Figures in Bengal Patachitra

A remarkable characteristic of 'Bengal Patachitra' is the faces of the human characters generally turn at three quarter angle, which is rare in other Indian vernacular art. Artisans also paint figures with complete side-profiles, depicted in Fig. 6a, b.

Human figures in 'Bengal Patachitra' are drawn in spontaneous flow of brushes. These painted human figures do not follow the iconographic descriptions of 'Shilpashastra' [13]. Use of different colours, like, yellow, orange, grey, green and blue as skin-tones is also present in this style of paintings.

4.1.5 Use of Patterns in Bengal Patachitra

Besides the use of flat colours artisans of 'Bengal Patachitra' also fill the area within a black outline with different verities of patterns to depict waves, grass lands, tree canopies, wardrobes, etc. Use of cross-hatches, angular hatches and repetitive dots to fill an area also creates different patterns (reference Fig. 7) in the composition.

Fig. 6 Faces turned at three quarter angle in Midnapur Patachitra and Kalighat Patachitra. *Source* **a** Painting of Moyna Chitrakar, "Sita's Ramayan", Tara Books. **b** Website of Peabody Essex Museum

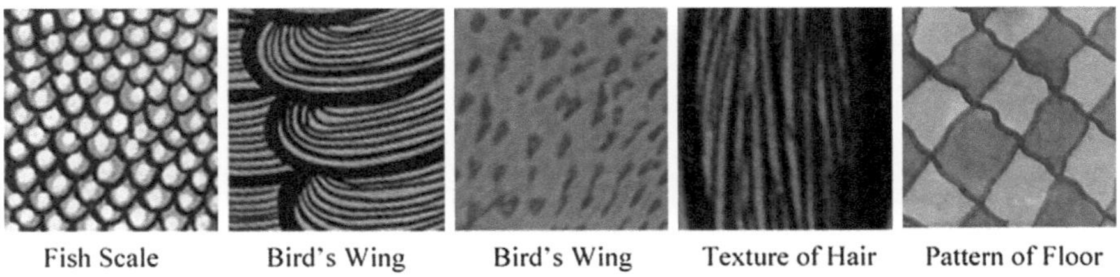

Fig. 7 Different patterns used in Midnapur Patachitra

4.1.6 Analysis of Potentiality of Adaptation into Animation

Visual analysis of the art style reveals the potentiality of being transformed into animation. The capability of being a style of an animated film from a style of a graphic narrative depends on some parameters. Considering these parameters 'Bengal Patachitra' is analyzed before adaptation into animation.

4.1.7 Recognisability of Characters

The characters designed for the story has to be identifiable throughout the movie and should have some recognizable features, which can make the character distinguishable. The chosen style should give the freedom to experiment with the character design without distorting the visual language of the style.

All the Indian vernacular art forms are idealized and consist of some kind of distortion. But the human characters in 'Bengal Patachitra' are much more recognisable if compared with 'Madhubani', 'Gond' or 'Warli' paintings (reference Fig. 8).

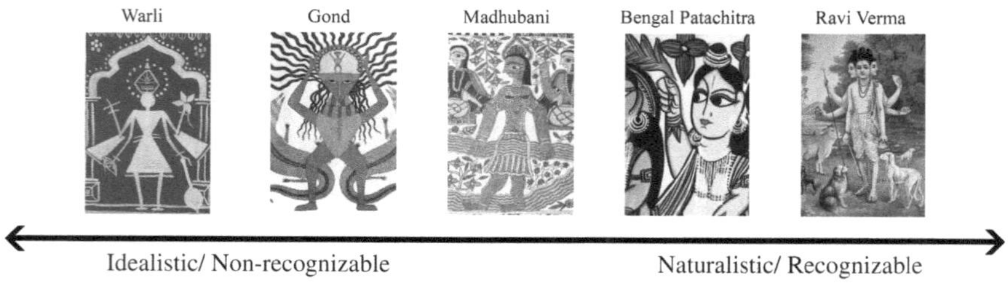

Fig. 8 Recognisability of human characters in different Indian vernacular arts and painting of Raja Ravi Verma. *Source* Author

4.1.8 Movement Capability of Characters

'Bengal Patachitra' can be adapted into 2D animation by drawing the character into different parts from the joints or by drawing each frame separately.

4.1.9 Potentiality of Using Different Camera Angles

Bengal Patachitrakar portrays faces which are turned at an angle (reference Fig. 6), though frontal and side faces are occasionally used. This unique feature of Bengal Patachitra portrays the character much prominently and provides the opportunity to use different camera angles.

4.1.10 Potentiality to Depict Modern Elements Without Losing the Visual Identity

'Bengal Patachitrakars' have already started experimenting with incorporating modern objects into their paintings to depict contemporary events (reference Fig. 2) [10–12].

4.1.11 Sense of Depth in 2D

As 'Oriental Perspective' with parallel guidelines is one of the characteristics of this painting, so to achieve sense of depth superimposition and placing further objects on top of nearer one is the only option. But sometimes further objects are also portrayed in smaller scale.

5 Adaptation and Final Design

In this stage characters are designed considering the design principles and elements of 'Bengal Patachitra' for a short contemporary story called "Matsyanyaya" (reference Fig. 10). The style of 'Kalighat Patachitra' and 'Midnapur Patachitra'

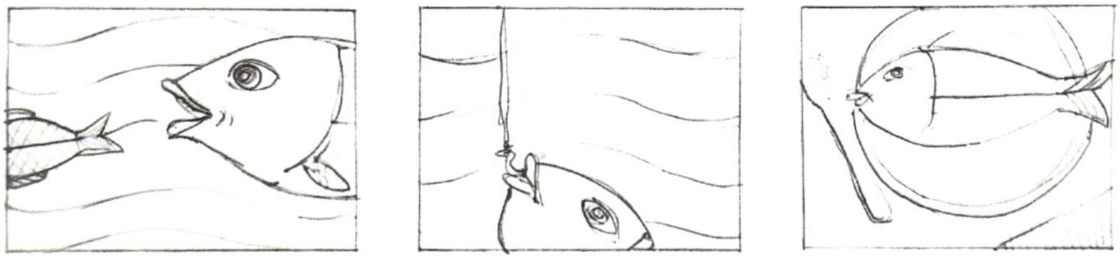

Fig. 9 Story-board for the animation

both influenced the created visual identity. A story-board (reference Fig. 9) is designed for a short animation film to explore the potentiality of adaptation of the art style.

To match with the features of 'Bengal Patachitra', the socio-cultural attributes of contemporary Bengal are revealed in the animated movie. Also contemporary houses and vehicles of Kolkata are incorporated to give the identity of a contemporary cityscape. To inculcate the intrinsic nature of the city Bengali typefaces are used to design the signboards and hoardings of the cityscape [14]. The colour palette of the animated movie is influenced from the organic colour tones of 'Midnapur patachitra'. During the translation of visual style the characteristic features of 'Bengal Patachitra' are incorporated in the animated movie. For example the thick black outline around a figure moves independently in the animated movie to give the feeling of free-flowing heavy bold out-lines added on top of the flat-coloured paintings of 'Midnapur Patachitra'. As floral border acts as a gutter space between the two different frames of a 'Jorano Pata' (reference Fig. 1), this similar floral border is used to blend two different shots of the animated movie (reference Fig. 10).

Fig. 10 Final animation, Matsyanyaya

6 Conclusion

6.1 Analysis

Though 'Bengal Patachitra' style is explored before in graphic novel like 'Sita's Ramayana', this form is not substantially explored in terms of animation. Some key features of this art form are very suitable for animation, which is described in previous chapter in details. 'Bengal Patachitra' style is already very popular and its popularity will help viewer to absorb the visual style of the animation.

Through the whole process, it can be derived that the Indian vernacular art has an immense potentiality to be identified and expressed a visual identity of India. As these are mainly designed for a storytelling purpose these art forms can be adaptable in animations as well as graphic narratives. The art style used here to design the final prototype is 'Bengal Patachitra'.

Similarly other different Indian art styles can be explored and can be adapted into animations. But some of these art forms have some special features which are difficult to incorporate into animations. For these art styles some different elements of visual design can be extracted and carefully blended with contemporary paradigm. But for these cases help from the real artisans is necessary.

6.2 Future Scope of Work

Using the similar process, the identified key characteristics of various art styles can be translated into different contemporary and popular paradigms other than animation. Graphic novels, new media applications, game design and any other visual art form can find inspiration from Indian art styles. Inspiration of identity design can be derived from more than one art style without distorting the visual language. But visual identity design should be created only after in-depth analysis of elements and principles of visual design of an art; a superficial visual copy will lead to the distortion of an existing art style.

It is important to understand a cultural practice with its historical and social aspects, without superficially adapting elements in design process. A thorough analysis of the existing visual design provides us with the clue to know the visual language to its core and make use of its essential sensibility with a shifted paradigm. This is important for identity design as well.

Acknowledgments The authors would like to thank Ms. Moyna Chitrakar and Mr. Dipak Barapanda, curator of Gurusaday Museum, Kolkata for their help and support. We are thankful to Indian Institute of Technology Kanpur for providing all the technical supports during research.

References

1. Subhramanian, K.G.: The Living Tradition: Perspectives on Modern Indian Art. Seagull Books, Calcutta (1987)
2. Havell, E.B.: Essays on Indian Art: Industry and Education. Natesan, Madras (1907)
3. Richardson, M.: Understanding the importance of eclecticism: K.G. Subramanyan and twentieth-century indian art. Southeast Rev. Asian Stud. **29**, 240–247 (2007)
4. Subhramanian, K.G.: Moving Focus: Essays on Indian Art. Lalit Kala Akademi, New Delhi (1978)
5. Subhramanian, K.G.: The Magic of Making: Essays on Art and Culture. Seagull Books, Calcutta (2007)
6. Sabnani, N.: The challenges of sleeping giant. Des. Issues **21**, 94–105 (2005)
7. Ranjan, A., Ranjan, M.P.: Handmade in India. Mapin, New Delhi (2007)
8. Barapanda, D.: Patua Sanskriti Parampara O Paribartan. Pustak Bipani, Kolkata (1999)
9. Gramin Vikas Seva Sanshtha: Evaluation Study of Tribal/Folk Arts and Culture in West Bengal, Orissa, Jharkhand, Chhatisgarh and Bihar. SER Division Planning Commission, Govt. of India, New Delhi (2011)
10. Arni, S., Chitrakar, M.: Sita's Ramayana. Tara Publications, Chennai (2011)
11. Wolf, G., Chitrakar, J.: The Enduring Ark. Tara Publications, Chennai (2013)
12. Chitrakar, M., Chitrakar, J.: Tsunami. Tara Publications, Chennai (2009)
13. Tagore, A.: Some Notes on Indian Artistic Anatomy. Indian Society of Oriental Art, Calcutta (1921)
14. Kolay, S.: Exploration of styles of visual representations in indian context. M.Des. Thesis, Design Programme, Indian Institute of Technology Kanpur (2013)

An Introduction to the Tendency of Online Cosmetic Advertisement Design

Yen-Ting Lai, Ching-Yuan Huang and Mu-Chien Chou

Abstract With the increasing popularity of the Internet, purchase patterns of the public have shown some innovations. This has encouraged advertising industry to march towards the web market, and establish models of communication with consumers. This study is intended to understand the focuses and elements of cosmetic web ads by widely collecting such advertising contents. Typically, in a cosmetic ad, the product endorser is often the subject of the advertisement. Therefore, this study is investigating such points as what role to choose, how demonstration of the difference role is distinguished, and how validity of the difference role is integrated. The current study is expecting to discover the classification of variables that contributed to the design tendency through surveys, and further provide cosmetics and advertising industry with introduction as reference. It is also expected that general rules of cosmetic web ads should be set out for visual designers of web pages, for preference of consumers, and even for applications in the teaching of webpage design.

Keywords Advertisement design · Cosmetics · Online advertisement

1 Introduction

Recently, the multi-media bring further possibilities to the advertisement industry. Take online advertisement for example, it contains the traits of effectiveness, variability and richness. And among online advertisement, banner ads are the most prevailing format.

A psychological study on recognition recall shows the result that "Generally, consumers have high recognition on banner ads" [1]. With the development of the Internet, the online advertisement industry has appealed more attention. Among all

Y.-T. Lai · C.-Y. Huang · M.-C. Chou (✉)
Department of Bio-Industry Communication and Development,
National Taiwan University, Taipei, Taiwan (ROC)
e-mail: choumc@ntu.edu.tw

A. Chakrabarti (ed.), *ICoRD'15 – Research into Design Across Boundaries Volume 1*,
Smart Innovation, Systems and Technologies 34, DOI 10.1007/978-81-322-2232-3_15

industries, cosmetics industry is undoubtedly one of the most highly-risen industries nowadays. According to the survey [2], total sales in the beauty and skincare industry were roughly $426 billion in 2011. We can see that skin care and makeup have become a necessary part for most of women in modern society; therefore, there are a large group of potential cosmetics buyers within the market. In accord to the trend that people are accustomed to purchasing online, launching advertisements onto websites has become a common marketing way for advertisers. Therefore, the issue which advertisers and consumers are eager to know is how to manage an attractive banner ads. According to definition on good banner ads from Bayles [3], the design of good banner ads must be compatible with consumer's needs. Hence, studies on advertisement graphics that trigger consumer's interests are really important in advertisement design. In line with the reasons above, the research gathers and organizes various types of designs on banner ads in the market. By using KJ method to categorize advertisement design elements, the research comes to the result of the key vision tendency of cosmetic banner ads. The results will provide a design tendency of banner ads for academic and industrial fields.

2 Literature Review

2.1 Cosmetic Banner Ads

Total global beauty sales increased by 14 % to $2,278,000,000 in 3 months [4]. According to the survey on consumers' purchasing behavior for cosmetics on the Internet, consumers who are used to surf more than five cosmetic websites in advance of making purchasing decisions account for 43 and 38 % of them browse through two to four cosmetic websites [5]. The phenomenon indicates that when cosmetic advertisers launch ads onto relative cosmetic websites, their target audience is often capable of absorbing effectively, which achieves goals of advertisements.

Such type of cosmetic advertisements has been the most popular advertisements form since 1994, when it first turned up. It is worth to mention that the form of banner ads is the most prevalent one both in America or Taiwan's Internet advertising industry [6]. Consequently, with an accurate eye on the research purpose, we choose the most frequently used banner ads as the discussing targets.

2.2 Graphic Design and Key Vision

The existence and performance of graphic designs and key vision play vital roles in one complete advertisement. According to Roth et al. [7], graphics include photography, charts and illustrations. Among them, photography, which is able to truly present meanings of ads, is the mainstream of contemporary ads [8]. Key vision is the most significant media message in every graphic for communicating with the audience. What mainly compose an ad's key vision are usually human beings (characters) or objects [9]. During the process of interacting with characters, face is one of the most accessible parts of a body. The field of facial cosmetics is quite extensive, and the ones seen the most commonly are foundation, eye makeup and lipsticks, which respectively represent the facial makeup of face, eyes, and lips. Additionally, usage of heavy or light facial makeup will be adjusted under different scenarios. From what are mentioned above, the research controls the design elements not inclusive in images themselves, and does a design tendency research specifically focusing on banners ads in the photography method.

2.3 KJ Method

KJ method is named after the abbreviation of its inventor, Professor Kawakita Jiro, who is one of the Japanese cultural anthropologists. When doing the national research on the Himalayan group, Professor Kawakita Jiro created one set of methodology which can be applied to qualitative researches. Originally, it used to apply to humane and geographic data survey. After constructing the methodology, Professor Kawakita Jiro spent more than 15 years on making the set of thoughts and methods systematically organized, and its accomplishment is the current well-known KJ method [10]. KJ method is one kind of research approaches relying on data similarity from a great deal of data to organize and generalize [11]. To face and solve problems and facts under unfamiliar conditions, KJ method takes advantage of intrinsically interactive net to categorize and combine, expecting to find out logic and a thread from complex and giant phenomena. Generally speaking, there are six main steps in KJ method. First, researchers should decide a concrete theme. Second, collect related and significant data that the research needs. Third, make cards and fill all collected information on them. Fourth, make the cards in the same category classified and rename them. Fifth, draw KJ pictures from the result of sorting out. Following the above steps, KJ method can make researchers do further research from a variety of angles. However, KJ method sometimes will be transformed into subjective judgment [10]. After using KJ method, the researcher will interview experts and invite them to check points for settling problems mentioned above, so that the researcher can start to analyze from the most objective aspect. Not only can KJ method simplify the data collected from the market, but it allows us to find out correspondingly relative variables to generalize and merge. Finally, the research will arrange properly-classified sample areas and discuss them furthermore. KJ method is the main approach of the research.

3 Market Survey

3.1 Data Collecting

The research collects the data from the following websites famous for women care: ELLE Taiwan (http://www.elle.com.tw/), iswiss beauty news (http://www.iswii.net/) and niusnews news (http://www.niusnews.com/), etc. According to Taiwan

Internet Usage Report [12], each of the most frequently used period of Internet in weekdays and weekends are "6 pm–12 pm" and "8 pm–9 pm". Moreover, a report indicates that the best-performed size of banner ads is 300 × 250 [5]. Therefore, the research mainly collects banner ads in the size of 300 × 250 from January to February in weekdays from 6 pm to 12 pm. All the samples were downloaded during 1/1/2014–1/2/2014.

3.2 Market Survey Method

The research categorizes the banner ads by KJ method, in view of the literatures mentioned above, the steps of the method are as follows: (1) Confirm the research theme: "A Study on the Design Tendency of Online Cosmetic Advertisement"; (2) Collect the banner ads that fit the above requirements; (3) Select the variables from key vision found in the collected banner ads and fill in the item cards; (4) Categorize and rename the key vision variables; (5) Present the classification results. To be objective, the research invites several experts to make appraisal of classification between step (4) and (5). It is so-called the "Expert Classification Method." The research visit a 40–50 years old makeup expert and a 50–60 years old photographer as the research experts. After a preliminary categorization by KJ method, the variables are discussed and revised. Then, the final classification of the market survey are made.

4 Market Survey Results

Because the researcher spends 2 months collecting samples and most advertisers just arrange advertisements scheduling for 1–2 months, the repeat rate of appearance of cosmetic banner ads on the Internet is high. Moreover, according to the best performing advertisements size from the literature, the size of 300 × 250 seems not to be the most popular type in current banner ads. Through the market survey, we found that the number of 160 × 160 sideline advertisements and 728 × 90 banner advertisements on the websites is quite a lot. In the slim-profit era, all enterprises chase after the most influential advertisements effects in the most economical and profit-maximization way. Consequently, the number of banner cosmetic ads which are put in the same period is not too many. According to the three factors, the research collects one hundred banner ads in every size to be the analytic base.

According to the market survey steps (Table 1), after setting a theme up and finishing collecting samples, the research follows KJ method to find out different variables of key vision, including a character, an object, eyes, face, lips, Western people, Eastern people, a character's exposure level, number or position of objects, position of slogan, content of a slogan. In the light of the above key vision

Table 1 Steps of KJ method

Steps of KJ method	Step 1	Step 2	Step 3
Objective	Find out different variables in key vision	Categorize and rename the variables	Display the classification result
Results	Characters, objects, eyes, face, lips. Eastern character, Western character, exposure proportion of the character, numbers of objects, objects in a row, objects in a column, slogan, slogan position	(1) The different composition in key vision: characters/objects (2) The difference of characters' significant parts: eyes/face/lips (3) The difference of character's skin colors: yellow/black/white (4) The difference of the exposure levels of characters: large/medium/small (5) The difference of the number of objects: one/two to three/four and more than four pieces (6) The different orders of objects: in a column/in a row	The cosmetic banner ads' key vision can divided into two parts: Character as a key vision, including eyes, facial, lips makeups and related care products Object as a key vision, including eyes, facial, lips makeups and related care products (1) The classification of character are the characters' level of makeups, skin colors and level of exposure (2) The classification of object are the objects' number, orders and levels of exposure

variables, because advertisements slogan should be distinguished from graphics to discuss, the research first delete two slogan-related variables before entering into the next step. After generalizing and categorizing the remaining variables, we gain the following classifications. The formation of key vision in ads' graphics are covered by the different composition of key vision—character/object, the difference of characters' significant parts—eyes/face/lips, the difference of characters' skin colors—Yellow/black/white, the difference of the exposure levels of characters—far/near, the difference of the number of objects- one/two to three/four and more than four pieces, the difference of order of objects—horizontal and vertical. When advertisements' key vision is a character, the makeup expert suggest that the research should directly divide horizontal axis into "a character" and "an object" two variables, and divide vertical axis into the sales of goods, including lips, eyes, facial makeups and related care products. Additionally, they advised that the difference of characters' significant body parts be expressed by three levels of their makeups, including heavy, medium and light, and also divide skin colors into yellow, white and black. After asking some professional photographers, they suggest that either a character or an object's key vision be all rigorously divided into the large, medium and small exposure levels. From the market survey and suggestions of specialists, cosmetic banner ads can mainly be divided into four categories, including eyes, lips, facial makeups and related care products, and key vision in banner ads mainly focus on "characters" and "objects".

According to KJ method and experts' interviews, the research concludes that in cosmetic banner ads, three variables in character-oriented key vision are makeup degrees, skin colors and levels of exposure; three variables in object-oriented key vision are numbers, orders and the levels of exposure. Furthermore, in character-oriented cosmetic banner ads, we can divide makeup degrees into heavy, medium and light, divide skin colors into yellow, white and black, and divide levels of exposure into large, medium and small by definition. The $3 \times 3 \times 3$ different variables can make total 27 crossed classifications. On the other hand, in character-oriented cosmetic banner ads, we can divide number into three levels, including one, two to three, four and more than four pieces, divide orders into a column, a row, mixed in a column and a row and irregular, and divide levels of exposure into large, medium and small by definition. The $3 \times 4 \times 3$ different variables can make total 36 crossed classifications. Above are the most popular current combination of design tendency.

5 Conclusions and Suggestions

On the basis of the market survey, the research concludes three results: (1) Cosmetic banner ads can be mainly categorized into four product types: lips makeup (7 %), eyes makeup (6 %), facial makeup (30 %), skincare (49 %), and the fifth category discovered after market survey: mixed products (8 %). And most of the banner designs can be categorized into two types of key vision: characters (40 %) and objects (60 %); (2) three variables in character-oriented key vision can be divided into three variables and three different levels: heavy (11 %), normal (17 %) and light (12 %) makeups; character with yellow (27 %), white (12 %) and black (0 %) skin colors; the exposure levels of characters: large (27 %), medium (9 %) and small (4 %). (3) three variables in object-oriented key vision can be divided into three variables and different levels: one (25 %), two to three (11 %) and four and more than for pieces (14 %); orders of the objects are in a column (47 %), in a row (0 %), mixed in a column and a row (3 %) or irregular (10 %); the exposure levels of objects are large (57 %), medium (3 %) and small (0 %). Table 2 points out the design tendency analysis of banner ads: Product types are

Table 2 Design tendency analysis of banner ads

Key vision	Code	Lips makeups	Eyes makeups	Facial makeups	Skincare products	Mixed products	Number	Percent	Total
A	Aax0	–	1	1	–	1	3	0.03	40
	Aax1	–	2	1	–	–	3	0.03	
	Aax2	–	–	–	–	–	–	–	
	Aay0	1	–	–	–	–	1	0.01	
	Aay1	1	–	–	–	1	2	0.02	
	Aay2	–	–	–	–	–	–	–	
	Aaz0	–	–	1	–	–	1	0.01	
	Aaz1	–	1	–	–	–	1	0.01	
	Aaz2	–	–	–	–	–	–	–	
	Subtotal						11	0.11	
	Abx0	3	1	1	–	–	5	0.05	
	Abx1	1	–	2	4	–	7	0.07	
	Abx2	–	–	–	–	–	–	–	
	Aby0	–	–	2	1	–	3	0.03	
	Aby1	–	–	–	–	1	1	0.01	
	Aby2	–	–	–	–	–	–	–	
	Abz0	–	–	–	–	–	–	–	
	Abz1	–	1	–	–	–	1	0.01	
	Abz2	–	–	–	–	–	–	–	
	Subtotal						17	0.17	
	Acx0	–	–	–	–	–	–	–	
	Acx1	–	–	3	6	–	9	0.09	
	Acx2	–	–	–	–	–	–	–	
	Acy0	–	–	–	–	–	–	–	
	Acy1	–	–	–	2	–	2	0.02	
	Acy2	–	–	–	–	–	–	–	
	Acz0	–	–	–	–	–	–	–	
	Acz1	–	–	–	1	–	1	1	
	Acz2	–	–	–	–	–	–	–	
	Subtotal						12	0.12	

(continued)

Table 2 (continued)

Key vision	Code	Lips makeups	Eyes makeups	Facial makeups	Skincare products	Mixed products	Number	Percent	Total
B	Bax0	–	–	6	18	–	24	0.24	60
	Bax1	–	–	2	8	–	10	0.1	
	Bax2	–	–	3	3	5	11	0.11	
	Bay0	–	–	1	–	–	1	0.1	
	Bay1	–	–	–	1	–	1	0.1	
	Bay2	–	–	–	–	–	–	–	
	Baz0	–	–	–	–	–	–	–	
	Baz1	–	–	–	–	–	–	–	
	Baz2	–	–	–	–	–	–	–	
	Subtotal						47	0.47	
	Bbx0	–	–	–	–	–	–	–	
	Bbx1	–	–	–	–	–	–	–	
	Bbx2	–	–	–	–	–	–	–	
	Bby0	–	–	–	–	–	–	–	
	Bby1	–	–	–	–	–	–	–	
	Bby2	–	–	–	–	–	–	–	
	Bbz0	–	–	–	–	–	–	–	
	Bbz1	–	–	–	–	–	–	–	
	Bbz2	–	–	–	–	–	–	–	
	Subtotal						0	0	
	Bcx0	–	–	–	–	–	–	–	
	Bcx1	–	–	–	–	–	–	–	
	Bcx2	1	–	1	–	–	2	0.2	
	Bcy0	–	–	–	–	–	–	–	
	Bcy1	–	–	–	–	–	–	–	
	Bcy2	–	–	–	1	–	1	0.1	
	Bcz0	–	–	–	–	–	–	–	
	Bcz1	–	–	–	–	–	–	–	
	Bcz2	–	–	–	–	–	–	–	
	Subtotal						3	0.03	
	Bdx0	–	–	3	2	–	5	0.5	
	Bdx1	–	–	3	2	–	5	0.5	
	Bdx2	–	–	–	–	–	–	–	
	Bdy0	–	–	–	–	–	–	–	
	Bdy1	–	–	–	–	–			
	Bdy2	–	–	–	–	–	–	–	
	Bdz0	–	–	–	–	–	–	–	
	Bdz1	–	–	–	–	–	–	–	
	Bdz2	–	–	–	–	–	–	–	
	Subtotal						10	0.1	
Total		7	6	30	49	8	100	–	100
Percent (%)		7	6	3	49	8	–	1	100

lips makeup, eyes makeup, facial makeup, skincare and mixed products. Coding are as follows: (A) Character-oriented key vision: (a) heavy makeup (b) normal makeup (c) light makeup; (x) large exposure (y) medium exposure (z) small exposure; (0) white skin (1) yellow skin (2) black skin; (B) Object-oriented key vision: (a) in a column (b) in a row (c) mixed in a column and a row (d) irregular; (x) large exposure (y) medium exposure (z) small exposure; (0) one object (1) two to three objects (2) four and more than four pieces.

In conclusion, the research finds that the design tendency on banner ads nowadays generally follows the variables classification mentioned above but differs in various levels in the light of the product's target audience and market purpose. Furthermore, facial-related products such as facial makeups and skin care products together account for 52 % of all types of cosmetics ads. As Table 2 shown, we can easily saw that the ads design tendency of facial-related products are mostly made up by Baxo: Object-oriented key vision, in a column, large exposure, one object. Besides, it is worth noting that lips and eyes makeups ads are all presented by the key vision of character-oriented, which would be more easier to demonstrate the makeups effects.

The research is just at the beginning age of online cosmetic advertisement's design tendency, which haven't found out a general design rule to follow but some conclusions from the survey. Thus, the following researchers can make permutation and combination on those classified variables, and design samples to conduct some experiments in order to find out the clicks and eye tracking effects on different graphic designs. It is expected that through setting up a general design rule of cosmetic banner ads' key vision, cosmetic and advertising industries can take research results as reference and even for application in the teaching of web page design.

References

1. Bayles, M.E., Chaparro, B.: Recall and recognition of static vs. animated banner advertisements. SAGE **45**(15), 1201–1204 (2001)
2. Euromonitor: Latest Research: Beauty and Characteral Care 2012 Now Live. Retrieved from http://www.thebeautycompany.co/downloads/Beyer_BeautyNumbers.pdf (2012)
3. Bayles, M.E.: Designing online banner advertisements: should we animate? In: Proceedings of SIGCHI International Conference on Human Factors in Computing Systems, 2002
4. Phelan, Haley: Beauty's Top 100. Retrieved from http://www.thebeautycompany.co/downloads/Beyer_BeautyNumbers.pdf (2012)
5. Google Think Insights: What's Trending in Display for Publishers? Retrieved from http://www.google.com/think/research-studies/whats-trending-in-display-for-publishers.html (2012)
6. Cho, C.-H.: Factors influencing clicking of banner ads on the WWW. J. CyberPsychology Behav. **6**(2), 201–215 (2003)
7. Roth, S.F., Kolojejchick, J., Mattis, J., Golstein, J.: Interactive graphic design using automatic presentation knowledge. In: Proceedings of SIGCHI International Conference on Human Factors, pp. 112–117 (1994)
8. Frascara, J.: Graphic Design: Fine Art or Social Science. The MIT Press, Cambridge (1988)

9. Lothia, R., Donthu, N., Hershberger, E.K.: The impact of content and design elements on banner advertising click-through rates. J. Adv. Res. **43**(4), 410–418 (2003)
10. Scupin, R.: The KJ method: a technique for analyzing data derived from Japanese ethnology. J. Hum. Organ. **56**(2), 233–237 (1997)
11. Karsak, E.E., Sozer, S., Alptekin, S.E.: Product planning in quality function deployment using a combined analytic network process and goal programming approach. J. Comput. Ind. Eng. **44**(1), 171–190 (2003)
12. Taiwan Network Information Center: Taiwan Internet Usage Survey. Retrieved from http://www.twnic.net.tw/NEWS4/130.pdf (2013)

Investigating Shape Comparison Tools for Benchmarking Differences in Product Appearance During Product Styling

Charlie Ranscombe, Philip Kinsella, Paul R. Stoddart and Gavin Melles

Abstract Product appearance is known to significantly influence market success of a product. Guidelines exist to direct designers as to whether a product should have a more similar or different appearance, in order to fit different marketing strategies. These guidelines are subjective and as such can be difficult for designers to apply when evaluating similarity/difference in the context of specific products being designed. This paper reports an exploratory study of the use of shape descriptors (previously used to search/match 3D digital models), to provide objective measures (benchmarks) for product shape in terms of similarity and difference. Following a case study of beer bottle shapes, it was shown that such descriptors can be used to provide measures for typical deviation in shape for certain product categories. Therefore these measures may form benchmarks for styling of new designs and communicating degree of difference in appearance.

Keywords Product styling · Measuring appearance · Shape descriptors · Measures for similarity in appearance

1 Introduction

Product appearance is widely regarded as having significant influence on the success of mass-market consumer products [1–3]. It follows that companies or brands choose the appearance of products ("style") products strategically in order to meet

C. Ranscombe (✉) · G. Melles
Faculty of Health, Arts and Design, Swinburne University of Technology, John Street, Hawthorn, VIC, Australia
e-mail: cranscombe@swin.edu.au

P. Kinsella · P.R. Stoddart
Faculty of Science, Engineering and Technology, Swinburne University of Technology, John Street, Hawthorn, VIC, Australia

A. Chakrabarti (ed.), *ICoRD'15 – Research into Design Across Boundaries Volume 1*, Smart Innovation, Systems and Technologies 34, DOI 10.1007/978-81-322-2232-3_16

marketing and product development goals. This activity is referred to as strategic styling [4]. Despite the increasing emphasis placed on the appearance/aesthetic properties of products, there are few approaches to aid designers in assessing and subsequently rationalising the most ideal, novel or brand appropriate appearance of products. Typically designers conduct qualitative reviews of existing and competing products that form inspiration/references used as benchmarks in the creation of new designs. However when proposing new designs, designers must rely on previous experience combined with these qualitative benchmarks when justifying the appearance of a product. This can make justification difficult when in the face of more objective attributes such as materials, manufacture and performance costs [5, 6].

The widespread use of modelling software in design and subsequent generation of digital models has attracted research interest in the field of model comparison and recognition. The intention of these studies is to search for models/files using model geometry (shape) as an input to the search, with an ultimate goal of streamlining the design process by making models easy to retrieve [7].

The research in this paper sets out to explore the use of shape comparison approaches to measure and compare shape/form in order to establish a quantitative understanding of shape across existing/competing categories of products. The objective for doing so is in providing a quantitative basis to assist in justifying styling decisions in new products relating to the degree that form/shape differs from other products on the market.

2 Background

This section presents a brief review of shape comparison tools for shape matching and shape searching, and previous research into supporting designers' styling decisions.

2.1 Shape Searching and Shape Matching Tools

Currently, images and 3D models are searched through the use of keywords and online catalogues. The sourced media are therefore dependent on the detail and appropriateness of the keywords, as well as the level of indexing utilized. Shape descriptors (SDs) aim to bridge this gap by extracting important visual features from the shape and representing them in a graph-like structure [7] that allows for numerical comparison and subsequent identification of the most similar shapes. In reviewing different SDs, primary concerns have related to their accuracy in distinguishing subtle differences, and being able to maintain accuracy in the presence of noise, translation/scaling of models, and model degradation [7–9]. Due to the focus on search and retrieval for the most similar shapes, little attention has been paid to the use of SDs in calculating average shapes or comparing deviation.

The focus in this study is on SDs defined on 2D contours, as this is the primary platform for testing SDs in the 2D domain. In particular centroid distance function (CDF), D1 and D2 shape descriptors have been utilized, for reasons of simplicity of computation and ease of comparison [8]. These analyses have different strengths in relating the descriptors back to shape, and thus additional study is needed to identify the most appropriate approach for use in a styling context.

2.2 Support for Strategic Styling

Person et al. [4] conducted research into product appearance with respect to the market context (referred to as dimensions) in which a new product will be launched. These contextual dimensions are defined as; current portfolio, succession of products and competing products. For each of these dimensions Person specifies opposing strategies of styling products to be more similar or more different in appearance. For the competing products dimension, of most interest here, greater similarity to other products is advised where a company may want to align its product to be recognised in the same context as a dominant and successful competitor. Greater differentiation is advised to ensure that a product is suitably unique compared with its competitors.

Person's framework makes a substantial contribution to product styling. This is firstly in formalising the dimensions that describe the market context of a product and secondly in supplying clear guidelines for styling products to be similar or different, in order to align them with a predefined market strategy. However, the nature of these guidelines is generic. In other words, guidelines can specify "more similar" or "more different", but do not go as far as specifying the degree of similarity/difference for a specific product context. As such the implementation of these guidelines (knowledge for designers as to what constitutes similar or different for a given design case) remains subjective to the designer. Thus there is still difficulty in justifying the way that a new design adheres to the strategy.

Ranscombe et al. [10] explored the issue of establishing appropriate degree of similarity in such styling judgements. In [10] a set of analyses to measure and compare appearance were presented, based on measuring and comparing the relative orientation and proportion of features, and the shape/contour of features with a form of CDF analysis. The analyses were applied in case studies of smartphones and vehicle fascias (front views), with results successfully providing quantitative data for typical/average features and the standard deviation from the average providing a measure for similarity and difference. While these analyses provide a basis for an approach to quantitatively assess degree of similarity (deviation from a mean shape/form), there is a need for further study to better understand the use of CDF in wider competing product scenarios.

Related to these approaches are a design tool termed "shape grammars", These are described as a set of shape transformation rules that have been used to generate concepts/designs in architecture, cars and motorcycles to name a few [5, 11, 12].

The significant difference between these approaches and those focused on in this study is in their application/integration to the design process. Shape grammars are a generative design tool where the concentration here is on tools that analyse designs rather than create them.

3 Research Aim

As discussed the objective behind this study is to provide designers with a tool to give objective basis to rationalise styling decisions during the design of new products. In particular to provide a way to justify adherence to the guidelines and the framework set out by Person et al. [4]. It follows that the aim of the study reported is to explore the use of D1 and D2 shape descriptors, in addition to the CDF shape analysis, for objectively benchmarking an average shape and the deviation from a mean shape seen in different product categories.

4 Method

The method adopted for this study is now discussed in terms of the analyses used, their application and the case study used to explore them.

4.1 Analyses Used and Procedure to Apply Them

D1 analysis is based on finding the centroid of the contour/shape and then calculating the radial length from the centroid to randomized points on the shape outline/contour. In this context, the centroid refers to the centre-of-mass of the 2D shape. A sample of 100,000 points on the contour is used to give an accurate impression of the shape. The distance between each sampled point and the centroid is then calculated and visualized using a histogram (1,000 bins).

The D2 analysis is based on calculating the distance between two randomly sampled points on the shape's contour. As with D1 analysis this is repeated on a sample size suitably large that it is deemed representative of the shape. For the purpose of this study 100,000 point comparisons were used. As with D1 analysis the data was visualized in histogram form (1,000 bins), plotting the distribution of distance between any two pairs of points.

As with D1 analysis, the CDF analysis is based around calculating the radial length from the shape centroid to the shape's contour. The major difference between this approach and D1 analysis is that a specific number of points is evenly distributed along the contour perimeter. The radial length from points to the centroid is recorded sequentially while moving from point to point along the contour. This is opposed to points on the perimeter being sampled at random as done in D1 and D2 analyses. Values for radial length are then recorded and plotted for each of the points on the contour. A total of 10,000 points was selected for this study.

In order to apply the SD analyses to the input shapes, the programming language Python was used with the OpenCV image-processing library. OpenCV provides several built-in functions for image processing, which allowed for simple calculation of overall dimensions, area, centroid and definition of shape/outline/contour. Tracing of the outline of the shape to the nearest pixel positions (x_i, y_j) allows for CDF, D1, and D2 to be calculated. Data for D1, D2, and CDF were then recorded in the form of an excel spreadsheet (.xlsx).

4.2 Case Study: Beer Bottle Shape

The case study to explore the use of the different analyses defined above was based on a selection of beer bottles. Beer bottles form the basis of this study for a number of reasons. Firstly these represent a mature mass-market product and as such a prime example of a product where appearance influences consumers purchasing decisions. It follows that beer bottles are also an example of a product where designers typically find it difficult to justify styling decisions in the face of increased costs. Furthermore, observations of the market show that for a number of brands bottle shape holds a significant amount of brand symbolism as evidenced by the consistent aesthetic changes that are made to the shape of bottles (as well as labels). Finally there is a rationale in using beer bottles relating to their geometric complexity. Using products with common features/architecture and subtle changes in shape allows assessment of the degree of similarity that can be identified by these tools. This is crucial as it is the more minor/subtle changes in form that can be more difficult to justify.

A sample of 28 bottles was selected from an Australian national alcohol retailer. This sample included brands that can be classed as Domestic (14), Imported (9) and Craft (5). Domestic beers are defined as those produced by established Australian brands. Import beers were defined based on the origin of the brand. Craft beers are categorized as being made in smaller volumes by local brands.

The shape of the beer bottles is defined from front elevation photographs. Photographs were scaled to reflect true size of the bottle. The outline of each bottle was traced by zooming the bottle photograph to over 300× and constructing the outline using a chain of Bezier curves. Some automatic tracing approaches were trialled; however, the bottle labelling along with reflections in the glass surface caused difficulties in accurately defining the true edges. Following the tracing, the bottle shapes were represented as white on a black background in an image 1,000 pixels by 2,700 pixels at resolution of 100 pixels/cm. The bottle shapes used are summarized in Fig. 1.

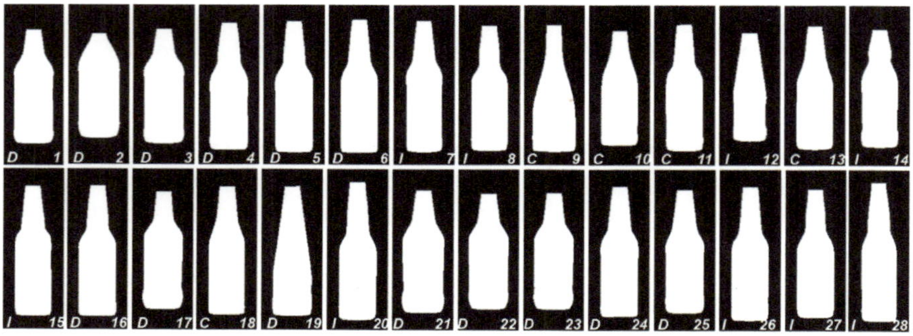

Fig. 1 Bottle shapes used in case study. The bottles are numbered 1–28, letters refer to the category of beer: *D* domestic, *I* import and *C* craft

5 Results

Results from applying the D1, D2 and CDF analyses are summarized in Figs. 2, 3 and 4, which show plots for the average shape descriptor and the associated standard deviation.

5.1 D1 Analysis

Referring to Fig. 2, it was deduced that the 1st peak represents the bottle width, the 2nd, the base and the third the height. Hence, the position of the second peak provides a description of the relative proportion of bottle shape. As the histograms are normalized D1 analyses are studied in terms of the relative proportions of the bottle's features (e.g. body, neck etc.). Referring to plots for standard deviation (Fig. 2, right) provides an indication of the areas in the plot that are consistent

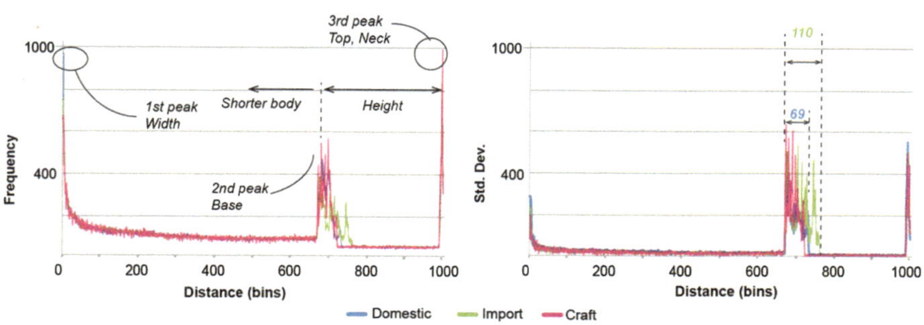

Fig. 2 Average histogram plots for each product category from normalised D1 analysis (*left*) and corresponding standard deviations (*right*)

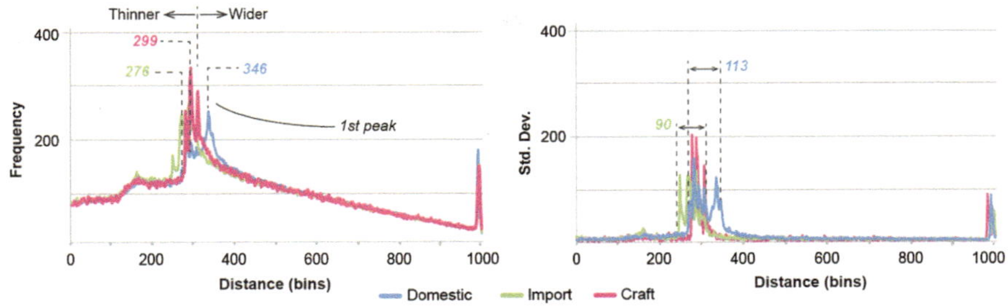

Fig. 3 Average plots for product categories from normalised D2 analysis (*left*) and corresponding plot for standard deviation (*right*)

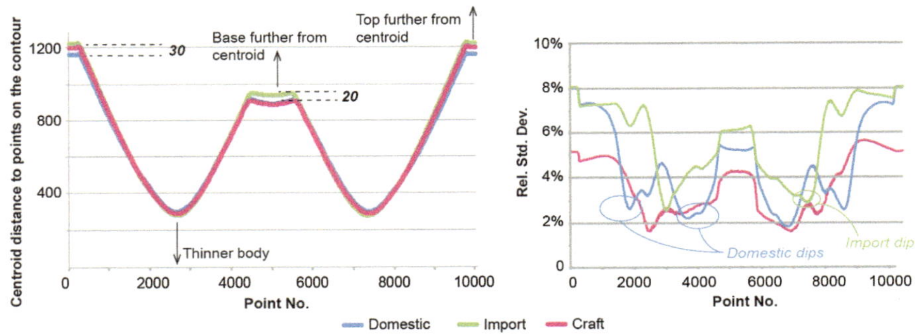

Fig. 4 Average plots for product categories from CDF analysis (*left*) and corresponding plot for relative standard deviation (*right*)

across different product categories and where major differences in shape occur. Thus it can be said that the range of significant deviation around the 2nd peak reflects the amount of variation in the relative proportion of body and the neck for a given product category.

For Import beers, deviation around the 2nd peak extends beyond that of the Domestic and Craft categories (Import deviates over a distance of 110 bins, versus 69 for Domestic and 70 for Craft). This indicates that the import bottle shape shows greater degree of difference in the relative body and neck proportions compared with the domestic and craft beer categories. More importantly, these values provide a limit past which the neck/shoulder proportion would break with what is currently on the market for a given product category.

5.2 D2 Analysis

As with D1 analysis, the D2 results give a typical shape plot for bottles. In this case, the 1st peak is indicative of the relative width of the bottle and the second peak reflects the length of the bottle. Taken together, these two peaks convey information about the aspect ratios of the bottles. Average plots indicate an overall skewing of the 1st peak, showing that domestic beers are typically wider and, since plots are normalized, also shorter than import beers. Craft beer shape lies between these two extremes. The 1st peak for domestic beer occurs at 343, versus 276 for import beers and 299 for craft beers.

Referring to Fig. 3, right, the extent/range of significant deviation around the 1st peak ranges from 110 for domestic beers, 90 for import beers and 70 for craft beers. This shows that as well as the average of typical aspect ratios being different for each category, the degree to which products on the market deviate from this also changes. Thus these values provide a measure of how far within, or outside this range a future design deviates.

5.3 CDF Analysis

The first observation of note in reviewing the plots for CDF analysis (Fig. 4) is that the variation in the plots is represented through the magnitude of peaks rather than shifts/skewing of the waveform. The absence of skewing results in an average that is more representative of the sample than was the case with D1 and D2 analyses.

Reviewing the plots of relative standard deviation for each product category (Fig. 4, right), it can be seen that the import beers have a clear dip in deviation reaching a minimum around points 2,750 and 7,100 on the contour. Conversely the domestic beers show a lower degree of deviation over a greater proportion of the contour (between points 1,700–3,800 and 6,000–8,100). Craft beers show a pattern between these extremes with overall lower deviation and a dip in deviation similar to that of the import beers. In terms of bottle shape this suggests import beer bottles vary more greatly in shape but have a specific point on the contour that is relatively consistent across this category. Domestic beers show a lower variation but over a greater range of points on the shape contour. As with the D1 and

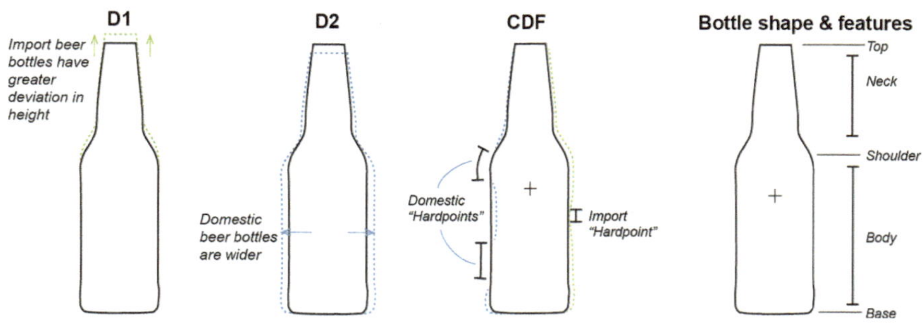

Fig. 5 Findings from applying D1, D2 and CDF shape descriptors to analysing bottle shape and a typical bottle defining the features discussed (*right*)

D2 analysis, these results give specific values for deviation in product categories, which can provide a benchmark for typical deviation. Furthermore CDF provides an indication of the location on the bottle where these sections of high deviation and low deviation (hardpoints) occur. Results from all three analyses are summarised and related back to bottle shape in Fig. 5.

6 Discussion

The results show the use of the different analyses in defining an average plot for shape and subsequently investigating the deviation across product categories to establish the current degree of difference/similarity in shape. The strengths and weaknesses of the different analyses are now discussed with respect to their utility in benchmarking shape.

6.1 Strengths and Weaknesses of Analyses in Product Benchmarking

From a review of plots it was seen that the various analyses highlight different attributes of shape. D1 analysis can be used to establish metrics for similarity/difference in the relative position of the side, base, and neck of the bottle. Similarly, D2 analysis provides metrics for the size of the neck versus the size of the body as well as giving an indication of the aspect ratio. One of the difficulties experienced when applying these analyses was in relating the resulting plots back to shape. It was only possible to gain an understanding of how a typical plot relates to shape by comparing plots from bottle shapes that are the most different. This does not prevent the calculation of averages and deviation against which to compare new designs (as demonstrated in the results). It does however reduce the ease with which this approach can be used to review substantially different shapes.

Data from the CDF plots gives more specific metrics in terms of differences in bottle height and specific points on the contour where shape differs the most within a given product category. Thus, conversely to D1 and D2, CDF analysis has strengths in that plots can be directly related to shape, as each point is plotted in sequence. In other words, the points and values for distance have not been randomized and binned to create a histogram.

While this is a strength in the study presented, it is acknowledged that the approach is only suitable if the contours are suitably similar. For example the addition of a feature such as a handle on one bottle would substantially increase the contour perimeter and thus skew the plot, making calculations of average and standard deviation inaccurate.

6.2 Utility of the Approach to Benchmark Degree of Similarity in Appearance

The first point to make is that despite the differing nature of the metric that the analyses contribute, they all can be used to provide objective measures for the degree of difference in shape across a given category. Relating to the creation of benchmarks for difference in appearance and Person's model [4] the degree of difference/deviation forms the starting point for being able to establish for a given context, exactly how different is different and how similar is similar. In terms of the design process, this has potential use in the form of a tool that designers may use to assess the degree of difference of new designs during the design process. Alternatively benchmark data may be used in the creation of more specific design briefs for styling. This can be in the form of metrics for specific elements of shape that are or are not to be changed, and to what degree.

The results from this study suggest metrics provided by D1 and D2 can be used to give an overall understanding of form in terms of relative proportions of key features. In the context of product benchmarking, this can give some objective measure relating more to overall impression of appearance. CDF analysis is more suited to specific/minor changes in shape as it offers the ability to concentrate on a specific section of a shape. This is demonstrated in the opposing peaks and troughs in deviation of import and domestic beers. These results would also indicate that D1 and D2 analyses are more appropriate in a context where the market appears to be highly diverse in shape, CDF has greater utility in markets with much less diverse forms, as seen in the case study presented here. It follows that D1 and D2 analysis may find use as primary analyses to establish an overall understanding of similarity, whereas CDF analysis would subsequently be applied to further investigate specific elements/aspects of shape that are similar or different.

7 Conclusions and Further Work

From results and discussion it is firstly concluded that CDF, D1 and D2 analyses provide a numerical description for deviation in shape and an average shape. It is also concluded that D1 and D2 can provide more holistic metrics for overall shape, but are less obviously relatable to form. CDF is more relatable to product form, but not necessarily applicable/comparable in highly variable or topologically differing shapes. CDF analysis was shown to be more appropriate in comparing shapes with a higher degree of similarity accounting for more subtle differences in scale and contour.

The major area for further work relating to the analyses is in developing a more generalizable understanding of the plots resulting from D1 and D2 analysis, as they are not directly relatable to form. There is also a need to extend the study further in applying analyses in different product contexts where appearance is more diverse. In particular there is a need to complement this study in looking at products with more complex appearance. Based on the findings in this study, there is an opportunity to explore the use of D1 and D2 as holistic measures for overall appearance (gestalt), and the use of CDF to focus on key features of appearance. As such, the findings from this study present a positive validation that these analyses can successfully establish quantitative measures for difference in appearance. This however only represents a proof of concept and hence the first step in further studies to understand the limits and the context of using this technique to benchmark differences in product appearance.

References

1. Bloch, P.H.: Seeking the ideal form: product design and consumer response. J. Mark. **59**(3), 16–29 (1995)
2. Crilly, N., Moultrie, J., Clarkson, P.J.: Seeing things: consumer response to the visual domain in product design. Des. Stud. **25**(6), 547–577 (2004)
3. Moulson, T., Sproles, G.: Styling strategy. Bus. Horiz. **43**(5), 45–52 (2000)
4. Person, O., et al.: Complementing intuition: insights on styling as a strategic tool. J. Mark. Manage. **23**(9–10), 901–916 (2007)
5. McCormack, J.P., Cagan, J., Vogel, C.M.: Speaking the Buick language: capturing, understanding, and exploring brand identity with shape grammars. Des. Stud. **25**(1), 1–29 (2004)
6. Warell, A.: Towards a theory-based method for evaluation of visual form syntactics. In: Xirouchakis, H. (ed.) TMCE, Tools and Methods for Competitive Engineering. Millpress, Lausanne (2004)
7. Kazmi, I.K., Lihua, Y., Jian Jun, Z.: A survey of 2D and 3D shape descriptors. In: 10th International Conference on Computer Graphics, Imaging and Visualization (CGIV 2013)
8. Osada, R., et al.: Matching 3D models with shape distributions. In: International Conference on Shape Modeling and Applications (SMI 2001)
9. Chen, D.-Y., et al.: On visual similarity based 3D model retrieval. Comput. Graph. Forum **22**(3), 223–232 (2003)
10. Ranscombe, C., Hicks, B., Mullineux, G.: A method for exploring similarities and visual references to brand in the appearance of mature mass-market products. Des. Stud. **33**(5), 496–520 (2012)
11. Pugliese, M., Cagan, J.: Capturing a rebel: modeling the Harley-Davidson brand through a motorcycle shape grammar. Res. Eng. Design **13**(3), 139–156 (2002)
12. Stiny, G.: Introduction to shape and shape grammars. Environ. Plan. **7**(3), 9 (1980)

Research in Visual Ethnography Focusing on Markets of Kanpur

Siddharth and Satyaki Roy

Abstract Documentation of visual culture makes up a valuable repository of information and is used for anthropological analysis and preservation of cultural heritage. The authors did the research following empirical methods, to understand human behaviour, patterns of exchange and communication, kinship and social structure, gender relations, socialisation, religion and local experience. Research includes absorbing visual elements using photography as a tool. This documentation and presentation project focussed on few major areas. The experience of being in semi urban markets from the perspective of visitors, the seller consumer interaction, the multitude of things on sale, and also on the stories of sellers themselves who make up the place in the system that it is. Data gathered was critically assessed and presented graphically in a visual form. Using collection of images, illustrations and text stating facts & insights associated with the images, a repository of 'visual essays' is made. This is a source of meaningful information for students or professionals researching about local semi urban markets, or to travellers coming to the city as for this research we have taken into consideration the local market places of Kanpur.

Keywords Qualitative research · Visual ethnography · Culture · Tradition

1 Introduction

1.1 Understanding the Idea

The importance and relevance of a project like this, lies in the much bigger field of Social Anthropology. According to Nilsson [1], Visual ethnography can stand alone or make up one piece of a larger study in social anthropology. He says, it is

Siddharth (✉) · S. Roy
Design Programme, IIT Kanpur, Kanpur, India
e-mail: siddhu@iitk.ac.in; sid065@gmail.com

A. Chakrabarti (ed.), *ICoRD'15 – Research into Design Across Boundaries Volume 1*,
Smart Innovation, Systems and Technologies 34, DOI 10.1007/978-81-322-2232-3_17

an excellent method used by international brands and research agencies. Whether the end result is to make sports equipment or fund social change, a well-executed visual ethnography project can help you make culturally relevant and informed decisions. Visual documentation or archival of visual culture is an indispensable part of a human-centered design research. A visual ethnographer has to move to both remote and populous areas according to the demands of the project to document the world we live in. With that background the researcher had to travel to different market areas to collect visual and other information relevant to the research. The focus was on unorganised sector, local markets which have their own unique brand value because of their equation with the buyers, the major elements of which are identity, basic approachability and trust.

On a more basic level, researching and archiving such important and unique elements of one's culture gives the chance to become better acquainted with Indian life in different contexts and scenarios. Every city in India has many markets and shopping areas, weekly or permanent, in fact they are fairly commonplace, serving thousands of people every day, and it was learned that therein lies the beauty and the deep interest for 'us'. Us, here range from, design students and professionals to psychologists, travellers, or laymen wanting to learn more about the chaotic order of Indian Society. Those sources encountered during the research, from discussions with people to learning the behind-the-scenes working of, for example, a small tea shop can challenge one's basic understanding of Indian local business in cultural and general point of view.

The workplace of the sellers and small businessmen in such street markets is so chaotic and unorganised that knowing and sharing some of their experiences becomes essential. The relevant insights gained during interviews and field visits by the author helps us in presenting a more comprehensive picture of the place and of the business transactions form the point of view of the sellers. Knowing more about the lives of the sellers itself was critical as besides being the imperative shareholder of the transactions they had multiple anecdotes to share being from different socio-cultural and economic origins and have extensive involvement with customers.

For analysis, using collection of images, illustrations and text stating facts and insights associated with the images, an attempt has been made to give it a whole meaning, to make a visual essay, to have a series of snapshots. The goal will be to have a good repository of information and insights, a documentation of happenstances, to piece together undercurrents under Indian societal chaos, using such bits and pieces for the sake of understanding human interactions on a new level.

It should be mentioned that when started the end product of this research was unclear. After collection of data from interviews and photographs a discursive picture was developed which later on took the form of a coffee table book showcasing semi urban market scenarios. While developing the representation of stories, illustrations and text was used along with the photographs to present author's observation about the place or the person as photograph alone could not do justice to the acumen of the story being told and the experience being shared which shall be clearer from the examples discussed below in part 3.

2 Design Process and Methodology

2.1 Process of Designing

The analysis achieved through research says that there is a need and comfortable space for exploration for ethnographic visual studies in Indian scenarios. Also a degree of freedom and new avenues for experimentation in the process of developing more and innovative visual montages in the field of visual ethnography is there. There are different parameters which play important roles in defining the identity and relevance of the creating such cultural archives like the captured image, the way of description and the visual style. For the authors, all the three parameters were subjectively defined on the basis of interactions and feedbacks received during the field visits.

2.1.1 Choosing an Area of Study

The markets of historical city of Kanpur which is now a bursting urban centre was chosen based on the following parameters:

Significance

Kanpur, which is our area of focus, previously known as Cawnpore, is the largest industrial city of Uttar Pradesh, believed by many to be the Economic and Industrial Capital of UP. It is also known as leather city as it contains one of the largest and finest tanneries in India and in South Asia. Widely known as "Manchester of the East", it is one of the oldest industrial townships of North India [2]. It has an area of over 450 km^2 with an approximate population of 3 million inhabitants in its area. It is the 75th largest city in the world [3].

Markets and Their Relevance

The success of such a project is very much defined by the significance of the information and stories being told through the project. The very definition of a market is any structure that allows buyers and sellers to exchange any type of goods, services and information. Physical retail markets, whether be local farmers' markets, shopping centres, market restaurants or shopping malls, to flourish and to function properly, will involve interaction and communication on many levels. The markets covered in Kanpur were Kalyanpur, Shivalaya, Hatiya and Khoya market. As the daily transaction in these places reach a huge number, both in terms of people visiting and in terms of daily monetary transactions. Also these places are a source of livelihood to thousands of people, if not more. The above mentioned market areas are much in 'vogue' locally and the happenstances are common phenomena and hence the experience seems like an intimate element of our shared culture.

Adaptability in the Chosen Medium

The medium of still photography been used in such a manner so that the inherent characteristics of photographs as tool of ethnographic research are enhanced. Photography is a time consuming but is a job which pays many dividends in research and aesthetic terms.

Markets in general and Indian markets in particular are rich in colour and are vibrant in nature because of their inherent quality of to and fro business and social interactions. These complex yet everyday interactions and the experience of being in the locale are the elements which were kept in mind while deciding the final format of the presentation. The visual richness of the such hustling bustling places, the sensory information one is bombarded with as one enters market areas, the social scenes, the business, the hundreds of people who come together every single day to make the place what it is, the old timers, the new visitors, and the collective and individual human element, everything is involved in a physical market area.

2.1.2 Choosing the Stories

Stories are loosely based on the following parameters.

Sense of completion or self sustaining: Since we are showing one instance to the viewer in each photograph, it becomes necessary that the snapshot captured when shown using chosen visual language should have a sense of completion to it.
Colour and vibrancy: The colour scheme of the visual style should be pleasing and eye-catching to the audience, following the principles of appropriate visual communication and hence the snapshot, the image should be visually vibrant and 'alive'.
Subjects: Such idea of conserving local visual culture obviously will involve a formidable human element which will require breaking down of information received during the field visit and linking it to the overall scenario. So, the chosen subjects and their stories should be able to provide enough capability for this be facilitated.
Picture as a whole: As we are dealing with static images in which the individual elements hold major importance, especially the element under discussion, it is also essential that the picture is complete as a whole.
Narrative potential: As the visual style includes textual information as well, its mode and content should match with the theme of the image.

2.1.3 Use of Illustrations as a Tool

The use of illustrations in the image itself added a new dimension to the whole picture. This was very helpful as the amount of information being rendered increased and the aesthetic quality also improved in terms of visual balance and focal points. It helped in further designing the whole page and explaining the intended idea to the viewer. An important point to note here about illustrations is that, in this particular project they play a number of different roles when accompanied by images and text. In some cases illustrations complete the story while in other cases they are used as interesting add-ons or little snippets of further information. Sometimes they tend to complete the picture as a story in totality while other times they are used just to enhance some idea already inherent in the picture. The idea here is that the illustrations have multiple advantages if used judiciously and that has been the authors' intention.

2.2 Developing the Idea

Once the above mentioned steps are carefully done with, we were able to work at the final look and feel of the visual ethnographic exercise which culminated as final presentation of the research as archives. There were different types of visual styles and techniques. While working at this area, and while explaining the experience of a place or a situation using fundamentals of visual and information communication, a unique style of depiction developed, owing its origin to the place or story under observation that made it all the more involving to the authors and in a general sense to the viewer as well.

2.2.1 Design

The font had to stand out and be readable and at the same time be non obtrusive, if used judiciously. It had to be informal, earthy and sketchy as with the mood of the vibrant local market areas keeping in mind the consistency with the subject material being said. After much deliberations, 'Cartoon relief' font was used which was freely available from internet, as that seemed to fit the bill most perfectly.

Illustrations They too had to undergo stages of development as the project progressed. These could, if used in excess spoil the overall result and if used in just the right amount work fascinatingly in bringing out the intended idea and as mentioned in the earlier chapter, making the unobservable visible.

Overall Look The final overall look of the whole image, which encompassed the picture in totality including title, illustrations and text if any, needed to impart the right idea in the viewers mind and at the same time giving the user's eye enough room to manoeuvre and not experience any form of visual tension. The amount of information has to be just right or we would run into trouble in the form of the image being too 'wordy' or text heavy and might divert attention from other more important elements of the picture.

3 The Visual Essays

3.1 Kalyanpur Market, Kanpur

Coordinates: 26.519121°N 80.249805°E

3.1.1 The Market in Particular

The area has one of the largest vegetable markets in the city due its proximity to the rural areas and nearby districts of western Uttar Pradesh. It is known as a vegetable market but with time the diversity of things sold here has grown enormously while in terms of cost it still caters well to demands of the original demography, which is lower to middle class.

3.1.2 The People

The hundreds of people who visit the place on a daily basis constitute a big percentage of the total population of Kanpur city. Mostly the customers come from nearby local areas to get their vegetables at the best rates possible. A large number of people also visit the place to buy meat, fish, other kitchen groceries, flour, rice etc. and also for cheap plastic things such as buckets, mugs, shaving kits, toys and numerous other knickknacks. The sellers are mostly small vendors who struggle daily to earn their livelihood. The general scene will always have rows of people with their things to sell on a push cart or on the ground itself using pieces of rag or plastic as floor, be it vegetables, fruits or little thingamajigs. The amazing thing is even though the number of sellers are huge as many lives depend on the business conducted here daily, the business doesn't seem to slow down at all. No matter the advent of shopping malls, which spring up frequently, or the blistering heat during summers or the bone chilling cold during winters in this open air market space.

The customers mostly belong to lower to lower middle class and people living only in the vicinity as the hygienic and aesthetic value of the place leaves a lot to be desired. Coming full circle though, many hotels and local restaurants also get their groceries wholesale from this place, which then do cater to all demography irrespective of class, making this place one very important vein for the life blood of the city.

3.1.3 The Pictures

Frequent visits to the place, resulted in a good collection of pictures which act as the backbone of our visual study. The three most important criteria that were kept in mind while trying to capture images were—the people involved, the things being sold, and the Social interaction that is taking place.

A good and thorough collection of pictures was taken which was of much help during the later stages of the design. A few as examples are attached along with. Each selected picture was also given a title which helps each snapshot stand out on its own (Fig. 1).

3.1.4 The Text

The pictures were supplemented with accompanying text which included the local stories and interviews with concerned people, gathered during the field visit.

3.1.5 The Illustrations

The illustrations which were added to each photograph were used earlier solely to enhance the picture's dominant mark. But later they evolved to fulfil a role bigger than that as illustrations gave freedom to enhance the point being made or to talk about something which was not visible at first and can likewise be used as postscript.

In Fig. 2, the picture was taken as the author's first introduction to the helter-skelter of street market of Kalyanpur. Even though the picture captures the sheer number of people on a chilling November morning, the authors felt that something else is needed to convey the experience of an unsuspecting visitor. The illustrations and the text convey that ancillary report.

Fig. 1 Sights of Kalyanpur Market

Fig. 2 First impressions of the Kalyanpur Street Market

Fig. 3 Affordable varied things on sale

Figure 3 shows the distinct things on display on this particular stall. The text added extra information detailing each article and its price. The idea was to detail that how such small businesses are gratifying many insistences. The things on sale were plastic knick knacks, including toys and kitchen utensils to shaving razors and toilet accessories and from comb to school supplies.

In similar ways a extensive visual anthology was developed for Kalyanpur and for three other semi urban market areas of Kanpur, Hatiya Mandi, Khoya Market and Shivalya Market.

3.2 Shivalaya Market

Coordinates: 26°28′24″N 80°21′9″E

Shivalaya is a small locality in the main city center of Kanpur. It is situated at about a distance of 20 kms from IIT Kanpur in the main city. It is a place famous for its quite a few temples and the market that has developed alongside them. Shivalya or Shiwala, as old books name the area, literally means, Home of Shiva.

Situated right next to imposing Z square mall, the biggest shopping mall in Uttar Pradesh, this old, simple, almost street market provides quite a contrast to the modernity and branded glitz of the mall.

The market is spread over not too big of an area, the whole market can be covered in an hour or two by walking. In fact, walking is the best option of experiencing the place in the system as the two parallel lanes which effectively define the market are quite congested, especially during shopping hours or during festivals. After the development of the city center and ironically, after the shopping mall was constructed, this area has become one of the overcrowded areas of Kanpur. The suburb of Shivalaya covers a large area and consists of several houses and temples. The proper Shivalaya area consists of mostly single storey houses situated in a congested manner.

3.2.1 The Market in Particular

The area has one of the most interesting markets in the city due to the kinds of products on sale here. It was known and was famous for only the temples it had and in earlier periods, the adjoined local vendors only sold prayer goods and items related to Hindu festivals. With time the diversity of things sold here has grown while in terms of cost and the myriad of things on sale, it still caters well to demands of the original demography i.e. middle class, which forms the biggest population of the city. Nowadays it is mostly frequented by denizens of the city looking for cheap and not so cheap jewellery and whole gamut of supplies needed during wedding ceremonies, from groom and bride attire to the prayer goods and worship items needed in religious congruence.

The area has quite a few temples for different gods and one mosque. The prominent temples are the ones for Shiva, Hanuman and Ma Chhinnamasta. Ma Chhinnamasta temple has the biggest and most striking temple complex. This particular temple opens only three days a year during the week of Navratri. The immense numbers of people who visit the place on a daily basis makes it one of the most frequently visited areas of the city. The customers, unlike Kalyanpur market, come from all over the city to do their shopping at the best rates possible. A large number of people who visit the place, are interested in getting footwear, clothing, bags, school and college supplies etc. and also for one of the best 'panipuri' in the city. The visitors also include numerous temple visitors.

Figure 4 displays devotees at one of the temples in Shivalaya. With market and advertising being such a integral component of our daily lives, it ironically and perhaps rightly depicts the faithful congregating in front of billboards. The illustration talk about the 'Rupa' brand in the detail.

3.3 Hatiya and Khoya Market

Coordinates: 26.475068°N 80.345979°E

The area is famous for its metal and iron goods. The market areas right next to it, Gayaprasad lane is famous for ladies cosmetics and Khoya market, which is covered in this project, is famous for its readymade sweets. These three areas form a sort of trio of market lanes having particular specialties.

Hatiya is a web of little lanes and by-lanes and the place is characterised by the sound of metal banging against metal and large daily customer traffic. As metal and iron business requires lot of physical labor, the place throngs with labourers of all kinds working as shop helpers and transporting metal goods to and fro. The lanes are too narrow and congested to allow passage of large or medium sized vehicles and hence people are seen either walking or on two wheelers. The market and the locality have been in place since decades now and most of the houses

Fig. 4 People praying in Shivalaya

belong to the times of British rule, though they are in fairly battered condition now. Even the shop owners are in the same business since 3 or 4 generations. As before the area is not too large and walking is the best option of treading the area for research. Although the area gets quite hot during summer days, yet it becomes bearable because of warm smiles of local people who often are forthcoming with glass of cold water for thirsty travellers and strangers to the area. Metal goods that are sold here in Hatiya market are of different kinds, ranging from metal and iron parts agricultural and farm equipments, construction equipments to kitchen utensils of all sizes for shops and for homes.

Khoya market, as the name suggest is known for milk based sweets. Though controversies have arisen in the past about the use of synthetic milk for manufacturing delicacies, this does not seem to have affected the business in a great amount as lots of whole sale supply of milk based goods and sweets take place daily form the locality. Khoya market is also the hotbed for wholesale supply of paper and plastic cups and utensils.

One very interesting thing to note here was the camaraderie between different shop owners and workers. May be it was because the lanes were too narrow and hence the shops were too close to not have possibilities of friendly banter. Everyone seemed to know everyone else and the shop keepers' major pastime was friendly discussions in groups and sharing jokes. The researcher while collecting pictures and interviews, himself was the butt of many jokes and starting point of conversations. As it was all in light and inquisitive spirit, that helped in breaking the ice between the observer and the observed subjects resulting in useful conversations.

Fig. 5 Hatiya Loha Mandi (Iron Market)

3.3.1 The People

The major categories of people visiting the area were, vendors from other areas buying large amount of metal goods at wholesale rate, customers buying metal goods and utensils for their homes and shops, and other vendors selling fruits and other food items. As the area is historically significant, a large number of people also have residence here. As with the case with other such areas having old housing complexes, families have been taking care of business and have been living in the same houses since generations.

Figure 5 shows a diminutive yet expansile shop in Hatiya. The shop was the front of a large depilated building which according to local legend housed the first Reserve Bank to be constructed in India after independence. The father son duo are the ancestral owners of the building and now look after this small business meanwhile still reminiscing about the sheen the place once had.

4 Conclusion

Decentralised, dynamic and social data provided in these kind of visual ethnographic exercises help create content strategy to be used when need arises. Such kinds of works that unobtrusively capture life as being lived provide a wealth of real information and data that can then be used to understand and interpret our shared visual culture unique to the time and place.

Visual ethnographic studies are not a new development, since historical times anthropologists have been recording cultural data for posterity, but times do keep changing, knowledge and information do keep evolving, society does keep on evolving. We should mention, that with information boom and easy accessibility of camera and other equipments to general public, people do make and keep repository of local visual cultural information but that is done in a largely informal context. Besides being essential form of qualitative research, these misapprehensions renders the task of doing formal visual ethnographic studies into local visual culture important if not crucial.

References

1. Nilsson, J.: Visual Ethnography by Jacob Langvad Nilsson. Retrieved 28 June 2013, from http://jacoblangvad.com/visual-ethnography/
2. JPS Associates (P) LTD: Final report: Kanpur city development plan. Retrieved 23 March 2013, from http://jnnurm.nic.in/wp-content/uploads/2010/12/CDP_Kanpur.pdf (2006)
3. City Mayors Foundation: The largest cities in the world and their mayors. Retrieved 19 Dec 2012, from http://www.citymayors.com/statistics/largest-cities-mayors-1.html (2011)

How to Impress Your User: Guideline for Designing the Product Impression

Constantin von Saucken, Andreas Wenzler and Udo Lindemann

Abstract The development of products requires considering multiple—partially contradictory—aspects (e.g. manufacture, usability, costs etc.). "Design for X" (X is replaced by the specific aspect) guidelines meet this challenge; however, there is a deficit of effective "Design for Emotion" approaches. We present a guideline with product parameters and design principles taken from engineering literature and enrich them with product impressions answering: How shall the product appear? Which emotions shall the design cause? We determine the dependencies between these elements and illustrate them in a figurative and usable way. By applying the guideline (engineering and industrial) designers can systematically vary parameters by applying adequate design principles to create a specific impression. This supports the emotional design of products caused by visual appearance.

Keywords Product design · User experience · Design for emotion · Design guideline

1 Introduction

1.1 Motivation

At its core the field of design is about creating a better life for people. Products provide functions in order to fulfill users' needs. Particularly in the field of engineering design, these functions were and still are mainly understood in a technical way: products need to be effective and efficient. The discipline of industrial design rather focuses on the visual appearance of products: being experts in sketching and

C. von Saucken (✉) · A. Wenzler · U. Lindemann
Institute of Product Development, Technische Universität München, Munich, Germany
e-mail: saucken@pe.mw.tum.de

A. Chakrabarti (ed.), *ICoRD'15 – Research into Design Across Boundaries Volume 1*,
Smart Innovation, Systems and Technologies 34, DOI 10.1007/978-81-322-2232-3_18

prototyping, they highly care for the users' needs for beauty, pleasure, meaning and emotions. Nowadays, products from different competitors offer predominantly the same technical functions (e.g. smartphones, cars). For those saturated markets, the visual impression of products and resulting emotions serve as a strong unique selling proposition.

We observe the following challenge in the tension field of engineering and industrial design: engineering designers are taught and used to apply guidelines (e.g. "Design for X" guidelines and the work of Pahl et al. [1]) allowing them to systematically design products with a certain focus. However, there is a lack of the mentioned issues of visual design and emotional product impression. On the other hand, industrial designers consider these topics, but rather unsystematically—sometimes due to the fear of killing creativity. This leads to the difficulty and sometimes inability of industrial designers to explain and defend their concepts and ideas. Furthermore, the results appear to be rather random depending on the designer's skill and mood.

1.2 Approach

Our goal is to combine the strengths of both disciplines: a systematic and thereby comprehensible approach in the form of design principles and corresponding product parameters (from engineering design) considering the rather "soft" issue of visual design and product impression on the user (from industrial design).

We gathered the most important 12 design parameters (e.g. product shape), 18 embodiment design principles (e.g. mimicry effect) in engineering design literature and 9 visual product impressions (e.g. robustness) in the field of industrial design. In the next step, we applied a matrix-based approach to investigate dependencies between these domains: Which product impression can be achieved by which design principle and which of these can cause which impression?

1.3 Result

The result is a design guideline that helps developers deciding for one or more targeted product impressions. The guideline proposes the application of corresponding design principles and finally gives hints which product parameters need to be varied by applying these principles. We paid careful attention that the guideline is easy to understand and intuitively explains its application by using a figurative sheet template with concrete examples without much textual descriptions.

The contribution of our guideline is to give engineering designers an idea which impressions can be achieved and give hints which parameters need to be adjusted by applying principles. At the same time, the guideline serves industrial designers as a tool to systematically adjust parameters by using principles allowing them to better explain how they achieve a targeted impression.

2 Fundamentals

The goal of this work is to support designers within a *product development process* in *designing for emotions* by systematically vary the product's *visual appearance*. We achieve this by presenting possible *product impressions* on the user, corresponding *design principles* to create these impressions and finally *product parameters* that need to be adjusted by the principles. In this section we define the fundamentals for our approach. Each of the highlighted terms is discussed separately in the following:

2.1 Product Development Process

By the term *product* we understand pure mechanical and mechatronic (integration of mechanical, electronic and software components) goods. *Processes* are necessary for the development of products to coordinate teams and individuals like mechanical engineers and industrial designers [2].

Established process models for product development like VDI 2221 [3] or Pahl et al. [1] propose elaborating systematically from abstract to concrete product properties: starting with product *specifications*, via *functions*, *principal solutions* to the concrete *layout*—the embodied product design. Our approach targets the most concrete product level: the embodiment design. It helps designing product parameters specifically, e.g. shape, color or material.

2.2 Design for Emotion

A big challenge in product development is the handling of numerous—partially contradictory—requirements on the product: Products need to be manufacturable, usable, cost-effective etc. at the same time [2]. Therefore, "Design for X" guidelines are established helping designers to refine their product concept regarding a specific aspect; the "X" is replaced by this aspect, e.g. "Design for Assembly".

In engineering design, we observe a deficit in the field of "Design for Emotion". Researchers like Desmet and Hekkert [4], Norman [5] and Jordan [6] offer descriptive models for emotional design. However, we made the experience in industrial projects, that the impact of emotional-driven approaches on engineering design is rather low.

Chakrabarti and Gupta [7] discuss the relations between *object* and *emotional features* as well as *primary emotions*. They show that images of animals, plants and products raise similar emotional reactions of observers—the visual appearance of products causes specific emotions. Nevertheless, their paper does not provide a proceeding for supporting designers in designing for emotions [7]. We want to achieve this with our approach.

2.3 Visual Appearance

"The visual sense is the most reliable of all senses" [8]. Humans perceive between 60 and 90 % of all incoming information visually. The eyes can distinguish finest contrasts in brightness and differences of colors. Furthermore, the product's visual appearance contains more information of its properties than the haptic, auditory, olfactory or gustatory one: material, shape, color, movements and three-dimensionality [8].

In order to create an emotional experience by embodiment design, the visual appearance offers the most opportunities for variation. For this reason, we restrict our supporting design guideline to visual principles and parameters.

2.4 Product Impression

A product fulfills different functions [9]: In addition to its *practical* function (a chair is for sitting, behavioral level [5]) it has an *aesthetic* function (cp. syntax, aesthetic experience [4], visceral level [5], physical pleasure [6]), a *symbolic* function (cp. semantics, experience of meaning [4], reflective level [5], social and ideological pleasure [6]) and an *indication* function (cp. usability, mental model [10]).

The product's visual appearance therefore plays an important role beyond the aesthetic function: It needs to communicate the status of its owner as well as brand values (symbolic function). Furthermore, it indicates the right product use and its properties: Humans make associations regarding a product's quality based on the embodied design (cp. [7]). For this reason, we aim at providing a guideline that helps designers creating product aesthetics and indicated properties by adequate design principles.

2.5 Design Principles

(Design) principles increase the likelihood to run a successful development process by influencing the designer's actions. They support the designer independently from his concrete problem. (Design) principles describe a prescriptive behavior and are often elementary parts of design methods [11]. Established engineering design literature offers a broad range of fundamental and design principles (e.g. [1]), in particular "Design for X" guidelines.

In our approach, design principles are the link between the product impression (*why* to design) and the product parameters (*what* to design): They describe *how* to design—how to change parameters in order to achieve an impression.

2.6 Product Parameters

Product parameters are the concrete properties of products a developer can vary directly (e.g. material, shape) or indirectly (e.g. costs, weight) [2]. They define the product's embodiment design and thereby fulfill functions. With a systematic variation of parameters a designer can change the product properties—goal-oriented in the sense of "Design for X" [2]. Including a collection of product parameters in our approach is essential to support developers as concretely as possible.

3 Method

This section firstly describes the process of collecting elements of the domains: product impressions, design principles and product parameters. Secondly, we apply a method from Structural Complexity Management—the Multiple Domain Matrix (MDM)—to determine dependencies between the three domains.

3.1 Collecting Impressions, Principles and Parameters

The collection of product impressions, design principles and product parameters was derived from literature. We took sources from engineering design [1, 2, 12] where we found product parameters and principles. Furthermore, we investigated publications from industrial design [13–15] focusing principles and impressions.

9 impressions, 18 principles and 12 parameters could be derived from literature, listed in Fig. 2. Furthermore, we defined relations between the three domains in preparation for the following MDM mapping. Figure 1 shows the meta model and dependencies: Design principles *create* product impressions and therefore *change* product parameters.

Fig. 1 Meta model of the domains

		Design Principles																		Product Impressions								
		Distance / Proximity	Similarity / Equality	Motion / Tension	Unit / Unity	Color / Contrast	Figure Ground Relati.	Common Destiny	Shape Structure	Continui. / Solid Curve	Lightness	Mimicry	Surface	Orientation Sensitivity	Conciseness / Simplic.	Proportion	Rhythm	Heaviness / Stability	Symmetry / Balance	Conspicuousness	Legibility	Weight	Complexity / Order	Strength / Dynamics	Robustness	Temperature	Softness / Hardness	Valence / Quality
Product Parameters	Contact Type				X				X													X	X		X			X
	Manufactr. Process		X		X	X			X		X	X	X					X		X	X	X	X	X	X	X	X	X
	Shape		X	X	X		X		X	X	X	X		X	X	X	X	X	X	X	X	X	X	X	X	X	X	X
	Size		X				X				X	X				X		X		X	X	X	X	X	X	X	X	X
	Compactness								X		X							X	X			X	X		X			X
	Connection Type							X	X													X	X	X	X		X	X
	Position / Direction	X	X		X			X	X		X	X		X	X	X	X	X	X	X	X	X	X	X	X	X	X	X
	Material		X			X	X				X	X	X					X		X	X	X	X	X	X	X	X	X
	Surface Treatment		X	X	X	X	X	X		X	X	X	X	X	X			X	X	X	X	X	X	X	X	X	X	X
	Connection Type		X		X				X		X							X		X		X	X		X	X	X	X
	Connect. Structure	X	X						X														X					X
	Number				X			X	X						X	X	X			X	X	X	X	X	X			X
Product Impressions	Conspicuousness					X	X					X		X	X													
	Legibility	X				X						X		X	X	X												
	Weight			X		X			X		X	X						X										
	Complexity / Order	X	X		X	X	X	X	X	X				X	X	X	X		X									
	Strength / Dynamics			X				X		X		X					X											
	Robustness				X	X			X		X		X			X		X	X									
	Temperature					X						X	X															
	Softness / Hardness					X			X			X	X															
	Valence / Quality		X		X				X			X	X	X		X	X		X									

X Set dependency
X Calculated dependency

Fig. 2 Multiple domain matrix (MDM) [16] of impressions, principles and parameters

3.2 Mapping Principles with Impressions and Parameters

The core element of our approach is setting dependencies between all elements of the three domains. This way, we may suggest principles changing parameters to achieve an impression. Setting these dependencies is a challenging task since there are theoretically 1,944 ($9 \times 18 \times 12$) combinations. For this reason, we illustrated the domains and links in a Multiple Domain Matrix (MDM) [16], shown in Fig. 2.

A MDM helps illustrating complex dependencies between elements of different domains. The method is a part of Structural Complexity Management and was developed in the context of product architecture to handle the multiple correlations of components, assemblies, functions and further domains. Corresponding tools provide functions for calculating indirect dependencies and for running analyses.

We set the links ("X") between principles and parameters (Which principle changes which parameter?) and between principles and impressions (Which principle can create which impression?). After that, the indirect dependencies between parameters and impressions were calculated automatically via principles: When an impression is created by a principle that again changes parameters, these two have a correlation (see Fig. 2).

4 Result

The goal of our approach is to support designers in industrial practice. Therefore, we designed a guideline as a handbook that figuratively describes all impressions, principles and parameters and depicts the correlations between these domains. Figures 3, 4 and 5 show extracts of each domain from the guideline. Section 4.2 describes how to apply the handbook; an example in Sect. 4.3 clarifies the approach.

4.1 Guideline for Achieving a Targeted Product Impression

For the design of the guideline we expressed the following requirements in order to be applicable in industrial practice. The guideline must:

- be general and solution-neutral,
- inspire designers in specific problem contexts,
- be intuitively usable and easy to understand,
- be visually attractive and
- give further literature references allowing users to deepen their knowledge.

Fig. 3 Sample for a product impression one-pager

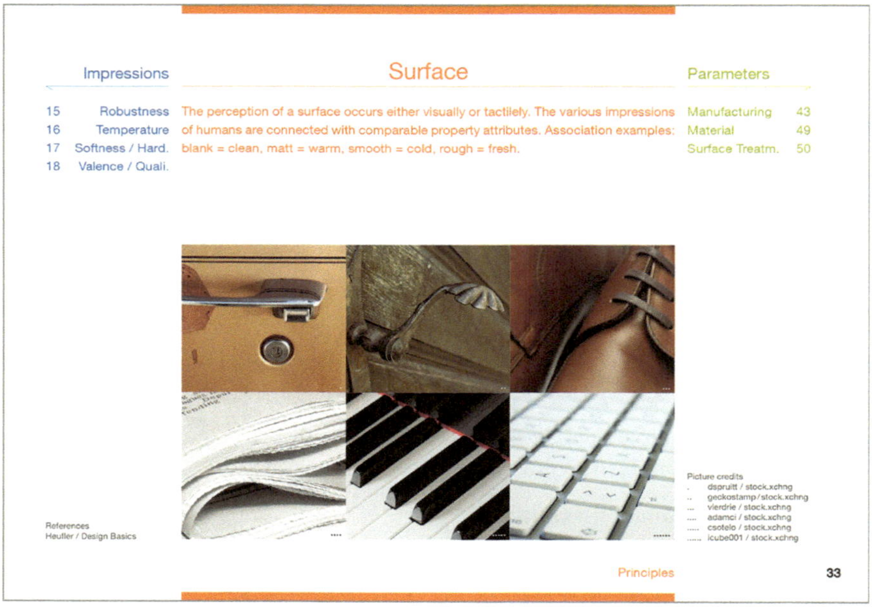

Fig. 4 Sample for a design principle one-pager

Fig. 5 Sample for a product parameter one-pager

The result is a handbook with one-pagers for each impression, principle and parameter as shown in Figs. 3, 4 and 5. The textual description is reduced to few lines; several figurative examples of products and natural objects clarify the range of opportunities. Furthermore, the corresponding dependent elements from other domains are shown. This way, a designer gets inspired by concrete illustrations, a general description describing the element and by corresponding links.

Finally, the one-pagers refer to literature for a detailed description of the element. The horizontal arrangement of the three domains, always appear in the same order, and a consistent color scheme support the orientation within the guideline.

This paper shows three samples of one-pagers; we can send the whole guideline via email. It is available in German and soon in English.

4.2 Application of the Guideline

The guideline is designed to support engineers and industrial designers in different stages of development process. *Engineers* are trained to work systematically, being able to explain their proceeding and decisions. They follow several design guidelines (cp. [3, 12]), norms and engineering catalogs. For them, this guideline provides a tool to comprehensibly vary product parameters by applying principles while being aware of the product's visual impression on the user.

At the same, *industrial designers* rather work intuitively in a "creative" way: They start with defining the targeted "character" of the product and its impression on the user. Therefore, they create moodboards to get a feeling for the user's context. Industrial designers create numerous variants of the embodied design by scribbling and rough prototyping—rather intuitively instead of systematically. The guideline provides a collection of principles and parameters fitting to their targeted product impression. It allows to work more systematically and provides a basis for explaining the design more rationally—in particular in collaboration with engineers.

Throughout the product development process, the guideline focuses the concrete embodiment design stage (cp. "define layout" [1, 3]). However, it can also serve as a basis for determining the product impression as emotional aspects in the specification sheet (in the early development stage [1, 3]). In particular engineers tend to care about the visual appearance only in late development stages—when most parameters are already fixed. We suggest dealing with those aspects already in an early stage by fixing the targeted visual impression as specifications.

4.3 Exemplary Application

With the following hypothetical use case we clarify the application and results of the guideline. A manufacturer for lamps wants to design a new product. The

Fig. 6 Conspicuous, warm and complex lamp design (thanks to aryan251 for the image)

design team has the task to create various concepts. As a starting point, the team agrees on a product parameter for variation. They decide for *surface treatment*, since in the past the lamp designs predominantly differed in *material* and *shape*.

The guideline suggests several design principles. In a workshop the design team decides to apply the principles of *color/contrast*, *conciseness/simplicity* and *rhythm*. This selection appears most promising for the kind of product. With these three principles the team gets suggestions for possible product impressions again: They find the highest overlap of the principles in the three impressions *conspicuousness*, *complexity* and *temperature*. The team decides to aim at designing a lamp that appears conspicuous, warm and complex by applying the mentioned principles to vary the surface treatment.

Figure 6 shows the result of this process. The lamp has a surface combining different materials and colors. The layout is irregular, but at the same time rhythmic. The lamp is conspicuous due to different colors and shapes of the glass splinters. The color range makes the lamp radiate a warm light.

5 Discussion

This paper presents a guideline with product impressions, design principles and product parameters from engineering and industrial design literature. It connects the elements of these three domains by means of a Multiple Domain Matrix. This way, the guideline specifically suggests applying adequate principles. The guideline is designed in a way that both engineers and industrial designers easily understand the elements (as one-pagers) and get inspired by concrete figurative examples.

We set the dependencies between the domains in a workshop based on our experience and intuition. The set links appear appropriate, but we will improve the proceeding for being scientifically more comprehensible. Therefore, existing products with a dominant impression could be analyzed regarding applied principles and parameters. Additionally, the links should be evaluated with designers from industrial practice.

The guideline was successfully applied by students in their academic theses. Nevertheless, we are aware that circumstances in industrial practice are different and that the guideline needs to be evaluated in this context. Therefore, we plan to implement an online version allowing for more examples using further media like videos and interactive links to corresponding online sources.

References

1. Pahl, G., Beitz, W., Feldhusen, J., Grote, K.-H.: Engineering Design. Springer, New York (2007)
2. Ponn, J., Lindemann, U.: Konzeptentwicklung und Gestaltung technischer Produkte. Springer, New York (2011)
3. VDI Guideline 2221: Systematic Approach to the Design of Technical Systems and Products. VDI (1987)
4. Desmet, P., Hekkert, P.: Framework of product experience. Int. J. Des. **1**, 13–23 (2007)
5. Norman, N.: Emotional Design. Basic Books, New York (2004)
6. Jordan, P.: Designing Pleasurable Products. Taylor and Francis, London (2000)
7. Chakrabarti, A., Gupta, A.: Design for emotions. In: International Conference on Engineering Design (ICED'07), Paper no. 148 (2007)
8. Steiner, P.: Sensory Branding. Gabler (2011)
9. Steffen, D.: Design als Produktsprache. Form (2000)
10. Norman, N.: The Design of Everyday Things. Basic Books, New York (2002)
11. Lindemann, U.: Methodische Entwicklung technischer Produkte. Springer, New York (2007)
12. VDI Guideline 2222: Engineering Design Methodology. VDI (1997)
13. Heufler, G.: Design Basics. Niggli (2004)
14. Tjalve, E.: Systematic Design of Industrial Products. IPU (2003)
15. Lidwell, W., Holden, K., Butler, J.: Universal Principles of Design. Rockport (2010)
16. Lindemann, U., Maurer, M., Braun, T.: Structural Complexity Management. Springer, New York (2009)

Creating Brand Recognition Through Product Design

Pablo Marcel de Arruda Torres

Abstract The main goal of brand is to establish a significant gap against competitors, essential to obtain a unique positioning. The products help brand to create recognition, reinforcing that positioning. Thus, brand perception must be supported by a coherent product, which can be deliberately created, transformed and managed to full this function. The objective of this paper is to show how product design can be used to create brand recognition and positioning to the marketplace. For this, the paper presents three case studies that demonstrates how global companies use product design to convey strategic messages and create brand perception. The third case still shows how the brand knowledge can lead to new product designs and projects in a practical way. Although the cases have different characteristics and peculiarities, they are similar in how the company's image is reflected on a planned and intentional way into new products.

Keywords Brand recognition · Product design · Corporate identity · Strategic design

1 Introduction

At the start of the production era, the handicraft was characterized by experience product expertise learning constructive excellence exclusive products customization and high costs. With the growth of the middle class, the demand for goods required a larger production capacity and lower costs: industrialization can be seen as a consequence of such factors. The need for a standard quality and a huge number of products resulted in serial production with emphasis on to the process and, in particular, the early stages of design the formalized and measured processes

P.M. de Arruda Torres (✉)
Researcher, Universidade Federal de Campina Grande, Brazil,
Second University of Napoli, Napoli, Italy
e-mail: pablotorres@ddi.ufcg.edu.br

A. Chakrabarti (ed.), *ICoRD'15 – Research into Design Across Boundaries Volume 1*,
Smart Innovation, Systems and Technologies 34, DOI 10.1007/978-81-322-2232-3_19

standardization, lower costs and quality control. The method of mass production has spread everywhere and increased competition: even complex products like cars are aligned to each market segment whose niches are divided by factors such as cost performance consumption maintenance and lifespan [1, p. 26].

From the 90s, several factors reconfigured the world production scenario. With the opening of the market and the emergence of the internet, the competition has become global, regardless of the geographical location [2–4]. Thus, with easy access to production technology and information, the technical characteristics of the products are increasingly similar in a growing number of areas, when not happen to have the same components. The constant standardization, the multiplication of products, market saturation and the pollution in the media undermine the effectiveness of brand messages [5, 6]. Product design as differentiation exerts special role while the brand becomes increasingly present and decisive for the purchase [7, p. 360].

The postmodern consumer is no longer dictated by the needs, but tries to produce symbols; objects tell us how to use them serving their symbolic representations and building the environment in which we live. Therefore, we may infer that consumers have become beings of unconscious desires, not only needs, indicating that people consume more than objects, but they consume unconscious desires [5, p. 116]. It is in this field between the intangible elements and physical elements that fits the brand recognition as strategic communication through the product design.

Borja Mozota [5, p. 110] states that products communicate a message and a combination of several individual products leads to a mix of communication which can be seen as a construction of meaning (connotation), which can be interpreted (denoted) by users. As well, it became clear that products should not be seen only from the point of view of satisfying needs or maximizing individual use, but also as a means of communication [7, p. 290].

Therefore, the paper has the objective to show how product design can be used to create brand recognition and positioning to the marketplace. For this, the paper presents three case studies that demonstrates how global companies use product design to convey strategic messages and create brand perception. The two initial cases are founded in the literature and emphasize the analysis of brand portfolios to create product design cues for future designs. The third case was developed by the author of this paper, showing the project of a new product beyond the brand analysis, in order to demonstrate the practice of product design for brand recognition.

2 Design and Brand Identity

Since the 90s, it is observable that design is fundamental to the task of developing products and designers are becoming increasingly communicative professionals in companies [7, p. 364]. The designer is like an interpreter of the company's personality, a strategist to guide your work with intentionality, and communicator to transmit and implement an identity in the public mind. Therefore, the communicative function can be considered implicit to the design action [8, p. 56]. In this sense, the design in its various forms, regarded as an element of communication and as a tool implemented in the business context, can only be considered strategic if used for strengthening and consolidation of the corporate brand.

In this sense we can understand the significance of brand as the central manifestation of the company's identity, involving key attributes in a "condensed" form, connoting specific meanings and activating a network of associations [9] with the functions to identify, describe, remember and create value [8, p. 34]. Thereby, the contact of the customer with any form of brand expression affects the brand experience, which should be planned with coherence and consistency to ensure positive consumption experiences [10, p. 130; 11, p. 152] and present a value proposition involving functional, emotional and self-expression benefits [8, p. 31]. The marks also serve like mirrors idealized in which individuals see illusory projected its self-image, in other words, the brand image is my own image [8, p. 41].

García Garrido [8, pp. 23–24] states "to build the image is to communicate, to express identity through all possible resources." According to the author, there will always be communication between organization and society, even if there is no intention, and all corporate action becomes your identity communication. In this sense, Borja de Mozota [5, p. 30] points out that design process is an identity process, differentiating the brand from the competitors and offering a key identifier from the company to the public. Therefore, we can conclude that brand is a concrete referent of the company's identity, which can be deliberately created, transformed and restructured according certain strategic intentions [7, p. 345, 9], reflecting a desired and intentionally planned image [8, p. 30].

Tirit Ekmekçi [12, p. 19] says that the basic meaning of identity is to distinguish one thing from other things; in the company context, the notion of identity predominantly refers to differentiation. Specifically directed to companies, organizations or community activities, the concept of identity means that the company's internal profile, namely know-how, expertise and entrepreneurial attitude, need to be equaled to the external profile through product design, communication and brand building [7, p. 345]. A brand, to be preferred within a category, must match up to or much resemble associations or attributes with the attributes determined by the market [5, p. 131]. The elements of identity should be consistent with the level of product offered by the company, because if the identity exceeds the product, the customer perceives the image as badly transmitted, while a lower identity weakens the value of the brand [10, p. 132]. In this way, brand communicates a personality through identity [11, p. 150], where communication reinforces the brand and brand recognition reinforces communication [10, p. 141].

3 Product as a Brand Message

In the literature, identity is divided into three dimensions: the client's identity, the company's identity and the product identity. The product identity is expressed by manufacturers who often want to control it through brand identity [12, p. 20]. This means that brands use products as a means to communicate your vision, because inevitably the preliminary meeting that the consumer has with the brand is performed through the product. How the user perception when confronted with the product depends of the design, it establishes increasingly the central aspect of the "corporate decisions". Wolfgang Sarasin called attention very early (1980) to the fact that in many cases the development of a strong corporate identity occurs from a development of a product identity profile. How the construction of the company's identity is first perceived from the outside, to the product design should be given special attention (BÜRDEK p. 346). Thus, the act of 'code' strategic associations of products is in the center of corporate interests [9].

Designers communicate with users through products, incorporating social and cultural meanings. Users communicate with other users through products, adopting a place in society and also modifying proposed use for companies and designers. In this way, product design is a notable manifestation of the brand identity, while the importance of the brand act evoking strategic associations through various ways of corporate communication in the contemporary market is evident [13].

When someone looks an object, a mental figure is constructed through associations and projections, activating a mental image by the visual image. Thus, the design that allows the brand perception directly affects consumer beliefs about the products belonging the brand portfolio, influencing the interpretation of functional and aesthetic informations [5, p. 110]. Karjalainen [9] says that products function like manifestations of the brand identity, evoking associations that, in an ideal situation, are strategically aligned to the message defined by the company.

How the product design establish 'connections' among symbolic worlds of users (or groups of users) and producers of symbols (companies), it is necessary understand about the process of encode and decode informations, which requires intensive handling of the communicative design functions [7, p. 325]. Thus, the product design must necessarily associate the brand with positive effects and pleasurable experiences because the intensity of an emotional reaction is function of the intrinsic elements of the perceived form (5, p. 113). This aspect has to be considered since some specific brand elements can become specific product elements. The intentional transmission of meanings through a design, therefore, requires specific product knowledge [9]. This question is more specifically about a 'defined' identity in terms of explicit features, and about an intentional communication through product design. That is why brand identity can be considered a strategic concept that fundamentally involves intentional actions and emphasizes product design such as messaging transmitter [9].

Figure 1 shows graphically the concept of semantic transformation developed by Karjalainen [13]. It presents important aspects of product design in terms of brand communication including semantic functions, typical features and design references. First, the accumulation of knowledge, processes, practices and methods used and tested before a specific design process constitutes the body of

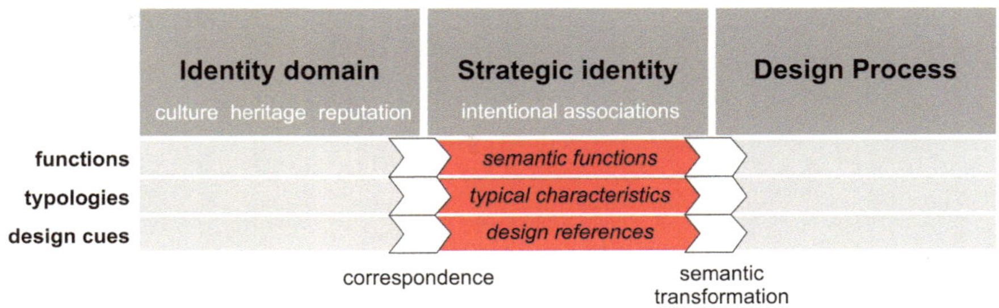

Fig. 1 Semantic transformation schemes. *Source* Karjalainen [13]

experiential knowledge. This knowledge is linked to the company's past, reputation, heritage and culture. Second, the strategic knowledge comes from recognition and internalization of the strategic company's objectives, or the use of product design in the creation of intentional meanings and associations. The process knowledge, in turn, relates to the design process itself. Therefore, the semantic transformation of strategic intentions into physical product design occurs in the interaction among these different types of knowledge.

4 Brand Recognition Through Product Design: Case Studies

The cases presented here illustrate in a practical way how brands may create recognition through product design. The first and second cases were founded in the literature and represent analysis of the brands and their products, developed by Karjalainen [13] and Tirit Ekmekçi [12] for Volvo (cars) and Bang and Olufsen (electronics), respectively. The third case shows a product development realized by the author of the paper for Alessi (home accessories) in order to demonstrate how the process can be used to translate design cues into brand recognition.

4.1 Volvo

Karjalainen [13] presented this case and approaches with the analysis of the products and the communicative aspects in two main levels (and between them): the 'personality' of the product and the physical manifestations, in other words, the characteristics of product design. In the Volvo design philosophy, the visual recognition for the brand was considered on a long term basis. Volvo had become known as the number one safety brand in the automotive industry, based on year-long emphasis and R&D spending on safety. Moreover, the Scandinavian heritage

Fig. 2 Volvo design features represented in the S60 model. *Source* Karjalainen [13]

was considered an inherent part of the Volvo brand, offering a vital basis for differentiate the brand from the others. However, it is important to preserve the strong heritage and Volvo brand recognition had achieved with earlier models.

Thus, Volvo defined a number of explicit design features, that include the frontal characteristic, with the soft nose and the Volvo logo diagonally, the V-shaped bonnet, the strong 'shoulder' line, the rear with distinctively carved backlight, the third side window, and the flowing line from roof to boot lid (Fig. 2). A number of other design features and details were carefully designed to enhance the emotional and premium appearance of the cars.

4.2 Bang and Olufsen

Tirit Ekmekçi [12] analyzed contents of internal and external brand identity and made a general framework of Bang and Olufsen, the danish brand of high-level sound systems. The brand structured the core principles guiding design and communication policies codified in the Seven Corporate Identity Components (CIC), were: authenticity, autovisuality, credibility, domesticity, essentiality, individuality and inventiveness.

The physical design elements and characteristics of Bang and Olufsen audio product range (brand specific design cues) were determined by giving specific importance to the brand specific meanings (semantic references) behind them. The coherency in the design qualities of products in a single product level and product portfolio level provides brand recognition through family resemblance by the previous experiences of the users with brand.

In terms of creating connections with beliefs and desires, metaphors, stories and myths that designers of the brand aimed to create around the audio products, they

Fig. 3 The metaphors, stories and myths created around Beosound 3,200 and the Beocenter 2 to create connections with beliefs and desires. *Source* Tirit Ekmekçi [12]

are lively, breathtaking in use, but also create connections with beliefs and desires of the users and provides an emotional interaction with the products. The coherency in the emotional side across the product range improves the possibilities of interpretation of implicit brand design cues, such as magical, breathtaking, exciting, surprising, striking, inspiring, inviting, enjoyable, desirable, communicative, domesticity, inventiveness, high-tech, in the way that the brand intents (Fig. 3).

4.3 Alessi

Alessi is an italian kitchen utensil company famous for its fun design home objects that make everyday items from plastic and metal designed by famous designers (Fig. 4). From the 1980s onward, Alessi has been particularly associated with the notion of designer objects. The company is committed to incorporating features from areas, such as art, handicraft, anthropology and psychoanalysis and is recognized by the incorporation of new ideas and strategies. By the contribution of these different cultures, the company created new market segments, many with much higher price. Most of the most memorable kettles, toothbrushes, graters and so on were Alessi products.

Fig. 4 Alessi ludic, colorful and functional objects

For that brand, ordinary tools and objects are projected in a post-modern style, with rounded forms and strong colors, sometimes different materials, creating funny and functional objects that have a great visual appeal and are connected with the emotional sense of the users, a company's 'trade mark'.

After realize the analysis of the brand, I started with the project for a new Alessi product, directed for an international competition promoted by the brand. The concept was that distant relationships are part of modern daily life. Friends, lovers, family… it is common to contact them through technology. But time passes and the day to meet those special people again finally arrives. Thus, the product is a timer called 'Time for Love' that symbolizes the distance, time passing and reencounter, using two rounded and stylized figures of a couple (Fig. 5). They are distant from each other, but the time is passing and with the sound of the bell, they are together! The materials are also elements of the brand identification, with a stand made of metal with a simple mechanism inside it, and the figures made of plastic. When the user wants to activate the mechanism of the timer, it is just get the figures away. The final result is an emotional and funny object, for those that are now waiting for the moment to see again a special people.

Fig. 5 'Time for Love' concept of kitchen timer designed to recognize the Italian brand Alessi (project developed by the author for an international competition)

5 Conclusion

The brand is currently the most valuable intangible good that a company can have; it needs to be managed into all points of customer contact for a positive perception around the brand. The product is the main point of contact between client and company, awakening desires and providing a positive brand experience. The company transmit predefined messages (brand values) to the target customers through communicative qualities of brand-specific product design, according the product strategy and the brand identity management. The role of product design in the brand recognition product is of a fundamental tool for companies that consider their image as a way to create value in the market. The companies studied in the cases of the paper present that attribute.

The first two cases present Nordic companies that have strategically employed design to create visual recognition for their brand values. In both cases, it was found that companies took a deliberate and planned effort to translate the core brand values of the company into visible and material design cues. In the Volvo case, the cues are more visible, with specific elements that are repeated along the company's product portfolio, where the repetition of features facilitates visual brand recognition by the market. Danish brand Bang and Olufsen presents another approach, which metaphors and storytelling are used to create fantastic products for a high-level market, in a some more free way, not so obvious if compared to Volvo, but inside a predetermined series of requirements presents in its Corporate Identity Components. In that case, the repetition of design cues is less in the elements of design, being conceptual and material.

The third case, developed by the author of the paper, presents the company's analysis and the design of a product to show how to use brand cues into new projects and products identified with the brand strategy. Here, the Italian brand Alessi was characterized in a different way from the others, which subjective concepts, like fun, style and high-market, are confronted with technical requirements, like materials, form and colors. That analysis made possible thinking a concept of product aligned with the Alessi brand, mixing storytelling with visible brand design cues.

Although the cases have different characteristics and peculiarities, they are similar in how the company's image is reflected on a planned and intentional way into products present in the portfolio. Thus, if a brand want to be recognized it has to specify a number of design principles and features to be used in the product portfolio.

Acknowledgments This work would not be possible to be realized without the institutional contributions and grants from CAPES—Coordenação de Apoio ao Pessoal do Ensino Superior (Brazil) and Second University of Naples (Italy), to whom I leave here my sincerely thanks.

References

1. Maiocchi, M., Pillan, M.: Design e Comunicazione. Alinea Editrice, Firenze (2009)
2. Beck, U.: Che cos'è la globalizzazione. Carrocci, Roma (1999)
3. Featherstone, M.: Cultura globale. Seam, Roma (1996)
4. Waters M.: Globalização. Celta Editora, Oeiras (1999)
5. Borja de Mozota, B.: Design management: using design to build brand value. Allworth Press, New York (2004)
6. Moraes, D.: Design e identidade local: o território como referência projetual em APL's moveleiros. In: Collection of advanced studies in design: Identity. Org.: Dijon de Moraes; Lia Krucken, Paul Reyes. vol. 4, pp. 13–34. EdUEMG, Barbacena (2010)
7. Bürdek, B.E.: Design: história, teoria e prática do design de produtos. Edgard Blücher, São Paulo (2006)
8. García Garrido, S.: Identidad, marca e imagen corporativa. Máster Internacional en Dirección de Comunicación. UCAM, Murcia (2013)
9. Karjalainen, T.-M.: Strategic design language—transforming brand identity into product design elements. In: Proceedings of the 10th International Product Development Management Conference, Brussels (Belgium), 10–11 June 2003
10. Bargellini, J.F.: Costruire um'azienda design–oriented: I 12 principi del design manegement. FrancoAngeli, Milano (2014)
11. Best, K.: The fundamentals of design manegement. AVA Publishing, London (2010)
12. Tirit Ekmekçi, H.: Strategic brand communication in product design. Master of Science Thesis in Industrial Design. Graduate School of Engineering and Sciences, Izmir Institute of Technology, Izmir (2007)
13. Karjalainen, T.-M.: Semantic transformation in design—communicating strategic brand identity through product design references. University of Art and Design Helsinki, Helsinki (2004)

Kundan Jewellery Design-Diverse Style of Form Clusters a Methodical Study of Their Characteristics

Parag K. Vyas

Abstract In vast gamut of Indian jewellery, *Kundan* is a traditional one with a unique character of its own that has remained unchanged over a period of time. Motifs, often identifiable by their individual names are used in this particular style. A semantic unit is the fundamental building block of an article of *Kundan* jewellery. These units are used in combination to make a form cluster. These form clusters are fitted around a curve to create an overall jewellery form like necklace or a bangle etc. These patterns are repetitive and appear through the body of jewellery as a coherent theme. Form clusters are made in a variety of ways and follow different construction details. Sometimes they are densely packed with many semantic units in a small area and sometimes they have a lot of space in with the, units just barely touching each other. These styles of form clusters can be differentiated from each other and can be classified for a better understanding of subject. This study elucidates three types of form clusters and discusses their respective characteristic form. This is a comparative study based on form to provide points of reference and mark out key aspects.

Keywords Kundan jewellery · Typification · Parametric design · Form clusters

1 Introduction

Jewellery in India is synonymous with bullion metals, preferably gold or silver. Largely, the quantity of gold held by a family was in sole possession of women, where they enjoyed absolute control over this material possession as shown in Fig. 1 [1: 162 and 222]. This gold was protected by a strict code of conduct in society. It can be understood from this example why this psychological affinity to gold runs so deep in Indian mindset.

P.K. Vyas (✉)
Grau Bär Designs, Indore, India
e-mail: paragvyas01@gmail.com

A. Chakrabarti (ed.), *ICoRD'15 – Research into Design Across Boundaries Volume 1*, Smart Innovation, Systems and Technologies 34, DOI 10.1007/978-81-322-2232-3_20

Fig. 1 Front and backside of *Kundan* jewellery

Fig. 2 *Kundan* image *Source* adopted from timeless jewels [16]

Kundan work is labor intensive, difficult to mechanize and that has remained unchanged over a period of time. Motifs are frequently mimicking nature [2: 118], liberally adopting from flora and fauna.

These units are fundamental building blocks and used in combination with each other make a form cluster. In turn, these form clusters make patterns, which are repetitive and appear through the body of jewellery as a coherent theme. There are manners of construction of these form clusters that are passed down from generation to generation.

To begin with a piece of jewellery, a form cluster is first made using diverse motifs. Adjustments are made to overall composition of the cluster, till a satisfactory visual effect is achieved. It may have one or more focal in centre, surrounded by others in an aesthetically pleasing manner. Alternatively, it is a symmetric construct with more than one foci (Fig. 2). These form clusters are fitted around a curve to get an overall form.

2 Review of Literature

The subject of *Kundan*, is not documented well and studied with the rigor it demands. In absence of such works forms and clusters are neither consistently recognized nor standardized. This study throws light on form clusters and propound on their comparison with each other. That can be adapted easily for a cluster-based approach to jewellery design.

Jewellery making has gone through a rapid transformation in recent years. With the advent of new tools and techniques, technology has taken a quantum leap which is well documented. There is moderate published material about jewellery making techniques and their various technological aspects [3, 4]. In other areas there are metallurgical research publications that cover alloy compositions such as depletion gilding [5].

Available literature was studied to get an idea of existing research works with focus on form based approach. It was found that the literature broadly falls under two broad categories. The first type has a depiction and pictorial documentation of existing articles, sometimes covering historical aspects [6: 270, 7: 107, 8: 21] in form of books. On the other hand, there is some published material on CAD as well as metallurgical aspects.

Some of these are briefly discussed as under giving high points of the research.

A Parametric Vowel Oriented CAD Paradigm to Produce Forming Components for Stretch Formed Jewellery [9: 137–145] it nicely touches Computer Aided Design (CAD) systems, reproducing traditional Jewellery. Further elaborations are made by [10: 811–821] which present a parametric vowel based jewellery modeller for designing and creating carved jewellery, theoretically.

An Aesthetics Driven Approach to Jewellery Design is taken by [11]. It puts forward a computer-based design tool to automate art form generation used in jewellery design, touching on Computer Design Support, Evolutionary Art, Fractal, and Jewellery Design. The paper again is highly theoretical in nature, with conceptual propositions.

In comparison to domain of jewellery, Research in the diamond industry is methodical in nature but limited to polishing for high reflectivity, weight and quality of cuts [12: 169].

There also seems a disjoint between shop floor practices and research, perhaps due to the secretive nature of the jewellery industry. This is a challenge to bridge the gap by a direct approach to *Kundansaaz* and in-depth interviews on the shop floor.

There are no ready catalogues of various forms and shapes used in jewellery. This study is a step forward in that direction with research insights.

3 Methodology

Methodology adopted for present research is a combination of in-depth interviews with *Kundansaazs*, small jewellery manufacturers, and field observations. These are later validated by domain experts and necessary corrections incorporated in study.

Conventional methods of questionnaires are of little use as in spite of of assurances of an academic nature of a study, rarely a clear insight can be gained by such attempts. Being pioneering attempts in documenting formal aspects of *Kundan* jewellery designs, this methodology is found to be of appropriate and in the line of views of different academicians, practitioners and jewellers.

For in depth understanding of form photographs were studied in a scientific manner. The methodology followed was a systematic labeling and tabulation of smallest semantic units contained in a particular form cluster [13] in form of a picture.

To methodically study photographs, renderings, and shop floor presentation, montages were meticulously traced out and converted into vector artworks. By way of analogy, a reverse engineering of the existing piece by tracing them out firmly, gave a foundation to understand the construction of units. This method allows minute details to be studied and interplay of units was revealed with clarity.

Study was made robust with a series of in depth interviews and time spent at the actual shop floor observing craftsmen working. Practicing jewellers and academicians validated the findings. Their valid suggestions were incorporated in study.

The insights gained about three types of clusters and their constructions are as elaborated in subsequent sections.

4 Diverse Styles and Their Comparison

Traditional *Kundan* has a manner of making form clusters by four parameters [14] to achieve a good *Gothavani*. This aspect can be described as a "satisfactory, aesthetically pleasing, composition of smallest semantic units" and a *Kundan*saaz may take a long time in this procedure until oneself and thereafter client is satisfied. The same theme carried to different articles creates "a set" a specific form cluster running as a common thread through them. The form clusters are frequently scaled up or down, or transformed yet retain essence and continuum.

There are types of such form clusters that follow a particular style, consistent within group yet distinctly different from the other groups. This part of study propounds on three of them namely, *Bharawdaar, Jaalikaam* and *Khila* clusters and gives insights of their specific construction details as follows.

4.1 Bharawdaar Form Clusters

These are form clusters, which have a dense appearance, due to close packing of units. These form clusters have, comparatively, a high number of smallest semantic units in propinquity per unit area. Characteristically, they have very little space left in between adjoining motifs and often their borders may overlap slightly with little counter space as shown in Fig. 3.

This type gives a 'heavy' appearance to articles. It is for this type of visual appeal that these are also termed as 'Bhari', (literally heavy). To emphasize in

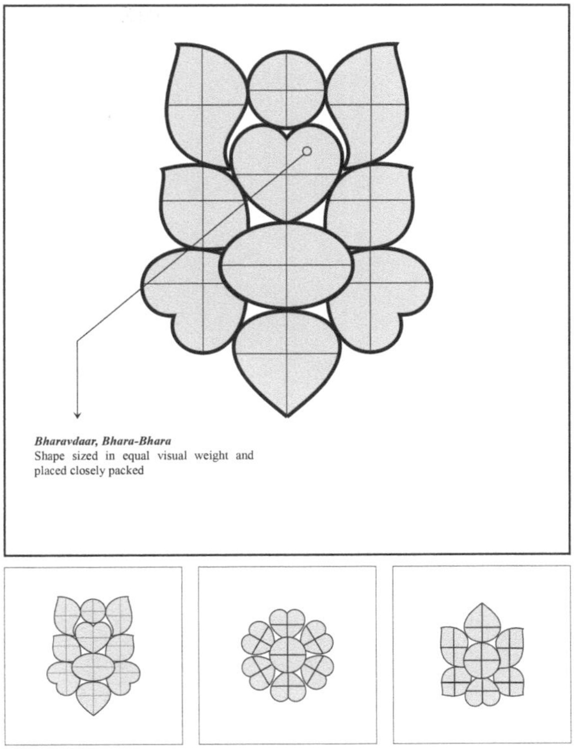

Fig. 3 Bharavdaar form clusters with little counter space in between

common parlance referred to as Bhara-Bhara, (repetition) suggestive of relatively heavy constructional and dense appearance.

These have one focal point or line, around which the emergence and arrangement of other units occur, further emphasizing focal unit.

The characteristics of this particular style, point-wise, can be listed as follows.

- Equal or equitable visual size of semantic units having approximately similar areas.
- One focal smallest semantic unit or units.
- Little counter space.
- High semantic unit density per unit area.
- Overlap of a few units.
- Fewer use of long stemmed units.
- High Axial symmetry.

4.2 Jaalikaam Form Clusters

In comparison, these form clusters have a light appearance. they have comparatively low number of smallest semantic units in propinquity per unit area. The motifs barely touch each other except on one or two well-positioned points to hold

Fig. 4 Jaalikaam form clusters with more counter space in between

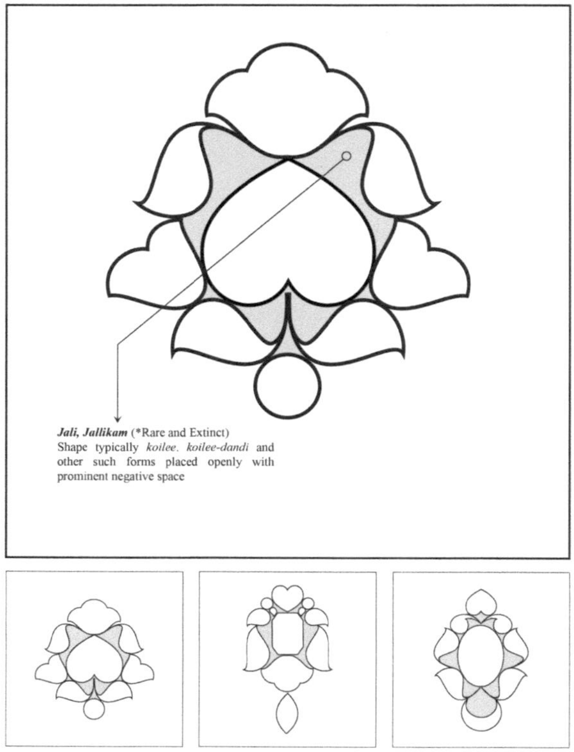

firmly. Typically, they have a large counter space in between adjoining motifs and often stylized motifs are used that have a 'Dandi' (a pronounced tail extending and thinning to an end). Overall, it has a latticed structure that has large open spaces with little metal veins running through them.

These types of clusters have light and see through appearance. For this type of visual appeal a lot of piercing work is required and in comparison, this is lightest both in appearance and weight per unit area.[1]

These clusters also have one focal large semantic unit around which emergence and arrangement of other units occurs, emphasizing the large unit and often a central stone (Fig. 4).

The characteristics of this particular style point wise can be listed as follows.

- Variable size of semantic units, visual weights balanced.
- One large focal smallest semantic unit, surrounded/supported by others.
- Large counter space.
- Low semantic unit density, with open spaces.
- Semantic units barely touching on select locations.

[1] This type of work is slowly becoming rare and extinct as it requires more efforts, skill and time. In absence of patronage, it does not encourage craftsmen to follow this style.

- Liberal use of long stemmed units.
- High Axial symmetry.
- Flourish and radiance in style.

4.3 *Khila Form Clusters*

These are the type of form clusters that have a radiating appearance, due to a radial arrangement of smallest semantic units in an area. These form clusters have diversely sized, medium number of smallest semantic units in propinquity per unit area as it can be seen in Fig. 5. Characteristically, they have space in between adjoining motifs thus creating a negative space, that itself seems to flow around creating a rhythmic visual motion.

This type gives an overall light and radiating appearance to articles using these predominantly. It is for this type of visual appeal; these are also termed as khila, or khula. To lay emphasis on appearance often the word is also called khula-khula, the repetition suggestive of relatively light constructional appearance.

These have one focal point or line, around which emergence and arrangement of other units occurs, poetically speaking, suggesting blossoming of a flower.

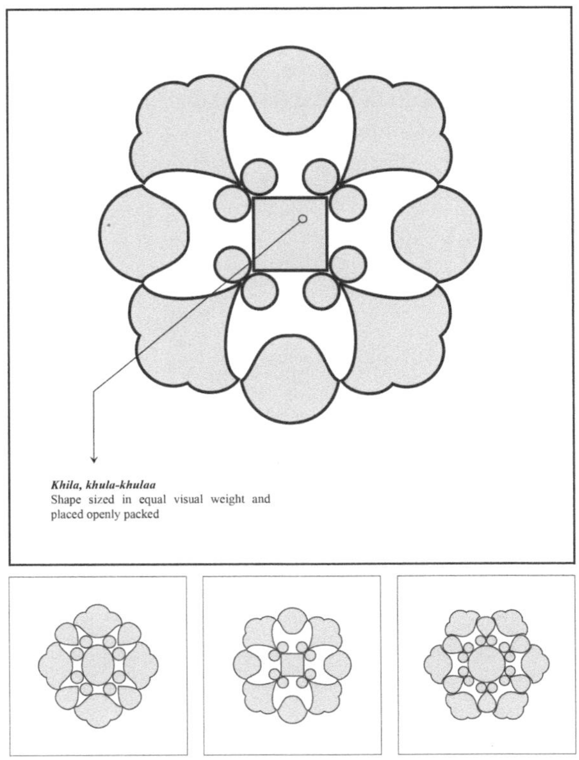

Fig. 5 Khila form clusters with pronounced counter space in between

Table 1 A comparison of attributes

S. No.	Attribute	Bhari	Jallikam	Khula	Remarks
1	Weight/unit area	Heaviest	Lightest	Medium	Typically heavy construction is preferred by a client who is making an investment in gold in form of jewellery. While a client who uses the article often prefers a khila cluster.[a] Only a rare connoisseur commissions a Jaalikaam work these days
2	Visible metal/unit area	Maximum	Minimal	Medium	
3	Skilled craftsmanship	Medium	Most skilled	Medium	
4	Labor intensive	Low	Most	Medium	
5	Rigorous	Low	Maximum	Medium	

[a]Weight and intricacy are important criterion for selection of jewellery. Generally labor for gold work is charged on basis of rupees per gram weight and intricacy of work. Intricate and delicate works, though less in overall weight are charged on a higher rate per gram basis. Upon liquidation of assets the labor, which is extrinsic value to jewellery is lost. Only precious metal value, which is intrinsic is retained. Therefore, jewellery which has an investment value in bullion, tends to become bulky. On the other hand, where a routine wear is expected, frequently worn articles are preferred owing to convenience of lighter weight

The characteristics of this particular style, point-wise, can be listed as follows.

- Diverse size of semantic units having large and small visual areas.
- One focal smallest semantic unit as a centre of attraction.
- Flowing negative space.
- Low to medium semantic unit density.
- Overlap of a few units.
- Mixed use of stout and long stemmed units.
- High rotational symmetry.

Above description of these particular styles clarifies and highlights the features of form in each of them. A quick summery of attributes is tabulated as follows (see Table 1).

5 Merits of Cluster Based Approach

With an understanding of a cluster based approach meaningful forms can be achieved as shown in Fig. 6. Methodically catalogued in study by [15], the smallest semantic units can be used in various forms and configurations to create a virtually limitless number of designs by changing various parameters.

Furthermore, these form clusters can be fitted over a circle to give a good judgment of size and can be scaled up or down to get the right size, in this manner working becomes mathematically precise and a parametric design is possible in domain of jewellery. A CAD interfacing becomes possibility, algorithms can be written for calculating weight of gold for that particular design.

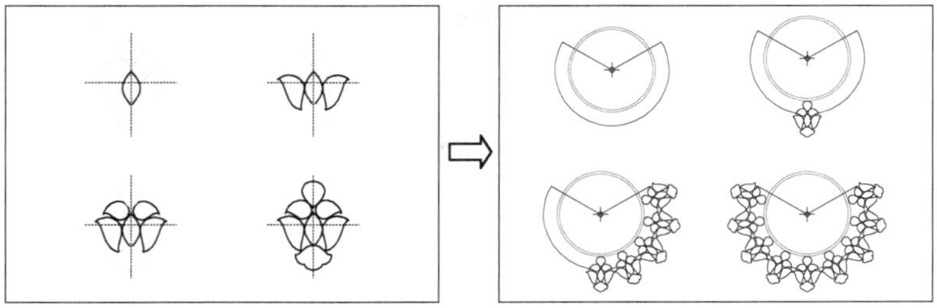

Fig. 6 Progressive stages in making a meaningful form cluster and fitting of a form cluster over a *circle* to get shape of a necklace

6 Implication of Work

The implications of work are beneficial both for the clients and designers alike. These can be used as a tool of expression and forward working for present and potential applications. Methodical approach as compared to traditional hand drawings becomes available for common use.

6.1 Managerial Implications for Clients

With portability of vector arts across diverse design software's and platforms following benefits are available for application and use.

- A methodical approach to design with understanding of form clusters.
- Exposure to a selective number of designs with a comparison amongst clusters will enable a client to select appropriate design giving them wider choice.
- Desirable changes can be incorporated at the design stage eliminating expensive reworking before setting.
- A selection can be defined and communicated more efficiently between client and designer.
- Designers can focus on details, workout specifications and estimate on paper without waiting for the gold structure to be ready by mathematical modeling.
- Work, time, and material estimates can be more precise with mathematical measurement of different variables which is plausible as opposed to an approach of trial and error in practice now.

6.2 Managerial Implications for Designers

- The study can provide a digital library of various basic units used with parametric control over form.

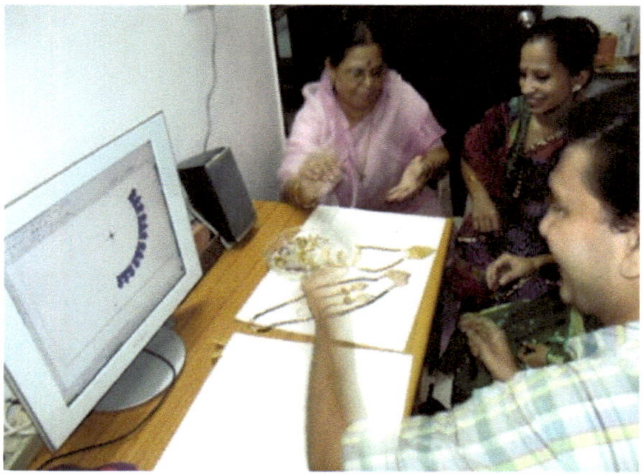

Fig. 7 Cluster approach in use by a designer *Source* Author

- This study serves as a foundation for preferential assessment of jewellery design for different groups of specific potential clients.

Jewellers and designers could observe (Fig. 7) and envisage to a certain degree, preferences of articles and control dimension for present as well as prospective clients.

7 Concluding Remarks

A methodical approach to jewellery with a parametric control over designs has a clear advantage over a trial and error tentative approach in practice today. As the study is expected to reduce substantial time by minimizing making selections firm, all stakeholders can equitably share benefit of this saving. This approach also allows specifications to be more articulate and precise in nature as compared to an indicative hand drawn sketch.

A new language can be devised by application of transformation to a semantic unit or a form cluster as a whole. The language can be adopted for other types of jewellery and life style products readily.

References

1. Untracht, O.: Traditional Jewellery of India. Thames & Hudson Ltd., London (1997)
2. Adorno, T.W.: Aesthetic Theory, 6th edn, p. 118. University of Minnesota press, USA (2008)
3. Ravi, B.: Rapid hard tooling process selection using QFD-AHP methodology. J. Manufact. Technol. Manag. **17**(3), 332–350 (2006). IIT Bombay
4. Wright, J.C., Corti, C.: Engineering approach to gold jewellery manufacture. In: 11th Santa fe symposium, Albuquerque, USA (1997)
5. Grimwade, M.: The surface enrichment of carat gold alloys—depletion gilding. In: 13th Santa Fe Symposium, Albuquerque, USA (1999)

6. Balakrishnan, U.R., Kumar, M.S.: Dance of the Peacock: Jewellery Traditions of India. India Book House Pvt. Ltd, India (2004)
7. Balakrishnan, U.R., Kumar, M.S.: Jewels of the Nizams. India Book House Pvt. Ltd, India (2006)
8. Sharma, R.D., Varadarajan, M.: Handcrafted Indian Enamel Jewellery. Lustre Press, Roli Books, India (2008)
9. Gulati, V., Tandon, P.: A parametric voxel oriented CAD paradigm to produce forming components for stretch formed jewelry. Comput. Aided Des. Appl. **4**(1–4), 137–145 (2007)
10. Gulati, V., Singh, H., Tandon, P.: A parametric voxel based unified modeler for creating carved jewelry. Comput. Aided Des. Appl. **5**(6), 811–821 (2008)
11. Wannarumon, S.: An aesthetics driven approach to jewelry design. Comput. Aided Des. Appl. **7**(4), 489–503 (2010)
12. Pagel Theisen, V.: Diamond Grading ABC: Handbook for Diamond Grading. Rubin & Son Bvba, Antwerpen (1993)
13. Vyas, P., Bapat, V.P.: Design approach to Kundan jewellery—development of a Tool to Study Preferential Likeness of Articles Using Mother Grid and Form Clusters in a Methodical Manner. Research Into Design, pp. 775–783. Research Publishing, India (2011)
14. Vyas, P., Bapat, V.P.: Identification and classification of semantic units used in formation of patterns in kundan jewellery, a methodical approach. In: Design Thoughts, IDC, IIT Bombay, pp. 59–72, Aug 2010
15. Vyas, P., Bapat, V.P.: Investigation of form clusters made of smallest semantic units and patterns they create as building blocks of kundan jewellery. In: Research in to Design, pp. 766–774 , Research Publishing, India (2011)
16. Timeless Jewels, vol. 7, issue3, p. 81

Sound Symbolism in India Comic Books

Subir Dey and Prasad Bokil

Abstract Comics are cultural artifacts that narrate stories through sequential images. With the turn of the century, these artifacts are being understood as a form of visual language. To understand the visual language of comics, it is imperative to study its core components: image and text. In this paper, the status of sound symbolic words used in Indian comic books is reviewed, which may lead to the analysis and creation of further sound symbolic words. It is found that the onomatopoeic words, created based on the typographic attributes like, alignment, weight, color, texture and size are more effective in conveying the meaning of a sound. Identification of these factors provides insights about the relationship of a sound to its textual representation. This study will prove helpful to the comic book designers by being an additional tool for communicating their story more efficiently.

Keywords Comics · Visual language · Sound symbolism · Expressive typography

1 Introduction

Comics are a cultural artifact that tells a story. Although comics sales dropped when internet and television era started, however it is growing at a steady pace in the 21st century. In India sales have taken a leap from 25 lakhs in 2011 to over a crore in 2013. Japan leads in terms of comics publication, where *manga* (Japanese comics) covers almost 27 % of the published materials. The readership has also shifted from young kids to mature adults and comics are now read by anyone between the ages of 15–40. Due to its popular cultural placement, comics have remained a subject for scholarly discussion for decades. There have been

S. Dey (✉) · P. Bokil
Department of Design, Indian Institute of Technology Guwahati, Guwahati, India
e-mail: subir@iitg.ernet.in

A. Chakrabarti (ed.), *ICoRD'15 – Research into Design Across Boundaries Volume 1*,
Smart Innovation, Systems and Technologies 34, DOI 10.1007/978-81-322-2232-3_21

numerous attempts to define comics and its semantic structure. The idea of comics as language existed since the 19th century. Pioneer artists like Jack 'King' Kirby and Osamu Tezuka and writer Frederik Schodt consider making comics as writing in images. On defining comics as a form of sequential art Eisner [1] and McCloud [2] have given their respective definitions. However with the turn of the 21st century comics is being enquired and argued as a form of visual language. The visual language of comics consists of two elements: image and text. This paper analyses the textual element of Hindi comics (Raj comics).

Sound symbolism in linguistic theory assumes that the relation between sound and meaning is arbitrary [3]. In Plato's *Cratylus*, Socrates debates with two pupils on the issue of whether the names for things are arbitrary or whether instead they are a natural reflection of the things named [4]. However, throughout the dialogue its debatable whether the relationship of sound and corresponding letters is arbitrary or not. Considering Socrates' view, it can be said that letters do have characteristic to depict softness, roundness and other properties. This paper makes an enquiry of the probable characteristics that might prove helpful in creating a relation between an action and its representation through written text.

2 Literature Review

Expressing meanings is what languages are all about. To understand how any particular language works, we need to understand how its individual elements work to fulfill its function as an intricate device for communicating meanings [5]. In the case of comic books, it is the visual language that communicates the story. Just like sequential units in speech comprises of the grammar, the sequential units of comics comprises of visual language. In a cultural domain, the visual language bonds with the written text and forms a common body of communication [6]. Therefore, it is essential to understand the parts of a visual language in order to understand the working of a comics.

Considering the sound symbolic words (SSW) as linguistic sign, it can be said by the principle of linguistic sign that the bond between the signifier and signified is arbitrary [7]. While Saussure concludes that linguistic signs are arbitrary, he also adds an exception to the onomatopoeic formation of words. Being small in category, these words are mainly neglected and used as *interjections*. This might be true in linguistic studies, but in comics the sound symbolic words remain an effective tool for communicating a message. This can be said based on the fact that more than thousand sound symbolic words exist in comic books.

The Oxford Online dictionary defines sound symbolism as the partial representation of the sense of a word by its sound, as in *bang*, *fizz*, and slide. In English, sound symbolism is a small and neglected category of words whereas Japanese language has an extensive list of sound symbolic words. The sound symbolic words can be classified in three main categories [8]:

(a) Phonomime or onomatopoeia: The words that mimic actual sounds.
(b) Phenomime: The words that depict non-auditory senses.
(c) Psychomime: The words that depict psychological or bodily feelings.

A well-known demonstration of the naturally biased mappings is [9] finding that, adults tend to relate meaningless word like ‘maluma’ to a curvy shape and ‘takete’ with a jagged shape. Another similar practical demonstration that sound–object correspondences are not completely arbitrary is that adults map nonsense words with rounded vowels (e.g. bouba) to rounded shapes and nonsense words with unrounded vowels (e.g. kiki) to angular shapes [10]. These studies show that there might be a relation between sound and shape. Linguistic studies mainly focus on spoken language whereas comics have to deal with written language. Though there is no correlation between the form of letterform and its pronunciation but there are evident efforts through comics to build such a relation.

3 Methodology

Raj comics is one of the oldest comic book publishers in India. It was founded in 1984, and primarily published in Hindi language. It was chosen for sample collection because the range and variation of sound symbolic words in Raj comics was found to be more extensive than in any other Indian comics. Also, Raj comics have remained in existence for more than three decades, so it was possible to observe the shift in SSWs across a long span of time.

For checking the frequency of SSWs appearing in comic books, two hundred comic books were selected. These comic books were published between 1992 and 2013. All the SSWs appearing in these comics were counted. The total number of SSWs was seven thousand two hundred. This also gives an average of thirty-six SSWs per comics. Considering the fact that most of the comics were thirty-two page issues, therefore it can be said that on an average each page contained at least one or two SSW. Selection of words was not limited to onomatopoeic words but words depicting attributes like shock and confusion also. However, phenomimetic and psychomimetic words were hardly identified. The words were categorized according to their appearance in comics. Different words conveying same meaning are put under one category of sound symbolic words. Since these words have already appeared in the comics, therefore reliability check was not necessary. The words were directly manually counted and categorized according to their context of appearance.

Table 1 shows the categorization of the words according to different sounds produced by either humans/animals/machines or objects. It categorizes the signifier words and its signified sound which covers all the words appearing across the two hundred comic books analyzed. Repeated words are removed from the table. A total of 28 different words were selected which represent different categories of sounds, like clash, impact, thought, etc.

Table 1 Sound symbolic words in Raj comics

Signified sound	Signifier word (written here in Latin script instead of *Devanagari*)
Kick/Punch/Slap	Tadaak, Dhaad, Thud, Dhadaak, Fadaak, Chataak, Thaak, Dhuum, Dhammm
Clash/Explosion/ Collision	Dhadaam, Thadaak, Bhadaak, Badaam, Dhadaam, Dhadaak, Dhammm, Badoom, Thamm, Dhapp, Varoom
Glass shatter	Chanaak, Khanaak, Chann, Chann-chanaak
Gas/Smoke	Skishh, Fwoosh, Fishhh
Bones breaking	Kad-kadaak, Kadach
Wall breaking	Kadad, Kraak, Kadachh, Khadaak, Kad-kad, Badaak
Liquid splash	Chapaak, Fachaak, Furrrr, Chapp-chapp, Fachh
Animals	Brokk, Kreyao, Boohoo, Gurrr, Kriii, Kriaaa, Kvaa-kvaa, Voouu, vuf-vuf
Birds	Chi-chi, Kaav, Kriiii
Bullet (big gun)	Chyoom, Dharrr, Boook, Barrrr, Jing-jing, Tra-tra, Boom, Bang-bang,Tad-tad
Bullet (small gun)	Dhay, Thay
Empty gun	Klik-klik
Weapon slash	Skinch, Khachh, Khachhak, Shak
Metal weapons	Thanaak, Karr, Kooomm, Khanaak, Tann
Air/Wind	Khoosh, Skrishhh, Foooo
Bubble burst	Plop
Machine/Engines	Huu-huu, Gharrr, Ghurrr, Dhadd, Charrr
Car breaks	Chiiiiiii, Krinch
Cloth/paper rip	Charrr
Electric shock	Kirrrr, Tiddd
Siren	Vao-vao, Vu-vu, Pio-pio, Boo-boo
Weapon throw	Saay
Choked	Gupp, Ghapp
Wormhole	Skiishhh
Switch/Knob	Khat
Shock/Surprise	Exclamation mark
Confusion	Question mark

4 Analysis and Observations

The SSWs are based on two variables. First is the verbalization of the sound itself. Second is the visualization of the verbalized sound in the form of a letter or text. While the former variable can be extensively studied and analyzed from a linguistic perspective, it is the latter part which is analyzed and explored in this paper. Figure 1 shows the steps for SSWs creation and highlights the relationship among the factors involved. It should be noted that sound is an invisible effect that is made visible in comics through visualization of its mimicking sound. It is identified and discussed in this paper the factors on which the visualization is developed.

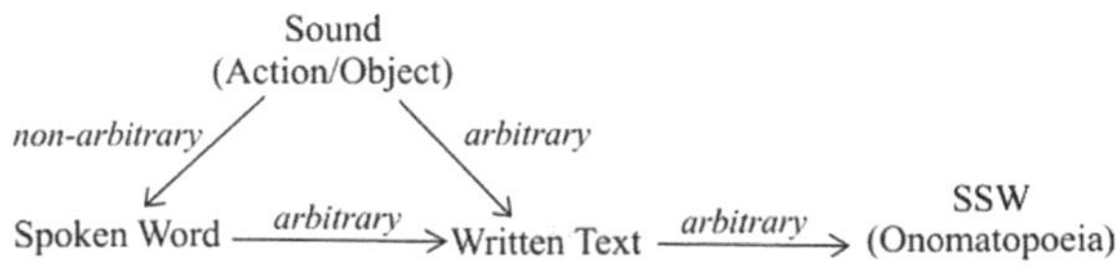

Fig. 1 Stages of SSW creation

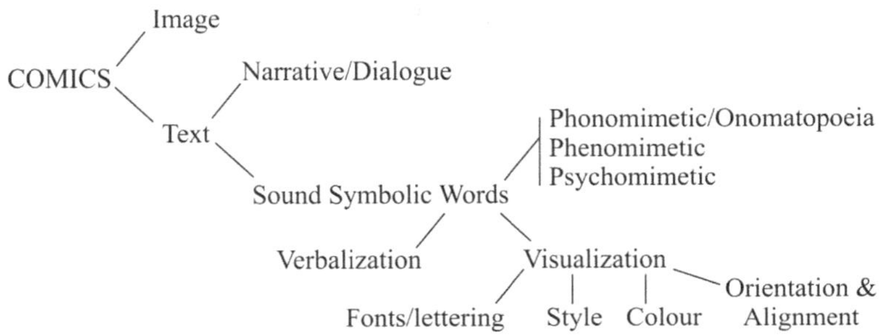

Fig. 2 Structural diagrams of syntactic elements in comics

Onomatopoeias are used across different languages. There might be similarity or differences among the representation of the same sound. In onomatopoeias, the relationship of the visualized text is to some extent related to the actual sound of the object. For instance, the written text of a baby crying is *ua-ua* in Hindi and *wah-wah* in English. Similarly, the written text of an explosion is *badaam* in Hindi and *badoom* in English. Likewise there are other similar words which actually resemble the sound and are same in characteristic across many languages. Differences are noticed mainly in the representation of the text in a visual format. Indian comic books have a range of words as phonomimetic words; however a gap is noticed when it comes to the representation of the phenomimetic and psychomimetic words. Since there's a lack in the other two categories therefore, this study could only be focused on the phonomimetic or onomatopoeic category. The structural diagram of the syntactic elements in comics is shown in Fig. 2.

The visualization of the sound is represented through variables like, style, fonts and letterings, color, texture, orientation and alignment. These are the semantic factors based on which the analysis is done. The concept of signifier and signified plays a vital role while creating the onomatopoeic words. In the case of Indian comics, however, it is found that these two factors have a minor role. The examples in Table 2 show the limited range of representation of onomatopoeia. Words are mostly created in a bold format and covered with a thick black outer stroke. Details of each variable are discussed separately.

4.1 Variables in the Visualization of SSW

In this section the important variables which embody the creation of sound symbolic words are discussed. The variables are analyzed as individual contributing factor but it was also found that these variables are interconnected in themselves.

Table 2 Examples of SSW with semantic features

Signified word	Signified sound	Context	Semantic features
	Collision/ Impact	Used for showing collision accompanied by directional lines from the source of impact	Bold, red, rectangular form for conveying loudness
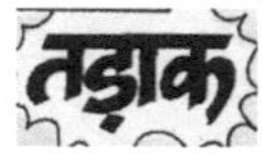	Punch/Kick	Used for a forceful kick. Other variations include 'Dhaad' and 'Thad'	Bold, angular edges, placed inside a jagged edge balloon
	Raging animal	Mostly used for Bulls and other big mammals	Bold, red to signify anger, repeated small 's' signifies an extended growl
	Birds	For small birds its 'Chi-chi', for scavengers its 'Kriaa'	Bold, red, rounded form for small birds, use of 's' for extended sound
	Bullet	Used for sound of revolver or beretta (small guns)	Bold, red, rectangular or square in form
	Water splash	Use not limited to water splash but any other form of liquid also (petrol, acid etc.)	Bold, yellow, blue or green depending on the liquid, vertical placing signifies a vertical splash
	Glass shatter	Used mainly for fragile objects specially glass	Bold, red, rounded, letters are separated to signify scattering of glass

The fonts are related to the style of the word. The colors are also context specific. The directional placement of the words helps in conveying the action in a more effective way. The overall shape of the word also determines the characteristics of the sound. Therefore, it is not only a single factor on which the signified sound depends, but a collective inclusion of different variables. The weightage of these factors are context specific and a hierarchy based on their significance is doubtful since they could be molded according to the context specified.

4.1.1 Style

The onomatopoeic words contribute to a significant part of the overall action/communication inside a panel. Arguably, it is an important element to convey a sense of sound in the reader's mind. In this context the style of the word becomes of much importance so that it justifies the characteristic of the represented sound. In order to relate the style of the word to its corresponding sound, the verbalization of the word is a determining factor. Therefore, first it should be identified that how does the sound of the word feels? Like the *kiki* and *bouba* test proves, some words give a sensation of being sharp and edgy (like zrrrt, krriiik, fizz etc.) in

Fig. 3 Differences in onomatopoeia in Hindi and American comics (*Source* Raj comics and Marvel comics)

nature, whereas some might convey a sense of being round and smooth (like blorb, glub, chug etc.). American comics use these factors to create range of styles for the SSWs. However in Indian comics a common type of model is used for almost every kind of sound, as shown in Fig. 3.

The analysis of the comic books provided insights about the pattern for creating the SSWs. In the comics samples from mid 90s it has been found that the onomatopoeic words are drawn rather than using software. This might prove to be a time consuming job when the frequency of these words needs to be increased. However in the comics from the last decade a consistency of the style can be noticed. Apparently, a template or software must be in use to create these words. Similar style is used for different words. In many instances, these words are just looked upon as interjections. Used with a thick black outer stroke, these words just appear to be forced upon the image instead of coherently merging with the action. From Fig. 3 it is evident that the characteristic of the sound in Hindi comics is not given much importance and two different words with different meanings are given the same visual treatment. Whereas in American comics the SSWs acts as a strong structural part of the image that emphasizes the overall impact of the action.

4.1.2 Fonts and Lettering

'Graphically' treated lettering acts as an extension of the imagery. It creates a mood for the story and implies sound [1]. Each comic series of the same character contains same style of fonts and letterings. Some styles are established for some characters which means, by identifying the font, the reader can guess who the main character of the comics is. For example, Fig. 4 shows the lettering style which is exclusively used for 'Doga' series. The lettering style is almost consistent in most of the issues.

Fig. 4 Same bold typeface used for different sounds (*Source* Raj comics)

The popular software ‘Comic Life’ created by www.plasq.com has an extended list of onomatopoeias which can be used for any beginner for making a comic. These words are compiled in a variety of styles and fonts. In Indian comics the application of such software is still a rarity. This opens up possibilities for the Indian comics industry to develop tools which will be helpful while using SSWs. Even if a compilation of templates of different words can be created, that’ll definitely prove as an additional tool for the comic book creators.

The fonts are mainly found in three variations: (i) Standard bold (ii) Italics (iii) Any other deviant. Unlike the standard bold type, italics and other deviants are a minor category. The bold typeface suggests loudness and sounds which cover a large area form the point of impact. At one hand bold typeface is used because of its attention getting character. On the other hand, too many bold types in a single panel might become illegible for the reader. Also, since there is not much variation in the styles of the letters so it makes the words as a forcefully incorporated element. First panel in Fig. 4 shows how bold typefaces in a single panel create legibility issue. Note how the same fonts are used for depicting the scream of the person as well as the breaking of the bones and also in depicting the sound of a big gun, a crow and electric shock.

4.1.3 Color

All the sample comic books are printed in color. That adds another variable for creating SSW i.e. the color. Although a dominant percentage of red is noticed among the words, occasional appearance of yellow, blue, green, white and black is also found. The dominance of red may be attributed to the fact that red is the most intense color and also stands for pain, scream, fear and loudness. The variable of color can be categorised in two groups: Monocolor and Multicolor. The onomatopoeic words are almost entirely are found in monocolors. Figure 5 shows the use of gradient colors and the bevel effect of onomatopoeia in American comics. The first and the last words signify the sound of glass shattered. The shattering of a glass is an instantaneous action. Also, glass is conventionally represented in blue color. The word ‘Kssh’ is thus logically filled with blue gradient (symbolizing the reflective nature of glass) with an uneven outer contour depicting the actual breaking of the glass. Whereas in the last panel, the Hindi word ‘Channak’ is simply filled with black without adding any characteristics of breaking glass. This creates an opportunity to create a database of SSWs that are context specific as well as resembles the character of the sound represented.

While different effects can be achieved through color explorations, keeping the SSW transparent creates further interesting effects. This effect mainly helps in signifying a huge environmental sound. Since filling the words will cover most of the image, so keeping it transparent seems to be an interesting and clever idea. Figure 6 shows sound of explosion with transparent text. The first panel has letters joined together whereas the rest of the images have distinct letters. This way the words can

Fig. 5 Different color styles used in American comics compared to a monocolor use in Hindi comics (*Source* Marvel comics and Raj comics)

Fig. 6 Transparent onomatopoeia used in American comics (*Source* Marvel comics and Raj comics)

even cover the whole page without disturbing the image, simultaneously adding a dramatic effect to it. Indian comics seem to lack on exploring in this aspect.

4.1.4 Orientation and Alignment

The conventional pattern for the positioning of the onomatopoeic words mostly depends upon the action being shown. Although the artist enjoys much freedom while placing the text, one important decisive factor to be considered is the nature of the sound. For instance, the sound of a kick would be different from that of a continuous siren of a police jeep. The former is more instantaneous and momentary, whereas the latter is environmental and echoed in the surroundings. Therefore, while positioning the sound of a kick, words placed near the point of impact would be more effective as compared to keeping the word in distance. Figure 7 shows some dynamic orientations and alignment of the onomatopoeic words use in American comics and Indian comics. With proper orientation, perspective acts as another important factor to construct a semantically strong structure of the word. In Indian comics use of perspective and distortion in onomatopoeic words is a limited. Among the samples, only in a single panel perspective is noticed (Fig. 7, last panel).

Fig. 7 Words in perspective and with distortion creates a dynamic effect (*Source* Marvel comics and Raj comics)

5 Conclusion

SSWs are important features of a comics' discourse. They help in creating additional environment for conveying the sounds and movements in an image. In this paper the structural elements of SSWs in Raj comics are analyzed and discussed. The findings based on the factors like style, colors, letterings and orientation shows the significance of SSWs in establishing a harmonious relation with image. This raises the idea to enquire the wordless comics that hardly uses any words compared to mainstream comics where onomatopoeia is an integral part of the action. Since this paper presents certain categorization factors from which SSW could be created, however a concern develops regarding the broader function of the SSWs. Research claims that a page is holistically understood with all its panels instead of understanding single panels. Then what role exactly the SSWs play in conveying the information? In Indian comics why there is lack in diversity of SSWs? Are SSWs made for only specific genre of comics? How the SSWs effect the reader's cognition process? What changes takes place in the absence of SSWs?

This study started as an analysis of SSWs in Indian comics, however since this study draws upon the existing printed comics, therefore the digital and motion comics remains areas for further study. With technological advancement, it becomes more demanding to seek the scope of SSWs in digital platform. This study was also limited to only onomatopoeic words used in Raj comics mainly because the other two categories (Phenomimetic and Psychomimetic) were not found. These lacunae posit the scope of research for image-text relationship in Indian comics and its equivalent media (such as story books, cartoons, infographics etc.).

References

1. Eisner, W.: Comics and Sequential Art. Poorhouse Press, Tamarac (1985)
2. McCloud, S.: Understanding Comics. Harper Perennial, New York (1994)
3. Ohala, J.J., et al. (eds.): Sound Symbolism. Cambridge University Press, Cambridge (1994)
4. Ohala, J.J.: Sound symbolism. In: 4th Seoul International Conference on Linguistics, pp. 98–103 (1997)
5. Goddard, C.: Semantic Analysis: A Practical Introduction. Oxford University Press, Oxford (2011)
6. Cohn, N.: Japanese Visual Language: The Structure of Manga. Continuum Books, London (2007)
7. Saussure, F.: Course in General Linguistics. McGraw-Hill Book Company, New York (1916)
8. Akita, K.: A grammar of sound-symbolic words in Japanese: theoretical approaches to iconic and lexical properties of Japanese Mimetics. Ph.D. thesis, Kobe University (2009)
9. Kohler, W.: Gestalt Psychology, 2nd edn. Liveright Publishing, New York (1947)
10. Hubbard, E.M., Ramachandran, V.S.: Synaesthesia—a window into perception, thought and language. J. Conscious. Stud. **8**, 3–34 (2001)

Anatomy of Bengali Letterforms: A Semiotic Study

Subhajit Chandra, Prasad Bokil and Darmalingam Udaya Kumar

Abstract The anatomy of letterforms defines the structural formation of letters. The study is based on semiotic approach. The methods used here are Syntagmatic and Paradigmatic analysis. The anatomy is developed through analysis based on the work on Latin letterforms from three different aspect which are structural grid lines, anatomical features and parameters. This syntagmatic analysis is yielded in identification of various structural features of letterforms like terminal, bowl, blob, stem, dot or nukta, ascender and descender. The analysis has been carried out using two techniques, repeated forms and unique forms of letters. The paradigmatic analysis discusses the comparative study of structure and feature of letterforms across different typefaces such as Lohit Bengali, Vrinda, Solaimanlipi and etc. The analysis offers distinct anatomical nomenclatures after analyzing paradigmatic transformations. Further the study categorizes the letterforms according to the appearance of common features.

Keywords Anatomy of letterforms · Bengali letterforms · Syntagmatic and paradigmatic analysis · Typeface design

1 Introduction

Over the last century, the anatomy of Latin script has been extensively studied by typographers and type researchers. Since the last two decades non-Latin scripts are getting more attention from the research community. 60 % or more of global population is dependent on non-Latin scripts including Indic scripts. It is globally used in education, politics, economics and cultural purposes. There are several print media like newspaper, hoarding, and poster are regularly getting printed using non-Latin scripts. Even non-Latin internet users are increasing day by day [1].

S. Chandra (✉) · P. Bokil · D. Udaya Kumar
Department of Design, Indian Institute of Technology, Guwahati, Assam, India
e-mail: c.subhajit@iitg.ernet.in

A. Chakrabarti (ed.), *ICoRD'15 – Research into Design Across Boundaries Volume 1*, Smart Innovation, Systems and Technologies 34, DOI 10.1007/978-81-322-2232-3_22

India is a multilingual country with various scripts. There are twenty-two official languages and eleven scripts in India [2, 3]. Bengali script is one of the most prominent Indic scripts used by 84 million Bengali speakers in India and 15 million in Bangladesh. The Bengali script is evolved from 'Siddham' script which is an offspring of 'Brahmi', the origin of all Indic scripts [4]. The letterforms of Indic scripts including Bengali are more structural and compositional complex [5, 6]. All Indic scripts are very different from each other with respect to their shape, proportion, height, width, stroke ratio and path of the stroke [5].

It is through the means of typefaces that any script can be printed or displayed for the purpose of communication. Bengali typeface design has elaborate history in print and publication over last 200 years [7]. Typeface design in Indic scripts involves knowledge of calligraphy and composition of the script. Understanding of the 'script composition grammar' of letterforms can certainly assist the type designer to design any typeface with better legibility and readability [4, 5]. The composition of Bengali script is not fully defined [8]. There is a scope for further investigation and design considerations. This paper focuses on grid and anatomical features of Bengali letterforms.

The enquiries on Bengali typeface anatomy are prepared based on two semiotic methods namely Syntagmatic and Paradigmatic analysis. The structural formation of a typeface is investigated using Syntagmatic analysis. The distinct shape of letter-parts are named by new terminologies or taken from Latin or non-Latin scripts based on appearance of the stroke characteristics. Most of the terminologies in Latin letterforms are based on animal anatomy like eye, ear, shoulder, leg, tail etc. [9]. Here, both plant and animal anatomical nomenclatures are used to identify letter-parts like stem, shoot, bud, knot, shoulder, leg, tail etc. The font used here for syntagmatic analysis is 'Lohit Bengali'.

The comparative study of anatomical features across various typefaces is conducted by using Paradigmatic analysis. The study is focused on the arrangement of structural features that vary in different typefaces. SolaimanLipi, Lohit Bengali, Vrinda and Rupali typeface are used for paradigmatic analysis. Further, categorization of letters is investigated according to common character and common structure.

2 Anatomy of Letterforms: Literature Review

The structural formation of a script is reliant on the tools used during the initial development of the script. The arrangement of letter-strokes reflect mediums like stone engraving, calligraphic brushing or Palm leaf lettering used to develop the particular typeface [4, 6, 10]. Many of Indic scripts share common mediums, though the 'script composition grammar' is different for each Indic scripts [3]. A grammar of anatomy formulates the structure of letterforms that assists type designers to design typefaces from a conception to its final letterform [5, 8]. The anatomy of Latin script and few of the non-Latin scripts is already well established [9].

The development of anatomy of letterforms is based on three distinct aspects. These are (1) Structural Grid Lines, (2) Anatomical Features and (3) Anatomical Parameters [2, 4–6, 9]. The structural grid lines, in practice, act as a ratio scale of height and proportion of alphabets [9]. The height to width ratio has an important role in designing of a typeface, targeting its use for a media [8]. Devanagari is the only Indic script where the use of structural grid lines is evident during design among the literature available on all Indic scripts, only Devanagari has a detailed discussion on structural grid lines [6]. Bengali script does not have such grids followed by practitioners; and it is sparsely discussed in literature [7].

Pelli et al. [11] assert that a letter identification is a recognition process of identifying its features. The Gestalt law of grouping plays a significant role in letter recognition by identifying combination, position and size of features [12, 13]. The letter-part that found in different letters repeatedly is the common feature and others are unique feature. Unlike Latin script [9], the anatomical features of majority of Indic scripts including Bengali are underdeveloped [8].

2.1 Latin Script

The research and development of Latin typeface suggest a well-defined anatomy of Latin letterforms [9, 10]. Most of the letters are combination of linear and curvilinear strokes. The structural complexities are fewer by using repeated comparable strokes and unique structural arrangement [9]. However, within the established forms there are still many possibilities for structural variation. The gridlines and anatomical features like Ascender, Bowl, Counter, Descender, Dot, Leg, Link, Loop, Shoulder, Spine, Spur, Stem, and Tail etc. (refer Fig. 1) are already defined, based on visual appearance of the letter-parts [9, 10].

2.2 Non-Latin Scripts

The anatomical foundations of non-Latin scripts such as Arabic, Chinese and Devanagari are already been developed [6, 14–17]. The significant works have been done in Arabic type design. The script has historical background of using calligraphic tools and written from right to left in repeated forms [14]. Horizontal toothy appearance is a specialty of this script [14, 15]. Similarly, Chinese letterforms are another example of calligraphic style of writing that successfully reproduces from print to digital displays. The letterforms are ideographic visual symbols that express emotion, narrative, motion and sentiment [16] and the central point of grid holds the visual balance of the letter. Figure 1 shows the grid system of Arabic and Chinese letterforms.

Devanagari script is used for writing Hindi, Marathi and other few languages and one of most explored Indic script. S.V. Bhagwat explains the anatomical

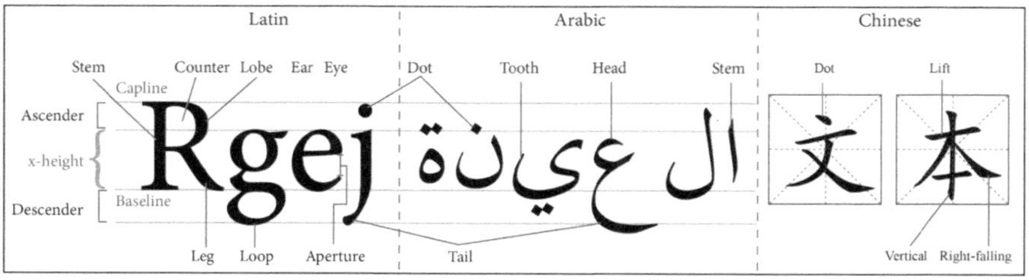

Fig. 1 Anatomy of Latin, Arabic and Chinese letterforms

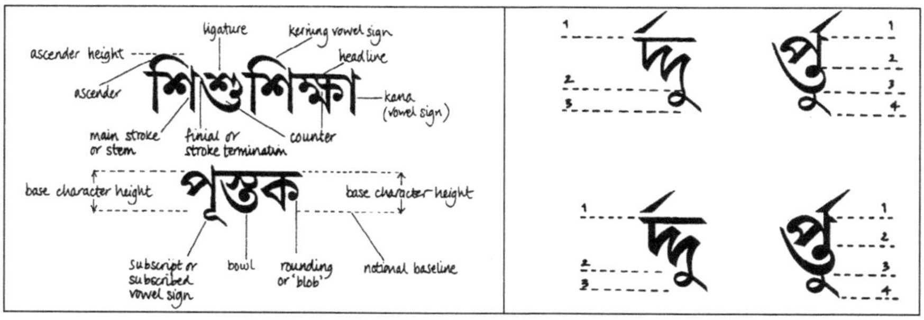

Fig. 2 Anatomy of Bengali. *Source* [7]

aspect of Devanagari script [17]. The letters are grouped according to appearance of common element of letterforms. Later, Naik [6] explains the grid system of Devanagari script based on Bhagwat's work [17].

2.2.1 The Bengali Script

The Bengali letterforms are sinuous. The positions of diacritics and juxtaposed letters (also known as conjuncts) suggest that the structure of the letterforms is complex. Ross [7] describes the basic grid lines and some of the anatomical features such as bowl, kana, rounding or blob, stem, etc. of the Bengali script (refer Fig. 2). Ross also identifies multi-tier grid system of the script in case of conjuncts.

2.3 Conclusion from Literature Study

The existing literature suggested that the structural grid lines are not fully defined in Bengali. Only base character height is identified by Fiona Ross. There are possibilities to identify more grid lines that segregate a letter vertically for better understanding.

The complexities of type design process and existing literatures indicate that there is a need of fine-tuning in the basic letter-parts of letterforms. Ross identifies only five features from five letters (as shown in Fig. 2). The research gap leads to an investigation on nomenclature of different letter-parts of all Bengali letters [8]. A standard anatomy helps to identify individual parts that lead to better understanding and improvement in field of type design.

3 Defining Anatomy of Bengali Script

The Bengali script consists of twelve vowel, thirty five consonants, the numbers and several punctuations. Apart from this, there are around five hundred conjuncts and few ligatures which are used for writing the language. Bengali script is an 'Abugida'. Every consonant ends with syllable of an inherent vowel. There is also a symbol 'hasanta' to quiet the sound of inherent vowel. Bengali is written from left to right. There is no uppercase or lowercase. So, there is no reference of x-height in Bengali script. There are four modifier signs such as Khanda-ta, Anusvara, Visarga and Chandra-bindu, used for contextual purposes [7]. The basic letterforms of Bengali are shown in Fig. 3.

Each vowel letter also has its own diacritic form and it is appeared with only consonant to adapt new sound of inherent vowel. The appearance of diacritics occur pre-glyph, post glyph, above-glyph and below-glyph with consonant. Some diacritics appear in variant forms or as ligature with specific consonant and their form is different than regular consonant-vowel forms [1, 7].

3.1 *Grid*

The grid system defines size and proportion of letters the grid model is an arrangement of virtual lines that constructs a vertical proportion of the letterforms. The grid model is prepared based on design practice to shape the Bengali letterforms taking the reference of existing literature [4–7, 17]. As shown in Fig. 4 the grid is mainly consists of 6 lines as

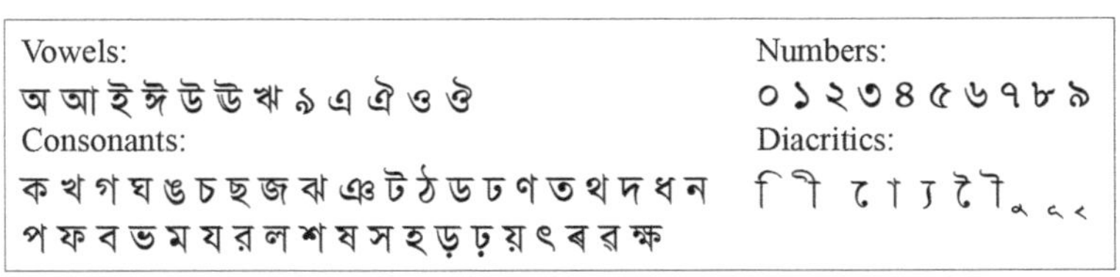

Fig. 3 Bengali Letterforms

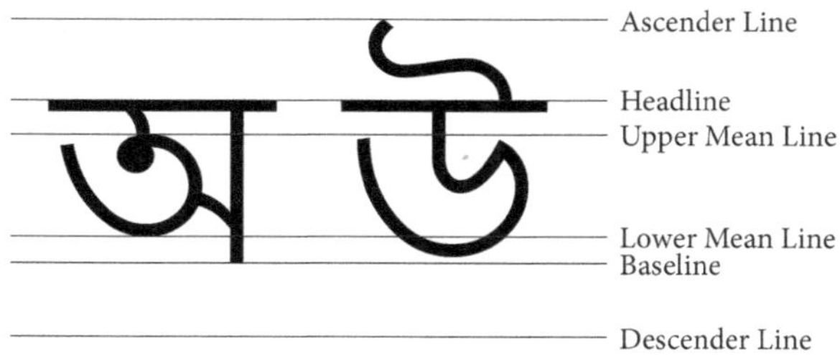

Fig. 4 Grid system of Bengali

1. Topmost Line/Ascender Line
2. Shiro-rekha/Headline
3. Initial Line/Shoulder Line/Upper Mean Line
4. Lower Mean Line
5. Lower Kana Line/Footline/Baseline
6. Extreme Bottom Line/Descender Line.

The distance between base line and headline is Base Character Height. Likewise, the distance between Headline to Ascender line is Ascender Height and Baseline to Descender Line is Descender Height. Headline is also known as Shiro-rekha, one of basic element of most of all Indic scripts. There is also two Mean line within Base Character Height, Upper Mean Line and Lower Mean Line. The identifiable body structure lies within the bound of upper to lower mean lines.

3.2 Anatomical Features

Syntagmatic analysis is a method to analyse the surface structure of any object. This method is used here to identify different anatomical features of a single typeface and to define its nomenclature. The analysis is carried out considering two facts, first the repeated forms among all letters and second the unique form of individual letter using prepared illustration as in Fig. 5. Then, a terminology is provided to each common forms that come out from the analysis of repeated form among letters. The process of feature analysis has done on vowels and consonants only using repeated forms that provides seventeen different identified features. Some of the similar features are segregated further like 'Blob' into 'Bud' and 'Knot' where Bud is connected at single end of a curve in letter but Knot is positioned at the joinery of two curves in letters 'A', 'E' and 'Ma'. Similarly, the 'Delta' feature is a combination or triangular formation of strokes in letterform as a main body element of letters 'Ka', 'Ba', 'Ra' and etc. There is a 'V'-like joinery named as 'Wedge' which is combination of a 'Stem' and 'Shoot' that started from the end of Stem in letters 'Ka', 'Kha', 'Tha' and etc.

The unique forms of individual letter are also identified and named accordingly such as 'Loop', 'Nose' and etc. Figure 6 is the detail analysis of letter 'A', 'Harsh-u' and 'Ka'. All letters are examined in same way and Table 1 is prepared with all possible nomenclature of vowel and consonant letters.

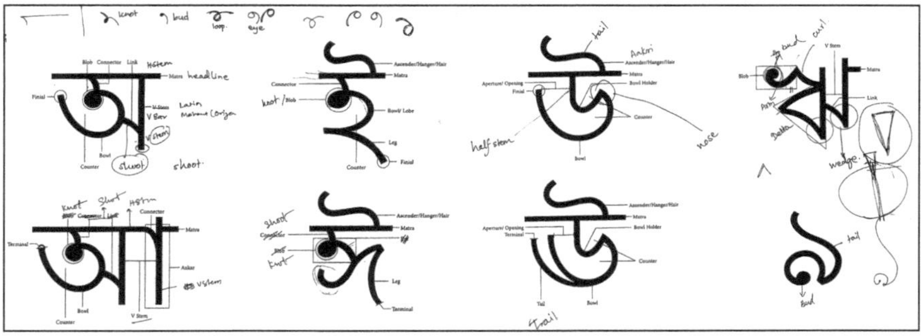

Fig. 5 Bengali Letter analysis

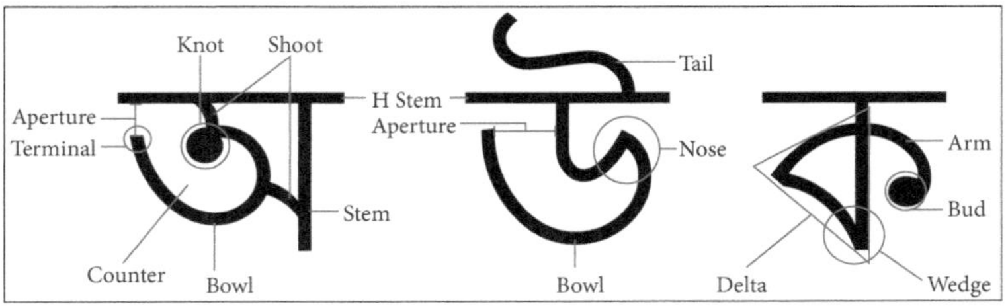

Fig. 6 Letter anatomy of 'A', 'Harsh-u' and 'Ka'

Table 1 Letter anatomy table

Terminology	Borrowed from	Description	Letterforms
Arm	Latin [9]	A curvilinear stroke within bound of 30 to 90 degree (approx.)	এ ঈ ক
Lobe	Latin [9]	A curvilinear stroke within bound of 90 to 180 degree (approx.)	ই গ ও
Bowl	Bengali [7]	A curvilinear stroke about 360 degree round	অ ত ড
Bud	*	A blob feature connected to Arm or Bowl or Lobe	অ ঋ ক খ ত
Knot	Bengali [8]	A blob feature connected to two continuous Arm or Bowl or Lobe	ম ই
Stem	Arabic [14], Bengali [7], Latin [9]	Vertical Bar	অ ক চ ন র ৱ
Half Stem	*	Short vertical Bar	উ ছ
Shoot	*	A stroke comes out from Stem or Half Stem	অ জ
Delta	*	Connected triangular stroke	ব ক ঝ
Tail	Arabic [14], Latin [9]	A stroke comes out from main letter part individually. Most of the Ascender is Tail in Bengali.	উ ই ঈ ঐ

(continued)

Table 1 (continued)

Terminology	Borrowed from	Description	Letterforms
Wedge	*	A 'V' shaped angle at bottom	অ ক থ
Loop	Arabic [14], Latin [9]	A round stroke with close counter	ঙ
Nose	*	A junction of two curves	উ ঊ ড
Dot or Bindu	Latin [9], Devanagari [6]	A Dot feature like in letter 'j'	র ড় য় ঢ়
Terminal	Latin [9]	Stroke end of main letter part	অ উ ত
Aperture	Latin [9]	Opening of Terminal	অ উ ত
Leg	Latin [9]	A stroke balancing the main body part	ই ঈ দ

* This term is introduced first time

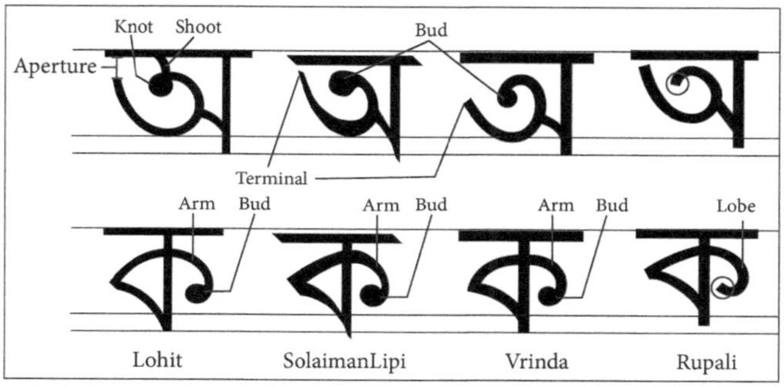

Fig. 7 Paradigmatic analysis of letter 'A' and 'Ka'

The paradigmatic analysis has been done on the chosen typefaces Lohit Bengali from RedHat Project, SolaimanLipi from OmniLab, Vrinda from Microsoft and Rupali from Ekushy Bangla. These typefaces are selected on the basis of variation of application context. Lohit Bengali is used for androids, Vrinda is used in PCs, Solaimanlipi and Rupali used for digital displays.

Figure 7 is the detailed study of letter 'A' and 'Ka' where letter 'Ka' (upper row) consists of Bowl and Stem. Only 'Ka' of Lohit typeface has Knot feature and rest of all have Bud due to absence of Shoot from the Headline. The Terminal cuts are distinct in each typefaces. The Aperture is also varying for each typefaces. Similarly the letter 'Ka' in Fig. 7 (lower row) of three typefaces Lohit, SolaimanLipi and Vrinda have Bud at the end of Arm. But in case of Rupali typeface, there is no Bud feature at the end of Arm. Here the Arm visually becomes like a 'Lobe'.

3.3 Anatomical Parameters

There are two anatomical parameters, stroke thickness and stress on stroke path are observed during analysis. These characteristics adopt from calligraphic style to typographic form. Most of the typefaces developed from manuscripts are high contrast. The thin to thick stroke significantly varies due to dominance of calligraphic tools. Here only SolaimanLipi typeface has the stress parameter. The stroke thickness and stress have significant role in letter legibility and the discussion is beyond the scope of this paper.

4 Categorization Based on Anatomy

The categorization has been done based on two parameters proposed by Mohanty (1998)–(1) common character and (2) common structure [5]. The groups of letter are prepared considering the appearance of common features or combination of features.

4.1 Common Character Parameter

Common character parameter identifies the groups of letter according to appearance of single feature within a typeface. Vertical Stem and Bowl are most common features of Bengali typefaces, encountered during feature analysis. The letters can be grouped based on these features in several ways—(1) Vertical stem at right side, (2) Vertical stem at middle, (3) Vertical stem at left side, (4) Hanging Bowl from top line/half stem, (5) Letters with leg and (6) Bowl at lower portion in Fig. 8.

Common Character Parameter:

Vertical Stem (Right Side)	Vertical Stem (Left Side)	Letters with leg
অ আ ঋ এ খ গ ঘ ণ থ ধ ন প ব ম য র ল শ ষ স য় ৰ ৱ	চ ছ ট ঢ ঢ় দ	ই হ ছ ঈ দ
Vertical Stem (Center)	**Bowl from Top Line/ Half stem**	**Bowl at Lower Portion**
ঐ ঞ ক ফ	উ ঊ ড ড় জ	অ আ ও ঔ ত ভ

Fig. 8 Grouping using common character parameter

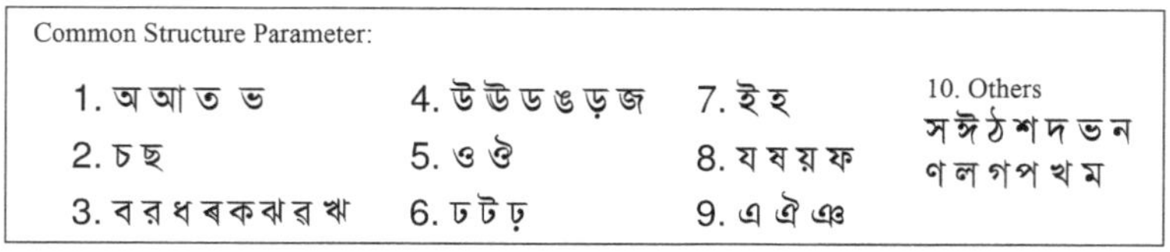

Fig. 9 Grouping using common structure parameter

4.2 Common Structural Parameter

Common structural parameter similarly provides several groups of letters in combination of strokes or features as a single unit. Figure 9 shows several groups consists of common structure letters.

5 Conclusion

The study offers a grid system for Bengali and a range of nomenclatures to identify different features that may help type designer to achieve rhythm and unity during design of a typeface. The horizontal to vertical ratio of letters can be achieved in practice by using the grid system [5]. The study proposes seventeen distinct features after analyzing only vowels and consonants over the five features that identified by Fiona Ross. The features and their position and shape can accompany to effective design of typeface that can solve the legibility and letter confusion to recognition issues [15, 18]. The features can also be used in OCR systems for detection of letters [19].

The study has been done only with vowels and consonants. The analysis of diacritics and conjuncts may provide more insight of grid system and features. The two semiotic analysis method have been used here to identify letter features. Further, the role of syntagmatic and paradigmatic transformation (such as addition and deletion or substitution and transposition) and the affordance of the letter shape can be discussed when a feature changes from one shape to another.

References

1. Ross, F., Shaw, G.: Non-Latin scripts: From Metal to Digital Type. St. Bride Foundation, London (2012)
2. Ghosh, P.K.: An Approach to Type Design and Text Composition in Indian Scripts. Department of Computer Science, Stanford University, Stanford (1983)
3. Sinha, R.M.K.: A Journey from Indian Scripts Processing to Indian Language Processing. IEEE Ann. Hist. Comput. **31**, 8–31 (2009)
4. Darmalingam, U.K.: Transformation of Tamil letterforms from Palm Leaf Manuscripts to Early Letterpress Printing. Ph.D. thesis, IDC IIT, Bombay (2010)

5. Mohanty, S.K.: The Formulation of Parameters for Type Design of Indian Scripts Based on Calligraphic Studies. Artistic Imaging and Digital Typography Lecture Notes in Computer Science, vol. 1375, pp. 157–166 (1998)
6. Naik, B.S.: Typography of Devanagari, vol. 1–3. Directorate of Language, Bombay (1971)
7. Ross, F.: The Printed Bengali Character and its Evolution. RoutledgeCurzon Publication, London (1999)
8. Ross, F.: Digital typeface design and font development for twenty-first century bangla language processing. In: Technical Challenges and Design Issues in Bangla Language Processing, pp. 1–15. IGI Global, Hershey, PA (2013)
9. Cheng, K.: Designing Type. Laurence King Publishing, London (2005)
10. Pflughaupt, L.: Letter by Letter: An Alphabetical Miscellany. Princeton Architectural Press, New York (2003)
11. Pelli, D.G., Burns, C.W., Farell, B., Moore-Page, D.C.: Feature detection and letter identification. Vision. Res. **46**(28), 4646–4674 (2006)
12. Sanocki, T.: Intra- and Interpattern Relations in Letter Recognition. J. Exp. Psychol. Hum. Percept. Perform. **17**(4), 924–941 (1991)
13. Pelli, D.G., Majaj, N.J., Raizman, N., Christian, C.J., Edward, K., Palomares, M.C.: Grouping in object recognition: the role of a Gestalt law in letter identification. Cogn. Neuropsychol. **26**(1), 36–49 (2006)
14. Abulab, S.D.: Anatomy of an Arabetic Type Design. Visible Lang. **42**(2), 189–200 (2008)
15. Chahine, N.:Reading Arabic: Legibility Studies for the Arabic Script. Ph.D. thesis, Leiden University, Leiden (2012)
16. Jiantang, H.: Chinese Characters. Cambridge University Press, New York (2012)
17. Dalvi, G.: Anatomy of Devanagari Typefaces. Des. Thoughts **1**(1), 30–36 (2009)
18. Fiset, D., Blais, C., Ethier-Majcher, C., Arguin, M., Bub, D., Gosselin, F.: Features for identification of uppercase and lowercase letters. Psychol. Sci. **19**(11), 1167–1168 (2008)
19. Sarwar, H., Rahman, M., Akter, N., Hossain, S., Ahmed, S., Chowdhury, M.R.: Selection of an optimal set of features for bengali character recognition. In: Technical Challenges and Design Issues in Bangla Language Processing, pp. 96–116. IGI Global, Hershey, PA (2013)

Redefining the Grid in Visual Design

Prasad Bokil

Abstract Grid is a well-known structural tool used for space organisation. Though there is a common consensus about the concept of grid it doesn't have a unique definition accepted by the design community. Creating a definition of grid will help to understand the scope of this concept, to define the research premise and ultimately to locate it in the design research domain. This paper tries to construct a definition of the grid in design. It demonstrates a methodology to construct a definition from collected consensus. The methodology used by Sarkar and Chakrabarti to develop a definition of creativity is taken for reference. The outcome is a theoretical definition of grid and its version as a working definition for graphic designers. After discussing the shortcomings of the definition created out of consensus a logical framework of knowledge representation is implemented to generate a definition. The FBS framework developed during last couple of decades is used to create this definition.

Keywords Grid · Graphic design · Definition · FBS framework

1 Introduction

Grid is a common tool in graphic design. The history has preserved and documented some grid-based architectures [1] and designs [2]. In the modern times graphic design has been redefined with design schools like Bauhaus and Ulm. With the advancement of printing technology, grid has gained importance in print and publication design. Grid as a design tool is extensively exploited in the domain of type design and typography. The first extensive writing on grids was done by Joseph Müller-Brockman in the form of a book—Grid Systems

P. Bokil (✉)
Department of Design, Indian Institute of Technology, Guwahati, India
e-mail: prasad.bokil@iitg.ernet.in

A. Chakrabarti (ed.), *ICoRD'15 – Research into Design Across Boundaries Volume 1*,
Smart Innovation, Systems and Technologies 34, DOI 10.1007/978-81-322-2232-3_23

[3]. It explains the role of grid in design, its advantages and demonstrates formulation of grid through design steps. This is followed by many examples of design layouts. The reader is expected to understand the process of design from the examples and design descriptions. Almost all literature published on grids [4–9] and about typography [10–12] follows the same model for knowledge transfer.

This model of knowledge transfer is based on direct internalisation of knowledge through observations. The available literature discusses the grid philosophically and then focuses on examples for practicalities; finally the readers are left to decode the structure and the involved thought process. The procedural knowledge is not externalised. There is a need to make this knowledge explicit, so that it can be communicated and shared easily by the design community. The current status of literature on grid shows incomplete knowledge externalisation, lack of knowledge representation and an inefficient model of knowledge transfer.

2 Definition of Grid

The ambiguity in articulation of grid starts with the definition of grid. Many design books just avoid defining grid; they use this term as a lay term. The literature does not give any common definition. There are many descriptions and definitions available, but none that covers the total concept of grid. For the concept of grid, expanded beyond the application of typography, the definition should also be more inclusive. The grid can be described with working experience or it can be explained to connect with epistemology. Grid etymologically means the mesh of horizontal and vertical lines, but practice has shown abundance of variations in the form and usage of grids. It is no more possible to restrain the term to a set of vertical and horizontal lines.

The objective of this paper is to construct a definition of grid which is valid in the design domain as well as outside it. Many experienced designers have talked about the use of grid for efficient designing, giving varying descriptions. However, in design practice it is the consensus which gives the notion of grid on the gross level, rather than a definition. It follows the general meaning of framework, guidelines or understructure which is then connected to its application in design, like effective layout, cohesive style, etc. But among these various descriptions it is necessary to find a common definition which will represent all possible grids. Descriptions of grids are acceptable in the design domain, but those are associated with personal opinions and lead to subjectivity. The ambiguity about the definition of grid needs to be eliminated in order to bring objectivity in research. Developing the definition is the effort towards objectivity which is required for validation and repeatability in the research work.

3 Methodology for Developing a Definition

This paper is an attempt to formulate the conceptual and working definition of grid which will encompass all possible forms and applications of grids. Developing a common definition involves the articulation of the concept of grid in design, shared among the designers. In graphic design, the grid is taken for granted as a tool; hence articulation of its definition is very sparse in the literature. The available descriptions are usually very trivial, sketchy and incomplete.

This research started with collection of all kinds of definitions and descriptions of grids given by design practitioners which were then analysed to formulate a common definition. The methodology used by Sarkar and Chakrabarti [13] to develop a definition of 'creativity' is taken for reference and the methods, 'Majority analysis and Relationship analysis', are borrowed and modified for this case. Majority Analysis is used to generate a common vocabulary to describe grid and Relationship Analysis has helped to create hierarchy and order among these articulations. The results from these two analyses have then been evaluated. The difference between Sarkar and Chakrabarti's work and this work is—(a) Creativity is a concept whereas grid is a tool in physical form, (b) Creativity is a theoretical concept; hence the data considered for their analysis was mainly gathered from researchers working in the related field whereas grid is a design tool in practice; hence the data was gathered from all design practitioners as well as design scholars, (c) Creativity being an abstract term the success of the definition depends on the agreement within the community on what is acceptable as the 'common' definition. The definition of grid also needs agreement, but at the outset, it needs to encompass all the existing practical forms and functional possibilities of grids and (d) since we are working with the description and not the concrete definitions we need to formulate the definition to satisfy the combined description.

3.1 Collection of Descriptions and Definitions of Grid

The descriptions of grid were collected by three sources—published literature, personal interviews and online questionnaire. All secondary data from the available literature was compiled and all primary data collected by interviews and questionnaires was collected together. The attempts to define the grid as well as the loose descriptions are included in the data pool. All this raw data is available at [14].

3.2 Primary Data Collection

The descriptions collected from interviews and questionnaires are tabulated together. Four types of responses came from designers when they were asked to define the grid—definition, description, simile and other information. All responses telling 'what the grid is' are labelled as definitions. Descriptions are loose definitions. They give information about the application, form and use of grid. Many similes are given by respondents where they have compared the grid to something else. All other information like comments, suggestions, references are put aside as other information. The data which does not assist in any way to define the grid is omitted. Out of more than a hundred descriptions only 46 are stated in the form of definition, and some of them are listed below. Although not in the form of definition, many descriptions are also taken into account for further analysis.

1. A grid is a backbone structure of a layout which defines the relative position of elements of design against each other.
2. A grid is a consistent system for placing objects.
3. A grid is a method of organising visual content in an orderly, consistent approach.
4. A grid is a tool to help organize information in a design, taking into consideration a harmony between page size, hierarchy of type and other elements.
5. A grid is something that is used to maintain the order in visual elements.
6. Grid to design is like a skeleton to the body. One does not see the skeleton but it holds the body together and gives it a form, a proportion and a discipline.

3.3 Secondary Data Collection

The descriptions of grid from all the literature available were collected. It mainly includes books and web sources. The language of publication is the limitation. This data is limited to the published material in English. Out of 16 descriptions collected from literature few are given below:

1. Grid is a predetermined understructure in publication design [4]
2. Grid is an organising principle in graphic design [8]
3. Grid is a geometric division of space [6]
4. Grid is an underlying structure used to guide the design and placements of elements on the page [15]
5. Grid breaks space or time in regular units [16]
6. Grid is an organisational tool which determines the internal division of the page, and the layout of the position of elements [11]

4 Analysis

There is a wide range of descriptions from an abstract one like—'most vivid manifestation of the will to order' [9] to very practical one like—'two-dimensional structure made up of a series of intersecting vertical and horizontal axes used to

Table 1 Sample analysis of definitions/descriptions of grid

	Noun	Adverb/ adjective	Verb	Application	Other key word	Function (purpose)
1	Guidelines	Easier	Creates order	Broacher, book, interiors, etc.	Reproduction, restrictions	Make task easy, creates order
2	Tool		Helps the eye to navigate	Visual design	Accuracy	Helps the eye
3	Approach	Consistent	Creates an order, hierarchy			Creates order and hierarchy
4	Series of lines		To organize or structure forms			Organize or structure forms
5	A method	Orderly, consistent	Organising visual content		Orderly, consistent approach	Organize visual content
6	A tool, supporting system		Helps to create an order			Order, clarity

structure content' [17]. The initial analysis of 127 descriptions (part of it is shown in Table 1) shows that there are typical noun phrases and verb phrases used to describe the grid. They are either ontological or teleological. Each definition is analysed for important nouns, verbs and adverbs or adjectives which give additional information about nouns or verbs. The information about the form and function of grid is also decoded. Few descriptions are presented as similes where the grid is compared to some other object, like Japanese bento box. Such similes cannot be considered as definitions but they give rich descriptions about the role of grid in design hence are tabulated separately. The applications mentioned in different descriptions and the important keywords are listed down. Most of the descriptions focus on the teleological aspect of grid. The grid is positioned according to its usage in design. The use of grid is tabulated under the function column. The list of applications shows that the concept of grid is usually defined in the area of typography.

These descriptions are further analysed in the next subsection to formulate a common definition with the phrases selected by the majority consensus. Such definition formed is then reviewed for better relationships among these phrases for more explicit definition.

4.1 Majority Analysis to Arrive at a Common Definition

Majority Analysis leads to a definition which is generated by the terms and phrases used by a majority of designers. A common definition should incorporate what is common across the existing definitions. The important phrases in the definitions are identified and then their occurrences are counted across all the definitions listed above. The occurrence of a phrase is not counted more than once for each description. The phrases are then grouped together semantically. There are four clusters observed within the common terms. They are concerned with 'taxonomy, purpose, function and context'. Table 2 shows the clusters of common phrases and the selected one for creating a common definition.

From the Table 2, the phrases selected from each cluster to make a common definition of grid can be assembled.

$$\text{The grid} = \text{structure} + \text{use} + \text{order \& consistency} + \text{visual} \tag{1}$$

Hence,

The grid can be defined as a structure used to bring order and consistency in visuals. (d1)

Majority Analysis had given a definition of grid (d1) which can be called as the most common definition. The form and application is not explicit hence can be

Table 2 Majority analysis cluster selection

Cluster semantics	Noun/verb phrase	Occurrence	Selected phrase	Comment
What is it? (taxonomy)	Tool	12	Structure	Tool and structure has equal statistical weight but contextually structure is more suitable. The tool is not the part of final product whereas structure is. It may or may not be visible but it can be perceived
	Structure	12		
	System	7		
	Lines	6		
	Guideline	2		
	Set of (…)	5		
	Method	3		
	Aid	3		
What does it do? (purpose)	Help	11	Use to	Help and use has equal count but the verb 'use' suggests the choice made by designer and it sounds more neutral than help. None of the published work uses 'help'
	Use + Used to	11		
	Organize	3		
	Maintain	6		
	Bring	5		
What is it for? (function)	Order	13	Order, consistency	All these are attributes and they are not overlapping each other. Order is used maximum. Consistency and homogeneity can be clubbed together
	Hierarchy	3		
	Consistency	6		
	Proportion	3		
	Harmony	2		
	Homogeneity	1		
Application context	Visual	13	Visual	Layout and page are phrases exclusive to typography. Selecting this word is focusing on visual domain
	Layout	8		
	Page	6		

expanded for a wide range of applications. It can be observed that the definition thus created is still not inclusive of all functions of grid; it only mentions order and consistency. Also, it is already established that the concept of grid in common consensus is not enough to encompass all possibilities. It is necessary to understand the relationships among the phrases used in the definition to build the hierarchy.

4.2 Relationship Analysis to Arrive at a Comprehensive Definition

In Majority Analysis the essential features are identified from the collection of definitions. But it does not take into account the relationships among these phrases and their hierarchy and importance in the definition itself. In Relationship Analysis, these phrases are analysed for relationship so that a more comprehensive definition can be formulated. After analysing all the definitions it is found that each definition consists of a few basic semantic units. They necessarily describe the function of grid, its form and/or locate it in the design application context. In the hierarchy, the form is the basic feature, then there is the function and the last is the context. The description of the form is sufficient for the grid to exist but the definition is complete after the inclusion of the function. The context can elaborate different shades of the concept of grid. We are defining the grid for the design community, hence within the designers it might not be required to add context. The hierarchy relationship of all the features of definition is shown in Fig. 1.

The Form is made up of two clusters- ontological form and physical form. In the definition it is necessary to state what the grid is. It is called as tool, structure, method, approach, system etc. Next is the physical form. Physically the grid

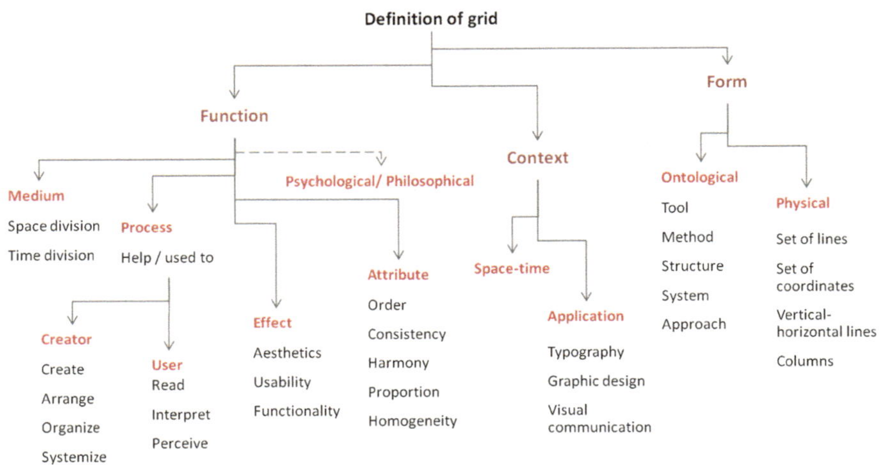

Fig. 1 Relationship analysis of definition of grid

is made up of different entities like lines, points etc. and the description of these entities, with relation to each other, is sufficient to identify the grid. The context is mentioned in some of the definitions. It is mainly connected to the type of the design application. Some of the definitions mention the typographic reference such as page layout. Many definitions just mention the visual medium and keep it open to include any application in visual communication.

Functional features of the definition have wide variations and levels. Four types of functional features are identified. Medium function is a feature which comments on the medium of grid. Process function is the cluster of the verbs phrases like 'helps (person) to (verb)'. It describes the process in which grids are involved. It is directed through either creator or user. For example—(a) helps designer to organize, (b) used by designer to bring order, (c) helps viewer in reading the hierarchy, etc. Sometimes the person is omitted and the process is assigned to the grid directly, like 'the grid determines the internal division of the page'. Attribute function gives the attributes of the designed visual as achieved by the use of grid. It can be a list of attributes like order, consistency, harmony, proportion, etc. A few definitions directly claim the effect of the grid as to improve aesthetics, usability or functionality. This is termed as effect function. In some of the cases, descriptions become abstract and describe psychological or philosophical aspect of grid like 'manifestation of will to order'.

From the above discussion, the definition of grid can be said to be composed of these features together as given below.

$$\begin{aligned}\text{Grid} = {} & (\text{Ontological form} + \text{Physical form}) \\ & + \begin{pmatrix}\text{Medium} + \text{Process [(Person) Action]} \\ + \text{Attribute} + \text{Effect} + \text{Psycho} - \text{Philosophical}\end{pmatrix} \\ & + (\text{Application context} + \text{Socio} - \text{cultural context}) \qquad (2)\end{aligned}$$

This definition after adding each feature becomes very clumsy. The attributes need not be included in this definition, as there can be many attributes which may vary from application to application. The definition should not depend on such a variable feature. Sometimes, grid is intentionally used to break the order or make the design heterogeneous. In such cases the definition with attribute function will become invalid. To avoid this situation, attribute features are removed from the definition. Also, the effect functions, such as improving the aesthetics or usability of designed visual, are very subjective features and depend on the skill and capability of designer and the nature of the content. Even though the expected effect couldn't be achieved the grid might be visible. Hence, the effect functions too are removed from the definition. Rearranging the phrases, replacing some of them with other phrases from the same category and after simplification the definition is as below:

> *Grid is a pre-determined foundational structure in the geometric form which creates the consciousness of space and guides the designer to create a visual representation and arrange the visual information in a demarcated space* (d2)

This is the working definition of grid for graphic design. This definition is contextualised to visual design in two or three dimensions. This is not valid for time division or grids dividing other media. To compare with the similar principle like grid from other contexts the definition needs to be more generic without any contextual restrictions. This exercise of developing a definition with Majority and Relationship Analyses of a collection of definitions is unable to yield an objective definition valid for all forms, functions and contexts of grid.

These methods of Majority and Relationship analysis are scientifically well defined but the success of their results depends on the nature of data. As commented before in graphic design there is a void in terms of objective articulation and scientific discourse. The data does not cover all the possible forms and functions of grid. Although this exercise did not yield an objective definition as required, it was fruitful for this research activity. It covered a large spectrum of potential key concepts for defining grid and also helped to identify the missing concepts.

5 Using FBS Framework for Grid Definition

The definitions (d1 and d2) were created by majority analysis and relationship analysis, but these definitions are subjective and dependent on the context, designer and/or effect on the visual. This subjectivity makes the representation ambiguous and open for multiple interpretations. To make the representation objective it is necessary to understand it through a logical framework rather than consensus. The FBS framework, originally proposed by Gero [18], has been proved very fruitful for design knowledge representation. Grid can be considered as a design created by designer to assist for further visual design. The two layered FBS representation of visual design using grid is proposed and discussed by Bokil and Ranade [19]. This framework can be used to derive an objective definition of grid. Let's understand the grid in terms of its variables and values of variables.

There can be many function variables expected from a designed grid and each variable can have different values. The function based on the effect on visual is subjective and so is the definition derived from it. It may vary from person to person and can be biased. Although the grid is claimed to bring order, discipline, consistency and homogeneity, these variables cannot be included in the definition. The definition of grid should hold true independent of its effect on the visual, because in spite of the grid the designer may fail to achieve the required effect.

Structure variables are essential to describe any grid but they are not of much use in defining the grid. The structural elements can be in the primitive form of point, line, curves or in the structure form of squares, circles, columns or can be more complex like combination of regular, irregular shapes. It is not feasible to include all possible descriptions of structure in the definition. The definition will become a long list of all that is represented by Eq. 2. It can be observed that in all definitions/descriptions listed above, emphasis is given on function and/or structure of the grid. In all articulations and externalisations there is hardly any mention of behaviour and the behaviour variables of the grid.

Nevertheless, the grid is characterised by its behaviour, whatever its structure and whether the function is achieved or not. Behaviour of grid actually connects the structure of visual to the structure of grid. Though there are a variety of behaviour variables of grid [19], the grid exists even if there is only one grid variable with only one possible value. It is possible to generalize various structural behaviour variables into a single behaviour description. In whatever form, the grid fixes the set of options from the continuation of infinite possible values for one or more behaviour variables of grid which can be used by the designer while designing the visual. Hence the characteristic behaviour of grid is the **discretisation** produced by its structure. Thus the definition emerges as:

Grid is a structure used to discretize any continuum (d3)

In graphic design, the definition is limited to space continuum but it can be extended for other applications. This definition is independent of structure variables and function variables and hence independent of form, content, context and design agents. It can cover all existing and potential forms and functions of grid.

6 Conclusion

This paper has successfully demonstrated a systematic way to formulate a definition of a commonly used term in design. Majority analysis and Relationship analysis has given an in depth understanding of the concept. The definitions created with these methods do not cover the complete scope of this concept due to the subjectivity of consensus and biased notion about this tool. Equation 2 resulted from Relationship analysis has lot of potential to produce any working definition of grid for a given context. The definition created based on FBS ontology is generic and simple. It holds true for all possible forms and functions of grid. FBS framework can help to identify non-variable aspect of any entity for its stable definition. The analysis presented in this paper and the definitions evolved through it are definitely useful for any further research based on grids.

References

1. Kramrisch, S.: The Hindu Temples. Motilal Banarasidass Publication, Delhi (1976)
2. Boner, A.: Principles of Composition in Hindu Sculpture. E. J. Brill, Leiden (1962)
3. Müller-Brockman, J.: Grid systems in Graphic Design. Arthur Niggli Ltd., Switzerland (1981)
4. Vignelli, M.:Grids: Their Meaning and Use for federal Designers, Vols. No. 036-000-00038-4. National Endowment for the Arts, USA (1976)
5. Hurlburt, A.: The Grid—A Modular System for The Design and Production of Newspapers, Magazines, and Books. Van Nostrand Reinhold Company, New York (1978)

6. Swann, A.: How to Understand and Use GRIDS. Quarto Publishing plc, USA (1889)
7. Bosshard, H.R.: The Typographic Grid. Verlag Niggli, Switzerland (2000)
8. Samara, T.: Making and Breaking The Grid—A Graphic Design Layout Workshop. Rockport Publishers Inc, Beverly (2005)
9. Vinh, K., Boulton, M.: Grids are Good. Retrieved 23 Sept 2010 (2007, March 10)
10. Ruder, E.: Typography, A manual of Design. Hastings House, Publishers Inc, Florida (1984)
11. Baines, P., Haslam, A.: Type and Typography. Watson Guptill Publications, New York (2005)
12. Elam, K.: Typographic Systems. Princeton Architectural Press, New York (2007)
13. Sarkar, P., Chakrabarti, A.: Studying engineering design creativity—developing a common definition and associated measures. In: Gero, J. (ed.) Studying Design Creativity. Springer, Berlin (2008)
14. Bokil, P.P.: Knowledge representation of grids in graphic design and its application for analogy-based design. Ph.D. thesis, Indian Institute of Technology Bombay, Mumbai (2013)
15. Clair, K.: A Typographic, a Primer to History, Techniques and Artistry. Wiley, New Jersey (1999)
16. Lupton, E.: Thinking with Type. Princeton Architectural Press, New York (2004)
17. Grid (Page Layout). Retrieved November 2011, from Wikipedia http://en.wikipedia.org/wiki/Grid_%28page_layout%29 (2011)
18. Gero, J.S.: Design prototypes: a knowledge representation scheme for design. AI Mag. **11**(4), 26–36 (1990)
19. Bokil, P., Ranade, S.: Function-behavior-structure representation of the grids in graphic design. In: Gero, J. (ed.) Design Computing and Cognition'12, pp. 533–552. Springer, Berlin (2014)

Investigating Frequently Used Stroke Sequence, of Handwritten Devanagari Letters, by Observing Second Script Learners

Santosh Kshirsagar and Ravi Poovaiah

Abstract Eleven scripts are used in multilingual India today. However, not enough research is found in the domain of handwriting acquisition. In this study, we investigate Devanagari handwriting activity by observing its second script learners. A text paragraph containing basic Devanagari letters, numbers and few consonant clusters, is given to a selection of students, who have learnt Devanagari as the second script. This activity is captured using a movie camera. A similar activity is conducted on a student who has learnt Devanagari as the first script. We separate key letters from the scanned manuscript for observation of shape and stroke sequence. To observe the direction and sequence of strokes, we separate video clips of key letters. In this paper, we present a detailed analysis of the key letter 'Ka'. It is observed that second script learners exhibit a variety in the types of stroke sequences used to produce the letter. Cross-checking the stroke sequence in a prescribed school book, provides us insights on the ideal stroke sequence.

Keywords Handwriting acquisition · Second script · Stroke sequence · Devanagari

1 Introduction

Needless to say, in the context of handwriting, an individual's first script will always have structural influence on the second learnt script [1]. As is the case in the first language, the learnt first language always has an influence on the second learnt language in grammar, dialect, pronunciation etc. In a multilingual country like India, we all are aware of this fact as our general observation supports it.

S. Kshirsagar (✉)
Sir J.J. Institute of Applied Art, IDC, Indian Institute of Technology, Mumbai, India
e-mail: kshirsagar.santosh@gmail.com

R. Poovaiah
Industrial Design Centre, Indian Institute of Technology, Mumbai, India

A. Chakrabarti (ed.), *ICoRD'15 – Research into Design Across Boundaries Volume 1*,
Smart Innovation, Systems and Technologies 34, DOI 10.1007/978-81-322-2232-3_24

At present, India uses eleven different scripts for 26 officially acknowledged languages in the country [2]. Devanagari is one of the scripts that are widely used in India and the rest of the world [3]. Devanagari is also used for Hindi, a language that has been endorsed for official correspondence by the Indian government [4]. Hence, along with every state's regional language, every Indian school-going child has to learn English and/or Hindi as a second language. This makes Devanagari a second script for many of the states. Extensive research has been conducted in order to understand first and second language acquisition [5–7]. However, the focus of such research has been on reading acquisition. On the other hand, minimal scientific research has been conducted on handwriting acquisition in any Indian scripts so far [8].

We can refer to handwriting research done in other countries. A great deal of scientific research has been conducted in the Latin/Roman script culture [9]. We get to see published research in the more generalisable areas like, how to hold a pencil or how the writer's finger grips the pencil [10], how to hold writing paper, and at what angle? [11], what should the seating posture be [12], and at what angle should the eyes be in order to meet the required focus?

Recent research covers cognitive, perceptual and motor skills in relation to handwriting acquisition [13]. Handwriting research is conducted with various objectives. There is a very thin line between handwriting as a product and as a process. Our focus is the activity or process of handwriting.

As aptly termed in palaeography, we are, in short, investigating the ductus[1] [14] followed by second script learners of handwriting. Every handwritten letter is learnt by practicing the stroke sequence, and these sequences provide the flow and rhythm to the handwriting activity [15]. This also affects the form of the letters. To separate the directionality of the stroke and stroke sequence within a letter is a challenge, since they are inextricably related to each other [1].

In this exercise, we shot the handwriting activity of the second script students using a video camera, and separated and counted the clips to decode the acquired stroke sequence. We separated select key letters from the scanned manuscript for observation of shape. To observe the direction and sequence of strokes, we separated video clips of the key letters. In this paper, we present a detailed analysis of the reproduction of the Devanagari key letter 'Ka'. We found variety in the types of stroke sequences used by the second script learners. The correlation with stroke sequence that is mentioned in a schoolbook helped us make a decision about frequently used stroke sequences.

---

[1] *Ductus: the overall, general "nature" of the production of a given script, defined in terms of the "number, sequence, and direction of the strokes used in forming each letter of the script's alphabet. A knowledge of ductus, which is more concerned with the dynamic than with the static aspect of letters, can be helpful in reading, dating, and placing scripts, but its most important service comes in explaining changes in the appearance of letters. It is largely ductus which determines where the inertial forces and strains generated by rapid writing will express themselves. But if ductus can help generate changes in appearance, changes in appearance can also generate changes in ductus."

2 Method

Ten students each, of fifth and seventh grade, from Telugu and Tamil medium schools, were randomly selected as subjects for observation. A total of $10 \times 4 = 40$ students participated. A paragraph was given to them for copying on an un-ruled paper. We video-recorded their handwriting activity. The letter 'Ka' was selected as a representative key letter for observation. The video clips and scanned manuscript of the letter 'Ka' reproduced by the subjects, was then cut and separated for observation.

2.1 Selection of the Students as Sample

It was assumed that students, who have learnt to write any script other than Devanagari as a first script, will carry some visual influence of that first script to Devanagari writing learnt as the second script [1].

After comparing all Indian scripts, we found that the Gujarati script is quite similar to Devanagari in basic appearance except for the latter's characteristic top line. Bengali has a prominent top line feature like Devanagari. Therefore, it was decided that students with a South Indian script background would be more appropriate as subjects for this enquiry. One more reason is that South Indian scripts are basically circular in visual appearance in comparison to the square-looking Devanagari script [16]. South Indian scripts were found to be radically different-looking from the Devanagari script. Moreover, government-run Telugu and Tamil municipal schools were more conveniently accessible in Mumbai than Malayalam and Kannada medium schools. These vernacular municipal schools teach Hindi or Marathi as the second language, which means that the Devanagari script is taught at the secondary level (since it is the state language of Mumbai, Maharashtra). Hence, Telugu and Tamil students who have learnt Devanagari as a second script, served as an appropriate sample for this project (Table 1).

2.2 Designed Text Paragraph for Testing (Fig. 1)

A paragraph was designed that included all the basic Devanagari 'Varna Mala' (basic vowels and consonants) along with a simple conjunct consonant cluster, and ending in numerals from 0 to 9. The paragraph also takes anticipated care to weave the letter 'Ka' ten times into the text in order to facilitate more focused micro-observation.

Table 1 Background experience of students served as a sample

No.	Count	School background	Grade	Experience writing Devanagari
1	10	Telugu medium municipal school	5th std.	1 year as a second script
2	10	Telugu medium municipal school	7th std.	3 years as a second script
3	10	Tamil medium municipal school	5th std.	1 year as a second script
4	10	Tamil medium municipal school	7th std.	3 years as a second script
5	01	Marathi medium school (served as a reference sample)	4th std.	3 years as a first script

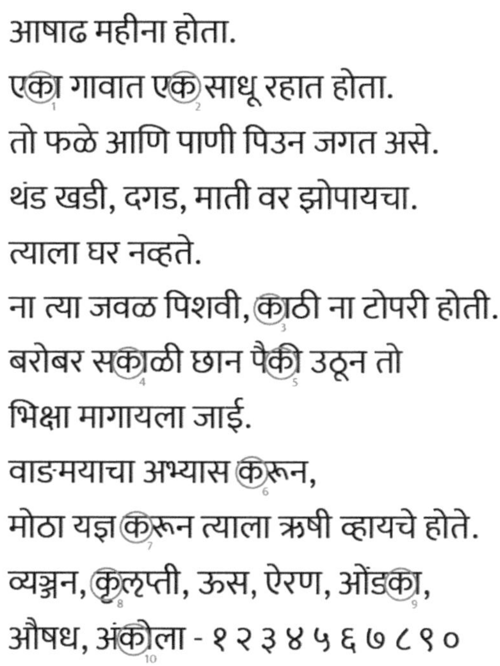

Fig. 1 Text paragraph includes 10 'Ka' letters

2.3 Selection of Devanagari Letter 'Ka' as Key Letter (Fig. 2)

Devanagari is written from left to right and has a continuous top line on every word in a sentence. Out of the 52 basic level* letters, 29 have a vertical bar (perpendicular to the top line) on the right side, fourteen have a short vertical bar, four letterforms have a middle vertical bar, and three begin with a circular stroke from the top line. Two letters are odd, i.e. they do not fall into any of the above groups. Apart from these geometric visual features, Devanagari also has compound clockwise and anti-clockwise curves with open and/or closed knots and some with a closed loop.

Fig. 2 Why letter 'Ka' as key letter

Among the basic level Devanagari letters, structurally, 'Ka' is a clear/countable combination of vertical and horizontal straight strokes with anti-clockwise and clockwise circular strokes. The letter 'Ka' is more frequently in use than even the letter 'La', the only similar looking letter existing in a basic level set. Also there are more complex letters with complex strokes but which are either less frequently used, or the strokes within them are less in use in the entire set of basic level Devanagari.

(*Basic level = only vowels and consonants [17].)

3 Procedure

We designed a text paragraph in Devanagari containing the letter 'Ka' 10 times, along with basic letters, numbers and a few consonant clusters. Ten students each, of fifth and seventh grade, from Telugu and Tamil medium schools, were randomly selected as subjects for observation. The paragraph was given to them for copying on un-ruled paper. We video recorded their handwriting activity. The letter 'Ka'

Table 2 Folder and file system for data management

Folder-1 = School name	Folder 1.a and 1.b	Folder 1.a.1 to 10	File names
Folder-1 = Tamil-5 = Tamil school fifth standard, Likewise Tam-7, Tel-5, Tel-7	Tam-Video	V-Tam-5-S-1 to 10	Ka-1 to Ka-10
	Tam-Manuscript	M-Tam-5-S-1 to 10	Ka-1 to Ka-10

Tam-5 Tamil fifth standard, *V* video, *M* manuscript, *Tam-5-S-1* Tamil fifth standard student one, *Ka-1* first attempt at producing the letter 'Ka'. Likewise *Telugu* Tel
For example /Tamil-5/Tam-Video/V-Tam-5-S-3/Ka-7

was selected as a representative key letter for observation. The movie clip data was then cut and separated. 10 handwritten Ka's, from each of the 40 students, produced 400 visually representative possibilities of the letter. Each individual possibility was sketched, and we derived 24 different types of 'Ka's based on shape and stroke sequence. After tabulating the data, we were able to ascertain the frequently used 'stroke sequence types' from the 24 types. We cross-checked these stroke sequence types with a reference student, who has learnt Devanagari as the first script. We further cross-checked the stroke sequence with a prescribed schoolbook for learning handwriting.

3.1 Organization of Data

Two types of data were generated: the first being a video of the activity, and the second, the scanning of the manuscripts. There was a need to create a short naming system to save each student's written page and then each of his/her separated letterforms. Ultimately we created several files and corresponding folders to carefully segregate them. Following is the discipline followed for organising the data.

3.2 Folder and File System for Data Management (Table 2)

Video cut clips folder name = Video, Scanned and cut manuscript = Manuscript.

4 Instrument for Analysis, Data Counting and Separation

The data that was received in manuscript format did not allow us to decode clearly the writing sequence used by the students. It was not possible to memorise and count many movie cut clips for categorisation purposes. From the entire data, every unique sequence possibility in handwritten 'Ka' was sketched first, by observing the cut video clips. Making a visually clear sketch of the 'Type of sequence' that is found while referring to individual video clips, has helped us as a data counting and categorising instrument. It has also helped us to define 'Type' in the context of 'stroke sequence in handwriting'. For clearer categorisation, we also created a definition of 'Type' in words (Table 3).

Table 3 Frequency of 'stroke sequence Type' used out of 345 actual handwritten 'Ka' samples

Stroke Order-Type	Telugu Std. 5	Telugu Std. 7	Tamil Std. 5	Tamil Std. 7
Type A 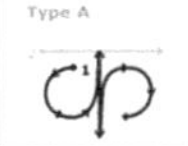	0	2	1	14
Type B 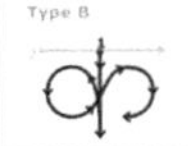	47	58	33	45
Type B.0.1 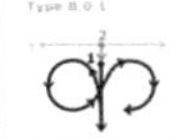	0	0	0	1
Type B.1 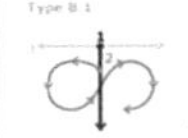	2	23	23	19
Type B.2 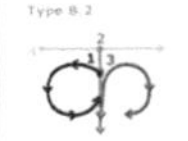	1	0	0	0
Type B.2.1 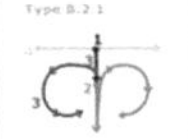	0	0	1	0
Type B.3 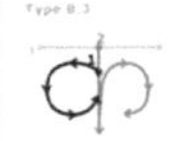	2	0	0	0
Type B.4 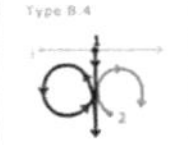	1	0	0	0
Type C 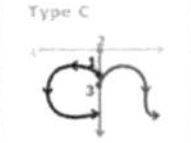	0	0	1	0
Type D 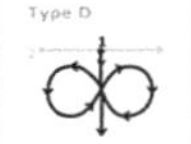	0	0	9	0
Type E 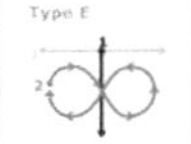	0	0	3	0
Type F 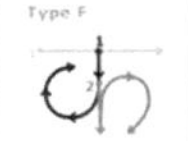	7	1	0	0
Type F.1 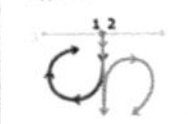	8	3	1	0
Type F.1.1 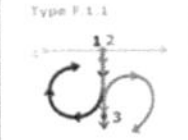	1	0	0	0
Type F2 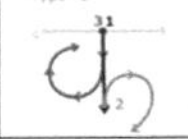	2	6	4	0
Type G 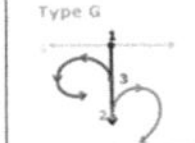	0	0	1	0
Type H 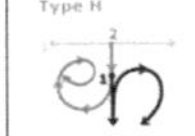	3	0	0	0
Type H.1 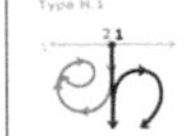	4	0	0	0
Type O 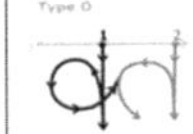 Type O 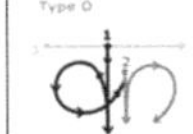 Type O 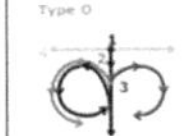 Type O 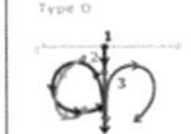 Type O 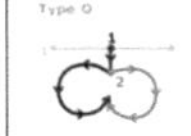 Type O 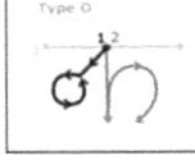	10	1	6	0

* The sketches of types in Table 2 have been made to understand merely the stroke sequences, and are not the actual shapes received from the students

4.1 Definition of Type

It simply means, a unique 'type of sequence of strokes' in a letter, found in handwriting. It is merely an identification code, assigned to every 'unique sequence combination of strokes used' that is found in a received handwritten data sample.

'Type', here, means such a hand drawn representation that shows the direction of strokes and their sequence, with clarity. Such an identified unique sequence is termed here as 'Type'. Even one such case of unique sequence identified from an entire data set is termed here as 'Type'. The following criteria were applied while deciding the definition of 'Type'.

Uniqueness in sequence of strokes + direction of stroke = 'Type'
Coding them, for example, as Type-A or Type B (Fig. 3).

4.2 Definition of Sub-type

A Type where the count of strokes and/or sequence of strokes is similar, but during the course of the activity, has repetition, addition or overlapping strokes, or an unnecessary pen-lift, leading to a 'dislocated look' in shape, is termed here as 'Sub-Type'. Although it does not differ or contradict in its sequence with the core Type, such a Type is termed here as 'Sub-Type'. For example: the 'Sub-Type' of 'A' is coded as A.1, A.1.1; and B, B.1 so on.

4.3 Definition of Odd Type

An Odd Type is a rarely seen sequence in comparison with the entire data received. May be it is an accident, or an attempt at correction of the non-intentional output. Hence, it remains un-fit to be included in any of the 'Type' categories that are seen in the received data.

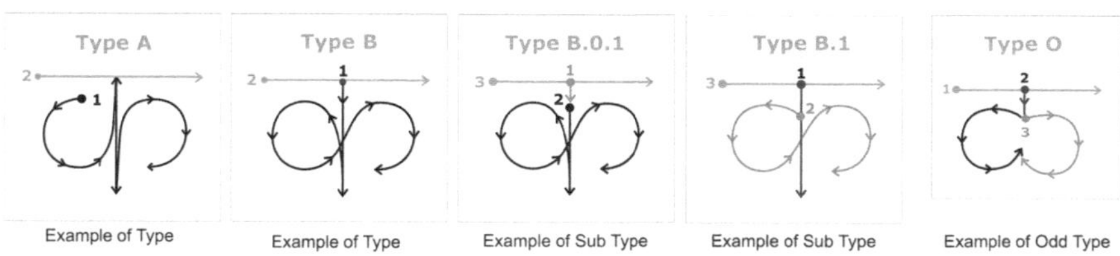

Fig. 3 Shows different sketched examples of 'sequence type', used as a code for counting data

5 Data Analysis

Due to various on-field errors by subjects, and technical errors in filming or file transfer, some clips have been found not useful for purposes of observation in research. Therefore, instead of 400 video clips, we received 345 clips in reality. Consequently, we took an average of the entire data that was actually received.

6 Result: Over All Observations of Data (Fig. 4)

Many of our observations are quite sensory in nature. One can read all the Devanagari letters that have been written by second script background students. When one conducts an overview of the data on the surface, one does not see radical differences while reading Devanagari letterforms from the received data. But one clearly notices differences in stroke sequence, while watching them in action, closely in the video clips.

The Devanagari letters written by Tamil or Telugu fifth and seventh standard children are quite big, and in various sizes as compared to normal Devanagari, as also in comparison to the reference sample. Most of the second script students' letters are round in shape.

Most of the second script students' overall writing is not perfectly parallel. Also, the letters do not hang from the horizontal top line adequately. This is perhaps a consequence of not being in the habit of drawing a top horizontal line in their first scripts.

The 'E' (के), 'Ai' (कै) matra that generally starts from the top left and ends at the bottom right obliquely, is found to be peculiar to the Devanagari script. However, the data of the second script students shows these accents produced the other way, which means that the 'E' (के), 'Ai' (कै) matra stroke starts from the bottom right and ends obliquely at the left top. Looking at the marching logic of the Devanagari script, every letter that concludes with 'E' (के), 'Ai' (कै) Matra always ends to the right, as it is then easier to move to the forthcoming letter.

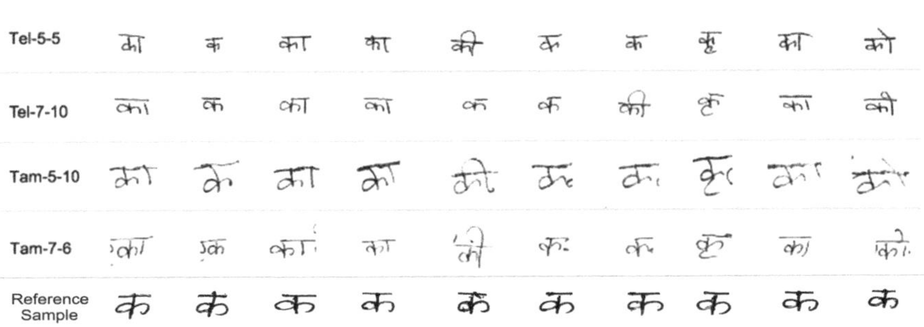

Fig. 4 Scan and sorted manuscript data for comparison

Overall, Devanagari letters are created using a left to right, top to bottom and clockwise direction of strokes. However, the second script students were found using a peculiar anti-clockwise movement while producing their round stroke letters.

It is also observed that the second script students' inter-letter spacing is quite inconsistent in comparison with that of the reference student. Also, the flow or rhythm of the top line that is achieved on the initial and ending letters of the word was found to be quite abrupt in the case of second script students. This is perhaps because in Telugu and Tamil each letter is written separately, and the word is separated by an extra space, but in Devanagari, the top line continues over the entire word. This means that the habit of writing each letter individually in the native script has revealed an influence on the second script learnt. The central axis of the letterform seems to change in most of the cases.

It is possible to mention many such overall observations here. But we are limiting our investigation by focusing on one key letter. This will make the observations more concise and enable us to conduct a productive enquiry on the sequence of strokes in Devanagari handwriting.

6.1 Observation of Most Frequently Used Sequences in Key Letter 'Ka'

The Type 'B' sequence is most frequently visible in 53 % of the entire collection, i.e. 345 'Ka's. The second level of this type of sequence was found in B.1, appearing in 19 % of the collection. Although this appears less frequently than Type B, it is very close to the sequence category as compared to 24 other categories. That adds up to nearly 73 % of similar category sequences that are found used by the subjects. Following are the qualitative observations made about B and B.1 types of sequences.

One commonality in these sequences is the drawing of the top horizontal stroke upon the completion of the letter. Completing every letter towards the right side with an exit point to right is a good sign. Since the Devanagari script flows from left to right, it helps to move quickly to the next letter.

Type B is made of two stroke/pen lifts, whereas Type B.1 uses three. Type B.1 is a bit more time-consuming compared to Type B owing to the additional stroke. In advanced stages it may slow down the speed of the learner. Hence, Type B is found to be more logical in comparison (Fig. 5).

The strokes in both these sequence types, in a flow, cross the vertical bar that is already drawn. In the long run, particularly when letters are written with greater speed, such crosses are harmful for readability as well as in producing a good and stable letter shape.

In Type B, the first stroke goes down and comes back up again using the technique of overlapping. This gives the writer some momentum and therefore can be conducive to achieving speed while writing. On the other hand, this overlap is unnecessary, and can result in complicating the shape of the letter in the long run.

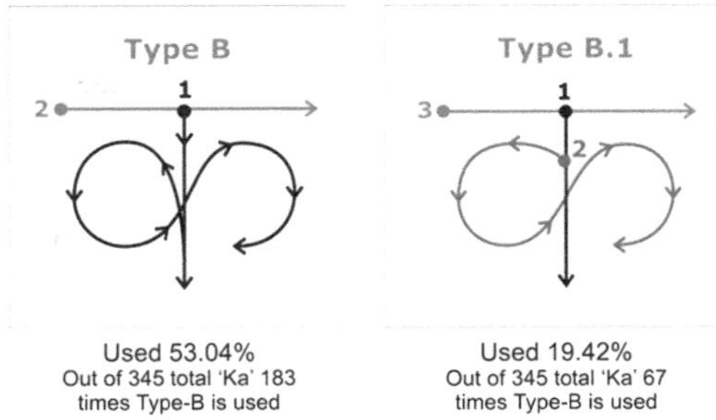

Fig. 5 Example of B and B.1 is considered here for their higher frequency of usage

Out of the 40 subject students, only nine students showed great consistency in completing all ten attempts of 'Ka' using Type B. This means that only 22.05 % of the sample showed consistency in the acquisition of the Type B sequence. Therefore, about 78 % results reveal inconsistencies in following one type of stroke sequence. The uncertainty regarding the sequence acquired is clearly seen in the second script students, as against the reference student, who has Devanagari as the first script, and followed the Type B sequence consistently across all ten attempts (Table 4).

However, we have used only one student here as a reference sample, and that provides us a very weak validation, as also any confidence to conclude that Type B is, in fact, an ideal stroke sequence. Therefore, we decided to further cross check

Table 4 Frequency of type used-in percentage

Type	Total use	Percentage
A	17	4.93
B	183	53.04
B.0.1	1	0.29
B.1	67	19.42
B.2	1	0.29
B.2.1	2	0.58
B.3	2	0.58
B.4	1	0.29
C	1	0.29
D	9	2.61
E	3	0.87
F	8	2.32
F.1	12	3.48
F.1.1	1	0.29
F.2	12	3.48
G	1	0.29
H	3	0.87
H.1	4	1.16
O	17	4.93
Total	345	100

with the book that is prescribed by the school in order to gain better insights into the matter of sequence in handwriting.

6.2 Comparison of Reference Student's Outcome with Prescribed Schoolbook

The prescribed schoolbook shows that the 'Type A' sequence (according to our definition of type) needs to be followed. This sequence is not followed by our reference sample, namely the student who has learnt Devanagari as the first script. Interestingly, it is followed by only one student out of the 40 s script learners (Table 5, *Tam-7-7*). Over all, across 345 attempts, the Type A sequence is followed only 14 times, and if we exclude the ten times by one student (Table 5, *Tam-7-7*), then it is followed only four times, in a rather irregular fashion, by one more student (Table 5, *Tam-7-5*). Therefore, we cannot state, with any confidence, that 'Type A' is also an ideal sequence to be followed as per the book.

7 Discussion

The outcome also suggests that although there are many types of sequences followed, the letters are fairly readable. And if letters are indeed legible, then the question arises as to why the stroke sequence needs to be followed at all. This is an extremely vital question, but it is also lopsided and reveals a general concern only towards the reader of the handwriting. When we focus our enquiry on the writer's point-of-view, then many such details are found extremely important.

Following a specific sequence when writing is essential, since it brings flow and ease into the handwriting [18]. The sequence of strokes definitely shows its impact on the shape of the letterform that we receive.

Following a certain sequence that is learnt initially (at an early stage of development and learning) also has an effect on the acquiring of speed in handwriting. In short, learning the correct stroke sequence can be described as a 'good habit' rather than as an essential habit for acquiring handwriting. 'Goodness' in this context is like 'correctness', and should not be considered as a requirement for making the letterform readable. Hence it becomes the first level of an aesthetic issue. It is more of a writer's issue than just that of a reader as the receiver of one's handwriting.

The stroke sequence in a handwritten letterform can be closely compared to the idea of pronunciation of the spoken word that leads one to formulate the diction of a person. But it is not as categorically followed as the spelling of the word. Should we consider a frequently used stroke sequence as correct? Is it further conducive to one's handwriting development? Or should we follow the sequence prescribed by the schoolbook? (Figs. 6 and 7). Looking at the result that we have received,

Table 5 'Type of stroke sequence' used by each student for each attempt

Code	ka 1	ka 2	ka 3	ka 4	ka 5	ka 6	ka 7	ka 8	ka 9	ka 10	Avg.
	TYP	TYP	TYP	TYP	TYP	TYP	TYP	TYP	TYP	TYP	
Telugu school											
5th Std.											
Tel-5-1	x	B	O	O	B.2	F.1	F.2	B.3	B	B	B & Mix
Tel-5-2	B	O	B	B	F.1	F	F	O	F	F	F & Mix
Tel-5-3	B	B	B	B	B	B	B	B	B	B	B
Tel-5-4	O	x	F.1.1	F.1.1	F	F.1	F.1	F	F.2	F.1	F
Tel-5-5	B	B	B	B	B	B	B	B	B	B	B
Tel-5-6	B	B	F.1	B	O	B.2.2	F.1	F.1	B	B	B & F
Tel-5-7	B	B	B	B	B	H.2	H.2.1	H.2	O	H.2	B & H
Tel-5-8	B.1	B.1	F.1	0	H	H	H	F	x	0	Mix
Tel-5-9	B.3	B	0	0	B.3	O	O	O	O	O	O
Tel-5-10	B	B	B	B	B	B	B	B	B	B	B
7th Std.											
Tel-7-1	x	x	F.2	B	F.2	F.2	B.1	F.2	F2	F.1	F
Tel-7-2	B	B	0	0	B	B	B	B	B	B	B
Tel-7-3	B	B	B	B	B	B	B	B	B.1	B.1	B
Tel-7-4	B.1	B.1	B.1	B.1	B.1	B	B	B	B	B.1	B
Tel-7-5	B	B.1	B.1	B.1	B	B	B.1	B.1	B	B	B
Tel-7-6	B.1	B	B	B	B	B	B	F.1	B	B	B
Tel-7-7	B	B	O	B	B	F.2	F.2	F	B.1	B	B
Tel-7-8	B.1	B.1	B.1	0	0	B	B.1	A	A	B.1	B
Tel-7-9	B.1	B.1	B	B	B.1	B	B	B	B.1	B	B
Tel-7-10	B	B	B	B	B	B	B	B	B	B	B

(continued)

Table 5 (continued)

Code	ka 1	ka 2	ka 3	ka 4	ka 5	ka 6	ka 7	ka 8	ka 9	ka 10	Avg.
	TYP	TYP	TYP	TYP	TYP	TYP	TYP	TYP	TYP	TYP	
Tamil school											
5th Std.											
Tam-5-1	x	x	O	B	B	B	B	B	B	B	B
Tam-5-2	B.1	B.1	B.1	B.1	B	B.1	B.1	B.1	B.1	B.1	B.1
Tam-5-3		B.1	0	0	B.1	B.1	B.1	B.1	B.1	B.1	B.1
Tam-5-4	B	B	O	B	O	O	O	B	B	B	B & O
Tam-5-5	B.1	B.1	B.1	C	B.1	0	0	B.1	B.1	B.1	B.1
Tam-5-6	D	D	D	D	O	D	D	D	D	D	D
Tam-5-7	B	B	B	B	B	B	B	A	B	B	B
Tam-5-8	x	x	x	x	x	x	x	x	x	x	x
Tam-5-9	E	E	F.1	E	F.2	G	F.2	F.2	B.2.1	F.2	Mix
Tam-5-10	B	B	B	B	B	B	B	B	B	B	B
7th Std.											
Tam-7-1	B	B.1	B	B.1	B	B	B	B	B	B	B
Tam-7-2	B	x	B	x	x	x	x	x	x	x	x
Tam-7-3	x	x	x	x	x	x	x	x	x	x	x
Tam-7-4	B.1	x	B.1	B.1	B.1	B.1	B	B.1	B.1	B.1	B.1
Tam-7-5	B	B	B	A	A	A	B	B	A	B	B & A
Tam-7-6	B	B	B	B	B	B	B	B	B	B	B
Tam-7-7	A	A	A	A	A	A	A	A	A	A	A
Tam-7-8	B.0.1	B	B	B	B	B	B	O	B	B	B
Tam-7-9	x	B	B	B	B	B	B	B	B	B.1	B
Tam-7-10	B.1	B.1	B.1	B.1	B.1	B	B.1	B	B.1	B.1	B.1
Reference sample											
	B	B	B	B	B	B	B	B	B	B	B

Abbreviations: *TYP* type of shape, *O* odd type, *x* errors in data capture/transfer

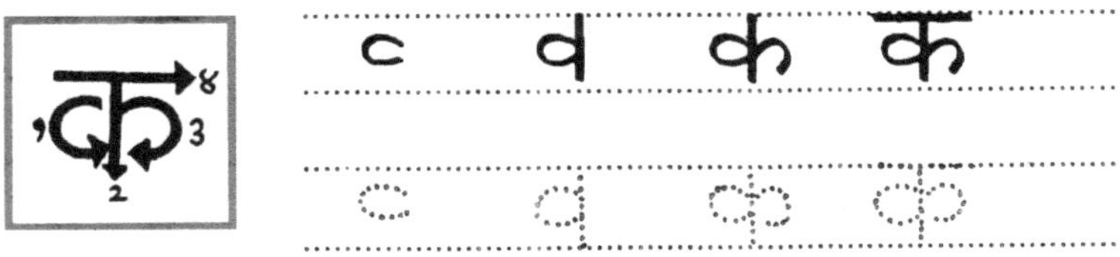

Fig. 6 Stroke sequence as shown in prescribed school book

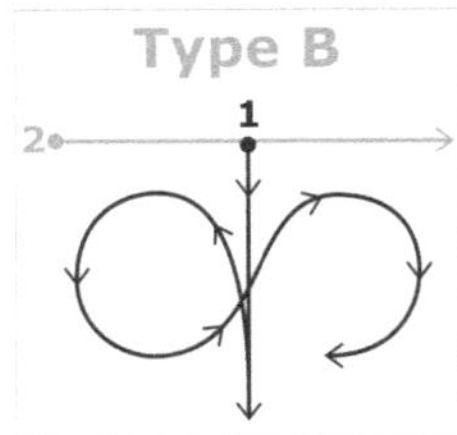

Fig. 7 As against outcome by students

we humbly challenge the existing stroke prescription of the schoolbook too. Are schoolbooks or privately published handwriting primers designed after conducting scientific research? Experts [19] have already noted that this issue has been largely ignored thus far.

The key lies in probing deeper, into the issue of acquisition of the first script, its pedagogy, and the design of the learning material.

8 Conclusion

Our investigation of the received data leaves us uncertain about any conclusions regarding Type B or Type A as an ideal sequence to be followed in Devanagari handwriting. However, this exercise has surely provided us the following realisations and insights:

(1) This method can serve as an instrument for analysis, data counting and separation.
(2) Type A appears to be the most logical stroke sequence with reference to the marching mannerism of the Devanagari script.
(3) A new point of enquiry emerges, i.e. to investigate Devanagari handwriting primers as a future scope.
(4) We can investigate second script learners to get more clarity in the investigation of the first script. In other words, our study validates Sassoon's original intuitive thought [1].
(5) In the context of research in handwriting, the stroke sequence is script specific in nature [20, 21].

9 Future Scope and Limitation of this Study

There is a need to investigate the model stroke sequence of Devanagari for handwriting teaching and learning.

There is a need for good/scientifically designed handwriting practice manual/ primers that will be useful for teachers.

There is a need to understand the handwriting factors that one should investigate, in early/developing age of first script students.

This project presents us with a limitation that of investigating the stroke sequence of a single Devanagari letter, which is not sufficient. Hence, one should probe deeper into the formation issues of first script handwriting acquisition in order to understand the cause and effect relationship that results in the formation of a correct stroke sequence.

Acknowledgments Thanks to K.C. Matunga Municipal Tamil School and K.D. Gayakwad Municipal Telugu School, Mumbai for allowing us to conduct this experiment in their schools. Thanks to Noopur Datye for making and adding finishing touches to the images. Thanks to Lavanya Varadrajan for proofreading the text.

References

1. Sassoon, R.: The Acquisition of a Second Writing System. Oxford: Intellect (1995)
2. Eighth Schedule occurs in articles 344(1), Constitution of Government of India 2004
3. http://listverse.com/2008/06/26/top-10-most-spoken-languages-in-the-world/
4. Eighth Schedule occurs in articles 351, Constitution of Government of India 2004
5. Jayaram, K.: Building Foundations for Reading and Writing. Paper presented at the USRN, JNU Seminar, 02. 30–31 Oct 2009
6. Menon, D.S.: Historical and Theoretical Approaches to Early Literacy. Report of the Consultation on Early Literacy, 25–28 Apr 2011. Submitted to SRTT-NRTT, 27. (2011)
7. Gupta, D.R.: Teaching the indic scripts: tradition and innovation. Language in India, strength for today and bright hope for tomorrow **12**(2), 425–441, (2012)
8. Nag, S., Snowling, M.J.: Reading in an alphasyllabary: implications for a language-universal theory of learning to read. 16th January 2014 (Pre-print version)
9. Walker, S.: Handwriting skills, models and methods. In: Edwards, V. (ed.) Encyclopaedia of Language and Education, vol. 2, pp. 97–105. Kluwer Academic Publishers, Dordrecht (1997)
10. Koziatek, S.M., Powell, N.J.: Pencil grips, legibility, and speed of fourth-graders' writing in cursive. Am. J. Occup. Ther. **57**(03), 284–288 (2003)
11. Yakimishyn, J.E., Magill-Evans, J.: Comparisons among tools, surface orientation, and pencil grasp for children 23 months of age. Am. J. Occup. Ther. **56**(5), 564–572 (2002)
12. Clayton, E.: A history of learning to write, p. 13. http://www.ejf.org.uk/Resources/ejhandw.pdf
13. Greer, T., Lockman, J.: Using writing instruments: in variances in young children and adults. Soc. Res. Child. Dev. Child. Dev. **69**(4), 888–902, (1998) (Blackwell Publishing)
14. John, J.J.: Latin Palaeography, Medieval Studies, 2nd edn., p. 8. As sited in—Reimer Stephen, R., Manuscript Studies Medieval and Early Modern IV. iv. Palaeography: Letter Formation, University of Alberta; Edmonton, Canada, http://www.ualberta.ca/~sreimer/ms-course/course/pal-ltrs.htm (1998)
15. Sassoon, R.: Joins in children's handwriting, and the effects of different models of teaching methods. Department of typography and graphic communication, University of Reading. Doctoral thesis, University of Reading, Department of Typography and Graphic Communication, UK, May 1988

16. Joshi, R.K.: Calligraphy the Art of Writing, CALTIS-84, p. 129, Special Issue on Calligraphy, Lettering and Typography of Indian Script, CALTIS-84 Seminar held on 11, 12, 13 February at Pragati Maidan, New Delhi, India
17. Palande, B.D.: Marathi First Book. Government Central Book Depot, Mumbai, Maharashtra, India (1867–1870)
18. LeH, T.: Smith, Six Basic Factors in Handwriting Classification. J. Crim. Law, Criminol., Police Sci. **44**(6), 810–816, (1954) (North-western University). Stable URL: http://www.jstor.org/stable/1139934
19. Vinter, A., Chartrel, E.: Effects of different types of learning on handwriting movements in young children. Learn. Instr. **20**, 476–486 (2010). www.elsevier.com/locate/learninstruc. (University of Bourgogne, Laboratoire d'Etude de l'Apprentissage et du De´veloppement (LEAD), Centre National de la Recherche Scientifique (CNRS) UMR 5022 Po^ le AAFE, Esplanade Erasme, BP 26513, 21065 Dijon Cedex, France)
20. Share, L.: On the anglo centricities of current reading research and practice-the perils of overreliance on an "outlier" orthography. Psychol. Bull. **134**(4), 584–615 (2008)
21. Smith, P.: Handwriting in the United Kingdom. Reading Teach. **41**(01), 27–31 (1987)

A Design Practice on Communicating Emotions Through Visual, Tactile and Auditory Simulations

Secil Ugur Yavuz, Monica Bordegoni and Marina Carulli

Abstract Emotional expression is an important human behaviour, which enriches communication. Sensory organs play crucial role in emotional perception. Today communication is mostly done via digital mediators, which dominantly address to vision excluding the other senses; therefore, communication becomes less affective. Wearable technology can appeal to sensory organs from very close distance due to its intimate interaction with human body. Hence, this technology can be used in order to make distant communication more affective by enabling multi-sensory interaction. This paper represents a user-centred design practice on wearable products that simulate sensorial feedbacks (tactile, visual and auditory) to express basic emotions. Three prototypes that transmit emotional messages were designed, built and tested to observe user behaviour. This paper discusses how user experience obtained through the user test can be taken further to design new communication products.

Keywords Emotion · Sensory perception · Wearable technology · Design practice

1 Introduction

Social interaction relies on the ability of expressing emotions and reacting to them. Emotions occur depending on the perceptual evaluation of the outer stimuli and the internal state. While the sensory system performs the perception of inner and outer stimuli, the motor systems trigger the action, and the cognitive system evaluates the experience of emotion. This natural process ends with an emotional expression that gives sense to communication.

S.U. Yavuz (✉) · M. Bordegoni · M. Carulli
Department of Mechanical Engineering, Politecnico Di Milano, Milan, Italy
e-mail: secil.ugur@mail.polimi.it

A. Chakrabarti (ed.), *ICoRD'15 – Research into Design Across Boundaries Volume 1*,
Smart Innovation, Systems and Technologies 34, DOI 10.1007/978-81-322-2232-3_25

Today, digital communication devices are the core means of communication. How people make contact with others has been changing depending on the abilities of the technology that is used as a mediator of communication. Mediators such as avatars, emoticons or Embodied Conversational Agents (ECAs) are the new "relationship enablers" that change how people interact with others [1]. Every communication medium has a different "affective bandwidth level" varying on how much affective information it is able to transmit [2]. While face-to-face interaction involves the whole human body, digitally mediated communication lacks physical contact that is an important part of emotional communication.

Lately there has been an interest in research on systems and technologies that can enable sensory organs generally overlooked in the digital communication in order to provide a more intimate and real-like communication. For instance, Wearable Technology that provides intimate and continuous connection with sensory organs is a new way of interaction with digital data through a whole body interaction. Wearable technology can extend the ability of how people communicate and share details about themselves in spite of the distance [3]. While wearable sensors can detect human gestures, actuators embedded into textiles can transmit messages in a multi-sensory way.

Through addressing the role of sensory perception in recognizing emotions, this paper presents a user-centered approach and design practice on communicating emotions through a wearable and tacit system. The wearable system consisting of three wearable prototypes was designed to transmit affective communication patterns between two people in three modalities: tactile, visual and auditory. Based on the user test results the paper discusses the user behavior towards the use of these mediators. This study aims to explore new communication modalities that are formed by human experience for developing new models of communication technologies.

2 Sensory Perception in Recognizing Emotion

In emotional recognition non-verbal language, such as facial expression, body gestures or tone of voice are important elements to figure out the other person's emotions. Each sensory organ filters these non-verbal messages, which are analyzed in the central nervous system in order to make sense of them. This sensory processing involves all sensory organs simultaneously.

As Darwin [4] mentioned, human movements, from facial expressions to respiratory actions are the main non-verbal expressions that reveal the state of the mind. Vision plays an important role in recognition of emotional expressions. The study of facial expression has been the most investigated subject of recognition of emotions [4–6]. Besides, body language is another essential visual cue that conveys emotions. Studies have been done on affective body gesture analysis that categorizes human movements based on spatial, temporal and sequential parameters [7, 8]. For instance, anger is mostly expressed through expansive and fast movements, whereas sadness is expressed through smaller and slower movements [9]. Andersson et al. [10] found that happiness was associated with a continuous movement, and anger was associated with shaking movement. Moreover, emotion recognition can arise due to the visual properties of a firm artefact, such as its shape, colour, etc. The emotional appraisal occurs through the judgment on the aesthetics of the artefact, the semantic interpretation, and the symbolic associations [11]. While aesthetical features can create primary emotions, through a cognitive processing secondary emotions can appear as well [12].

Besides the visual cues voices, noises and music can also convey emotions [13, 14]. Emotions are expressed through voice by applying different amount of stress to the words [15]. Patterns of pitch and loudness are used to convey specific emotions [16]. While joyous speech shows higher pitch range, sadness results into lower pitch [17].

Furthermore, touch is a very emphatic sense that can elicit emotions. Interpersonal touch is an intimate and immediate way of expressing emotions [18]. The way people touch each other differs depending on the meaning they want to convey. Touch can communicate hedonic tone of emotion, such as warmth and intimacy or pain and disturbance [18, 19]. Hertenstein et al.'s [18] research on touch behaviors associated with emotions has showed that sympathy was associated with stroking and patting, anger with hitting and squeezing, disgust with pushing, fear trembling, and love with stroking. Besides its communicative role, touch has also a positive effect on human wellbeing. Affectionate touch, such as hug, can reduce stress and create calming effect [20].

3 Wearable Technology for Intimate Communication of Emotions

Today connection is the most important phenomenon of our lives. Internet and smart mobile phones have been providing an ease to be connected to people over distance and share personal information. Emerging technologies have been focusing on mediating intimate relationships that are mostly based on emotions [21]. According to Gaver [22], technologies can bridge the emotional distance through providing non-verbal displays that evoke or communicate emotions. These non-verbal messages have been generally transmitted through immediate sensory channels, such as sound or touch. For instance, Shake2Talk was designed in order to send sounds and tactile sensations to one another through simple gesture interactions [23]. On the other hand, ComTouch is a device that converts hand pressure into vibrational signals in order to complement auditory communication with tactile inputs [24]. Keep in Touch, which is a fabric touch screen designed to support intimacy for couples in long distance relationships combines the visual and tactile senses together by allowing the users to communicate through touch, gestures, and body language [25].

Until recently wearable technology has been mostly used for healthcare and personal tracking. However, due to its intimate relation with human body it has been seen as a new medium for interpersonal communication. While it can easily detect bodily gestures in order to involve the human body more into the interaction, it can also evoke awareness of a remotely located person through haptic signals. Thanks to Affective Haptics that can “elicit, enhance, or influence the emotional state of people by means of touch” [26] it is possible to integrate tactile sensation into communication.

For instance, Hug Over a Distance is a jacket that enables users to give a hug to the other person, while physically separated [21]. Heart2Heart is another example of wearable electronic garment that emotionally links two individuals by simulating the pressure and warmth of physical embrace and heart beat pulses [27]. On the other hand, Schiphorst [28] designed sensual skirts—called Exhale—that were simulating the rhythm of breath for sharing affective non-verbal data. Wearable technology can also be used as an emotion mediator during a virtual communication. For instance, iFeel_IM! is a vest shaped device that can generate pressure, vibration and temperature simulations in 3D virtual world Second Life in order to enhance user’s emotional experience [26].

These wearable technology examples show that there is a tendency towards a new type of intimate communication, which involves body gestures and multi-sensory non-verbal messages to transmit emotional data over distance. Starting from this notion, the paper represents a design practice, which examines the user behavior towards this type of wearable devices. Through focusing on user experience, the paper brings out findings that can throw light on design of new communication products.

4 Experimental Project

4.1 Research Question and Hypothesis

This study questions how wearable technology can support emotional communication by enhancing the role of human body in interaction through involving not only vision but also auditory and tactile sensory organs and how this means of communication open up new user behaviors and create new communication patterns. Findings of the previous study on communicating emotions through wearable technology shows that people prefer intimate communication rather public exposure [29], thus as a continuation this study aims at exploring one-to-one communication modalities. In order to answer the research questions, the study concentrates on prototyping and testing affective communication tools transmitting non-verbal emotional messages through multi-sensory stimulations. The idea is not only transmitting non-verbal codes that can be decoded as emotional messages, but also inducing emotions on the other person to create an emphatic communication. The envisioned prototypes can translate users gestures into non-verbal instant messages and transmit them in form of tactile, auditory and visual signals.

4.2 Prototypes

The project consists of three sets that can transmit emotions through long distance and create a non-verbal intimate communication pattern between two people. Emotion occurs in a process that consists of sensory system (vision, audition, taste, smell and touch), motor system (action, proprioception) and cognition (mental state, memories, motivation). Inspired by this natural process, each set of EMbody was constructed within a system that is made of sensors, processors and actuators. Each set consists of two wearable devices that are worn by two distant people. The system works with input and output data that are operated by a processor (Arduino) (Fig. 1). Pressure and touch sensors detect the gestures of the wearer, then the processor processes this data and transmits it to the other device's actuator, where the output appears as tactile, auditory or visual signals. The output messages are generated by embedded LEDs, micro motors, vibro-motors and buzzers embedded into the wearable prototypes. Due to the fact that self-perception of visceral activities is an important fact of emotional experience [30, 31], heartbeat rhythm was used in two prototypes (Sound Pad and Skin Deep) in order to transmit emotional messages.

The first prototype is Sound Pad, shoulder pads transmitting auditory heartbeat messages between two people (Fig. 2). The prototype enables the user to send two kinds of sounds with low and high pitch. When touching the chest, it sends a higher pitch heartbeat sound envisioned as affection-love message and when touching the shoulder it sends lower pitch heartbeat sound envisioned as sadness. While sending the messages, the wearer makes certain gestures (tightening while holding the shoulders, patting while holding the chest) that are translated into emotional audio messages. Buzzers, which generate the heartbeat sound in high and low pitch, were embedded into the pads on the shoulder area in order to provide an intimate audio-perception.

The second prototype is Skin Deep, set of a collar and a bracelet transmitting emotional messages in tactile and visual signals (Fig. 2). When the wearer touches the prototype (chest on the collar, wrist on the bracelet), the other prototype blinks

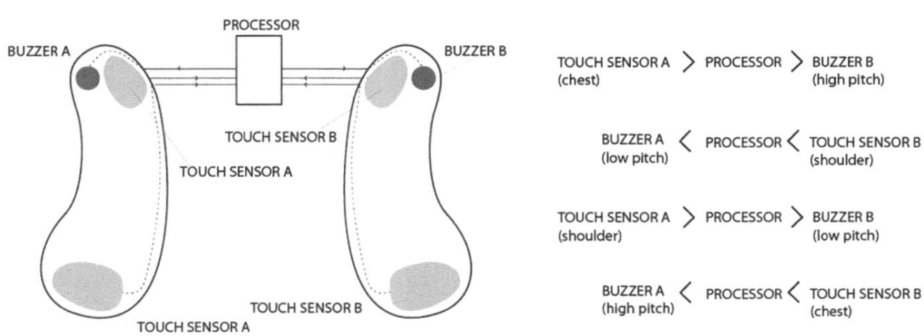

Fig. 1 The architecture of the system—sound pad

fig. 2 Sound pad on the *left*, skin deep in the *middle*, hand muff on the *right*

and starts vibrating as heartbeat envisioned as sender's heartbeats. Heartbeat speed is envisioned as the parameter for the arousal level of the emotion felt by the wearer. While the vibro-motor embedded in the textile is simulating heartbeat sensation, the LEDs represent the same information in visual form as blinking. The tactile sensation is perceived in two different places, chest and the artery. These wearable prototypes provide an intimate communication, which allows the wearers to touch and see their inner rhythms.

The last prototype is Hand Muff that can transmit affective gestures into tactile messages (Fig. 2). The idea is to create an intimate emotional communication, where the hand gestures are not visible from outside and hidden in a cozy object that enables the user to receive and send tactile messages. The Hand Muff can translate two gestures: tapping movements into playful light touch and constant pressure into firm affective touch. Each Muff has pressure sensors in side, where the hands are located and these sensors detect the typology of touch applied on (tapping/constant pressure). The tactile messages are created by a hexagon mechanism embedded in the Muff and activated by a small servo-motor. The mechanism closes continuously to create the firm affective touch; opens and closes rapidly to create playful light touch.

4.3 User Test and Findings

After building the prototypes, a user test was done in order to observe communication patterns between two people during the interaction. The test was done by 12 participants: six with engineering backgrounds, six with design backgrounds. The far distance situation was simulated in a laboratory setting, where the participants sat back to back without any visual contact. After a brief description of the concept, they were asked to communicate their emotions to the other person by using the prototypes without explaining the gesture coding that was envisioned while designing the prototypes. The test was analyzed qualitatively and quantitatively through a real-time observation and a questionnaire conducted after the test.

The observations done during the test gave valuable information about user behavior and led to find new opportunities to further develop the prototypes. The first experience that they had with the prototypes was problematic. They spent 10–15 min to understand how the system works. Due to the fact that this kind of novel products are unfamiliar to the user, it needs some time to get to know. Furthermore, receiving messages back from the other person made them understand that their message was received. A feedback mechanism is useful to be sure if the message is received.

It was observed that receiving a reaction after a message sent can be perceived as the presence of the other person. When an object reacts to the user, it creates interactivity, and therefore in this setting it makes the user believe that there is another person sending messages on the other side. Mimicking the other's messages was another user behavior, which happened after a time. For instance, when a participant sent a joyful message, the other person also sent the same message. And they repeated that for a while, until one of them changed the message. This shows that mirroring emotions in a face-to-face communication can also happen in a mediated communication setting. Therefore, while designing mediators for communication natural human behaviors should be the basis of the design approach, and designing alienated products that force users to do un-natural behaviors may not work in the practice.

On the other hand, it was observed that participants created their own language codes between each other. For instance, while using the sound pad they created different melodies by tapping the sensors. This shows that although the products are designed for a certain purpose, they can be utilized in another way, which the designer can overlook while designing. Therefore, it justifies that user tests are crucial and helpful steps of the design process, where the designer can develop new ideas through observations. This observation shows that the products should give more freedom and personalization option to the user in order to express emotions to the other person.

The other user behavior was closing eyes, while interacting with prototypes that gave auditory and tactile feedbacks. Although Skin Deep prototype provides both visual and tactile messages, some of the participants closed their eyes and concentrated on only the tactile message. The same user behavior was observed also in the participants that were using the Sound Pad. This observation shows that tactile and auditory feedbacks can induce concentration on emotional perception.

When the participants finished their experience with the prototypes, they were asked a questionnaire in order to define how the sensory messages were perceived and how effective they were. Each device induced different emotions on the participants. Most of the participants (66 %) reported that the higher pitch-heartbeat signal obtained by Sound Pads was funny and joyful; on the other hand the lower pitch was annoying and associated with negative emotions. The firm pressure sensation obtained by Hand Muffs was mostly (75 %) associated with relaxation, warmth and affection; besides the light touch was reported playful by 58 % of the participants as if it was tickling their hands. On the other hand, vibro-tactile heartbeat sensation in Skin Deep was mostly (66 %) associated with excitement and aroused emotion. Most of the participants (66 %) reported that they felt more intimately connected to the person with Skin Deep and they felt more able to communicate various emotions with Hand Muffs (83 %). Participants gave rankings for each sensorial experience with a scale from 0 to 10. Touch was rated with 7 points, sound with 6.3 points and vision had the lowest ranking with 3.7 points. The ranking and the comments of the participants showed that tactile based prototypes would be more effective to communicate emotions in long distance.

User experience goes beyond the usability and is based on different dynamic aspects, such as user's personality, physical environment and product meaning [32]. The background of the participants can affect the results of the user experience. Participants with engineering background reported that wearing the prototypes in public would seem weird; and therefore, although they found the experience socially and physically very pleasurable, they would not appreciate with their appearance to others. On the other hand, the participants with design background reported that these objects could be nice to wear around. The results show that although these products can create an interesting way of communication with others, people might need intimacy while communicating their emotions. And especially when these products are worn on the body, they are not any more functional objects, but personal identities. Therefore, while designing wearable technology this fact should be taken into consideration. Intimate communication products should be less salient that cannot be visually perceived by others. On the other hand, the environment, where these products are used, also affects the user experience. Some of the participants informed that they would use these products at home in a private setting. Therefore, the environment, where the experience happens, is one of the aspects that determine the product design. Besides, the meaning of product can affect the user experience. For instance, Hand Muff has a soft shape resembling a cushion; therefore it created an intimate interaction with user and was found relaxing while in use. Some of the participants reported that this product could be used to calm down people. This shows that the significance of a product can change the user's perception and affects the way she/he experiences the product.

5 Conclusion

The human body is an important matter, which functions as a sensorial filter and at the same time an expressive display pouring the inside out. Cognition and physical body work in a synergy in order to induce and express emotions. While people are shifting from traditional human-human interaction into new avenues of communication through digital communication mediums, which lacks for physical interaction, there is a need for new solutions to recreate the balance between the physical body and cognition. This paper argues that this can be possible by enhancing the role of human body in communication through involving not only vision but also auditory and tactile sensory organs. Instead of using one sensory channel the design practice that is presented in this paper combines the visual, tactile and auditory signals in order not only to communicate, but also induce emotions. This practice is an example of how affective communication can be done through a multimodal sensorial interaction and what kind of impacts this can create in the human-human communication. The observations during the user experience test gave useful ideas on people's behaviors towards this new phenomenon and how this kind of products can be further developed in a more efficient way.

One of the main findings is that tactile interaction is an inevitable solution for mediated emotional communication. While today many people have been suffering from "touch hunger" [33], this new way of communicating emotions can help people to sustain healthy relationships with others, enhance their wellbeing and moreover, this technology can be used as a therapeutic solution for people with psychological disorders. Furthermore, the findings show that this type of products requiring intimacy should be worn on the body in a salient way and convey emotions to an intimate partner. Due to the fact that these novel products are unfamiliar to users they require an adaptation period, which can also lead to new ways of usage that can be overlooked by the designer and designers should bear natural human behaviors in mind while designing these products. Moreover, the results show that the physical characteristics of this type of products can influence the perception of the user and effect the way the user gives meaning to it.

This study tries to underline the role of designer and user-experience in developing new ways of using technology in order to support human-human communication. Today we are in a technologically influenced era, where there is a need for user-centered approach in order to create solutions that are driven by human experience. While new technologies are shaping the way people naturally behave, the designers' role should be to create new interaction avenues that can enhance social, physiological and psychological wellbeing of the user. As Dunne and Raby [34] mentions, through critical design the designer generates new notions of use and pushes the limit of experiences. This paper claims that design can find human-centered solutions through reshaping the technology and using it not merely as a tool but an invisible power that enhances the human body in a physical sphere rather than digital.

References

1. Konijn, E.A., Utz, S., Tanis, M., Barnes, S.B.: Mediated Interpersonal Communication. Abingdon. Routledge, Oxon (2008)
2. Picard, R.W.: Affective Computing. The MIT Press, Cambridge (1997)
3. Labs, P.S.F.K.: The future of wearable tech. Des. Stud. **19**, 103–117 (1998)
4. Darwin, C.: The Expression of the Emotions in Man and Animals, 3rd edn. Oxford University Press, New York (1998)
5. Russell, J.A.: Is there universal recognition of emotion from facial expression? A review of the cross-cultural studies. Psychol. Bull. **115**, 102–141 (1994)
6. Ekman, P., Friesen, W.V., Hager, J.C.: Facial Action Coding System. A Human Face, Salt Lake City, UT (2002)

7. Glowinsky, D., Camurri, A., Volpe, G., Chiarra, N., Cowie, R., McMahon, E., Jaimovich, J., Knapp, R.B.: Using induction and multimodal assessment to understand the role of emotion in musical performance. In: Proceedings of WS Emotion in HCI—Designing for People, pp. 8–12. Fraunhof, Stuttgart (2008)
8. Shan, S.G.C., McOwan, P.W.: Beyond facial expressions: learning human emotion from body gestures. In: British Machine Vision Conference (2007)
9. Wallbott, H.G.: Bodily expression of emotion. Eur. J. Soc. Psychol. **28**(6), 879–896 (1998)
10. Andersson, G., Höök, K., Mourao, D., Paiva, A., Costa, M.: Using a wizard of Oz study to inform the design of SenToy. In: Proceedings of Designing Interactive Systems DIS'02, ACM, London (2002)
11. Crilly, N., Moultrie, J., Clarkson, P.J.: Seeing things: consumer response to the visual domain in product design. Des. Stud. **25**(6), 547–577 (2004)
12. Rafaeli, A., Vilnai-Yavetz, I.: Emotion as a connection of physical artifacts and organizations. Organ. Sci. **15**(6), 671–686 (2004)
13. Meyer, L.B.: Emotion and Meaning in Music. University of Chicago Press, Chicago (1961)
14. Zwicker, E.F.: Psyhoacoustics Facts and Models. Springer, Berlin (1990)
15. Khulage, A.A., Pathak, B.V.: Analysis of Speech Under Stress Using Linear Techniques and Non-Linear Techniques for Emotion Recognition System. CoRR (2012)
16. Scherer, K.: Vocal communication of emotion: a review of research paradigms. Speech Commun. **40**, 227–256 (2003)
17. Mohammadi, G., Vinciarelli, A.: Towards a technology of nonverbal communication: vocal behavior in social and affective phenomena. In: Gokcay, D., Yildirim, G. (eds.) Affective Computing and Interaction: Psychological, Cognitive and Neuroscientific Perspectives, pp. 133–156. IGI, Hershey (2010)
18. Hertenstein, M.J., Keltner, D., App, B., Bulleit, B., Jaskolka, A.: Touch communicates distinct emotions. Emotion **6**, 528–533 (2006)
19. Jones, S.E., Yarbrough, A.E.: A naturalistic study of the meanings of touch. Commun. Monogr. **52**, 19–56 (1985)
20. Grandin, T.: Calming effects of deep touch pressure in patients with autistic disorder, college students, and animals. J. Child Adoloscent psychopharmacol. **2**, 63–70 (1992)
21. Vetere, F., Gibbs, M.R., Kjeldskov, J., Howard, S., Mueller, F., Pedell, S., Mecoles, K., Bunyan, M.: Mediating intimacy: designing technologies to support strong-tie relationships. In: Proceedings of the SIGCHI Conference on Human Factors in Computing Systems, pp. 471–480 (2005)
22. Gaver, B.: Provocative Awareness. Comput. Support. Coop. Work **11**, 475–493 (2002)
23. Brown, L.M., Williamson, J.: Shake2Talk: Multimodal Messaging for Interpersonal Communication HAID 2007. LNCS, pp. 44–55 (2007)
24. Chang, A., O'Modhrain, S., Jacob, R., Gunther, E., Ishii, H.: ComTouch: design of a vibrotactile communication device. In: Proceedings of DIS 2002, pp. 312–320. ACM (2002)
25. Motamedi, N.: The aesthetics of touch in interaction design. In: Proceedings of the 2007 Conference on Designing Pleasurable Products and Interfaces, pp. 455–460. ACM (2007)
26. Tsetserukou, D., Neviarouskaya, A.: iFeel IM: augmenting emotions during online communication. IEEE Comput. Graphics Appl. **30**, 72–80 (2010)
27. Grimmer, N.: Heart2Heart. In: Student Design Competition Intel'01. http://www.baychi.org/calendar/20010508/#1 (2001)
28. Wilde, D., Schiphorst, T., Klooster, S.: Move to design/design to move: a conversation about designing for the body. Interactions **18**(4), 22–27 (2011)
29. Ugur, S., Bordegoni, M., Carulli, M., Mangiarotti, R., Wenveen, S.: Embodiment of emotions through wearable technology. In: Proceedings of ASME'10, pp. 28–31. Washington DC (2011)
30. Cacioppo, J.T., Berntson, G.G., Klein, D.J.: What is an emotion? The role of somatovisceral afference, with special emphasis on soma- tovisceral illusions. Rev. Pers. Soc. Psychol. **14**, 63–98 (1993)

31. Schachter, S., Singer, J.: Cognitive, social, and physiological determinants of emotional state. Psychol. Rev. **69**, 379–399 (1962)
32. Jääskö, V., Mattelmäki, T.: Observing and probing. In: Proceedings of International Conference on Designing Pleasurable Products and Interfaces, pp. 126–131. ACM Press, New York (2003)
33. Field, T.: Touch. MIT Press, Cambridge (2001)
34. Dunne, A., Raby, F.: Design Noir: The Secret Life of Electronic Objects. Birkhäuser, Basel (2001)

Title Design in Bollywood Film Posters: A Semiotic Analysis

Mohammad Shahid, Prasad Bokil and Darmalingam Udaya Kumar

Abstract Indian cinema comprises films produced across India in different languages. Bollywood, a Hindi cinema industry based in Mumbai, is the biggest entertainment industry in India. It shows a diverse and dynamic pattern of visual culture. One can experience this fluid panorama of visual culture through film posters. Over the decades, film posters have been one of the major medium for film publicity. It features images and text to create firsthand visual experience to its audience. As a key element of film poster, title design plays an important role in connoting the theme of the film. It has evolved under influence of various factors like, socio-cultural context, technology and usages. The impact of this development can be seen in terms of technique, composition, color and expressive typography. This paper aims to study and analyze the title design in Bollywood film posters using semiotics framework. Analysis is done based on its structural parameter, relationship with the theme of film, context of use and production technique.

Keywords Bollywood · Film posters · Semiotics · Title design

1 Introduction

Indian cinema, produced across India, has cinematic culture of different states. It shows a very diverse pattern of visual culture. This could be because of its multilingual and multicultural nature. Bollywood is one of the dominating Indian film industry based in Mumbai [1].

In the history of film production in India, different mediums have been implemented for film advertisement. During early period of the cinema, the mode of publicity has been dominated by print media in the form of newspaper advertisements, handbills, lobby cards, publicity booklets, posters and hoardings.

M. Shahid (✉) · P. Bokil · D. Udaya Kumar
Department of Design, Indian Institute of Technology Guwahati, Guwahati, India
e-mail: m.shahid@iitg.ernet.in

A. Chakrabarti (ed.), *ICoRD'15 – Research into Design Across Boundaries Volume 1*,
Smart Innovation, Systems and Technologies 34, DOI 10.1007/978-81-322-2232-3_26

Throughout the timeline film posters have been one of the important medium for film publicity. Like other advertising material, film poster too responds to its environment. From time to time it gets influenced by different art movements and socio-cultural changes [2, 3]. Bollywood film posters also perceived as a public urban icon which offers a visual experience of changing social and emotional standards to its audience [3].

Film posters being the most significant form of publicity is a symbolic visual representation of film in two dimensions where it condenses all the value and theme of a film in a single static plane [4, 5]. It features images and text to create firsthand visual experience to its audience. Keeping the mass audience in mind, the use of textual content is very strategic. Because of the regional language problem and low literacy levels in majority of the audiences, posters show minimum textual content to cater all [3]. The textual content generally includes film title, tagline, credit block, and names of leading characters. Title design plays an important role in suggesting the theme of the film. Due to advancement in the technology and other influential factors, title design has seen gradual changes in terms of form, style, texture, colour, composition, perspective and typeface.

Film posters in India have reflected the diversity of its audiences in terms of culture, religion, class and language. Though the films titles are mostly remains in Hindi across the timeline, one can observe the changes in film title from trilingual script i.e. Latin, Devanagari and Urdu to mainly Latin due to multiplex paradigm [2]. Also the changes in literacy rate can be observed through the changes from decorative lettering to more sophisticated and modern typography.

This paper has aimed to understand the role and development of title design in Bollywood film posters by using semiotic framework. Through this analysis the relation of title with respect to layout of the poster, letterform, film genre, treatments and poster production techniques will be discussed.

2 Methodology

Human beings always try to interpret thing as per their understanding and background knowledge. We can't take reality for granted and define it objectively. Semiotics teaches us that reality is a system of signs [6]. Studying semiotics can help us to be more aware of reality as a construction and of the roles played by everyone in constructing it.

Semiotics is a study of signs and according to Saussurean model; the sign is the whole that results from the association of signifier and the signified. This relationship is called as 'signification' and the value confirmed by a sign depends on its relationship with other sign within the system [6]. The meaning making and persuasion of Bollywood films posters depends on relationship generated by different components of the film posters. Titles design in these posters are analysed separately by syntactic, semantic and pragmatic approach.

Table 1 Prominent trends in letterform structure across timeline

Number variables	1940s	1950s	1960s	1970s	1980s	1990s	2000s
	6	18	18	19	18	20	20
Style	*Bold sans-serif*	*Bold mixed*	*Bold serif*	*Bold mixed*	*Bold mixed*	*Bold serif*	*Bold mixed*
Strokes	*Thick monolinear*	*Thick monolinear*	*Thick low contrast*	*Thick low contrast*	*Thick low contrast*	*Thick low contrast*	*Thick low contrast*
Texture	*No*	*No*	*No*	*No*	*Yes*	*Yes*	*Yes*
Outline	*Yes*	*Yes*	*No*	*Yes*	*Yes*	*Yes*	*No*
Shadow	*No*	*Yes*	*No*	*Yes*	*Yes*	*Yes*	*No*
Perspective	*No*	*No*	*No*	*Yes*	*Yes*	*Mixed*	*No*
Colour	*Black, Red, Yellow, White*	*Black, Red, Yellow, White*	*Black, Red, Yellow, Blue White*	*Black, Red, Yellow, Blue Green White*	*Black, Red, Yellow, Blue White*	*Black, Red, Yellow, Blue White*	*Black, Red, Yellow, Green White*

For this study 129 posters have been selected based on the popularity of the film at the box office during each decade from 1940s to 2000s. This study covers the analysis of title design considering letterforms, title composition, external typo elements and poster layout. As shown in Table 1 different variables like style (normal/bold/italic), stroke, texture, outline, shadow, perspective and colour are proposed for the analysis of letterform structure across the decades.

Syntactics is a study of relationship among signs in formal structure. It is useful to understand structural relationship among the parts of sign. This section utilizes the syntactic principle to see the relationship of title design with respect to poster layout, letterform structure and kind of treatment.

Semantics is a study of meaning created by signs in a system, where they interact with others signs. This approach has been used to analyze title design to see how effective is the film title in overall meaning in relationship with film story and genre wise type classification.

Pragmatics is study of relationship between signs and sign-using agents. Here context contributes to the meaning and interpretation of particular design. This approach helps to see the effect of poster productions technique, display positions, and display technology on title design in films posters.

3 Syntactic Analysis

Poster layout, letterform structure and decorative elements play a key role in the overall title design and its expressiveness. In this section title design is examined based on its structural parameters like layout and typography.

Fig. 1 Title position with respect to poster layout *Source* Posters [2, 9]

3.1 Poster Layout

Title has occupied different positions in poster layout. Figure 1 shows three possible position of title in film posters. Out of 129 posters taken for analysis, 61 % shows title at bottom whereas 28 % at top and 11 % in the middle. Positioning of title was also influenced by the visual hierarchy principles. In maximum cases central characters has been considered as a primary element of the poster [2, 7]. This put title at secondary position in term of importance. So in most of the cases central space is occupied by the leading character's visuals whereas title comes next to it. When it comes in centre of the poster, in 73 % cases it is horizontal whereas 17 % shows diagonal orientation.

3.2 Letterform

In a title design, individual letters act as a building block where its structure and style contribute a lot in suggesting the film's theme. The denotation of individual letters is a straightforward task. Anyone who knows the written language can read and understand the literal meaning of the title but the constructive meaning can only be generated through the understanding of message hidden somewhere in the style and structure of the individual fonts [8]. When we analyses individual letters across the title, it gives an interesting statistics. Approx 63 % posters shows title in uppercase whereas 23 % in title case, 3 % in lowercase and 11 % with mix lettering.

Various parameters of letterforms and the values acquired by them over the period of 70 years are tabulated in Table 1. Different rends and shifts can be noticed in these tabulated variables. At the beginning of the studio era very little experimentation has been done in terms of texture and colour of lettering. Most of the alphabets are plain and monocoloured, basically bright colours like yellow, green and red has been used. Film distributors always try to make posters for the masses so that it could ensure a sufficient return on investment in film production. As obvious, poster artists have used bright colour and glossiness effect to attract the viewer [7]. Darker tone has been used for outline and shadow purpose. Most

Fig. 2 Decorative elements used in letterform design

of the cases letters are decorated with shadow or outline to create more emphasis. The perspective look dominates in the film produced during 1950–1980.

Title design shows variation in individual fonts in term of stroke thickness, size, and counter spaces up to 1970s. One strong reason behind this trend could be the poster making technique. Section 5.1 has covered the poster making technique and its effect on characteristics of letterform. Use of texture, colour and manipulation of letterform enables the title to convey the meaning more effectively.

3.3 Use of Decorative Elements

Posters are made to attract the viewer at first sight [4]. They are also competing with the other visuals across the city wall to draw attention [2]. This competitive streetscape culture leads designer to give some ornamentation to title so that it can stand out of the poster. Titles in 96 % posters show the use of decorative elements to make it more prominent and appealing. Out of 124, 89 (72 %) titles show decoration in terms of outline and shadow whereas 35 (28 %) titles show decoration by manipulating text and image as shown in Fig. 2. Films like *Anand* (1970), *Aradhana* (1969), *Upkar* (1967), *Ghulami* (1985), *Dilwale Dulhania Le Jayenge* (1995) are some example in second category. Contrast colours are prominent in title design throughout the timeline.

4 Semantic Analysis

Semantics is a study of meaning. It deals with the generation of meaning from any sign. This section enquires the relationship of title to the theme or story of the film.

4.1 Title Design Based on Meaning of Title

Title design in posters in the early period of Bollywood cinema shows very minimal variations in letterforms irrespective of the meaning of the title. Letters have

Fig. 3 Letter form design reflecting the theme of the film

Fig. 4 Film titles where letterforms directly relate with the meaning of the title

been designed randomly like in *Bandhan* (1940), *Ram Rajya* (1943) or *Anmol Ghadi* (1946) showing individual artistic skill and style. Some of the exceptions from this era were *Barassat* (1949), *Aan* (1952), *jewel Thief* (1967) *and Aradhana* (1969). Figure 3 shows the title *Barasaat* in which artist has used rain strokes to give the feeling of rain whereas in *Aan* poster letterform is in bold, 3D perspective placed in the background in such a way that it reflects the meaning of '*Aan*' or pride. But this kind of exploration was rare till 1960s where only 13 % title designs show relationship with the meaning of the title.

Beginning of 1970s saw a dramatic increase in the use of expressive typography in the title design. As shown in Fig. 4, poster of films like *Mera Nam Joker* (1970), *Bobby* (1973), *Andhi* (1975), *Sholay* (1975), and *Shatranj Ke Khiladi* (1977) are some example of this trend. In title *Gadar*, the letter has been designed in such a way that it reflects the literal meaning of the title whereas in *Mera Naam Joker* the form of letters has been designed to give a sense of comical and jovial mood. It is interesting to see the letterform in the title *Shatranj Ke Khiladi* where each later is designed to look like 'Mohra' (64 square pieces in chess), exactly relating with the chess game. In similar way the fluffy form and colour of letters in *Bobby*, flowing strokes of letters in *Aandhi*, and flame colour with cracks in *Sholay* reflect the meaning of the title in first sight.

This trend continued during 1980s and 1990s with some prominent example like *Disco Dancer* (1982), *Coolie* (1983), *Rzia Sultan* (1983), *Ghulami* (1985), *Tezaab* (1988), *Agneepath* (1990), 1942 *a Love Story* (1994), *Rangeela* (1995) and *Satya* (1998). Like in *Razia Sultan,* the title and letterform are made in such a way that it looks like a fort in the background. This goes well with the heroic nature of *Razia Sultan* and reflects her power in the poster.

By the end of 1990s and beginning of 2000s design studios were taken hold of poster making in the Bollywood film industry. Advancement in technology and availability of different medium gave more opportunities in designing film's titles. Designers cleverly exploited text-image relationship to make title more attractive and meaningful. When it comes to the whole title composition and its meaning, in most of the cases it succeed in conveying the gist of the film. Looking at the poster of first full length Indian film *Raja Harishchandra* (1913), by Dadasaheb Phalke, the title itself is giving a sense that the film is all about the legend Raja Harishchandra, a story from Indian epic.

4.2 Title Design Based on Film Story

It is difficult to relate the letterform with the meaning of the story. Very few movies title has been designed according to the meaning of the story. To reflect the story, exploration was done mainly with letterform structure and colour. *Khiladi* (1992) is one example where letter 'A' is replaced by human figure holding gun in hand to reflect the action in the film irrespective of the literal meaning of the title which is player.

Most of the Indian films are melodramatic with full of emotions, drama, action and more. This makes difficult for an artist to design a film poster which can represent all parameters of the film [2]. So in most of the cases the letterforms depict only a part of the total theme of the film. As shown in Fig. 5, letterforms in the title of *Disco Dancer* (1982) reflect the theme of the film which is musical and romance. The use of led bulb to create type gives a sense that this is a film revolved around a disco dancer but this is one part of the film. The other part is action, revenge, and family drama.

4.3 Title Design with External Semantic Elements

Film posters are meant to invite the viewer to watch the film. Thus it is important that posters should show the highlights of the film into prominence [2]. Till 1960s Bollywood film poster were less explored in terms of use of external typo elements. In most of the cases title designs were plain and variations were limited to orientation, texture and forms of the letters and in some cases with perspective

Fig. 5 Letter form going with central theme of the film

Fig. 6 Use of external typo elements in title design

style. From mid 1970s, designers started using some external typo elements or "image as a type" for title design. A symbolic language was developed to communicate with illiterate people [2]. *Zanjeer* (1973) in Fig. 6 is one of the examples which shows use of shackles around letter 'Z' to signify sense of tying.

This style comes randomly in many posters throughout the timeline. Images were interestingly used as a type in many movie like *Teesari Manzil, Mera Naam Joker, Khiladi, Zanjeer, Coolie, Ghulami, Dilwale Dulhania Le Jayenge, Border, LOC, Iqbal, Raja Hindustani, Rab Ne Bana Di Jodi* and many more. This introduction of images as a type plays an important role in making posters more persuasive and powerful in terms of conveying message of the film.

4.4 Letterforms with Respect to Genre of the Film

Having the melodramatic nature, it is hard to categories most of the Bollywood films into a specific genre. The basic genres in which most of the films fall under are social, romantic, comedy, action and thriller. The industry started with film based on epic story like *Raja Harishchandra* (1913), *Alam Ara* (1931) and later dominated by romantic and action films. The "Golden period" of Hindi cinema (1940s–1960s) gives some of the most critically acclaimed films of all time featuring the social themes mainly dealing with working class urban life in India. Films like *Awaara* (1951), *Naya Daur* (1957), *Shree* 420 (1955), *Pyaasa* (1957) and *Mother India* (1957) are few of them. The print technology was limited and result of it, the main emphasis was given to keep titles plain and simple. Some exceptions like *Mugha-e-Azam,* where artist has used 3D perspective and shadow to create an illusion of fort suggesting a historical genre.

In the 1970s India was going through social and economical changes and this influenced the Bollywood film making significantly [10]. Industry came up with more of action and violent films in terms of commercial cinema. Films like *Sholay* (1975), *Deewar* (1975) and *Muquaddar Ka Sikandar* (1978) is prominent examples of this cinema. In most of the cases letterforms have followed similar pattern, like use of outline, 3D perspective and shadow to create emphasis irrespective of genre of the film. Action films which established the trend of using 3D style decorated with shadow and outline gel well with the film genre. This type of lettering dominated up to 1980s and became a trend throughout the decades independent of film genre.

5 Pragmatic Analysis

This section enquires about the title design in relation with the context. It enquires how designing of the title in film poster has been influenced by the techniques of poster productions, display positions or platform, and technology.

5.1 Poster Making Techniques

In the beginning of Bollywood film industry, only hand painting was in use for making posters and lithography was the only technology for mass production. These limitations in a way restricted the possibilities in terms of font exploration, whereas in other way gave freedom to poster artist to use their style and imagination. Because different hands were involved, no specific pattern has been followed till 1970s. Cut-paste technique was introduced in early 1970s which has been overcome by computer in 1990s. Table 2 summarizes the dominating poster making technique and their characteristics with an example.

Table 2 Bollywood film poster production techniques

Year	Dominant production technique	Characteristics	Example
1920s–1960s	Hand painting	Broad visible brush strokes with a striking array of colors and typography. Variation in typeface stroke thickness, width, form, weight and letter-spacing	 (1950)
1970s–1990s	Cut-paste	Artist cut the images of the actors out from still photos and then pastes them in a collage fashion on to a canvas board with a hand painted background	 (1986)
1990s–till now	Digital	More standardized. It provides wide scope of exploration. Use of available fonts and less time requirement provide the scope for correction	 (2006)

5.2 Display Position

A display position is one of the deciding factors in terms of title design. What would be the distance of viewer from the poster, whether it will be used for day light display or night, kinds of viewer (rural or urban), whether it is alone or competing with surrounding environment full of other forms of advertisement, all these parameters are deciding factors in title design. These constrain makes designer to think about the size of fonts, visibility, and contrast in the poster. Extensive use of bold, sans-serif (evident from Table 1) and 63 % uppercase letterform reflects the designers' concern. Letterforms in most of the cases are shadowed and decorated with outline in contrast colour so that it can stand out of the poster.

Till late 1960s, when there was no regulatory rule about the poster display platform, there were plenty of spaces in the society to use for the poster display. So posters used to came everywhere, above eye level, bellow eye level, far from the viewer as hoardings or just side of the viewer on the sidewall of the street. This might be the reason behind random positioning of title in the film poster. Since late 1970s, when guidelines on outdoor advertisement were made by the Indian government, posters designs become more standardized and in majority of the cases titles started appearing in the bottom of the film posters. Out of 64 titles analyzed from late 1970s–2000s, 47 (73 %) show title at bottom, 11 (18 %) at middle and 6(9 %) at top.

6 Conclusion

This study deals with title design in the context of Bollywood film posters. Findings reveal that variation in the structural elements in title design has lot of syntactic, semantic and pragmatic influences. Syntactic influences in structural variations include stroke thickness, weight, texture, colour and letter-spacing. In most of the cases (96 %), titles are decorated with outline and shadow to create more emphasis. Semantic influence has been adapted by the use of decoration and external semantic elements. A symbolic meaning has been created by manipulating letters and use of colour. It evolved gradually and usage of external typo elements became evident in the later period as out of 35 titles showing image-text manipulation, 31 (89 %) are from 1960s onwards. External typo elements like use of Coolie badges in the place of letter 'O' in film *Coolie* (1983), enhances this ability by making it simpler for common viewer to read and understand the hidden meaning behind the title. Expressive typography maximizes persuading ability of film poster. There are also pragmatics aspects involved in the title design variation like positioning of title in the film poster and context of usage. Advancement of technology has its effect on the title characteristics across the timeline.

The sample size used for this study is not statistically significant to generalize any arguments. But it definitely provides enough qualitative insights to predict few trends and various causes responsible for those trends. For future study it will be interesting to regenerate these patterns by considering more number of posters from each decade. Future study can also explore the relationship of film title with other elements in the film poster.

References

1. Matusitz, J., Payano, P.: Globalisation of popular culture: from hollywood to bollywood. South Asia Res. **32**, 123–138 (2012)
2. Pinto, J., Sippy, S.: Bollywood Posters. Thames & Hudson, New York (2008)
3. Haggards, S.: Mass media and the visual arts in twentieth-century South Asia: Indian Film Posters 1947-present. South Asia Res. **8,** 71 (1988)
4. Mazumdar, R.: The Bombay film poster. Seminar **525**, 33–41 (2003)
5. Uberoi, P.: The pain of love and the love of pain. In Blamey, D., D'souza R,. (eds.) Living picture: perspective on film poster in India (pp. 79–88), London, Open Editions (2005)
6. Chandler, D.: Semiotics the Basics. Routledge, Oxon (2002)
7. Pinney, C.: Notes on the epidemiology of allure. In Blamey, D., D'souza, R,. (eds.) Living picture (pp. 45–54), London (2005)
8. Leeuwen, T.V.: Semiotics and Iconography. In: Jewitt, T.V. (ed) Sage Publications, New York (2001)
9. Ausaja, S.M.M.: Bollywood in Posters. Om Books International, New Delhi (2009)
10. Kothari, S.: Prasar Bharati. http://newsonair.nic.in, http://newsonair.nic.in/100-YEARS-OF-INDIAN-CINEMA.asp. Accessed April 1 2014

Processes, Methods and Knowledge Creation in Jewellery Design Practice

Noor Adila Mohd Rajili, Andre Liem, Elin Olander and Anders Warell

Abstract This study aims to examine the practice of jewellery design, i.e. how practitioners describe their work and how they create the knowledge needed for their practice. The paper reports on an interview study with jewellery designers, which was analysed using a grounded theory approach. The study indicates that jewellery design practice is defined by the relationship between three main factors; the role of the designer, the type of jewellery, and the type of material used in the process. These factors are in turn affected by differences in (1) approaches used in the design process (2) external and internal factors influencing designers' ways of working, and (3) ways of establishing knowledge. The study provides new insights, which add to the understanding of how designers comprehend their practice and create the knowledge they need for their work.

Keywords Jewellery · Jewellery practitioner · Design process · Knowledge

1 Introduction

Jewellery designers are usually specialised in some aspect of design and craft, such as metal smithing. The "designer's role is to create jewellery that is visually appealing and highly wearable" [1]. In creating jewellery, practitioners of jewellery design make use of a range of principles and techniques in their work. The principles involve different stages such as finding inspiration, exploring and

N.A. Mohd Rajili (✉) · E. Olander · A. Warell
Department of Design Sciences, Division of Industrial Design, Lund University, Lund, Sweden
e-mail: adila.mohd_rajili@design.lth.se

A. Liem
Department of Product Design, Norwegian University of Science and Technology, Trondheim, Norway

A. Chakrabarti (ed.), *ICoRD'15 – Research into Design Across Boundaries Volume 1*, Smart Innovation, Systems and Technologies 34, DOI 10.1007/978-81-322-2232-3_27

generating ideas, and evaluating and translating ideas from sketches to end product [2]. While the stages of design work are established and taught in jewellery education, how jewellery designers actually reason in their work and how they create the knowledge needed in the process is fairly unexplored. Therefore, there is a need to expand the understanding of how knowledge is created in the practice of jewellery design.

In jewellery practice, designing and making processes are highly integrated in contrast to other design disciplines [3, 4]. Mäkinen [3] found that jewellery practitioners included the phase of making into the design process when describing their work. This indicates that their design processes are different from how design processes are described in general [5, 6], which in part may be attributed to the material characteristics of the designed artefacts and the close relation to the arts. Furthermore, jewellery designers use different competence strategies for different design projects, which indicate a need within the profession to develop strategies to secure the right competence for each project. In this context, competence strategies refer to individual design competencies, which consist of knowledge, skills and talents [3]. For example, a price-based strategy can be described as the price of jewellery based on the materials and techniques used and the cost of labour, while a differentiation-based strategy comprises the appearance of the design and the materials used. Jewellery designers use various design-related activities in the designing and making phases [3]. In their work, craft knowledge is associated with action-centred skills and know-how. Such processes require practitioners to use reflection on action [7] where backtracking and reassessment of earlier decisions is driving the design process forward. Hence, reflection is naturally and iteratively embedded in the process irrespective of the tools, materials and techniques used when designing and making jewellery [4].

As a field of education and practice, jewellery design rests on a solid craft and art-based foundation. There is a lack of a theoretical perspective of the field, which is important in order to increase our understanding of the discipline, its practices and impact. This raises our curiosity about understanding jewellery designers' working process. How designers' think [5] and reason [8, 9] in the process of making in other design disciplines has been a focus for design research for some time, however there are very few attempts to understand jewellery designers' reasoning and processes of making.

Hence, up-to-date knowledge about jewellery designers' work processes is quite sparse. Some exceptions exist, however. Wuytens and Williams [4] conducted a study, which aimed at enhancing the understanding of typical traits of jewellery practice and how practitioners create pieces of jewellery by describing the use of processes and methods as well as different ways of using reflection upon action in design reasoning. However, in their study [4] do not give any attention to knowledge creation in their practice. Hence, the rationale of jewellery designers' work remains unexplored.

Therefore, the focus of this study is to investigate jewellery designers' understanding of their work and how they describe their practice. A literature review complemented with first hand empirical data was conducted aiming to answer following three research questions: (1) What types of practices do jewellery designers employ in their work? (2) How do jewellery designers reason about their work and how do they describe the rationale for their practice? (3) How do jewellery designers create the knowledge needed for their design activities? In the following sections, related work, methods of inquiry and results addressing these questions are presented.

1.1 Related Studies

Wuytens and Willems [4] present a conceptual framework describing the variation within design processes that are used by jewellery designers. Since reflection-on-action is the foundation for their view of the design process, instead of describing the process in steps or phases as it usually is done for example in engineering design [6], they [4] have focused on explaining how designers make decisions regarding various design parameters within their design processes using reflexive approaches in their work. They have found that jewellery designers employ two different strategies in their work; 'design parameter reflection' and 'cluster reflection', where 'design parameter reflections' imposes constraints on a design by looking backward and looking forward, 'design cluster reflections' imposes a hierarchically structured order onto the design. Both of these strategies are based on the designer backtracking from an earlier design decision in order to reassess that decision and further progress the process. According to Wuytens and Willems, their way of describing the design process explicitly visualizes diversity within different design processes, which is useful in educational situations. Although Wuytens and Willems's [4] work is a good starting point for understanding jewellery designers working processes, they do not consider designers' knowledge creation needed for their design processes.

Other studies in jewellery design have focused on how certain aspects of jewellery design are supported by methods and tools, such as form specification using computer aided design (CAD) software tools [10, 11], including rapid prototyping (RP) [12]. Legg [13] emphasises the makers' approach to material and process, which highlights the transformation of sensory experience into tangible objects; however the scope of Legg's [13] studies is restricted to contemporary jewellery practice. Siu and Dilnot [14] indicate that although these digital techniques exist and provide clear benefits for certain aspects of the process, jewellers in small firms still prefer manual methods, which is confirmed by Mäkinen [3].

Based on the review of previous studies, knowledge about the jewellery design process is fragmented as identified studies tend to focus on designers' use of specific tools or methods, rather than discussion the rationale of their work. However, understanding designers' processes is challenging, since they may occur unconsciously [15], without explicit awareness. Therefore this paper focuses on how jewellery practitioners understand their work, in order to gain better insights into the practices of design and jewellery making. This will contribute to illustrate and characterise their activities including capabilities, skills, knowledge, and experiences.

2 Method

Four jewellery designers from Sweden and Denmark, all women ranging from 44 to 68 years of age, were interviewed using semi-structured interviews. The participants positioned themselves in different categories with respect to their working situation and jewellery work they produced; such as an artist designer, a goldsmith, a designer maker or marketing designer. The participants were selected using a "snowball" sampling method [16] based on availability and personal networks. Additional interviews with more designers are currently carried out as part of the on-going project.

The interviews were conducted at various venues including a conference, in a studio and at one interviewee's own office. All interviews were conducted by one and the same of the authors; each interview took approximately 1.5 h. Each interview was audio recorded and complemented by note-taking [17], and afterwards transcribed word by word. The interviews were guided by thematic questions grouped in three sections; (1) the practitioners' demographics (background information), (2) experience and knowledge and (3) design inspiration. Before each interview session, the participants were informed about the purpose of the interview study as well as interview ethics such as anonymity. After the interview session, the participants also had an opportunity to add information that they felt had not been discussed in the interview but was relevant to the topic of conversation. Additionally, they were also encouraged to ask questions before, during or after the interview session to avoid misunderstandings.

3 Analysis and Result

The interviews were analysed using a grounded theory approach [18–21]. The analysis involves establishing categories, properties, and concepts [18] as well as identifying the interrelationships between concerns identified in the interviews.

Coding processes are essential in grounded theory analysis. According Charmaz [19], "coding is the pivotal link between collecting data and developing an emergent theory to explain the data". Accordingly, coding is used to identify what has happened and what was the meaning represented by the data. The analysis used in this study has followed the three analytical steps of grounded theory; open coding, axial coding and selective coding [18].

In open coding, concepts and categories were identified in the data. This was done by interpreting the meaning by identifying, naming, categorizing and describing the phenomena [19] which were discovered in the transcriptions. All authors have read the transcription independently and thereafter given coding suggestions for the participants' answers. One of the authors subsequently completed all of the coding. While reading, the codes were gradually developed in order to find categories.

In the beginning the raw data was studied line by line in order to identify the participants' descriptions of feelings, actions and thoughts related to the themes of the interviews. Two different types of codes; in vivo codes and constructed codes [19, 20], have been used in the coding process. Furthermore, the open codes were assigned and grouped into categories. The open coding resulted in a large number of categories related to jewellery design practice. In this paper, only a selection of the most relevant categories are presented.

In axial coding, each sentence was grouped according to their open code contributing to a clear overview of conceptually similar axial codes. These resulting axial codes were then combined into larger analytical categories. In this stage, the links between different categories in open codes were found, for instance, 'experience in working with jewellery', 'different ways of working' and 'norms of working with jewellery' formed the higher category 'exploring jewellery designers activities related to experience and ways of working'. By using these grouping techniques, the open codes were condensed and aggregated to a manageable quantity of relevant axial codes. In this paper, three selected core categories are presented.

Finally, **selective coding** was done after the grouping stage was completed. In selective coding, data from the open and axial coding were examined in a comparative and reflective manner to explain the inner meaning of the participants' descriptions. During the different stages of analysis, a diagram (Fig. 1) was used in order to organize different levels of codes. This diagram facilitates the process of perceiving different concepts but also to distinguish relationships between the codes [18].

Although a range of design activities are implemented by various designers in their design processes, meaning there is a huge diversity in 'how your work is carried out', there are several common and important factors that define the practice. Firstly, all designers worked with different concepts of jewellery, which indicates that there are different types of jewellery and that these are related to the different roles the designers engage in. One designer may use different types of roles. Most respondents described their profession by the type of jewellery they produced, as exemplified by the following statement: *So, if I just call myself a maker, it depends for people to decide what kind of work it is, from what you need and it's for me, I split my career between two concepts of jewellery that I have made. The design work which is commercial jewellery and then I made my art jewellery which is totally liberated from any demands except from my intentions.*

Secondly, the respondents emphasised the use of different types of materials for producing jewellery. How they select materials for their jewellery indicates the type of jewellery they produce and further determines their role in the practice such as whether they work as an artist, a designer or jeweller. Thirdly, there are two distinctively different types of jewellery design; commercial jewellery and art jewellery. The latter uses a variety of materials such as metal, wood, plastic and glass when they make art jewellery, whereas commercial jewellery uses precious materials such as gold and silver as well as gemstones. When creating commercial jewellery; designers often deal with a customer or a company where the piece

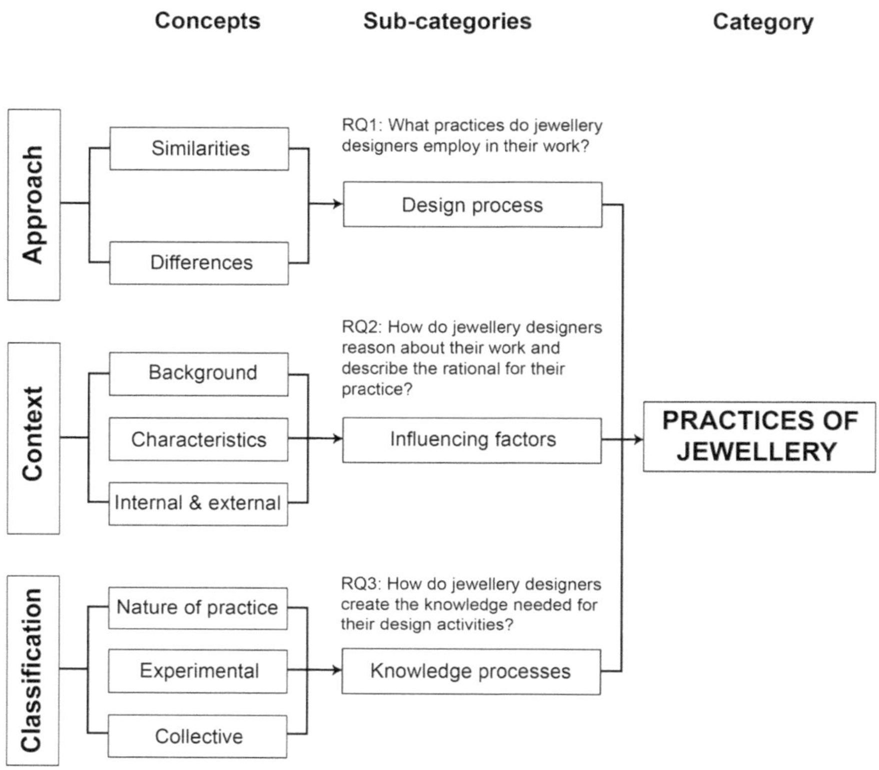

Fig. 1 Illustration of emerging category from axial coding

should represent or express the client, whereas when they create art jewellery, the piece is typically a representation of their own values and expression.

3.1 Approaches for Conducting Design Work

Although all participants have different backgrounds, they have many similar experiences in connection to how they conduct their work. *I'm working very similar to the designer in the process of creating because I am working with, like you see the models and… I see the similarity with the designer but I'm designing for my own workshop.*

When the designers talk about the different activities in relation to their design processes it becomes clear that they use four different approaches when accomplishing them. All designers work from (1) an intuitively driven process, if not through the whole process at least in some parts of the process. Hence, they describe a process where design decision are based on their intuition aiming for expressing own value and feelings trough their jewellery. There are indications that when designers work in teams they use (2) a more structured approach. Such an approach is more tied to that the design decisions are following a specific order of steps or phases connected to predetermined design process. "*The way we do all of our design processes are so structured is not by dreaming of shape and then you shape*". Most of the designers utilize (3) an experimental approach when making design decisions in relation to material, both regarding which type of material to use and to explore what possibilities different material possesses. Finally, (4) collaborative approaches are used (a) for understanding customer requirements in relation to commercial jewellery, and (b) for consultation with other practitioners when executing very complex designs.

3.2 Factors Influencing the Way of Working

Differences in design output are closely related to the designer's background and personal characteristics such as commercial designers and artist designers. When participants described their work, they mentioned different aspects which can be divided into external and internal factors. In the case of jewellery designers, external factors of the designers can be described as a customers, market, colleague and context; social, behavioural and cultural values, whereas internal factors related to characteristics of the designers, which include the designers' emotional and aesthetic values, own motivation and interest. Following is an example on the external factors influences; *Sometimes that happens that the customer comes with something and is the starting point for process, of course and then I think that, oh, I will like to use this more than only for this customer, I can use this idea.*

3.3 Establishing Ways of Knowing

When participants describe their working processes they simultaneously describe their relationship to knowledge creation. In these stories it becomes obvious that knowledge creation is an integrated part of the jewellery designers' core competence. Three different ways of establishing knowledge are identified. All of these correspond to the nature of jewellery practise in a natural way and may be considered embedded into the design process. The most frequently described way of establishing new knowledge, 'reflection upon action', is in agreement with existing literature, indicating that activities of reflection and doing are highly integrated. *being very analytic and then go back to working with your hands then look at what you've been doing and analysing it and taking it further. …it allows you to kind of digest your thought and it's there in the back of your head all time anyway and so it's kind of leads you anyway towards what you would like to express.*

The other two ways seem to be related to what type of approach the designer uses for their design process. Hence if they use an experimental approach their knowledge creation is based on learning principles such as 'learning by doing' or 'trial and error'. One participant states that she likes to challenge herself in order not to get stuck in a certain way of working. Another designer describes that she uses a 'library of different shapes and forms'. For every new piece of jewellery, she makes a prototype for example in silver, which is later used as a shape reference for future objects. The last way of creating knowledge is related to the collective approach, where the knowledge needed is acquired from a third part, like company or user.

4 Discussion

The main focus of the present study has been on understanding how jewellery designers describe their work and how they create knowledge needed for their design processes. The study indicates that jewellery design practice can be characterised as being defined by the relationship between three main factors; the role of the designer, the type of jewellery, and the type of material used in the process. These factors are in turn affected by differences in (1) approaches used in the design process (2) External and internal factors influencing designers' ways of working, and (3) ways of establishing knowledge.

These findings are summarized and illustrated in Fig. 2. In particular, the model highlights the numerous ways in which jewellery designers' work with jewellery and reasoning within their design processes. This particular finding should be seen

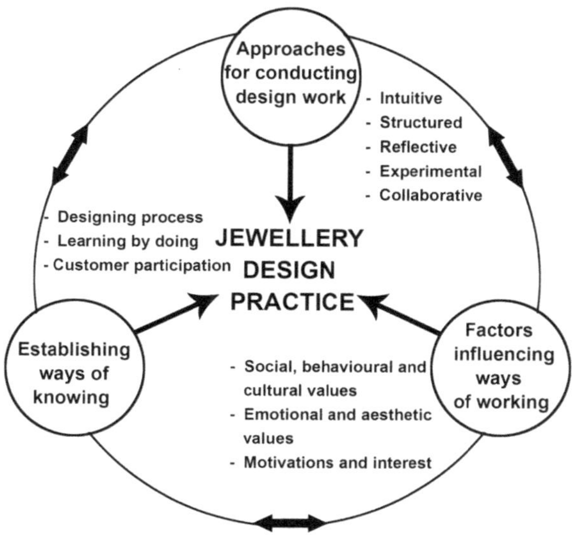

Fig. 2 Jewellery design practice as defined by different approaches used in the design process

in contrast to earlier studies, which emphasise singular aspects of jewellery processes, e.g. the use of tools, as described before [11–13].

In jewellery practices, it seems that most jewellery practitioners have more than one role in relation to their profession, this is valid both in the designing and production phase [3]. The findings from this study confirm the use of multiple roles in jewellery practice. The roles are determined by their characteristics, hence the knowledge and experience used in different roles are connected with action-centred skills and know-how and thereby a different concepts of jewellery mirror different ways of working.

In relation to the approaches for conducting design work, this study indicates that a structured way of working with jewellery is used by commercial designers working in design teams. The principal of structured way of working with jewellery indicate the use of stages similar to a general approach to design. In general, the main working process used by the design team is identified by various designing stages and strategies.

Design reasoning [7] is typically understood in other fields of design as understanding "how design happens" [8]. This is because their design process are determined by relationships between the different roles they engage in, the different types of jewellery they make, and the different types of material they use. Some of the designers' stories indicated that they, to some extent, adopt a strong tendency towards using the hermeneutic perspective, which aims at obtaining "an understanding of the designed product which is related to contexts, values and functions" [22]. Similarly, in this study, jewellery designers associated contexts and values with their creative thought.

As a whole, this study broadens the knowledge on jewellery design by introducing a model describing the various factors affecting the design process. Additionally, the findings support previous research [23, 24] arguing that intuitive and experimentally driven approaches are being employed by jewellery designers in their practice.

The study confirms that the practice of jewellery has similarities with other craft based design such as ceramics and furniture design [25]. These similarities can be exemplified by that designers in these fields integrate intuitive approach and experimental practices in their design process.

This study contributes by clarifying and positioning the profession of jewellery design practice, which will improve the understanding of the discipline. The findings of this initial study have provided a background for more in-depth studies of how jewellery designers reason about their jewellery making practice and have the potential to contribute to a better understanding of the role of the profession in the jewellery field.

5 Limitations and Future Research

The present study has some limitations that need to be addressed in future research. This paper presents a first analysis of the findings from a still ongoing study. In the study, eight jewellery designers were interviewed, however only four interviews have been thoroughly analysed and included in this paper. A more in-depth analysis will be done in the near future and reported continuous.

In this study, we present indications of how designers talk about their design activities, however, how jewellery designers create knowledge needed for their design processes is still under investigation. In order to discover this issue, the authors propose a direction of this research by conducting a situated case-based research using observation study.

Future research should aim at validating the proposed concept model (Fig. 2) describing the nature of jewellery design practice. Since this exploration is about knowledge creation and its relation to working and thinking, there are several aspects that can be developed further.

Acknowledgments The authors wish to thank the participants in the interview for contributing to the findings of this study. I would also like to acknowledge the financial support of the Ministry of *Higher Education of Malaysia* and *Universiti Teknologi MARA* for the duration of my PhD Studies.

References

1. Galton, E.: Basics Fashion Design 10: Jewellery Design: From Fashion to Fine Jewellery. AVA Academia, Lausanne (2012)
2. McGrath, J.: The complete jewellery making course; principles, practice and techniques: a beginner's course for aspiring jewellery makers, Page one, (2007)
3. Mäkinen, H.: Product design as a core competence in a design-oriented industry. Adv. Appl Bus. Strategy **9**, 103–126 (2005)
4. Wuytens, K., Willems, B.: Diversity in the design processes of studio jewelers. EKSIG: Experimental knowledge, method & methodology. Availabe at: http://www.academia.edu/885664/Diversity_in_the_design_processes_of_studio_jewellers (2009)
5. Lawson, B.: How Designers Think: the Design Process Demystified. Routledge, London (1997)
6. Cross, N.: Engineering Design Methods: Strategies for Product Design. Wiley, New York (2008)
7. Schön, D.A.: The Reflective Practitioner: how Professionals Think in Action, vol. 5126. Basic books, New York (1983)
8. Goldschmidt, G., Weil, M.: Contents and structure in design reasoning. Design issues, pp. 85–100, (1998)
9. Rittel, H.W.J.: The Reasoning of Designers. IGP, Montreal (1987)
10. Molinari, L.C., Megazzini, M.C., Bemporad, E.: The role of CAD/CAM in the modern jewellery business. Gold. Technol. **23**, 3–7 (1998)
11. Wannarumon, S., Bohez, E.L.: Rapid prototyping and tooling technology in jewelry CAD. Comput. Aided Des. Appl. **1**(1–4), 569–575 (2004)
12. Hohkraut, U.: Rapid prototyping and jewelry design. In: Annals of DAAAM and Proceedings (2010)
13. Legg, B.: Materiality of place: an investigation into the makers approach to material and process as a reflection of place within northern European contemporary jewellery practice. PhD thesis, University of Edinburgh (2013)
14. Siu, N.W., Dilnot, C.: The challenge of the codification of tacit knowledge in designing and making: a case study of CAD systems in the Hong Kong jewellery industry. Autom. constr. **10**(6), 701–714 (2001)

15. Wallace, J., Dearden, A.: Digital jewellery as experience. In: Future Interaction Design pp. 193–216, Springer London (2005)
16. Goodman, L.A.: Snowball sampling. Ann. Math. Stat. **32**(1), 148–170 (1961)
17. Steinar, K.: Doing Interviews. Sage, CA (2007)
18. Corbin, J., Strauss, A. (eds.): Basics of Qualitative Research: Techniques and Procedures for Developing Grounded Theory. Sage, CA (2008)
19. Charmaz, K.: Constructing Grounded Theory: a Practical Guide Through Qualitative Analysis. Pine Forge Press, CA (2006)
20. Glaser, B.G., Strauss, A.L.: The Discovery of Grounded Theory: Strategies for Qualitative Research. Transaction Publishers, New Jersey (2009)
21. Seale, C., Gobo, G., Gubrium, J.F., Silverman, D. (eds.): Qualitative Research Practice. Sage, CA (2004)
22. Lie, U.: Framing an eclectic practice; historical models and narratives of product design as professional work. Doctoral dissertation, Norwegian University of Science and Technology, Trondheim (2011)
23. Amadeu, F.: Creativity and emergingknowledge: Intuitive practice in design and crafts. Trans-techresearch.net (2012)
24. Pereira, Á. Tschimmel, K.: The design of narrative jewelry as a perception-in-action process. In: Proceedings of the 2nd International Conference on Design Creativity, Glasgow: The Design Society, vol. 97, p. 106, (2012)
25. Mohd Rajili, N.A., Olander, E., Warell, A.: Characteristics of jewellery design: An initial review. In:Conference Paper Submitted for ICoRD'15 (2014)

Aesthetic Design Optimization of a Sustainable Human Cum Electric Powered Vehicle Through Implication of Golden Ratio Perspectives

Sachin Mishra and Pradeep Kumar

Abstract The present research envisages the aesthetic design optimization strategy employed to achieve an aesthetically pleasing sustainable design of a Human cum Electric Powered Vehicle. The perspective of golden ratio is implicated as a tool for predicting, analyzing and iterating the design for an aesthetic and sustainable design. The vehicle—Vajrayana, is a two seater, tricycle with convertible rooftop cum wind shield type aerodynamic body. This paper is concerned, first with the problem of designing a streamlined vehicle shape having a convertible rooftop cum wind shield type aerodynamic body, followed by a sustainable artistic approach, to approximate, various iteration with golden ratio-especially in the form of the golden rectangle. The final design is simulated to validate that golden ratio, with the concept of sustainability and aerodynamics can be tailored to give pleasant vehicle design and overall performance.

Keywords Golden ratio · Sustainable · Aerodynamics · Streamline

1 Introduction

Green vehicles are assumed to be equipped with a low and limited power generating units like human power, fuel cells, battery etc. A highly efficient aerodynamic body structure is required to reduce drag force [1] and thus power consumption. For a sustainable design of vehicle, an aesthetically pleasing look is required with efficient overall performance. The concept of streamline design enhanced with aesthetic optimization through the implication of golden ratio caters the above requirement.

S. Mishra (✉) · P. Kumar
School of Engineering, Gautam Buddha University, Greater Noida, India
e-mail: sachinmishra92me@gmail.com

A. Chakrabarti (ed.), *ICoRD'15 – Research into Design Across Boundaries Volume 1*,
Smart Innovation, Systems and Technologies 34, DOI 10.1007/978-81-322-2232-3_28

The Vehicle—Vajrayana, as shown in Fig. 1, is an ergonomically designed human cum battery powered hybrid vehicle. The concept of sustainability, equipped with aesthetics and aerodynamics is implicated to design, analyze and simulate the aerodynamic body. Golden ratio or divine proportion is a ratio of 1.618—a naturally occurring number called Phi (φ) [2–7]. This mathematical pattern is evident in the nature at many places. Its ease on our perception is justified by its recurrence.

A golden rectangle [2, 3, 7] as shown in Fig. 2 is one whose side lengths are in the golden ratio, 1: $(1 + \sqrt{5})/2$ or approximately 1:1.618 as the value of term '$(1 + \sqrt{5})/2$' is equal to 1.618 approximately. A distinctive feature of this shape is that when a square section is removed, the remainder obtained is another golden rectangle.

This paper is concerned, first with the problem of designing a streamlined vehicle shape having a convertible rooftop cum wind shield type aerodynamic body, followed by a sustainable artistic approach to approximate various iterations with golden rectangle using the software PhiMatrix [8]. Computational Fluid Dynamics

Fig. 1 Vehicle Vajrayana design

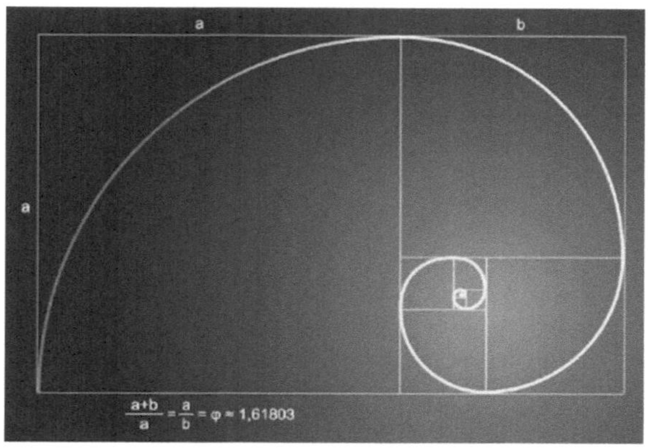

Fig. 2 Golden rectangles and Phi (φ)

[1] is utilized as a tool for predicting and analyzing fluid flow across the body. Ansys Fluent [9] is used as CFD tool to predict, simulate and improve the aerodynamic behavior of vehicle's body considering the coefficient of drag, pressure contour and velocity contours as key factors.

2 Vehicle Design

Vajrayana is a green convertible car which is an ergonomically and aerodynamically designed human powered cum battery powered hybrid vehicle. It is designed and fabricated to enhance performance, maneuverability, aesthetics and comfort while catering the safety norms. It is designed to accommodate 95 % population by considering the various factors of man machine interaction.

2.1 *Conceptual Design*

The vehicle is designed using Autodesk Inventor software [10] as a prototype hybrid car to participate in Efficycle Competition [11] organized by SAE India. As per the rules and regulations of the competition, the challenge was to design and build a three wheeler human cum battery powered vehicle to accommodate two persons.

The outer body is designed to be aesthetically pleasing with an aerodynamic streamlined shape. Figure 3 shows the workflow drawn using MS Word software [12].

2.2 *Sketching*

The outer body creative sketches were made for side view, front view and top view as per the vehicle design. The two proposed designs are shown in Fig. 4. As the side view have the most of the effect on streamline and aerodynamics, various iterations of side views are further analysed and improved manually.

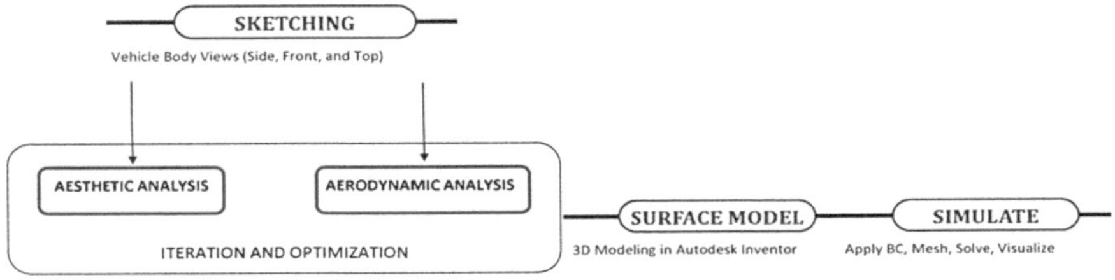

Fig. 3 Outer body design and optimization workflow in Microsoft Word software

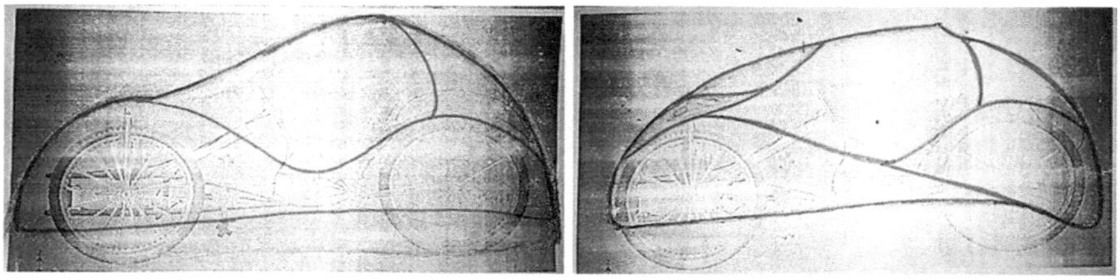

Fig. 4 Hand drawn sketches of proposed vehicle body

Fig. 5 Iteration 1 in Autodesk Inventor

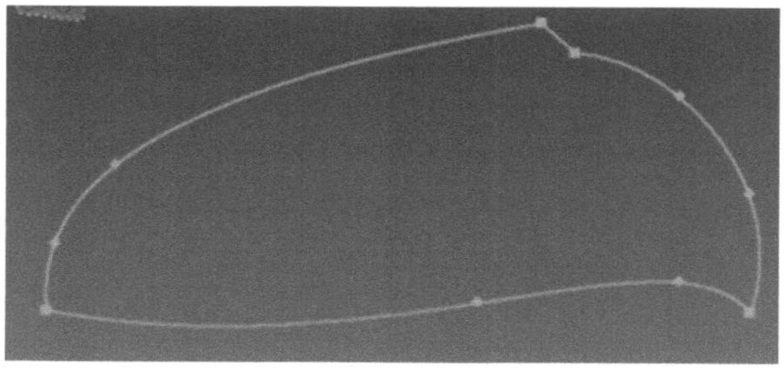

As the second proposal is aesthetically pleasing and also provides a streamline shape when compared to the first design, it was selected for further improvement to produce the side view of the vehicle. Also, the aerodynamic streamline effect is much more a function of side view as compared to top view and front view.

2.3 Iteration and Optimization

Various iterations were performed to modify the shape of side view profile for the selected proposal. The design was improved by analysing the aerodynamic behaviour and streamlines of air and modifying the shape through the implication of golden ratio.

2.3.1 Iteration 1

The side view drawn in Autodesk Inventor [10], as shown in Fig. 5 is as per the proposed and selected vehicle design sketch, shown in Fig. 4.

Aerodynamic analysis using CFD Software Ansys Fluent [9] is shown in Fig. 6. Velocity contour suggest that with an input of 12m/s velocity, maximum velocity of 18.7 m/s is developed at the top of the notch. Various input parameters, Analysis output and modifications in curve profile are shown in Table 1.

Fig. 6 Velocity contour for Iteration 1

Table 1 Analysis summary for Iteration 1

Aerodynamics	Value	Aesthetics	Details
Analysis type	2D	Golden ratio	Not implicated
Wind speed	12 m/s	Problem area	Top notch, outer curve
Avg. drag coefficient	0.59	Modification need	Remove notch, apply golden rectangle for curve fitting
Maximum velocity	18.7 m/s		

Analyzing the velocity contour, there is a large separation shown by dark blue color at the back side which is not desirable. Further, for comparison, average coefficient of drag (C_d) is observed which is coming out to be 0.59 through the convergence of 30 iterations in Ansys fluent calculation. This design profile acts as the base for further modification to improve the aerodynamics and aesthetics using curve fitting through the implication of golden rectangle.

2.3.2 Iteration 2

The side view in this iteration is a modified view through the removal of notch and by the implication of golden rectangle as shown in Fig. 7. The point A as shown, divides the frontal curve and rear curve, thus its horizontal position is important both, aesthetically as well as aerodynamically. The vehicle side view look is dependent on it and the streamline changes with its position.

Curve fitting was done to improve the shape so that, point A represents the golden rectangle's vertical line. The frontal curve lies in the left rectangle and the rear hatchback curve in the right side rectangle.

As shown in Fig. 8, velocity contour is modified in this iteration with a maximum velocity of 15.3 m/s developed at the top point A. The large separation as in previous case is now reduced, shown by dark blue color at the back side which shows the improvement in streamline shape. The average coefficient of drag is coming out to be 0.44. Table 2 shows the Analysis summary and modifications for Iteration 2. Still the modification is needed to reduce the separation at back bottom and C_d Value.

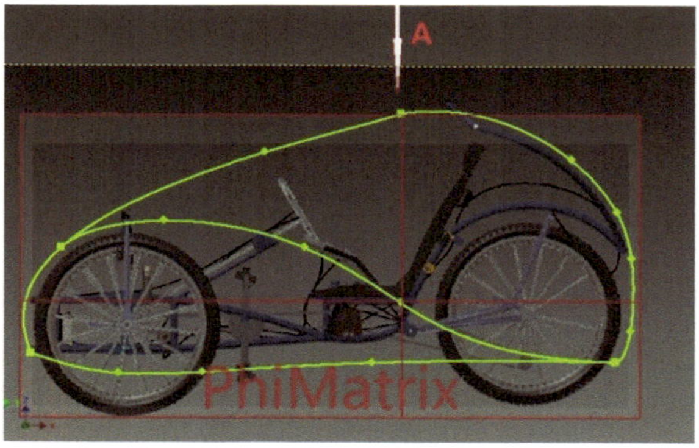

Fig. 7 Side view Iteration 2 with PhiMatrix

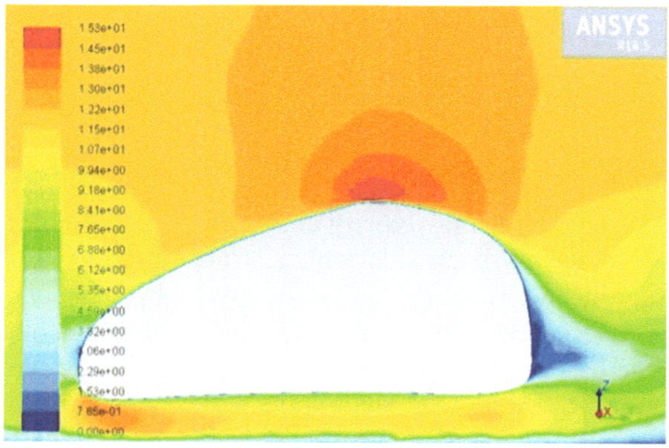

Fig. 8 Velocity contour for Iteration 2

Table 2 Analysis summary for Iteration 2

Aerodynamics	Value	Aesthetics	Details
Analysis type	2D	Golden ratio	Implicated
Wind speed	12 m/s	Modification done	Top notch removed, outer curve fitting for Pt. A by golden rectangle
Avg. drag coefficient	0.44	Problem area	Separation at back side
Maximum velocity	15.3 m/s	Modification need	Apply golden ratio for curve fitting and reduce separation

2.3.3 Iteration 3

This iteration is made by the modification of Point B in Inventor [10] as shown in Fig. 9 with respect to the vehicle height. Point B is acting as the deciding factor for end separation of air from the vehicle and also the bottom back look, thus the position is fixed as per the golden rectangle. The inward movement of rear curve at the bottom starts at point B so that the upper curve lies in upper rectangle and the rest in lower.

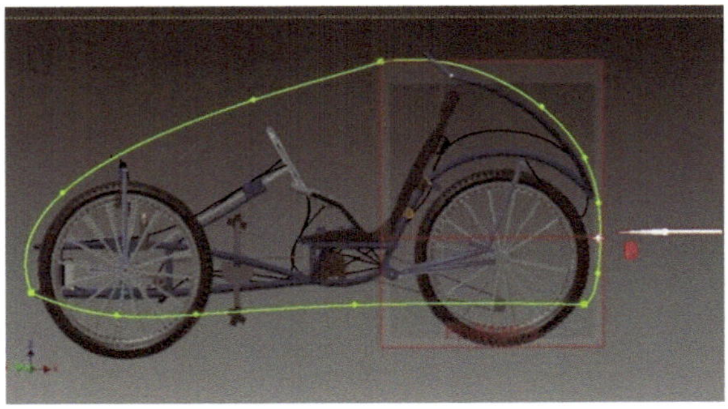

Fig. 9 Side view Iteration 3 using PhiMatrix

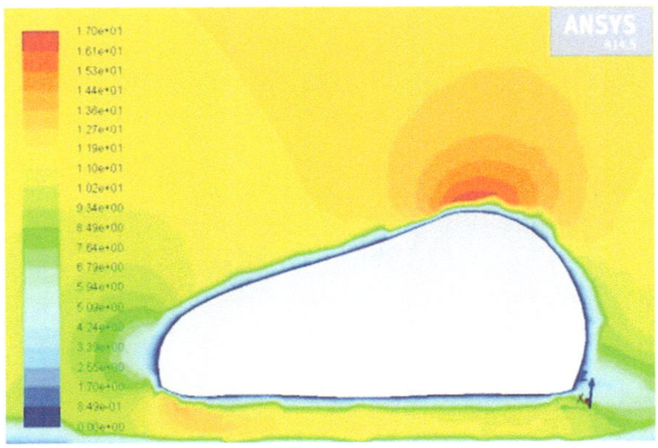

Fig. 10 Velocity contour for Iteration 3

Table 3 Analysis summary for Iteration 3

Aerodynamics	Value	Aesthetics	Details
Analysis type	2D	Golden ratio	Implicated
Wind speed	12 m/s	Modification done	Rear curve fitting for Point B by golden rectangle
Avg. drag coefficient	0.23	Problem area	None
Maximum velocity	17 m/s	Modification need	None

Velocity contour in Fig. 10 shows that the separation is reduced to almost negligible area and the curve is behaving like perfect streamline shape. The average coefficient of drag is coming out to be 0.23. This shape is final optimized curve profile with no further modification needed.

The summary of analysis for this optimized design is shown in Table 3. As the velocity contour is representing no large separation zone and also no vortex is generated at the back, the design can be said to be optimized.

The final output is now further utilized as the side view for modeling the vehicle 3D body design. With the minimization in separation area and C_d value, the curve profile presents an aerodynamically cum aesthetically optimized outer shape for the vehicle. The generated profile is analyzed at various wind speed and verified for its behavior.

3 Surface Modeling

The side view as per the Iteration 3 above is further improved for aesthetics by curve fitting of internal curves as shown in Fig. 11. The curve for side window at Point C, D and E are fitted as per the driver view requirement and golden ratio.

The curves will be further extruded to model the complete three dimensional body. The front view, as shown in Fig. 12 is designed using the golden rectangle to decide the curve fitting through the Point F and G. Point F and F' lies on the first horizontal line height from the bottom and G lies on the middle of top line. The top view is designed as per the vehicles outlines shown in Fig. 13.

Fig. 11 Side view internal curve fitting with PhiMatrix

Fig. 12 Front view optimization

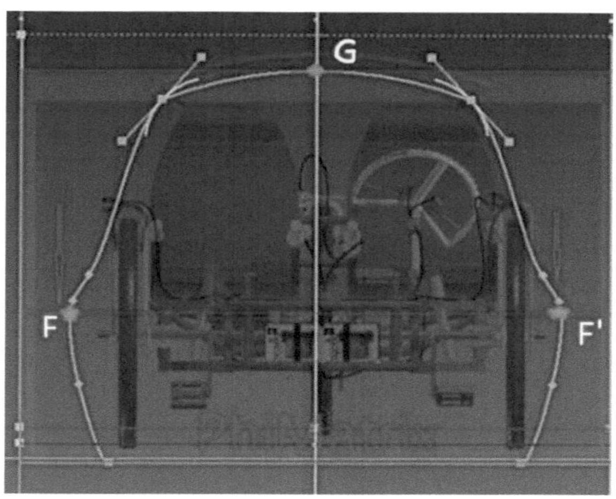

Fig. 13 Top view profile

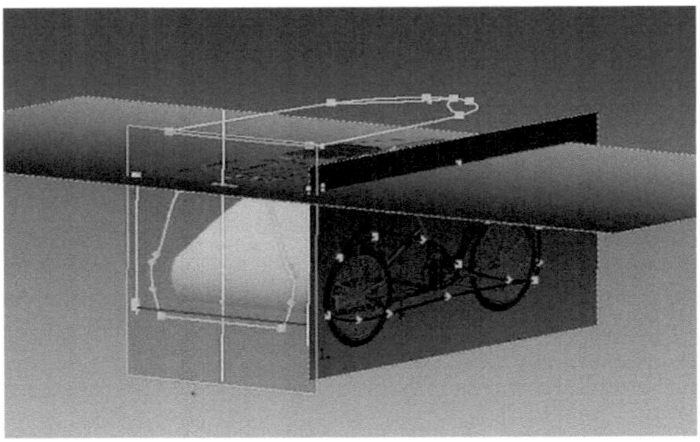

Fig. 14 Front view improvement in Autodesk Inventor

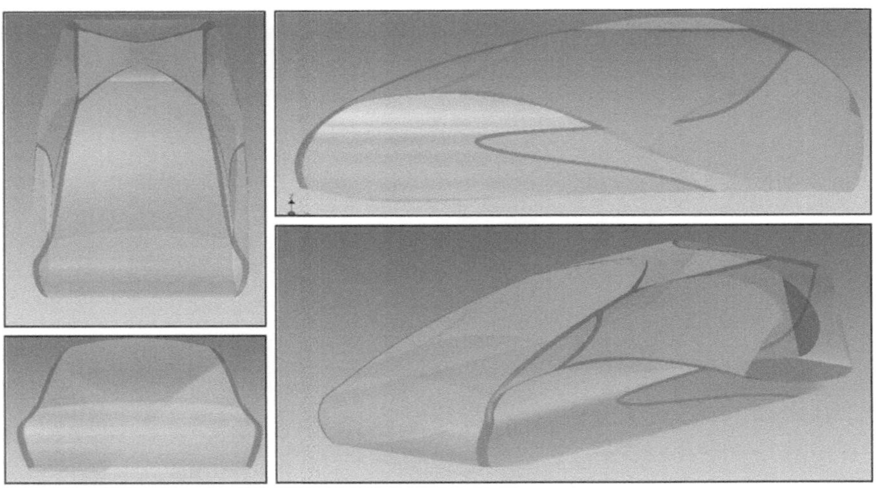

Fig. 15 Final output vehicle body design in Autodesk Inventor

The three views, side, front and top views are used to model the 3D Body using surface modeling concepts in Autodesk Inventor [10] as shown in Fig. 14. The overall curve fitting provides an aesthetically pleasing look and fits on mechanical design concepts as per the standards of safety and ergonomics.

The final optimized design as per the implementation of golden ratio and aerodynamic principles are shown in Fig. 15.

4 Simulation and Visualization

The final output 3D Model is analyzed for aerodynamic behavior using CFD Tool Ansys Fluent [9]. The input parameters and output results for analysis is shown in Table 4. The body is simulated for Velocity Contour, Static Pressure, Drag Coefficient and Path lines. The half body is analyzed with symmetry across the YZ plane. By using the Boolean operation, the half body space is reduced to no air zone and then meshing of air is done for CFD Analysis.

The pathline of velocity magnitude in Fig. 16 shows a maximum velocity of 21.9 m/s generated on the top side. Figure 17 shows the velocity contour at YZ symmetry plane. A maximum static pressure of 87.3 Pa is generated in the front of the body as shown in Fig. 18. Average Coefficient of Drag is coming out to be 0.29 which is considered to be very good for an aerodynamic body. The Turbulent Kinetic Energy is shown in Fig. 19 having a maximum value of 11.4 m^2/s^2.

Table 4 CFD aerodynamic analysis of final body

Input parameters	Value	Output results	Details
Analysis type	3D	Frontal area	0.78 m^2
Wind speed	12 m/s	Max. velocity	21.9 m/s
Material used	Air	Max. static pressure	87.3 Pa
Meshing	Tetrahedral	Drag coefficient	0.29

Fig. 16 Pathline of velocity magnitude

Fig. 17 Velocity contour at symmetry plane

Fig. 18 Contour of static pressure in Fluent

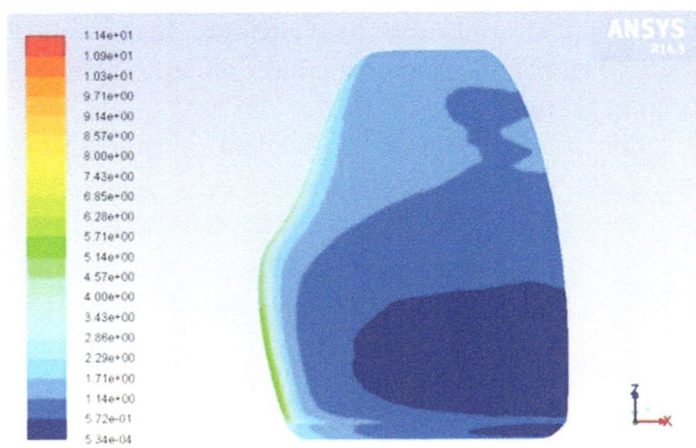

Fig. 19 Turbulent kinetic energy contour

5 Results and Discussion

The vehicle is designed to improve the overall aerodynamic performance as well as aesthetic performance using the CFD tool Ansys Fluent and PhiMatrix Software.

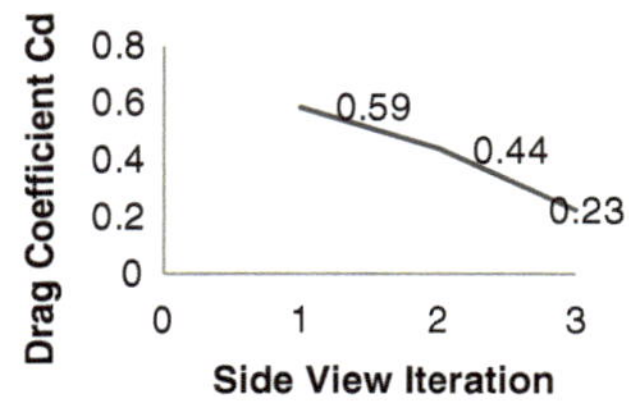

Fig. 20 C_d variation side view iteration

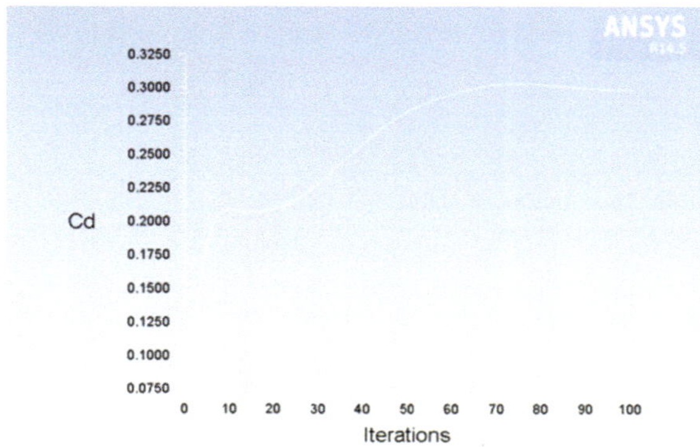

Fig. 21 C_d convergence curve in Ansys Fluent

As per the aerodynamic analysis shown for various iterations of side view, the average value of coefficient of drag (Cd) is plotted with respect to iterations as shown in Fig. 20, and in the final iteration, the lowest value of Cd is observed, which becomes the criteria for selection of this curve profile in modeling. Iteration 3 is obtained as a result of intense implication of golden rectangle concept, thus it is well shown that the concept of golden rectangle can be used to generate aerodynamic profiles.

The average value of coefficient of drag (C_d) is coming out to be 0.29 for the final 3D body. The variation and convergence curve is shown in Fig. 21. As the value is below 0.3, the designed vehicle body shows an excellent aerodynamic behavior. As a result of blending aerodynamics and golden ratio, an aesthetically pleasing streamline body is designed for the vehicle.

6 Conclusion

The aesthetic and aerodynamic behavior of the vehicle Vajrayana is enhanced by employing the concept of golden ratio and CFD together from sketching to modeling. The role of CFD is well played by the tools to predict, simulate and visualize the flow across the body. The application of PhiMatrix is an effective tool for generating aerodynamic and aesthetic curve profile.

The presented research validated the role of golden rectangle concept for enhancing the aerodynamics and aesthetic optimization of green vehicles. The study successfully presents the example that aerodynamics with the concept of golden ratio can be tailored to give better vehicle look and overall performance. Further research includes the development of mathematical correlation between aerodynamics and golden ratio.

References

1. Versteeg, H.K., Malalasekera, W.: An Introduction to Computational Fluid Dynamics—the Finite Volume Method, 2nd edn. Pearson Education Limited, New Jersey (2007)
2. Livio, Mario: The Golden Ratio: the Story of Phi. Broadway, New York (2002)
3. Mario Livio.: The golden ratio and aesthetics. Plus Magazine, November (2002b)
4. Akhtaruzzaman, M., Shafie, A.A., Ahsan, T., Alam, M.S., Raihan, S.M., Hasan, M.K., Haider, M.B.: Golden ratio, the Phi, and its geometrical substantiation. In: Research and Development (SCOReD), 2011 IEEE Student Conference on, pp. 425–430. IEEE, (2011)
5. Akhtaruzzaman, M., Shafie, A.A: An attempt to develop a Biped Intelligent Machine BIM-UIA. In: Mechatronics (ICOM), 2011 4th International Conference On, pp. 1–7. IEEE, (2011)
6. Schewe, G.: Reynolds-number effects in flow around more-or-less bluff bodies. J. Wind Eng. Ind. Aerodyn. **89**(14), 1267–1289 (2001)
7. Akhtaruzzaman, Md, Shafie, Amir A.: Geometrical substantiation of Phi, the golden ratio and the baroque of nature, architecture, design and engineering. Int. J. Arts **1**(1), 1–22 (2011)
8. Phimatrix software. http://www.phimatrix.com/
9. Ansys fluent software. http://www.ansys.com/
10. Autodesk inventor software. http://www.autodesk.com/
11. SAE India efficycle competition. http://effi.saenis.org/
12. Microsoft Word Software. http://office.microsoft.com/en-in/word/

Cognitive Theories of Product Emotion and Their Applications in Emotional Product Design

Anirban Chowdhury, Swathi Matta Reddy, Debkumar Chakrabarti and Sougata Karmakar

Abstract Product attributes which can elicit positive or negative emotions among users, determine product acceptance in market. Different cognitive theories of emotion are found in literature for emotional product design. These theories explain the various aspects of underlying phenomena of emotion generation based on product attributes and their impact on product choice. In present paper, six popular cognitive theories of emotion (viz. 'Kansei Engineering', 'Theories of product personality', 'Pleasure model', 'Product Appraisal Model', 'Emotional Design Model', and, 'Technology as Experience Framework') have been reviewed, stating their potential benefits and limitations. An integrated theoretical framework of emotional product design has been proposed by incorporating different potential approaches of earlier mentioned cognitive theories of emotion. It is expected that developed framework would be helpful to industrial designers, ergonomist, product innovation managers and cognitive scientists towards better emotional product design.

Keywords Cognitive ergonomics · Design · Emotion · Pleasure · Product

1 Introduction

Cognition directed emotion is found in many daily life scenarios/situations, though there are different causes of emotion other than cognition. Actually, cognition and emotions are closely related with each other; and cognition of an object/stimulus may lead to generation of different basic emotions. For example, fear may be evoked due to paying sudden attention to a stranger with a gun in a dark empty road.

A. Chowdhury · S.M. Reddy · D. Chakrabarti · S. Karmakar (✉)
Ergonomics Laboratory, Department of Design, Indian Institute of Technology (IIT) Guwahati, Guwahati 781039, Assam, India
e-mail: karmakar.sougata@gmail.com

A. Chakrabarti (ed.), *ICoRD'15 – Research into Design Across Boundaries Volume 1*, Smart Innovation, Systems and Technologies 34, DOI 10.1007/978-81-322-2232-3_29

In other case, one student may be pleased being rewarded for top score in a design course. Similar to these incidents, it is possible to generate positive emotions among consumers or users through product design. For the last three decades, designers and human factor researchers are more concerned about formulating design theories based on human cognition and emotion. Some practices regarding application of cognitive theories of emotions in design related industries are in full swing.

Industrial designers/engineers are trying to build different strategies for emotional product design but till date only few cognitive theories of emotion are presently being followed for product, service and interaction designs. Though many cognitive and non-cognitive theories are available in literature; in present paper total six cognitive theories/models of emotional product design which are very relevant to product/service design have been described. These includes 'Kansei Engineering', 'Product Personality', 'Pleasure Model', 'Emotional Design Model', 'Basic Model of Product Appraisal', and, 'Technology as Experience Framework'. Though each established cognitive theory of product emotion has its own potentials, none of the theory is able to explain the underlying process of emotion generation due to design/features of each and every product. Therefore, there is still need to study of cognition and emotion related theories to establish an integrated theoretical framework for better emotional product/service design outcome. In this context, detailed understanding of cognitive theories of emotion is necessary. In the present paper an attempt has been undertaken to bring together available information on cognitive theories of product emotion and to highlight their potentials in product/service design. A theoretical framework has also been developed for better emotional and pleasurable product/service design. Before going into details about cognitive theories of emotion, it is better to know basics of cognition and emotion.

2 Understanding Cognition and Emotion

Cognition is the mental process which includes series of sensory events such as perception, attention, use of short term memory/working memory, exploration of long term memory, information processing, identification and classification of objects, reasoning, decision making etc. [1]. The concept and meanings of cognition may be stated from following perspectives:

- Understanding about objects/events/surrounding environment
- Perception and consequent acceptance of meaning of objects
- Thought/thinking process
- Reasoning and interpretation
- Initiator of emotion(s)

Emotion which is defined as the different mental and bodily changes with certain features, has subjective, behavioral, cognitive and internal/peripheral physiological components [1, 2]. Cognitive processes are associated with emotional responses. Emotions either consciously or unconsciously may vary from time to time, place to place and situation to situation even in the same place [2]. In his great philosophical path breaking work "On the Origin of species", Darwin [3] first proposed that there are a limited number of basic and universal emotions. Ekman and Friesen [4] stated that there are six basic emotions corresponding to anger, disgust, fear, happiness, sadness, and surprise. Emotion which is actually one type of 'affective state' can have different aspects which are as follows:

- Emotion may be both conscious and unconscious
- It originates from different parts of the brain
- It combines both mental and bodily changes
- It affects attention and information processing in predictable way
- Emotion may influence motivation and behavior

3 Causes of Emotion

Different emotions occur by different stimulus. Causes of emotion can be categorized into two broad classes based on their origins: cognitive and non-cognitive causes [2].

3.1 Cognitive Causes

Cognitive causes of emotion refer to how a thought causes emotion. Cognitive scientists have widely accepted that thoughts can cause emotions but it depends on what kind of thought it is [2]. Many thought such as 'rose is red', 'ice is cold' etc. may have little impact on most people. Thoughts which evoke emotion are likely to be evaluative thoughts. These thoughts assess things in a way that reflects our attitudes towards them. For example, one may not have any emotional response to the thought 'kitchen knife is sharp', but emotion will likely follow the thought that 'kitchen knife is dangerous'. The concept of more dangerous or less dangerous is directly related to one's perception of safety.

3.2 Non-cognitive Causes

It is assumed that perceptual states may evoke/induce emotions by affecting our appraisal process. In most of the cases perception requires interpretation before emotion arises. On contrary, in some cases perceptual states may not require interpretation to elicit an emotion. This second phenomenon is considered as non-cognitive cause of emotion. For example, occurrence of a sudden high pitch sound rapidly generates an emotion. Such stimulus would seem gratuitous to postulate an intervening thought. Similar incidence happens of when fear response is evoked misidentifying coiled rope as snake on the road. In such situation, visual information that is not yet reached to neocortex (where visual object recognition takes place) via the optic nerves is transmitted to amygdala to generate fear [5]. Thus, fear may already be experienced before our realization that we are not in danger.

Though there are so many other cognitive and no-cognitive causes of emotions, in the present article authors' primary focus is to discuss about cognitive theories of emotion which are very relevant to design practice.

4 Cognitive Theories of Emotion and Their Applications in Product Design

4.1 Kansei Engineering

The concept of Kansei Engineering (KE) was established by Nagamachi [6]. 'Kansei' is the Japanese word that covers several English meanings such as sensitivity, sense, sensibility, feeling, aesthetics, emotion, affection, and intuition. All these words are associated with mental responses to external stimuli, often summarily referred to as feelings. Idea and goal of KE is similar to pleasurable product design. According to Jordan [7], KE is roughly translated as 'Pleasure Engineering' . KE helps the investigator to understand the relationship between formal and experiential properties of a product. KE is also helpful for gaining insights about expectations of users and consequent product properties through these intended user benefits. According to Nagamachi [6, 8], KE follows four basic steps. First step starts with collecting appropriate Kansei words or adjectives related to product of interest from targeted user population. Second step involves establishment of correlation between product features/attributes and Kansei words. In third step, a data bank of these correlations is searched for Kansei words. These words in turn are represented with semantic differential scales, and, analyzed, typically using factor analysis to reduce often large numbers to a manageable set of words. Approaches taken in this step is similar to the semantic product design. Fourth and last step follows evaluation of new design with potential users in terms of Kansei words to establish how close the tested product is to the ideal product. Nagamachi [9] described two different directional approaches/flows of KE . These are 'from design to diagnoses' and 'from context to design'. The first approach involves manipulation of individual attribute of a product in order to test users' overall responses towards the product. In the second approach, qualitative data about products are gathered via field observations and then establishment of relationships between formal properties of design and the user benefits associated with the products. With the help of KE, products could be engineered by the designers based on semantic meanings of the products to improve sales, usability, and user's satisfaction. However, KE cannot predict feelings of consumers directly and unable to explain underlying phenomena in the brain due to product emotion but it is still helpful to explain socially constructed phenomena (for instance, various social factors related to product features and product emotion).

4.2 Theories of Product Personality

Product personality assignment is an approach which was developed by 'Philip Design' [10]. The concept of product personality comes from semantic meaning(s) of the product and it is closely associated with product semantics (the study of relationships between signs and the objects/products they refer to or signify). These symbolic meanings may refer to physical product itself and the product can be described by human personality characteristics which are termed as 'product personality' [7, 10]. Consumers not only consider product in terms of their functionality but also often thinks about product's personality like human beings [11, 16]. Product personality has prominent influence on product choice while functionality and usability across products are same. For example, a person can be happy, cute, and honest; similarly a product may appears like happy, cute, and honest. Although, personality of a product is reflected mainly through visual attributes of the product, it may also be perceived through physical interaction [12]. There are some proposed models of product personality which conclude the role of product personality on product choice [11, 13–15]. Personality congruence model well explains why people would like to select a product based on product personality [15]. This model states that product personality influences product choice due to personality congruence effect of consumers but not due to user congruence effect. According to this model, product personality depends on product attributes and if perceived personality of a product matches with personality of the consumer then consumer would like to choose the product. It is necessary to state that product personality assignment is one of the important techniques for pleasurable product design [7]. In a recent study, it is reported that different usability dimensions are associated with different product personality characteristics [16] and users selects a product that represents comparatively more positive personality characters such as lively, interesting, childish etc. Designers may manipulate characteristics of a product using personality assignment technique to make it pleasurable one to achieve more market acceptance than its competitive products with similar usability.

4.3 Pleasure Model

This model was proposed by Jordan [7]. This alternative affective model focuses more on pleasurable aspects of our interactions with products. This model considers all the potential benefits that a product can deliver. Pleasure model of Jordan, proposes four conceptually distinct types of pleasure [7]. These include (a) Physio-pleasure; (b) Socio-pleasure; (c) Psycho-pleasure and (d) Ideo-pleasure. Physio-pleasure is related with bodily aspects which are connected to sensory experiences about product, e.g. tactual pleasure of holding a mobile phone. Socio-pleasure refers to enjoyments derived from relationships with others. Products can facilitate social interactions in different ways. For example, a coffee maker provides a service which may act as a central focus of attraction of small social gathering 'Coffee evening'. Psycho-pleasure is associated with cognitive and emotional reactions. A product may need a particular level of cognitive ability to use it and the product attributes/interaction style may elicit emotional response based on the experience about the product. Ideo-pleasure refers to people's understanding of their own values. For example, a product made up of biodegradable materials is associated with value of environmental responsibilities of user.

Similar to 'Maslow's hierarchy of needs', Jordan [7] highlighted about a hierarchy of consumer needs. In this hierarchy Jordan placed pleasure in the third level after usability and functionality as per user's priority. According to Jordan, while consumers/users take a purchase decision, they are firstly concerned about functionality, then usability and lastly pleasure. While functionality and usability of many products are similar, pleasure plays important role in product selection. Though pleasure model does not able to explain how pleasures happen in biological and behavioral level, but it helps designers to think about ways of expression of different kind of pleasure through product attributes/interaction styles. Another drawback of this model is that it does not prescribe how to incorporate all types of pleasures into a single product. This model talks much about product acceptance on the basis of pleasure; but role of other emotion (e.g. fear, anger, surprise etc.) in product choice is still less clear.

4.4 Basic Model of Product Appraisal

A model of product appraisal process was developed by Desmet and coworkers [17, 18]. This model is similar to Frijda's 'action readiness account' [19] and chiefly based on appraisal theory of Ortony et al. [20]. According to this model, product appraisal depends on product attributes and the judgment of product design against a concern; and, the appraisal leads to emotion (Fig. 1). For example, if attitude (concern) towards a product (stimulus) is positive, appeal of that product (appraisal) may lead to emotion (joy/pleasure/happiness).

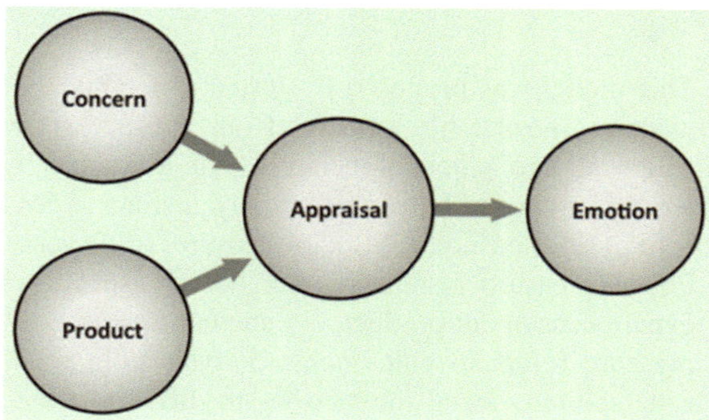

Fig. 1 Model of product appraisals. *Source* This figure is adapted from [17]

In this model a holistic approach has been taken for explanation towards generation of emotions due to product through appraisal process. Therefore, this model will be beneficial for designers to explain how a product may elicit emotion among users.

4.5 Emotional Design Model

This model was established by Norman and his colleagues, Andrew Ortony and William Revelle [21]. It explains how emotion and behavior are determined by different levels in the brain. According to this model, there are three levels of brain organizations which are related to emotion regulation (Fig. 2). First level which is known as visceral level, is associated with immediate communication in response to change in surrounding environment. Second level -behavioral level involved in regulation/control of our everyday behavior. At the highest level, brain further processes and anticipates the signals of behavioral level. This third level is called reflective level.

According to Norman's model [21, 22], visceral level responds rapidly, making judgments about what is good or bad, safe or dangerous, pleasurable or disgusting etc. It also triggers different emotional responses against stimuli, such as fear, joy, anger and sadness which are combination of physiological and behavioral responses. For example if someone finds a large hairy spider in the bathroom, he/she experiences fear and screams. These screaming or running away are due to the response from the behavioral level wherefrom most human activities are controlled. In reflective level people decides how to control emotion and how to respond further towards the stimulus.

Several explanations have been made on the basis of this model e.g. how we respond to stressful or pleasurable situations. The central theme of this model is that our affective state be it positive or negative, depends on how we think about the situation. For example, when we are tensed, the emotional response is to focus on the problem at hand and try to resolve it. In such situation, our bodies respond

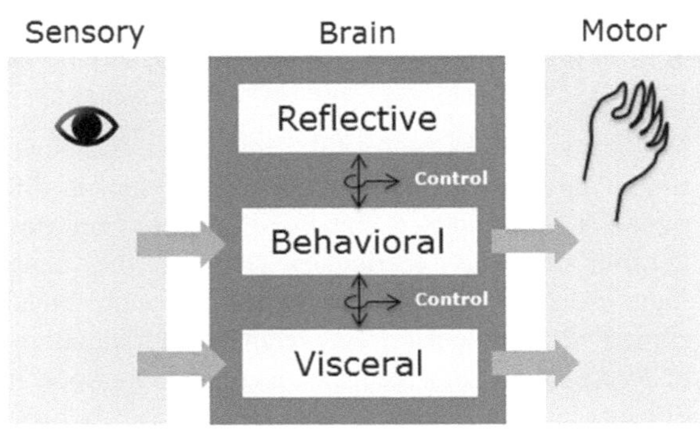

Fig. 2 Model of emotional design. *Source* This figure is adapted from [21]

Table 1 Examples of affects in relation to different sensory modalities

Affect	Positive	Negative
Visual	Smiling faces	Frowning disgusted faces
Touch	Rounded objects	Sharp objects
Hearing	Soothing/harmonious sounds	Harsh/discordant sound
Smell	Sweet smells	Smell of spoiled egg
Taste	Sweet taste	Bitter taste

Source Information presented in this table was adapted from [21]

by enhancing our muscles tone and sweating. On contrary, when people are happy due to wining of their favorite team in a cricket match, emotional responses are to lough, cheer, and jump about. Human body relaxes in such moments.

Following above discussions, few questions arise in our mind regarding application of this theory in product design. (A) Should designers create products according to varying emotional states of the users/consumers? (B) Is it realistic to conceptualize this type of product?

Creating product according to varying emotional states or conceptualization of such product is less practical but Norman's model is applicable in context specific design practice. Norman mentioned that a product which is intended to be used during leisure time or moment of fun, designers need not to worry about amount and type of information coming from product interface but they should concentrate on how to make the product more enjoyable. On contrary, he argued that for designing interface for serious task e.g. car driving, designers need to pay attention to all information required to perform that task and the interface should be clearly visible and unambiguous in feedback. Most acceptable application of his model is how to design products in relation to the three levels of information processing. Visceral design is widely applied for providing product's look, feel and sound good. Designers can use aesthetic techniques to make product emotional e.g. designers may use clean lines, balance, colour, shapes and texture for this purpose. According to this model, some affects related to our senses and their valances have been presented in Table 1 which may be useful for product/interface design.

4.6 Technology as Experience Framework

McCarthy and Wright [23] proposed a framework regarding user experience to explain how user feels about the product in different phases of product experience. They proposed that there are four core threads which are responsible for the user's holistic experiences. These are the 'sensual thread', 'emotional thread', 'compositional thread', and 'spatio-temporal thread'. The 'sensual thread' is concerned with our sensory engagement with situation and is similar to visceral level of Norman's model [21]. For examples, sensory engagement during interactivity of users with cell phones, computer games, chat-room etc. The 'emotional threads' involve common human emotions, such as sorrow, anger, happiness etc. which are influenced during product use. For example, a person may feel anger while facing problems in typing due to faulty key board. The 'compositional thread' refers to narrative part of product experience; as it unfolds the way a person make sense of them. For example, if consumers experience frustration during product purchase process due to e-retailer's website design related problems, they may not purchase products from the same website though they like the products. The 'spatio-temporal thread' is concerned with space and time in which our experiences take place and effect of various factors which influence space and time. For example, people avoid crowd and prefer quiet place/ambience for important discussion. Space designers may think about such situation for their design implementation.

5 Theoretical Framework for Designing Emotional Product

An integrated framework has been developed after identification of advantages and disadvantages about theories discussed in the above sections for designing of emotionally appealing product. Benefits and drawbacks of above described six theories presented in Table 2. In proposed framework different potential and useful components

Table 2 Examples of affects in relation to different sensory modalities

Theories	Advantages	Disadvantages
Kansei engineering	Relates product appearance with different emotional and semantic meanings of product	Not able to explain how emotion is generated by product
Product personality theory	Relates product appearance/ usability with different emotional and human personality related semantic meanings	Though it is able to explain consumer congruence effect but still unable to explain how emotion is generated
Pleasure model	Explains different sources of product pleasure e.g. physiological, psychological, ideological and sociological	Not able to explain cognition and emotion related behavioral information processing
Product appraisal model	Can explain how context and emotional product attribute generate emotion through appraisal	Less empirical evidences are found on directly related to product design and validating this model
Emotional design model	Can explain psycho-physiological basis of emotional information processing and product choice	Less empirical evidences are found on directly related to product design and validating this model
Technology as experience framework	Highlights on way to overcome different threats of user experiences in real time scenarios	Not able to explain cognition and emotion related behavioral information processing

of emotional product design have been included from various cognitive theories of emotion. Designers may take following considerations while designing emotional products (see Fig. 3). This framework highlights in details about both pre-design and methodological considerations to make a product more emotionally attached.

Various pre-design concerns of the proposed framework would help designers to consider about different aspects for designing emotional product. Prior to emotional product design, designers need to consider context of design, feasibility of value addition through emotional attributes of the product, different sources of pleasant user experiences, suitable way of emotion expression through product characteristics involving one or more sensory modalities, and socio-economic feasibility for unambiguous emotional product purchase decision. When, designers are able to presume and conceptualize product/service design features based on all these pre-design consideration, they may proceed to subsequent development of prototypes.

After developing probable prototypes, user evaluation can be performed using methodological concerns. Methods and techniques mentioned in Fig. 3 can be followed for prototype evaluation. In addition, evaluation of emotional product design may further be extended with some advanced neurophysiological techniques (such as eye-tracking, facial EMG, GSR, and MRI [24–27]). In this context it is worthy to mention that 'Somatic Marker Hypothesis' [28] is also very relevant for identification of neurophysiological basis of emotional product choice decision. After whole evaluation of prototypes, designers can take decision for redesigning of product or they can proceed for manufacturing. Thus, the proposed framework would be beneficial for designer to get overall idea about emotional product design and steps to be followed for the same.

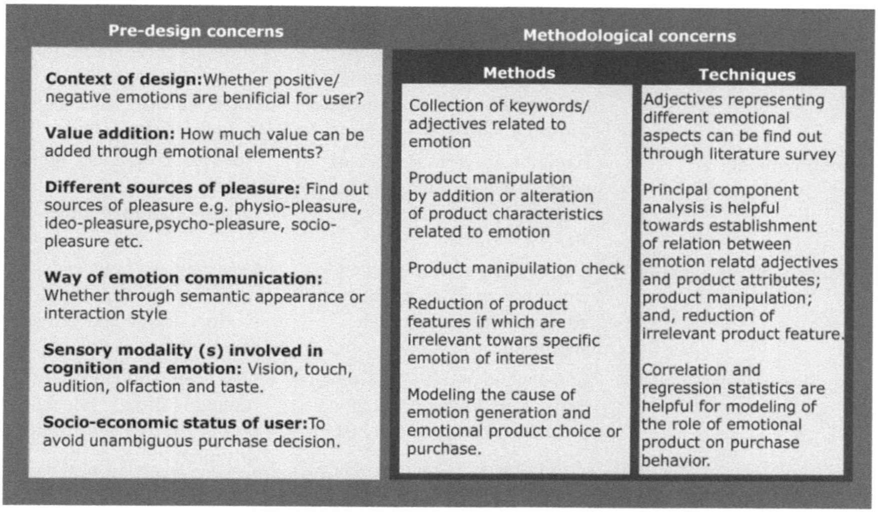

Fig. 3 Framework to design emotional product

6 Summary and Conclusion

Present paper covers six well established models for emotional product design. Each model and theory has its unique potential for emotional product design. First three models (viz. Kansei Engineering [6, 9], Product Personality Assignment [7, 10] and Pleasure Model [7]) described in this paper are mainly related to semantic product design but these models are not able to explain emotional signal processing and other bodily changes. On contrary, 'Basic Model of Product Appraisal' [17] and 'Emotional Product Design Model' [21] are more effective to explain the underlying cognitive and emotion processing in the brain. Technology as experience framework highlights some important aspect of product design such as trendiness and some other socio-economic aspects that users perceive in the market while purchasing a product. Based on advantages and disadvantages of aforesaid six models an integrated framework has been developed for product design with improved emotional appeal. Therefore, it can be concluded that theoretical framework for emotional product design which has been come out from present paper by incorporation of valuable inputs from important cognitive theories of emotional product design, will be helpful for visual communication designers, interaction designers, ergonomists, product innovation manager and cognitive scientists for designing and developing emotional products.

References

1. Andrade, J., May, J.: Cognitive Psychology. BIOS Scientific Publishers, Oxford (2007)
2. Prinz, J.: Emotion. In: Frankish, K., Ramsey, W.M. (eds.) The Cambridge Hand Book of Cognitive Science, pp. 193–211. Cambridge University Press, Cambridge (2012)
3. Darwin, C.: The Expression of Emotions in Man and Animals. John Murray, London (1872)
4. Ekman, P., Friesen, W.: Constant across cultures in the face and emotion. J. Pers. Soc. Psychol. **17**, 124–129 (1971)
5. LeDoux, G.E.: The Emotional Brain. Simon & Schuster, New York (1996)
6. Nagamachi, M.: Kansei engineering: a new ergonomic consumer oriented technology for product development. Int. J. Ind. Ergon. **15**, 3–11 (1995)
7. Jordan, P.W.: Designing Pleasurable Products. CRC Press, Boca Raton (2000)
8. Nagamachi, M.: Kansei Engineering. Kaibundo Publishing, Tokyo (1996)
9. Nagamachi, M.: The Story of Kansei Engineering. Kaibundo Publishing, Tokyo (1995)
10. Jordan, P.W.: Products as personalities. In Robertson, S.A. (ed.) Contemporary Ergonomics, pp. 73–78 (1997)
11. Govers, P., Hekkert, P., Schoormans, J.P.L.: Happy, cute and tough can designers create a product personality that consumers understand?. In: McDonagh, D., Hekkert, P., Erp, J.V., Gyi, D. (eds.) Design and Emotion, pp. 345–349 (2004)
12. Desmet, P.M.A., Nicolás, J.C.O., Schoormans, J.P.L.: Product personality in physical interaction. Des. Stud. **29**, 458–477 (2008)
13. Dumitrescu, A.: A Model of product personality. In: 4th European Computing Conference, World Scientific and Engineering Academy and Society (WSEAS), pp 88–93 (2010)
14. Govers, P.C.M., Schoormans, J.P.L.: Product personality and its influence on consumer preference. J. Consum. Mark. **22**, 189–197 (2005)
15. McDonagha, D., Bruseberg, A., Haslam, C.: Visual product evaluation: exploring users' emotional relationships with products. Appl. Ergon. **33**, 231–240 (2002)

16. Chowdhury, A., Reddy, S. M., Karmakar, S., Ghosh, S., Chakrabarti, D.: Is perception of product personality related with product usability? A cognitive ergonomics perspective. In: Parimalam, P., Premalatha, M.R., Banumathi, P. (eds.) Ergonomics for Enchanced Productivity, pp 177–182 (2013)
17. Desmet, P.M.A.: Designing emotions. PhD thesis, Department of Industrial Design Engineering, Delft University of Technology (2002)
18. Desmet, P.M.A., Hekkert, P.: The basis of product emotions. In: Green, W., Jordan, P. (eds.), Pleasure with Products Beyond Usability, pp 60–68 (2002)
19. Frijda, N.H.: The Emotions. Cambridge University Press, Cambridge (1986)
20. Ortony, A., Clore, G.L., Collins, A.: The Cognitive Structure of Emotions. Cambridge University Press, Cambridge (1988)
21. Norman, D.: Emotional Design: Why we Love (or Hate) Everyday Things. Basic BooK, London (2004)
22. Boehner, K., DePaula, R., Dourish, P., Sengers, P.: How emotion is made and measured. Int. J. Hum. Comput. Stud. **65**, 275–291 (2007)
23. McCarthy, J., Wright, P.: Technology as Experience. MIT Press, Cambridge (2004)
24. Mele, M.L., Federici, S.: Gaze and eye-tracking solutions for psychological research. Cogn. Process. **13**(1), S261–S265 (2012)
25. Miesler, L., Leder, H., Herrmann, A.: Isn't it cute: an evolutionary perspective of baby-schema effects in visual product designs. Int. J. Des. **5**(3), 17–30 (2011)
26. Laparra-Hernández, J., Belda-Lois, J.M., Medina, E., Campos, N., Poveda, R.: EMG and GSR signals for evaluating user's perception of different types of ceramic flooring. Int. J. Indus. Ergon. **39**, 326–332 (2009)
27. Balconi, M., Bortolotti, A., Gonzaga, L.: Emotional face recognition, EMG response, and medial prefrontal activity in empathic behaviour. Neurosci. Res. **71**, 251–259 (2011)
28. Damasio, A.R.: The somatic marker hypothesis and the possible functions of the prefrontal cortex. Philos. Trans. Royal Soc. Lond. B **351**, 1413–1420 (1996)

Designing Meaning to Change Consumer Behaviour: An Exploration

Gregor Waltersdorfer, Kilian Gericke and Luciënne Blessing

Abstract The focus of our research is to support designers in fostering a more sustainable behaviour of consumers by creating meaning in products and services. The paper describes the results of a literature study into the process of meaning making and the mechanisms through which meaning affects consumer behaviour. Meaning is defined as a mental representation of possible relationships. An initial model, the Meaning-Behaviour Model, is presented, integrating the mechanisms found in literature. Five possible interventions, derived from the model, show how designers can use meaning as a lever to foster enduring behavioural change. The paper contributes to the discussion of introducing meaning through design by exploring the link between meaning and behaviour.

Keywords Meaning · Behaviour · Experience · Sustainability

1 Introduction

40 years after the first report of the Club of Rome on the "Limits to Growth" humanity has to cope more than ever with numerous challenges, such as environmental degradation, resource overconsumption, climate change, and dependency on fossil fuels [1]. These challenges are caused by lifestyle and consumer behaviour, mainly in developed countries [2], but affecting the whole world. Companies need to actively address consumer behaviour during the design of products and services in order to face these challenges. The emphasis of our research is on changing the behaviour of the consumer, as buyer and as user, to decrease the

G. Waltersdorfer (✉) · K. Gericke · L. Blessing
Research Unit in Engineering Science, University of Luxembourg, Luxembourg, Luxembourg
e-mail: gregor.waltersdorfer@uni.lu

A. Chakrabarti (ed.), *ICoRD'15 – Research into Design Across Boundaries Volume 1*, Smart Innovation, Systems and Technologies 34, DOI 10.1007/978-81-322-2232-3_30

environmental impact of their products and services throughout their entire life-cycles. First attempts have already been made by focusing on technology, behavioural economics and psychology [3, 4].

Our research follows the proposal of Oehlberg et al. [5] to foster sustainable behaviour by creating *meaning* in products and services. Motives for consumption stem "in large part from the meaning of consumption objects and the value that meaning provides" [6]. Jackson [7] examines that consumption can be seen in part as a pursuit of meaning, which is constructed to "give both our personal lives and our society a sense of significance and purpose". But it is the consumerism, denoted as " 'excessive' individual consumption", which humans have to rethink in facing the global challenges [8].

Cultural psychology sees human as meaning makers, who have an actual need for meaning [9] and actively seek meaning instead of only constructing it [10]. Apart from some notable exceptions such as [11], the impact of meaning on behaviour is rarely addressed in design research.

After an introduction to the role of design in creating meaning (Sect. 2) this paper presents the results of a literature review into the process of meaning making (Sect. 3) and the mechanisms through which the meaning of products and services affect consumer behaviour (Sect. 4). The results are integrated in an initial Meaning-Behaviour Model (Sect. 5). The model allows the identification of possible interventions stimulating behavioural change by design, which are discussed in Sect. 6.

2 Meaning Creation by Design

According to Park [12], "meaning is a mental representation of *possible relationships*" connecting things, events, people, places and times and allows to get an understanding of all kind of experiences [13]. Crilly et al. [14] emphasise that people can assign different meanings to the same thing.

Describing the process of meaning creation by design we draw on the idea of design as communication [14]. The designers' intention is transferred through the product and interpreted by the consumer, who may give a behavioural response [15]. Kazmierczak [16] speaks of *intended, constructed, received and re-constructed meaning*. Krippendorff [17] labels the reconstructed meaning as *second-order understanding* that designers must anticipate. This process highlights the difficulty of creating meaning that is reconstructed as intended. For example, while designing a laptop, the designers intend to convey high quality (intended meaning) by choosing a unibody aluminium casing rather than plastic casing (constructed meaning). The consumer could perceive this solution (received meaning) indeed as high quality (correctly re-constructed meaning) or as less sustainable due to its energy-intense and wasteful production (incorrect re-construction, yet a possible meaning).

3 Meaning Making

In order to support designers in creating meaning, first an understanding is needed of how meaning is reconstructed by consumers. Meaning is made upon experiences through interacting with products or services. It is driven by the need of humans for meaning [9]. The processes of meaning making to cope with stressful events in life have been extensively discussed, e.g. by Park [12]. Following her broad definition of meaning (see Sect. 2), we assume that these processes can be transferred to other experiences, such as those with products and services. In this section we describe the process of meaning making, using Park as main source.

3.1 Definition of Meaning Making

Meaning making can be defined as the comparison of initially appraised and global meaning [12]. *Appraised meaning*, loosely based on cognitive appraisal, refers to situational intuitive and immediate evaluation of an experience in a particular environmental encounter. *Global meaning* "refers to individuals' general orienting systems, consisting of beliefs, goals, and subjective feelings" and self-view, and is more stable than appraised meaning. When meaning making involves reviewing the past it can be described as a learning process [18], which everyone develops in unique ways [19].

Meaning making aims at searching for *comprehensibility*, which is described as making an event "fit with a system of accepted rules or theories", and subsequently searching for *significance* by "determining the value or worth" of an event [12].

3.2 Process of Meaning Making

Making meaning of an event can be considered as a two-stage process: (1) appraising meaning, and (2) comparing appraised with global meaning [12]. In semiotics these two stages are called *denotation* and *connotation*, and in consumer research *identification* and *interpretation* [20].

Appraised meaning may be "instantaneously determined" by attributing why an event occurred and determining its implications through emotional and cognitive processing [12]. This stage is also called *sense-making*[1] and may be re-examined

[1] According to Krippendorff [17] "meanings are invoked by sense".

through continuous revision [20, 21]. The second stage involves the comparison of the appraised meaning with a person's global meaning to determine if and how the experience fits the personal orienting systems.

Ultimately, meaning connects the self, people, things, events, expectations, the past, the present [12], places [22] and, things beyond them all [23].

The comparison of appraised and global meaning may result in a perceived discrepancy—also called self-discrepancy [9], violation of meaning [13], or cognitive dissonance [7], e.g. due to conflicts between attitudes, beliefs, values or goals of the self or others [7]. The potentially resulting personal distress can drive more deliberate meaning making efforts [12], also called meaning maintenance [13], in order to solve these conflicts and to stay self-consistent [24].

Proulx and Inzlicht [13] describe five ways, Five A's, of "*meaning maintenance*": *assimilation, accommodation, affirmation, abstraction and assembly.* Assimilation involves the adaptation of the appraised meaning to the global meaning; accommodation involves the adaptation of the global to the appraised meaning. The other options are: to remain at the previous understanding and avoid the source of conflict (affirmation), to find something familiar to the discrepancy in the external environment in order to obtain understanding (abstraction), or to create a completely new understanding independent of the global meaning (assembly).

DeGrandpre sees meaning making as a dialectical process, in which meaning is refined through reviewing a "behavioural episode" [18], i.e. a closed loop process. Wright et al. [25] divide meaning making of past experiences into: *reflecting* (an inner recounting); *appropriating* (relating experiences to the self); and *recounting* (involving others in the review of past experiences). Reflecting and recounting can be seen as reviewing the appraised meaning, and accordingly appropriating as reviewing the global meaning of a person.

4 The Influence of Meaning on Human Behaviour

In this section, the mechanisms underlying the link between meaning and behaviour are analysed. Referring to Holzkamp, Brockmeier states that meaning only indicates "possibilities for action" and not as a determinant [26]: one can always abandon one's efforts of making meaning. Since meaning is not a direct determinant of behaviour, it is often left out of the discussion, as Darnton's extensive overview of behavioural models indicates [27]. Therefore, the question is how meaning can influence human behaviour and which schools of thought consider the effects of meaning on behaviour.

4.1 Meaning and Motivation

According to DeGrandpre [18], the *motivational qualities of meaning* can "guide individual actions". For Quigley and Tymon [28], who investigate career self-management, meaningfulness—next to choice, competence and progress—is one of four components of intrinsic motivation, and "the feeling of meaningfulness occurs when an individual is progressing on a path that they believe is worth their time and energy". In view of the self-determination theory of Ryan and Deci [29] intrinsic motivation: comes from unsatisfied needs for autonomy, competence and relatedness; directly leads to according behaviour; and provides higher levels of motivation, performance and well-being than extrinsic motivation. They emphasise that meaning facilitates the internalization of extrinsic motivation, when it comes with support for autonomy and relatedness. For example, the motivational quality of meaning can be observed with bicycle lovers, diving into every detail of their bike (belief of worth the time), repairing it on their own (need for competence) and taking exceedingly care of it (relatedness).

4.2 Meaning and Adaptive Attitude

Focusing on organizational change, van den Heuvel et al. [21] describe the role of meaning making by employees, as "the ability to integrate challenging or ambiguous situations into a framework of personal meaning using conscious, value-based reflection".

Meaning making can be considered as a personal resource, allowing employees "to remain resilient when confronted with organisational changes". In these circumstances, meaning making forms adaptive attitudes, "such as willingness, openness and readiness to change" [30]. This can lead to "a positive behavioural intention towards the implementation of modifications in an organisation's structure, work, or administrative processes (…)" [21]. Since the reflection upon challenging and ambiguous situations can also occur outside the work environment, we assume that these findings can be transferred to interactions with products and services. For example, consumers with affinity to new technologies are open to these technologies and adopt them quickly, as they are easily integrated into their frameworks of personal meaning.

4.3 Meaning and the Self-concept

Sirgy [24] defines the self-concept as the "totality of the individual's thoughts and feelings having reference to himself as an object", and distinguishes actual (present) and ideal (future) self. Drawing upon several scholars, Sirgy concludes that the self-concept directs behaviour "toward the protection and enhancement of self-concept", which can be observed in consumer behaviour based on the symbolic meaning of products and services.

Wright et al. [31] discuss product symbolism with the consumer self-concept as the "extended self". Consumers define themselves based on product meaning. According to Sirgy's self-congruity theory, recognising product symbolism is the ability to draw personal meaning out of a product [31]. For example a vegetarian, discovering a new tofu variety, interprets the product as a symbol fitting one's diet and therefore connects (assigns meaning) to it, which can lead to its purchase.

In case a discrepancy between appraised and global meaning, as described in Sect. 3.2, is only *anticipated* [32], one may act to prevent "an aversive sense of meaninglessness" [13]. For example, to avoid potential discrepancy between behaviour and the self [24] or the actual and ideal self [7], people buy products that provide an image, that is consistent with the self.

4.4 Meaning and the Social Identity

Jackson [7], drawing on symbolic interactionism [33], argues that the self-concept also has a social dimension, since the self is negotiated by social conversation. This results in shared values, attitudes and beliefs of groups. Consequently the self and the social identity cannot always be clearly separated.

Symbolic interactionism [33] also provides the ground for the discussion of a "socially oriented self" and its product relations. The social conversation is what connects people and objects, since objects can be part of the conversation [7]. In this way, objects become familiar and therefore meaningful. "People respond to material artefacts on the basis of the symbolic meanings that these artefacts carry" [7]: they become stimuli for action [33]. People not only set their actions through meaning making, they also review them in the lights of the objects, the people engaged in the social conversation, and the socially formed identity [33].

One can distinguish two social groups: the in-group, to which the person belongs or wishes to belong, and the out-group [7]. For example, driving a SUV may symbolise success for the in-group, but may be less accepted in the out-group, because of its resource intensity.

4.5 Meaning and Reviewing

The dialectical character of meaning making, as described in [18], is supported by Alea and Bluck [34], who discuss the search for meaning as reviewing one's past. They argue that making meaning of past experiences directs present and future behaviour, by using the "directive function of autobiographical memory". Kahneman and Riis [35] speak of two selves when it comes to memory: the experiencing and the remembering self. The first is built during immediate introspection, whereas the second involves retrospection and is dominant in reviewing past experiences. The difference between making meaning of recent and past experiences is that the latter may be biased due to the dominance of the remembering self [35].

DeGrandpre [18] argues that the dialectic of meaning making can act as a reinforcement of behaviour, increasing the probability that the behaviour is executed. For example, the shape of the doorknob may require the reflection on past experiences in order to understand whether to push or pull.

5 Meaning-Behaviour Model

In this section, the process of meaning making and the mechanisms behind the effects of meaning on behaviour, as described in Sects. 2–4, are summarised and integrated into one, initial model, the Meaning-Behaviour Model, shown in Fig. 1.

Consumers experience products and services through interaction (Sect. 2). Driven by the need for meaning, consumers go through the meaning making process (Sect. 3), in order to get an understanding of the experience. In a first step they attribute a reason to that experience and determine its implications, which results in an initially appraised meaning. In a second step this appraised meaning is compared to the global meaning in order to determine if and how the experience fits the global meaning, which represents individuals' general orienting systems and self-views. This comparison may lead to a perceived discrepancy between those two meanings. In case the discrepancy is not accepted, the meaning needs to be maintained by either adapting the appraised or the global meaning. The discrepancy can also be accepted, for example if someone is forced to act

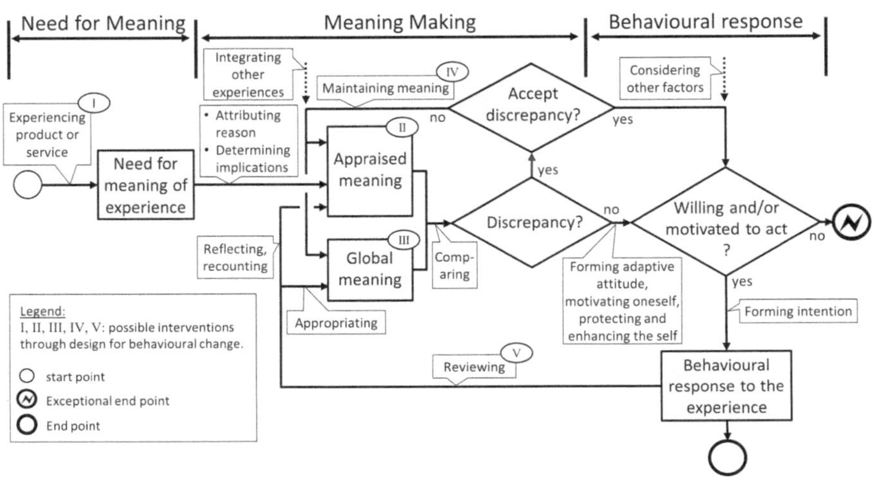

Fig. 1 Initial Meaning-Behaviour model

against his/her global meaning. Meaning making can also involve reviewing past experiences after a behavioural episode, by appropriating them to the global meaning or reflecting and recounting the appraised meaning.

As a result, the meaning, both the initially appraised and the meaning inferred from comparing appraised with global meaning, can be made of an experience or a product or service. The meaning is the built mental relationship to the self, other people, things, events, expectations, the past, the present, places, and things beyond them all.

There are several mechanisms how meaning can influence consumer behaviour, once there are no perceived discrepancies (Sect. 4):

- Regardless to what the relationship may be established, the meaning of an experience may influence behaviour:
  - As one of four components of intrinsic motivation coming from unsatisfied needs for autonomy, competence and relatedness, also facilitating the internalisation of extrinsic motivation.
  - Through the formation of an adaptive attitude towards a changing environment.
- If meaning relates an experience to the self (actual or ideal self), as part of the individuals' global meaning, consumers may want to protect and enhance their self-concepts at any time, also if a discrepancy of meanings is only anticipated, and behave accordingly through products and services as symbols.
- If the relationship is built to other people by negotiating the self through social conversation, consumers may respond to products, services or an experience based on the symbols they represent to other people.
- If meaning making establishes a relationship of past experiences, such as behavioural responses, through reviewing, it can guide present or future behaviour and even reinforce it.

Since meaning only indicates possibilities for action, the consumer may not be willing or motivated to act. Either the process ends here, or the consumer forms an intention to give a behavioural response to the experience. The process of meaning making may be repeated by reviewing.

6 Discussion

Our focus is on creating meaning for consumers in order to encourage enduring behavioural change. Following the initial Meaning-Behaviour Model, behavioural change can be the consequence of various interventions, as indicated in Fig. 1:

I. In case of a new experience, a new behavioural response may be evoked.
II. In case any of the relations resulting from appraising meaning is different from the usual relation.
III. In case different beliefs, goals, subjective feelings or self-views, as part of the individuals' global meaning, are addressed.

IV. In case of intentionally caused discrepancies between appraised and global meaning, conflicts need to be resolved, potentially resulting in changed meaning through maintenance, which in turn may affect behaviour.
V. In case of reviewing past experiences, future behaviour may be influenced.

DeGrandpre [18] argues that the dialectic of meaning making can also act as reinforcement of past behaviour, when it is affirmed by meaning making. This is in line with the findings of van den Heuvel et al. [30], indicating that meaning has the potential to make a behavioural change stick, if the change was made by the person itself as this results in more stable relations. This leads to the assumption that designers can foster enduring behavioural change by creating meaning.

7 Conclusion

Seeing design as communication and therefore contributing to the creation of meaning, this paper discussed the effect of meaning on behavioural change. We explored the process of meaning making, which occurs when consumer reconstruct the meaning intended by the designer. The mechanisms underlying the effect of meaning on behaviour were examined, resulting in an initial Meaning-Behaviour Model. Literature suggests that meaning can influence behaviour in several ways, which—in our opinion—can all be addressed by designers. Hence, meaning can be considered as a high-potential lever for designers to create products and services that can stimulate lasting behavioural change.

A first attempt has been made by Lai [36], who combined symbolic interactionism with interaction design, but much more research is needed. We will focus on verifying our model, on investigating the link between types of relation and types of behavioural response, and on developing guidelines that allow the design of products and services that are effective in changing behaviour and in maintaining this behaviour.

Acknowledgments The authors thank the Fonds Nationale de la Recherche Luxembourg for funding this research. Furthermore, the authors thank Renee Wever for the valuable feedback on the interventions and the reviewers for their useful comments.

References

1. Randers, J.: 2052: A Global Forecast for the Next Forty Years. Chelsea Green Publishing, Burlington (2012)
2. DESA: World economic and social survey 2013: sustainable development challenges, New York (2013)
3. Boks, C.: Design for sustainable behaviour research challenges. In: Design for Innovative Value Towards a Sustainable Society: Proceedings of EcoDesign 2011: 7th International Symposium on Environmentally Conscious Design and Inverse Manufacturing (2012)
4. Lockton, D.J.G.: Design with Intent: A Design Pattern Toolkit for Environmental and Social Behaviour Change. Brunel University, Uxbridge (2013)

5. Oehlberg, L.A., Aipperspach, R. Jeffery, S.R.: Sustainability through meaning. In: Proceedings of the Workshop on Ubiquitous Sustainability at Ubicomp, July (2007)
6. Richins, M.L.: Valuing things: the public and private meanings of possessions. J. Consum. Res. **21**(3), 504–521 (1994)
7. Jackson, T.: Motivating sustainable consumption. A review of evidence on consumer behaviour and behavioural change. A report to the sustainable development research network (2005)
8. Crocker, R.: From access to excess: consumerism, 'compulsory' consumption and behaviour change. In: Crocker, R., Lehmann, S. (eds.) Motivating Change: Sustainable Design and Behaviour in the Built Environment. Routledge, London (2013)
9. Heine, S.J., Proulx, T., Vohs, K.D.: The meaning maintenance model: on the coherence of social motivations. Pers. Soc. Psychol. Rev. **10**(2), 88–110 (2006)
10. Wong, P.T.P.: Viktor Frankl's meaning-seeking model and positive psychology. In: Batthyany, A., Russo-Netzer, P. (eds.) Meaning in Existential and Positive Psychology. Springer, New York (2014)
11. Chapman, J.: Design for (Emotional) durability. Des. Issues. **25**(4) (2009)
12. Park, C.L.: Making sense of the meaning literature: an integrative review of meaning making and its effects on adjustment to stressful life events. Psychol. Bull. **136**(2), 257–301 (2010)
13. Proulx, T., Inzlicht, M.: The five 'A's of meaning maintenance: finding meaning in the theories of sense-making. Psychol. Inq. **23**(4), 317–335 (2012)
14. Crilly, N., Good, D., Matravers, D., Clarkson, P.J.: Design as communication: exploring the validity and utility of relating intention to interpretation. Des. Stud. **29**(5), 425–457 (2008)
15. Crilly, N., Moultrie, J., Clarkson, P.J.: Seeing things: consumer response to the visual domain in product design. Des. Stud. **25**(6), 547–577 (2004)
16. Kazmierczak, E.T.: Design as meaning making: from making things to the design of thinking. Des. Issues **19**(2), 45–59 (2003)
17. Krippendorff, K.: The Semantic Turn: A New Foundation for Design. CRC Press, Taylor & Francis Group, Boca Raton, FL (2006)
18. DeGrandpre, R.J.: A science of meaning: can behaviorism bring meaning to psychological science? Am. Psychol. **55**(7), 721–739 (2000)
19. Kegan, R.: Making meaning: the constructive-development approach to persons and practice. Pers. Guid. J. **58**(5), 373–380 (1980)
20. Puntoni, S., Schroeder, J.E., Ritson, M.: Meaning matters. J. Advert. **39**(2), 51–64 (2010)
21. van den Heuvel, M., Demerouti, E., Schreurs, B.H.J., Bakker, A.B., Schaufeli, W.B.: Does meaning-making help during organizational change?: development and validation of a new scale. Career Dev. Int. **14**(6), 508–533 (2009)
22. Manzo, L.C.: Beyond house and haven: toward a revisioning of emotional relationships with places. J. Environ. Psychol. **23**(1), 47–61 (2003)
23. Vis, J.-A., Boynton, H.M.: Spirituality and transcendent meaning making: possibilities for enhancing posttraumatic growth. J. Relig. Spiritual. Soc. Work Soc. Thought **27**(1–2), 69–86 (2008)
24. Sirgy, M.J.: Self-concept in critical review consumer behavior: a critical review. J. Consum. Res. **9**(3), 287–300 (1982)
25. Wright, P., McCarthy, J., Meekison, L.: Making sense of experience. In: Blythe, M.A., Overbeeke, K., Monk, A.F., Wright, P.C. (eds.) Funology: From Usability to Enjoyment, vol. 3, pp. 43–53. Kluwer Academic Publishers, Dordrecht (2003)
26. Brockmeier, J.: Reaching for meaning: human agency and the narrative imagination. Theory Psychol. **19**(2), 213–233 (2009)
27. Darnton, A.: An Overview of Behaviour Change Models and their Uses (2008)
28. Quigley, N.R., Tymon Jr, W.G.: Toward an integrated model of intrinsic motivation and career self-management. Career Dev. Int. **11**(6), 522–543 (2006)
29. Ryan, R.M., Deci, E.L.: Intrinsic and extrinsic motivations: classic definitions and new directions. Contemp. Educ. Psychol. **25**(1), 54–67 (2000)

30. van den Heuvel, M., Demerouti, E., Bakker, A.B., Schaufeli, W.B.: Adapting to change: the value of change information and meaning-making. J. Vocat. Behav. **83**(1) (2013)
31. Wright, N.D., Claiborne, C.B., Sirgy, M.J.: The Effects of product symbolism on consumer self-concept. Adv. Consum. Res. **19**(1), 311–318 (1992)
32. Higgins, E.T.: Self-discrepancy: a theory relating self and affect. Psychol. Rev. **94**(3), 319–340 (1987)
33. Lee, D.H.: Symbolic interactionism: some implications for consumer self-concept and product symbolism research. Adv. Consum. Res. **17**(1), 386–393 (1990)
34. Alea, N., Bluck, S.: When does meaning making predict subjective well-being? Examining young and older adults in two cultures. Memory **21**(1) (2013)
35. Kahneman, D., Riis, J.: Living, and thinking about it: two perspectives on life. In: Huppert, F.A., Baylis, N., Keverne, B. (eds.) The science of well-being. Oxford University Press, Oxford (2005)
36. Lai, M.T.: Symbolic interactions as inspirations. In: Design 2014—Proceedings (2014)

Craft: A Narrative Barometer for Interior-Architecture Specific Focus on Stone Crafts in Religious Buildings of India

Smriti Saraswat and Gaurav Raheja

Abstract According to existing literature, the design culture is prominently dominated by two different approaches—'non-rhetoric design' and 'story-telling design'. This paper discusses the latter, with focus on craft and narrative (the analysis of narrative is an important branch of semiotics). Craft has the potential of becoming a strong narrative barometer for Design (here, Interior-Architecture). Narrative Barometer refers to a medium that has a potential to measure; create; carry forward and deliver narratives or Narrativity. Moreover, every craft has a narrative framework and a narrative structure (grammar) underlying it that contributes to design semantics and visual language in Interior-Architecture. It is simpler to study the narrative framework and structure in the literary narratives. But, in the non-literary narratives like paintings, relief works, sculptures, folk performances, crafts; it is difficult to do study the same. This research focuses on the non-literary narratives, specifically stone crafts in the religious buildings of India. Very few efforts have been made to study the narrative framework and narrative structure within crafts and design. This paper is an attempt to translate various models and codes deduced from different theoretical perspectives into Interior Architecture, with narratives being the centre and stone crafts being the medium. It aims to contribute to an inter-disciplinary research on Narratives, Narratology, Craft, Design, and Interior-Architecture.

Keywords Design · Interior-Architecture · Craft · Stone crafts · Narrative · Narrative barometer · Narrative structure · Religious buildings · India

---

S. Saraswat (✉) · G. Raheja
Department of Architecture and Planning, IIT Roorkee, Roorkee, India
e-mail: smritisaraswat@gmail.com

A. Chakrabarti (ed.), *ICoRD'15 – Research into Design Across Boundaries Volume 1*,
Smart Innovation, Systems and Technologies 34, DOI 10.1007/978-81-322-2232-3_31

1 Need and Importance of Such an Interdisciplinary Research

The motivation for such a research was primarily guided by authors' natural interest in craft (it refers to "making" of something and the "process" involved in it) and interior-architecture (a cohesive term encompassing the outer envelope as well as the internal environment of a building or a built-form). It was further strengthened by the observation that not much effort had been made in the past to discuss the Narrativity and narrative framework of these crafts, although remarkable work has been done on symbolism, semiotics and story-telling in interior-architecture. Equipped with a generic understanding of the narrative and craft based on the available definitions; theories; types such as textual, spatial, visual, oral; manifestations such as poetic, literary and non-literary; modes; and other aspects such as formal, structuralist, and post-structuralist; authors pursued their interest with an aim to bridge the possibly missing perspectives linking the narrativity of craft to design semantics, and visual language as applied in interior-architecture (but derived from craft).

The significance of the narrative was established in early Indian theories on poetics when a formal grammar and structure of the narrative process was established leading to a very important breakthrough in the study of human discourses. A similar contribution, much later in the western poetics, led to the study of narratives as a formal discipline, with formal methods and deeper analytical constructs.

The reigning principle of these narratives is that of an intent, a belief, and a philosophy that is organized as either one or multiple stories, but what is fascinating, is how they all fall into a unified, single narration, in the form of a collective art, as in a painting, or any other. And it is this narration that has a voice, a tense, an order, and many narrators. According to Seema Khanwalkar (quoted in Saraswat, unpublished post-graduate thesis 2010 that is now accepted by Lap Lambert Academic Publishing to be published as a book), "Much of our built environment, architecture and cities are planned on the basis of intent, belief, philosophy and narrative. But, this intent or discourse is hardly acknowledged or researched due to disciplinary boundaries." This paper is an attempt to break these boundaries, and do some trans-disciplinary research focusing on craft, narratives and interior-architecture. It is an important contribution to the study of Narratives, Narratology, Craft, Interior-Architecture and Design. It establishes and highlights the need to conduct more such studies that can help to understand the relationship and bond between humans, and the environment they create in built forms.

From the perspective of Pedagogy as well, this research holds great importance. One of the objectives of the Design Manifesto 2014 (released by MHRD, India in consultation with varied reputed institutions of the country) focuses on creating inter-disciplinary platforms for collaborative approaches to think about real world problems. And, one of its goals emphasizes on imagining, designing and implementing systems that are responsive to context, including markets, governance, social and cultural processes and existing knowledge levels and bases. In light of the above, such an inter-disciplinary research promises to create a new paradigm in the field of design pedagogy and profession.

1.1 Methodology

From a methodological perspective, it is a complex task to bring together the narrative intent and the stone craft discourse, as well as provide linkages with the visual language as applied in the Interior-Architecture. Going back and forth between the data and the theories, it was realized that there is no one rigorous theoretical model, on which this research could be based. In fact, this research attempts to translate various models and codes deduced from different theoretical readings into Interior Architecture, narratives being the centre and stone crafts being the medium. Methodology for this research has been derived from an eclectic theoretical base. It is primarily deductive in nature, and involves the following steps:

a) Foremost, a historiographical account of the literature was prepared. Major theoretical readings (which may or may not have any direct or indirect influence, first level or higher level application) include:

    1. *Art History, Indian Art and Indian Architecture* (Kapila Vatsayayan; Ananda Coomaraswamy)
    2. *Crafts and Craftspersons* (Kamaladevi Chattopadhyay; Jaya Jaitly, Jyotindra Jain; Richard Sennett)
    3. *Narrative* (H. Porter Abott and Paul Cobley)
    4. *Structuralism and Post-structuralism* (Roland Barthes)
    5. *Indian Narratology* (K. Ayyappa Paniker)
    6. *Narratives in Buddhist Art in Indian context* (Vidya Dahejia)
    7. *Function and Sign: The Semiotics of Architecture* (Umberto Eco)
    8. *Architecture as a Narrative Medium* (Joseph Tabbi)
    9. *Semiotics for Beginners* (Daniel Chandler)
    10. *Semiotics* (Seema Khanwalkar)
    11. *Stone Crafts of India (Neelam Chibber)*
    12. *Narratives in Interior-Architecture: Stone Crafts, a Medium* (Smriti Saraswat and Seema Khanwalkar)
    13. *Space-Making Elements and Crafts* (Yatin Pandya, Jay Thakkar and research done at Design Innovation and Craft Resource Centre at CEPT University in Ahmedabad)
    14. *Narrative structure in Crafts* (research by Smriti Saraswat, done at Design Innovation and Craft Resource Centre, CEPT University, Ahmedabad)

(b) Few concepts that are common across these theoretical readings were identified:

    1. *Narrative*
    2. *Communication*

3. *Sense of space and time*
4. *Structure*
5. *Craft*
6. *Design*
7. *Interior-Architecture*

(c) Stone Crafts were specifically studied through historical examples and case-studies, especially the examples of built-forms that have exquisite stone carvings and inlay work (discussed in subsequent sections). All the examples chosen confirmed to the above identified concepts to varied degrees. These studies reflected that we never see the story directly, but always pick it up through the narrative discourse, the role of which is played by stone craft in the buildings. It led to the understanding how crafts become integral to the purpose of a story. Further, it was observed that a story was always 'constructed'. In varied examples of stone architecture, it was constructed through the medium of stone crafts. Therefore, it became evident that when there is a story, it becomes equally important that it is narrated. At this stage, authors could reach a hypothesis—stone craft becomes the narrative barometer for creating; seeing; telling; carrying forward and delivering the story of a building.

1.2 Craft, a Narrative Barometer for Interior—Architecture

It is established that narrative is a Universal phenomenon; and exists everywhere, even in Interior Architecture [1–3]. Architecture in the real world has a lot to offer than basic need of shelter—it talks about the time, the era it belongs to, it is indicative of its culture and its people—architecture has a story to tell. Looking back at History gives a fair understanding of how all Interior-Architecture is permeated with narratives. Further, the basic understanding of Indian Narratology reflects upon the rich story telling tradition of India, which does not restrict only to the literature, drama and poetry, but also extends to the Interior Architecture of India—relief sculptures, carvings, paintings (Figs. 1 and 2), murals and the elements of space-making [4]. But, the stories exist in their "telling". There are no stories if they are not told. "Life is jammed with events. But call them what you will—stories, latent stories, virtual stories, untold stories—what is interesting is what happens when they are told, or re-told or staged or filmed, or mediated in whatever way. In other words, we are interested in narrative, the first rule of which is that it leaves its mark on the stories it tells [1]." This is where the need of a medium arises. Stone craft has the potential of being a powerful medium to measure; create; carry forward and deliver narratives or Narrativity. Moreover, every craft has a narrative framework and a narrative structure (grammar) underlying it that contributes to design semantics and visual language in Interior-Architecture. "An initial definition of "craft" means objects made for use. "Craft" also refers

Fig. 1 Ritual designs in the caves of Lascaux in Southern France. *Source* http://en.wikipedia.org/wiki/Lascaux

Fig. 2 *Bhimbetka* rock shelters with narrative paintings. *Source* http://en.wikipedia.org/wiki/Bhimbetka_rock_shelters

to the manual dexterity and artistic skill required in working with materials and in creating objects that can be ornamental, functional, or both [Rosy Greenlees, Director, Crafts Council]." Every craft is a piece of communication and can be like a word adding to the overall syntax of the structure (in this case, interior-architecture). "Presence of craftsmanship is virtually inseparable from the design and realization of the material world [5]."

Every architecture has a story to tell, and every craft has an inherent possibility of being integral to the grammar of the structure, forming a discourse, and structuring or creating a narrative, just like human language [6–8]. "A phenomenological consideration of our relationship with architectural objects tells us that we commonly do experience architecture as communication, even while recognizing its functionality [9]." It is important to understand craft narratives in the context of the belief systems that form a backdrop and are the driving force for the people as an interesting interplay between craft and the cultural milieu. "Design is permeated with narratives, because it is constituted within a field of discourses: formal, psychological, ideological, and theoretical. Every design is defined by and gives definition to social narratives that influence the behaviour associated with design spaces or objects [10]." "As soon as there is a society, every usage is converted into a sign of itself [11]."

1.3 Stone Craft, a Narrative Barometer for Interior—Architecture

The story of stone is the story of civilization. Stone is one such material, which has witnessed all the eras and narrates a timeline. It was available in abundance, and its use could be found all across the places in various products of daily use, objects of arts and crafts, and architecture, all of which tell a story of their respective epoch. Stone has the potential of narrating such a vast history because it is long lasting and could easily stand the test of times. Earlier, stone craft and architecture were considered the two sides of the same coin, but with the passage of time, this relationship has changed. Stone found extensive use right from village constructions to urban dwellings, palaces to temples, traditional sculptures, folk products and masonry for urban areas.

Stone has been used significantly as a building material. In India, one specific reason for its acceptance other than the availability is the material property and quality of stone suitable for carving. Large number of built forms can be traced in the areas which had easy access to the raw material. In fact, it was the wisdom in the '*shilpi*' tradition [12] that allowed the craftsmen to understand the composition of stones, their material properties, have expertise in quarrying techniques and exploit the potential to the best. "Craftspeople contribute their knowledge of local materials, giving form to human experience. Their visual and tactile connections with the land and with cultural ideas help to define the sense of place. The role of craftspeople in reviving cultural integrity and stimulating development and independence among small groups and communities throughout the world is profound [13]." Moreover, this vast repertory of information and knowledge was not just by experiences passed on from generations, but a completely scientific approach. A lot of hands and brains are involved in bringing stone form the quarry to the finished form. It is a time consuming process involving quarrying, cutting of stones, carrying loads and dressing stones.

Stone craft is not that predominant as it used to be. The new materials in the market and shift in the trends, have led to its decline in present times.

> "The Indian subcontinent has stones of all varieties, shapes and sizes. From expensive marble to cheap granite, the sheer abundance of this raw material is a compelling reason

to revive its use. Unless we recognize and respect this, stone craft and skills will quietly pass into history [14]."

In various interior-architecture case studies and examples, one can find narratives of life, values, society, environment, culture, religion, space and many more aspects. These are expressed in tangible form through visual and textual media, in the form of reliefs, sculptures, inscriptions, carvings and inlays. Craft reflects storytelling power of a building, with narrative as its life sustaining force. Stone crafts like stone carving and stone inlay become a narrative medium that negotiate between what is seen and what is experienced. In Khajuraho, we find narratives of society and life style. *Rani-ki-Vav* emphasizes on duty, *shringar*, jewellery, costumes, hair-styles (there are more than thousand hair styles that can be seen on the band exactly within the eye level, as soon as one steps down) and postures. Madurai temple looks inside religion. It is a reflection of that place and culture, and a narrative of particular religion. "Narrative is a linguistic structure designed to facilitate cultural cross-overs [15]." Craft is a medium to manifest this. "The character (*bhava*) of the form created inspiration and the divine vision was born. The action for this was the science of carving (*bhedanavidya*). In order to manifest the character of the image, the sculptor was very careful in creating the forms, so some of the images were peaceful (saumya) and some fierce (ugra). The face of a Buddha image is always peaceful but the faces of *nrsimha* and *Durga* are always terrible [16]." In City Palace at Udaipur and Laxmi Vilas Palace at Mysore, one can observe open-ended narratives with multiple narrators. The eclectic crafts tell multiple stories, and communicate diverse aspects. The placement of a single story in a chain of stories is a very natural form of narrative art in India. Even today in postmodernist novels and short stories too, one finds the effort of the narrator to locate each story in the 'story of story-telling'. The approach of understanding the multiple narratives and their manifestations in crafts can be the key to realizing the potential and significance of craft traditions in historic and present context.

In Aristotelian terms, stories have a beginning, middle and an end. But, in a Post-Modern sense, Narratives like—modern sagas, myth cycles, comic strips and TV series seem to have no proper end. In fact, the ending is postponed. To draw analogy with this research, it is fascinating how the craftspersons (skilled and non-skilled) have mastered their crafts, and kept the stories in process, while constructing various built-forms that have taken many decades to finish. In many built-forms or buildings, the motifs being chosen are quite varied with lot of diverse influences. The craftspersons do not allow the process to be easier. Their imagination, memory, "the Aesthetic Principle of Strangeness", "The Aesthetics of Absorption", all contribute in delaying the process and making the architecture timeless. Stone crafts; with their imbibed principles of space-making, craftspersons' skills, intricate detailing done by hand and the tacit knowledge involved; have rendered timelessness to interior-architecture throughout history [17–20].

Religious Buildings in India build totally through narratives. There are no major working drawings; no proper system of organization; no rationale; they build up through stories told by craftpersons and crafts, and it is important to understand this process (example—*Radhasoami Samadh*, Soami Bagh, Agra; Meenakshi Temple; Konark Temple). Therefore, they have a potential of being few best examples for the study of narratives. The idea of worshipping a supreme creator, a guru or an idol, is universal, because we as human beings share this across the civilizations. The basic idea of worship, devotion and meditation is understood alike by all human beings around the world. Religious buildings in India have been always constructed in stone. They provide a platform for the convergence of people, common point of reference being the "narrative universe".

Stone Crafts have carried forward some of human civilization's best known narratives. Throughout the Indian history, there have been so many examples of temples, mosques, churches and other religious buildings in stone, therefore, people can relate to the idea of universality through stone crafts. The idea of construction using hands, and laboriously detailing the idols, motifs and the buildings, as an act of devotion, has been a corner stone of most of the religious buildings [21]. People all around the world tend to address spirituality in a better way through stone. *Sewa*, which means service, is integral to every religion. The idea of contributing to institutionalize a belief system is a part and parcel of all the religious practices. It is this idea of *Sewa*, that turns craftspersons into devotees and encourage them to get associated with the construction of the building as an act of devotion, irrespective of the fact whether they are skilled or unskilled, which religion or caste they come from, and at what stage they join the construction.

The selection of right material is crucial for Interior Architecture. Architects and craftspersons did reach to find out the properties and qualities in stone that perfected their skills. Stone is considered purer than any other material (especially for religious buildings); easy to achieve geometry; wide range of colors (material property—natural as well as taken); does not crack and scatter easily; and, wide range of carving and inlay is possible in stone

Vidya Dahejia's work on visual narratives (Fig. 3) of India, especially conducted in the Buddhist context is also significant for this research.

"From the start of the century before Christ, Indian monuments display a wide range of sculpted and painted narratives portraying the legends of Buddha [22, 23]." Her article addresses the technique of presentation of these stories and analyses the structure of visual narratives as opposed to their content. "Seven distinct modes of narration have been identified, and it is suggested that the artists might have considered them to be equally acceptable alternatives since they frequently occur together on the same monument [22, 23]." She has also discussed the need and intention for Buddhist Narratives in her study on India's visual narratives. There must have been various factors that governed the choice of narratives:

- To link the gifts with the donors
- To set a parallel between the story of donor's choice and the one already set up
- Certain narratives added to the value of the monument
- Carved or painted guardians were considered auspicious at the entrances.
- It was considered sacred to decorate these monuments with *Jataka* stories

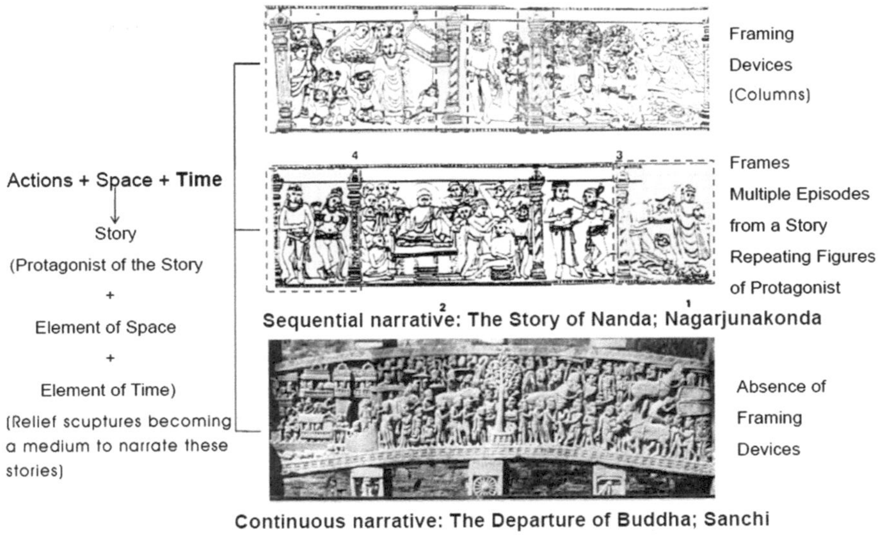

Fig. 3 *Jataka* stories and visual narratives. *Source* Images and research by Vidya Dahejia, representation by Smriti Saraswat

- Few narratives were placed beyond the reach of easy observations. This suggests that the motivation was not just the aspect of viewing, but also simply decorative.
- Narratives were also needed to add validity and significance to the ceremonies.

In light of the above discussions, it is clear that stone has been used throughout history in India as a medium of communication and a way of seeing. Especially, Indian temples in stone are the living examples of narratives. "In India, the history of making stone images for worship and decoration, begins with the Harappan Civilisation and is continuing till these days excepting a few centuries from 1500–600 B.C. From the Mauryan times onwards, we have a continuous history of making stone images mainly up to the 13th century A.D., with its various centres from Gandhar to Bengal and Kashmir to the far south of India [24]." Hence, it is proved that stone craft becomes a narrative barometer for creating; seeing; telling; carrying forward and delivering the story of a building.

References

1. Abott, H.P.: Cambridge Introduction to Narrative. University Press, Cambridge (2002)
2. Cobley, P.: Narrative. Routledge, London (2001)
3. Tabbi, J.: "Architecture as a Narrative Medium", A Review of: Beatriz Colomina, "Privacy and Publicity". The MIT Press, Cambridge and London (1996/1994)
4. Paniker, K.A.: Indian Narratology. Indira Gandhi National Centre for the Arts, New Delhi in association with Sterling Publishers Private Limited, New Delhi (2003)

5. Frampton, K., in the review of Adamson, G.: The Craft Reader. Berg Publishers, Oxford (2009)
6. Wright, B.: Working Theory of Communication. http://occasionallywright.typepad.com/occasionally_wright/2004/09/working_theory_.html. 15 June 14, 21:46, searched through Search Engine Google
7. Chandler, D.: Semiotics for the Beginners. http://www.aber.ac.uk/media/Documents/S4B/sem01.html. 5 May 14, 17:34, searched through Google
8. Mehrotra, L., Vajpayee, R. (eds.): Communication Through The Ages—An Indian Perspective. Aryan Books International, new Delhi in association with Media Centre for Research and Development, Gurgaon (2009)
9. Gottdiener, M. (ed.): Semiotics. Part VI—Semiotics of Non-Verbal and Complex Systems, vol. 3. Sage Publications, London, Thousand Oaks, New Delhi, p. 242 (2003)
10. Porter, L., Sotela, S.: Design by Narrative. http://www.ierg.net/confs/2004/Proceedings/Porter_Sotelo.pdf. 25 Sept 09, 19:24, searched through Google Search Engine
11. Barthes, R. quoted in Eco, U.: Function and sign: semiotics of architecture. In: Gottdiener, M. (ed.) and others Semiotics. Part VI—Semiotics of Non-Verbal and Complex Systems, vol. 3, p. 243. Sage Publications, London, Thousand Oaks, New Delhi (2003)
12. Jaitly, J.: The Craft Traditions of India. Lustre Press Pvt. Ltd, New Delhi (1990)
13. Maker and Meaning: Craft and society. In: Proceedings of the Seminar, Tamil Nadu, India (1999)
14. Chhiber, N.: Stone Craft of India, vol. 1 and vol. 2, Crafts Council of India, Chennai, with assistance from Department of Culture and Development Commissioner, Handicrafts, Govt. of India (2002)
15. Nair, R.B.: Narrative Gravity: Conversation, Cognition, Culture. Oxford University Press, New York (2002)
16. Boner, A., Sharma, S.R., Soumer, B. (ed. and Trans.): Vastusutra Upanisad, 3rd edn, pp. 59–66. Motilal Banarsidass, Delhi (1986). In: Roy, B.P. (ed.) Technique of Making Stone Images in India. In: Sinha, C.P. (ed.) Art, Archaeology and Cultural History of India (U.N. Roy Felicitation Volume), Part I, pp. 34–35. B.R. Publishing Corporation Ltd., Delhi (2006)
17. Pandya, Y.: Concepts of Space Making in Traditional Indian Architecture. Mapin Publishing Pvt Ltd., Ahmedabad (2005)
18. Pandya, Y.: Elements of Space Making. Mapin Publishing Pvt Ltd., Ahmedabad (2007)
19. Pandya, Y.: Art as a Spatial Narrative in Architecture. Architecture + Design (June, 2009)
20. Tillotson, G.H.R.: Paradigms of Indian Architecture: Space and Time in Representation and Design. Curzon press, Richmond (1998)
21. Saraswat, S.: Craft, a Narrative Medium: Enhancing Interior-Architecture, to be published as a book by Lap Lambert Academic Publishing (manuscript accepted)
22. Dahejia, V.: India's visual narratives: dominance of space over time. In: Tillotson, G.H.R. (ed.) Paradigms of Indian Architecture: Space and Time in Representation and Design. Curzon press, Richmond (1998)
23. Dahejia, V.: On Modes of Visual Narration in Early Buddhist Art, vol. 72, 3rd edn. The Art Bulletin, College Art Association (Sept, 1990)
24. Roy, B.P.: Technique of making stone images in India. in: Sinha, C.P. (ed.) Art, Archaeology and Cultural History of India (U.N. Roy Felicitation Volume), Part I, p. 33. B.R. Publishing Corporation Ltd., Delhi (2006)

Comparison of Indian and Central European Shape Contour Meaning Comprehension

Vanja Čok, Mihael Kline, Nikola Vukašinović and Jože Duhovnik

Abstract Since every culture has their own specifics, there is often the question, whether customers from diverse cultures perceive shape characteristics equally or differently? This study explore how two different cultural groups (Indian vs. Central European) of participants experience or response to predefined four types of product dimensions—aesthetic, functional, social and future. We assume that the Central-European and Indian students differ by their perception of different shapes. To gather the necessary statistical data, we prepared questionnaires wherein the samples were assessed using the semantic differential technique and a five-level Likert scale. The responses were then analyzed using Factor analysis. After conducting statistical analysis using Factor analysis, two main affective dimensions were gained, which show how differentiate Central-European and Indian participants. The methodology presented in this study can help designers to identify a pattern of consumer shape categorization.

Keywords Central European · South Asian (Indian) · Shape contour · Product characteristics

1 Introduction

Today products rarely break and mostly do what they promise. Therefore, it is not surprising that consumers increasingly make brand choices based on aesthetic value and distinctiveness of visual design [1–3]. Emotions elicited by products

V. Čok (✉) · N. Vukašinović · J. Duhovnik
LECAD Lab, Faculty of Mechanical Engineering,
University of Ljubljana, Ljubljana, Slovenia
e-mail: vanja.cok@lecad.fs.uni-lj.si

M. Kline
Chair of Marketing Communications and Public Relation,
Faculty of Social Sciences, University of Ljubljana, Ljubljana, Slovenia

A. Chakrabarti (ed.), *ICoRD'15 – Research into Design Across Boundaries Volume 1*,
Smart Innovation, Systems and Technologies 34, DOI 10.1007/978-81-322-2232-3_32

are changeably (i.e. diverse people can have different feelings towards the same product). However, universal patterns can be identified in the underlying process of how these emotions are elicited [4]. Customer's cultural background may be the reason for distinguished emotional value perceptions through the product [5]. Designers with a task to develop a new product for Asian or European market have to be aware of possible differences in customer perception, influenced by their culture, habits, religion etc. Costumer's background is affected by several factors and it is difficult to identify whether the difference is caused by cultural insight or not.

However, collaboration between Western and Eastern industrial partners is growing. Following the current business trends the need to understand the role and potential of differences in customers' product perception has appeared.

1.1 Cultural Aspect into Product Design

Several discoveries witness that there exist relative intensity differences within emotions among cultures [6, 7]. Thus, those differences are related to stable and meaningful dimensions of cultural variability [6]. Interesting study done by Russell [8] shows that the product proportions appreciated by US consumers were regarded as unattractive by Japanese consumers.

However, regarding Matsumoto [6] studies, Americans have rated expressions more intensely than Asians, particularly Japanese. Afterwards, this effect was limited only to ratings of external display, since Japanese gave higher ratings than Americans when rating presumed internal experience. There are other dimensions within which the consumer cultural influences come at forefront. Many factors and circumstances are affecting customers and their perception towards particular product form. Aside, environment may have ingrained cultural characteristics and values that influence the extent to which certain design characteristics are appreciated [9]. In his survey Hofstede included over fifty countries and created a set of dimensions by which national culture could be defined. According to Hofstede [10] five dimensions of national culture are: power-distance, individualism, toughness, uncertainty avoidance, and long-term orientation. Researchers rely on those dimensions when searching for cultural differences between different countries [11]. It was discovered that acts of confrontation, such as debate and contradiction, are not encouraged in East-Asian culture [12]. Western tradition was constructed differently, wherein individualism and debate are encouraged [13]. Purani and Shukla have made a comparison between India and Britain, which includes collectivist and individualist cultural contexts [14]. Their results suggest that British consumers' overall luxury value perceptions are significantly influenced by their utilitarian/functional value perceptions. They discovered that consumers in Western, individualistic markets increasingly look for luxury goods that are useful and practical. British consumers consider their luxury value perception more important than South Asian consumers. According to their opinion, South Asian consumers are modest, which could explain why hedonic values may not be significant [15]. Globalization also confronts companies to decide between 'global' or 'local' featured design of products [16]. However we can't skip the fact that every individual has his own sensibility in regards to product aesthetics.

Bloch et al. [14] believed that "Consumers with high visual aesthetics centrality are also likely to believe that fine design is valuable to society generally and to believe that the quality of life for everyone is affected by the quality of the designed environment". Berlyne [17] and Crilly et al. [18] confirmed that some people are endowed with more aesthetic taste than others. Responses to design aesthetics of objects have valence as well as intensity. Product appearance can elicit different emotional responses.

Certain forms or product designs will generate positive responses in a particular consumer while others evoke negative reactions [14]. There exist different models that explain role of emotions in product design [19]. Russell described emotions with Circumplex Model of Affect, Fig. 1. He developed a four-dimensional system of emotions in regards to arousal (*activation-deactivation*) and in regards to bipolarity of affect (*unpleasant-pleasant*) [20, 21]. The model proposed by Desmet uses appraisal theory as the basis for explanation how products elicit emotions through addressing someone's concerns [19]. Thus, negative affect causes avoidance while positive affect attracts people. According to these components we made a study using bipolar adjectives, which represent negative and positive affect, Fig. 1.

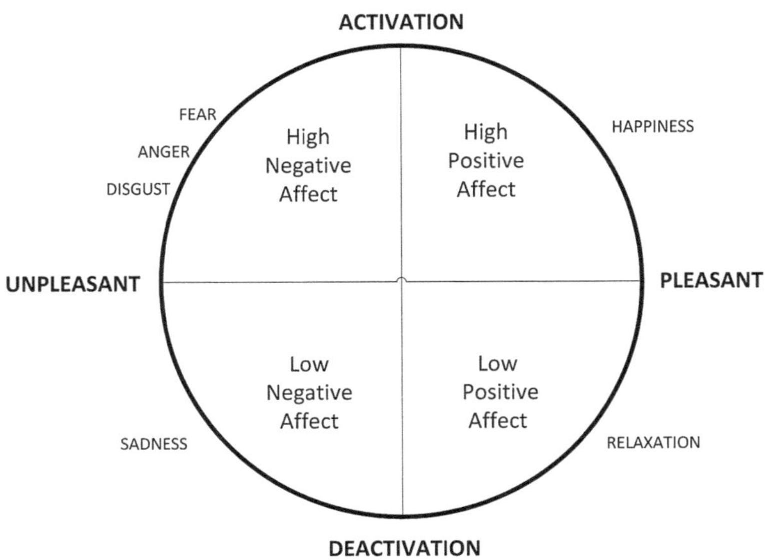

Fig. 1 Adapted Circumplex Model of affect (from [20])

2 Design of Experiment

2.1 Research Goal and Method

We were interested if South Asians (Indians) associate shape samples with the same meaning of bipolar adjectives as C. Europeans. The research procedure followed the next steps: (1) task and goal definition; (2) selection of population; (3) collection of representative bipolar adjectives; (4) collection and selection of product designs and concepts sketches (converted to abstraction), (5) categorization, (6) evaluation (questionnaires), (7) statistical analysis (Factor analysis and descriptive statistics), (8) synthesis.

2.2 Subjects and Samples

Subject of this research were 137 engineering students from India (Chennai n = 13, Bangalore n = 18, Mumbai n = 24) and Central Europe (Hungary n = 28, Croatia n = 28, Slovenia n = 28). They were undergraduate, master and PhD students. (33 females and 104 males) aged between 25 and 40. Their educational background was: mechanical engineering, industrial design engineering and product design. The research was performed using paper questionnaires so all students had the same evaluation forms.

The main idea of collecting samples procedure was to go from concrete product or sketch to abstraction. Since at early design phases we deal with market analysis and initial concepts (sketches) it is necessary to test potential consumer emotional responses towards existing and non-existing product samples in order to get new solutions. Therefore, we created 10 samples of contour shapes that were represented with black contour on white background in order to avoid colour influence, Fig. 2. In this experiment we intentionally did not associate the shape

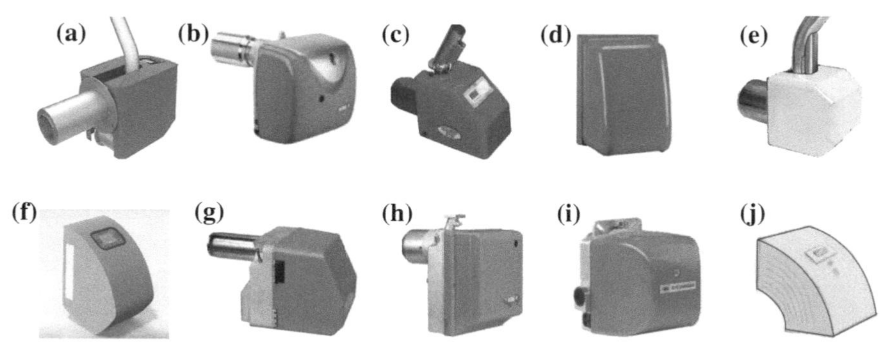

Fig. 2 Samples of pellet burners as a basis for form characteristic

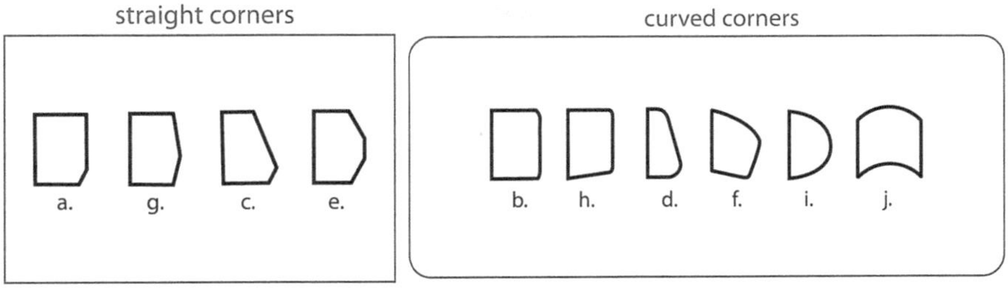

Fig. 3 Shape categorization based on *straight* or *curved* corners and edges

with any shape symbolism. For that reason non-consumer, industrial products were selected. Pellet burners for central heating systems were selected as a shape model, since most of the customers don't identify any pre-defined emotional value in them, while on the other hand they can be found in various shapes. Shape contour characteristics (2D) were therefore taken from pellet burners designs (3D) and 2 from designer's sketches. For clearer data interpretation we divided samples into two shape categories based on angularity of their shapes, Fig. 3. Criteria for separation were straight or curved edges and corners.

2.3 Observed Product Characteristics

We have defined 4 product characteristics that were observed during this experiment, Fig. 4. Adjectives were collected from different sources: brainstorming, previous studies and literature preview. They were arranged according to product dimension categories. It was important that all adjectives used in the questionnaire are common, simple and daily used in order to be understood in the same way by all participants. All students received the same identical questionnaire in English language.

In questionnaire 5-point scale was used to rate: shape aesthetic attribute (1 = very ugly, 5 = very beautiful); shape functional attribute (1 = very uncomfortable, 5 = very comfortable); shape social attribute (1 = very feminine, 5 = very masculine); shape functional attributes (1 = very unreliable, 5 = very reliable), (1 = very complex, 5 = very simple); and shape future attributes (very unfuturistic–very futuristic), Fig. 4.

3 Results and Data Interpretation

3.1 Factor Analysis (Survey on Affective Dimensions)

Factor analysis was used to recognise and describe Indian and C. European affective patterns. Its goal is to extract the important information from the data table and to express this information as a set of new orthogonal variables called

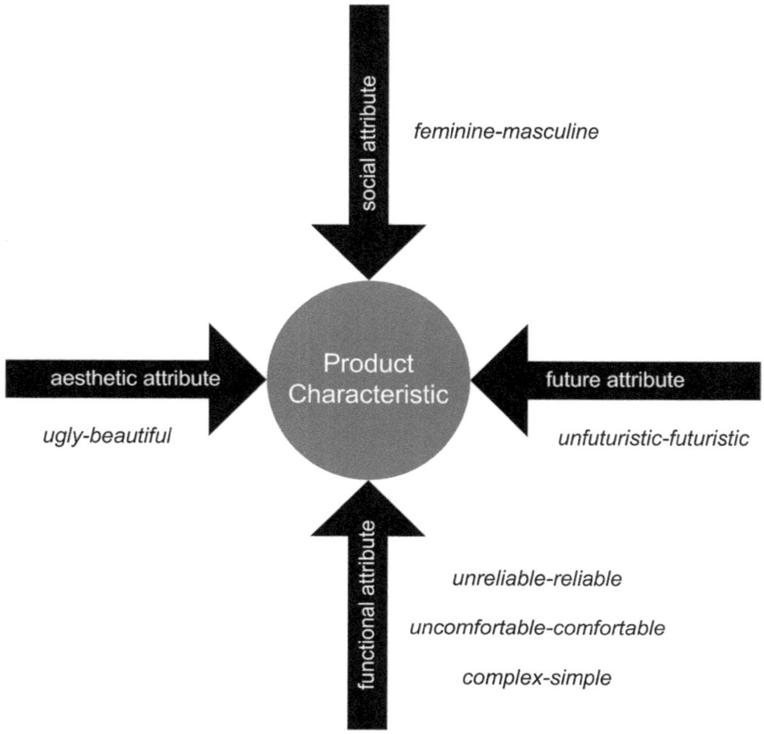

Fig. 4 Model of defined product characteristics

principal components. *By reducing a data set from a group of interrelated variables to a smaller set of factors, Factor analysis achieves parsimony by explaining the maximum amount of common variance in a correlation matrix using the smallest number of explanatory constructs* [22].

First, we were interested whether Indians associate shape samples with the same bipolar adjectives as C. Europeans. Performing Factor analysis we identified two main principal components that represent two emotional dimensions. We named these two components as *affective components* since they include bipolar adjectives that express aesthetic, functional, social and future product features. The values are given in Table 1 and are illustrated in Fig. 5. Bold values correspond to the variables (given as bipolar adjectives) which have significant impact on particular affective component. After the check of eigenvalues and factor contribution first two factors (AC1 + AC2) both for India and Central Europe had the highest contribution (India; 81.7 %) and (C. Europe; 87.9 %).

For **Central Europe** Kaiser-Meyer-Olkin Measure of Sampling Adequacy was 0.604 and Bartlett's Test of Sphericity was significant with 0.000. We used an extraction method Principal Component analysis where components were extracted. Component 1 had 52.2 % of variance while component 2 had 35.67 %

Table 1 Component matrix of AC for Central Europe and India

Bipolar adjectives	Central Europe		S. Asia (India)	
	AC1 (52, 19 %) *classical*	AC2 (35, 70 %) *attractiveness*	AC1 (54, 21 %) *attractiveness*	AC2 (27, 46 %) *classical*
Reliable–unreliable	**0.94**		**0.89**	0.38
Simple–complex	**0.90**		**0.80**	0.38
Futuristic–unfuturistic	**−0.89**	0.32	**0.57**	
Masculine–feminine	**0.80**			**0.93**
Comfortable–uncomfortable		**0.96**	**0.94**	−0.55
Beautiful–ugly		**0.95**	**0.77**	−0.42

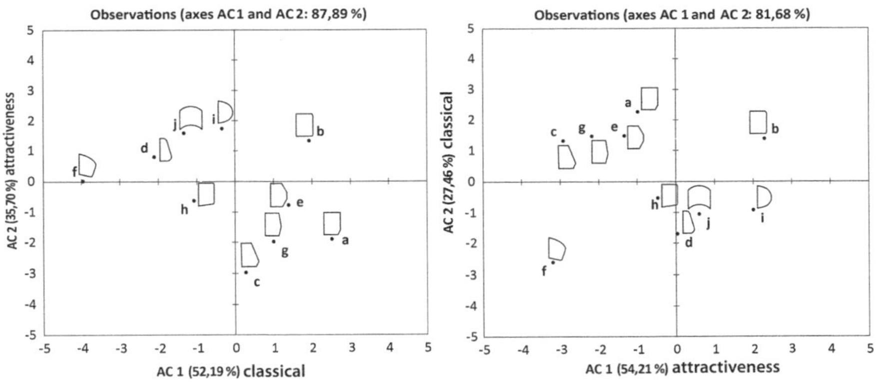

Fig. 5 Affective components (1, 2) and bipolar adjectives, Europe (*left*), India (*right*)

of variance. Together they had 87.9 % of cumulative contribution or total variance explained. Thus, affective component 1 with 52.19 % has the largest contribution. The content of AC1 has four most influential bipolar adjectives: *reliable*, *simple*, *masculine and unfuturistic*. Affective component 2 (AC2) with 35.70 % has three influential bipolar adjectives *beautiful* and *comfortable*, Fig. 5. Hence, we have two factor structures since some variables are present both, in AC1 and AC2.

For **India** Kaiser-Meyer-Olkin Measure of Sampling Adequacy was 0.620 and Bartlett's Test of Sphericity was significant with 0.000. We used an extraction method Principal Component analysis where components were extracted. Affective Component 1 (AC1) had 54, 2 % of variance while Affective Component 2 (AC2) had 27.5 % of variance. Together they had 81.7 % of cumulative contribution or total variance explained. The content of AC1 with 54.21 % has five most influential bipolar adjectives: *simple*, *reliable*, *futuristic*, *comfortable and beautiful*. AC2 with 27.46 % has most influential bipolar adjective *masculine*, Table 1.

Two affective components were named *attractiveness* and *classical*, based on the corresponding adjectives (variables). The meaning of affective dimensions between India and Central Europe was different: India with AC1 *attractiveness* and AC2 *classical* while Central European AC1 *classical* and AC2 *attractiveness*, Table 1. According to affective dimensions we selected the most extreme shape stimuli by factor scores. For clearer insight into results we create table of stimuli and affective components.

Relations between the most influential bipolar adjectives from Fig. 5 were represented graphically, to give the clearest overview. Figures 5 show the FA chart that represents a relationship between Factor 1 (AC1) and Factor 2 (AC2) which both have the largest contribution.

Generally, C. Europeans associate Functional (simple, reliable), Social (masculine) and Future (unfuturistic) dimensions with shapes having straight corners. Indians differ in a shape association. They associate Functional (simple, reliable, comfortable), Aesthetic (beautiful), Future (futuristic) with *shape B* which is assumed to have curved corners while *shape A* which is angular was associated with Social (masculine) dimension. According to Gordon [23] square oriented shapes give an impression of solidity and strength. Consequently, the straightness of its lines and the practical simplicity of its construction gives an impression of primitive and rugged thing. In contrast, curves are in general felt to be more beautiful than straight lines [23].

3.2 Straight Versus Curved Corners Shapes Contours (C. Europe Vs. S. Asia)

Surprisingly, the results show that Central Europeans typically provided more distinctive marks, while the Indians provided more neutral responses, Figs. 6 and 7.

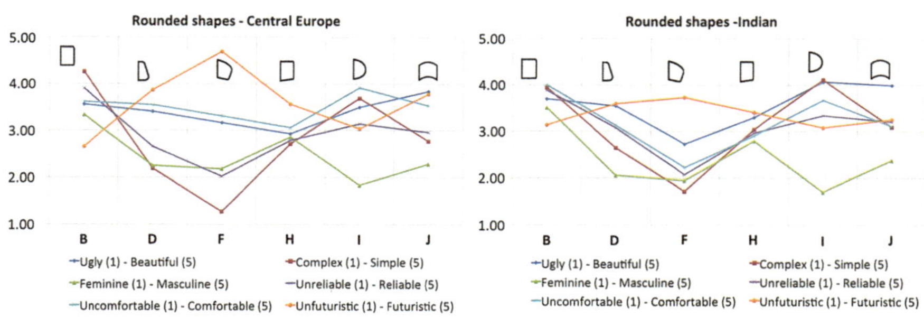

Fig. 6 *Curved* corners shapes samples means based on the bipolar adjectives

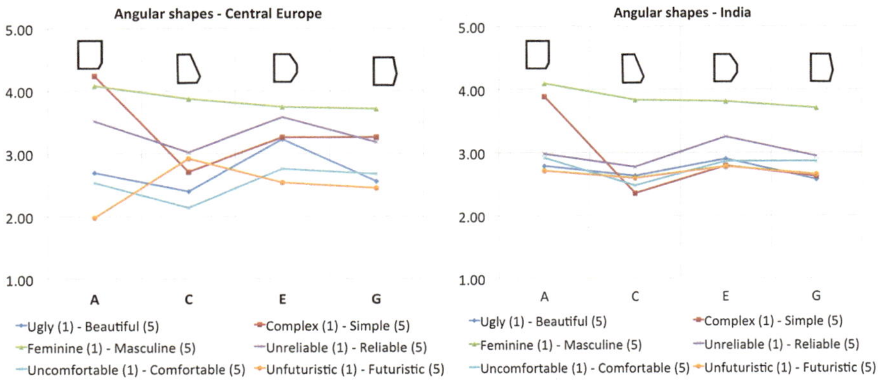

Fig. 7 *Straight* corners shapes samples means based on the bipolar adjectives

3.3 *Curved Corners Shapes Contours*

Shape B received the highest response values (Fig. 6). The participants differ at bipolar adjective *unfuturistic–futuristic.* C. Europeans find *shape B* to be unfuturistic (M = 2.65) while Indians were neutral. The lowest response values received *shape F.* Again, the groups of participants estimated it differently. C. Europeans find it *beautiful*, *complex*, *feminine*, *unreliable*, *comfortable* and *futuristic*. Indians find it *ugly*, *complex*, *feminine*, *unreliable*, *uncomfortable* and *futuristic* (Fig. 6).

3.4 *Straight Corners Shapes Contours*

Highest values from straight corners shape category received *shape E* (Fig. 7). C. Europeans perceive it as *beautiful*, *simple*, *masculine*, *reliable but uncomfortable* and *unfuturistic.* Whereas, Indians perceive it as *complex*, *masculine*, *reliable* and *unfuturistic*, but were neutral at adjectives *ugly–beautiful* and *uncomfortable–comfortable*. *Shape C* get the lowest values. Thus, Indians perceive it as *ugly*, *complex*, *unreliable*, *uncomfortable*, *unfuturistic and as masculine* (Fig. 7).

4 Conclusion

In this paper we investigated the influence of different product shapes at participants meaning comprehension. The case study has been performed using semantic differential technique to receive feedback for user's shape preferences and their emotional responses. With Factor analysis we identified two main affective components for Indian and Central European student population, which represented main intangible product dimensions: aesthetic, functional, social, future attributes. The results showed that Indian and European participants shape preferences are very similar. However, there were some research limitations. Since a survey match only students in engineering sciences we need to consider a population bias. Although, based on the research results we can conclude that both Indians and Central Europeans perceive curved corners shapes as attractive while straight corners shapes were perceived as classical. Straight corner shapes were also perceived as *masculine* while curved corners shapes were in most cases perceived as *feminine*. We received more profound information about single bipolar adjective estimation. Indian participants tended to give lightly higher scores to curved corners shapes while Central Europeans value higher straight corners shapes. We found that in shape perceptions between Indians and Central Europeans there are minor differences. The reason for minor differences in shape perception between Indians and Central Europeans could be an influence of Globalization. Since many scientific papers witness that fast growing Indian economy and consumerism has become an Indian value [24]. However, deeper research and monitoring during longer time period are necessary to give the final answers, as cultural and personal preferences are constantly changing phenomena.

Acknowledgments We would like to express gratitude to all the colleagues who were involved into this survey. Special thanks to Indian institute of Technology Madras (Prof. Dr. V. Raghu Prakash), Indian institute of Science, Bangalore (Prof. B. Gurumoorthy), Industrial Design Centre, IIT Bombay (Ms. Susmita Sharma), Budapest University of Technology and Economics (Prof. Balazs Vidovics), Mechanical Engineering Faculty in Slavonski Brod University of Osijek (Prof. Dr. Milan Kljajin), Faculty of Mechanical Engineering Ljubljana. This survey has been done with financial support of Ministry of Higher Education, Science and Technology.

References

1. Schütte, S.: Engineering emotional values in product design kansei engineering in development, Linköping, Linköping Studies in Science and Technology, Dissertation 951 (2005)
2. Schmitt, B.H., Simonson, A.: Marketing Aesthetics: The Strategic Management of Brands, Identity and Image. Free Press, New York
3. Dumaine, B.: Design That Sells and Sells and ..." Fortune, (March 11), pp. 86–94 (1991)
4. Desmet, P., Hekkert, P., Hillen, M., Desmet, P.M.A., Hekkert, P., Hillen, M.G.: Values and emotions; an empirical investigation in the relationship between emotional responses to products and human values. In: Proceedings of the Fifth European academy of design conference (2003)
5. Shukla, P., Purani, K.: Comparing the importance of luxury value perceptions in cross-national contexts. J. Bus. Res. **65**(10), 1417–1424 (2012)
6. Matsumoto, D., Fazilet, K., Kooken, K.: American-Japanese cultural differences in judgements of expression intensity and subjective experience. Cogn. Emot. **13**(2), 201–218 (1999)
7. Ekman, P., Sorenson, E.R., Friesen, W.V.: Pan-cultural elements in facial displays of emotion. Science **164**, 86–88 (1969)
8. Russell, P.: The aesthetics of rectangle proportion: effects of judgment scale and context. Am. J. Psychology. **113**(1), 27–42 (2000)
9. Jordan P.W.: Designing Pleasurable Product. Taylor and Francis, United Kingdom (2000)
10. Hofstede, G.: Cultures and Organisations. Harper Collins, London (1994)

11. Shukla, P., Purani, K.: Comparing the importance of luxury value perceptions in cross-national contexts. J. Bus. Res. **65**(10), 1417–1424 (2012)
12. Zhang, Y., Feick, L., Prince, L.J.: The Impact of self-construal on aesthetic preference for angular versus rounded shapes. Soc. Pers. Soc. Psychol. PSPB **32**(6), 794–805 (2006)
13. Nisbett, R.E.: The Geography of Thought: How Asians and Westerners think differently... and Why. Free Press, New York (2003)
14. Bloch P.H., Brunel F.F., Arnold T.J.: Individual differences in the centrality of visual product aesthetics: concept and measurement. J. Consum. Res. **29**(4), 551–565 (2003)
15. Douglas, M., Isherwood, B.: The World Of Goods: Towards an Anthropology of Consumption. Routledge, New York (1996)
16. Diehl, J.C.: Globalization and cross cultural product design. In: International Design Conference-Design. Dubrovnik, Croatia (2006)
17. Berlyne, D.E.: Aesthetics and Psychobiology. Appleton-Century-Crofts, New York (1971)
18. Crilly, N., Moultrie, J., Clarkson, P.J.: Seeing things: consumer response to the visual domain in product design. Des. Stud. **25**(6) 547–577 (2004)
19. Desmet, P.M.A., Hekkert, P.: Framework of product experience. Int. J. Des. **1**(1), 13–23 (2007)
20. Russell, J.A., Caroll, J.M.: On the bipolarity of positive and negative affect. Psychol. Bull. **125**(1), 3–30 (1999)
21. Russell, J.A.: A circumplex model of affect. J. Pers. Soc. Psychol. **39**(6), 1161–1178 (1980)
22. Field, A.: Discovering statistics using SPSS, 3rd edn. SAGE Publications Ltd, Thousand Oaks (2009)
23. Gordon, K.: Esthetics. Henry Holt, New York (1909), retrieved from https://archive.org/details/esthetics00gorduoft
24. Derné, S.: The (limited) effect of cultural globalization in India: implications of culture theory (2005)

Aesthetic Design Process: Descriptive Design Research and Ways Forward

Santosh Jagtap and Sachin Jagtap

Abstract Consumer response to designed products has a profound effect on how products are interpreted, approached and used. Product design is crucial in determining this consumer response. Research in this field has been centered on studying the relationship between product features and subjective responses of users and consumers to those features. The subject of aesthetic or styling design process has been relatively neglected despite the important role of this process in fulfilling intended consumer response through product design. In this paper, we present a review of descriptive design research on aesthetic design process, and highlight limitations of this research. We also suggest opportunities for further descriptive research on the subject of aesthetic design process.

Keywords Product aesthetics · Product design · Descriptive design research · Design process

1 Introduction

1.1 Consumer Response to Designed Products

Consumer or user response to a designed product can be classified into categories such as 'aesthetic', 'semantic', 'symbolic', etc. [1–3]. Aesthetic response is about the perception of (un)attractiveness in products. Semantic response relates to the interpretation of a product's function, the mode of using the product, etc. Symbolic response is about the associations between the product-characteristics and owner- or

S. Jagtap (✉)
Industrial Design, Department of Design Sciences, Lund University, Lund, Sweden
e-mail: snjagtap22@gmail.com; santhosh.jagtap@design.lth.se

S. Jagtap
Maharashtra State Board of Secondary and Higher Secondary Education, Latur, India

A. Chakrabarti (ed.), *ICoRD'15 – Research into Design Across Boundaries Volume 1*,
Smart Innovation, Systems and Technologies 34, DOI 10.1007/978-81-322-2232-3_33

user-identity. Consumer response to a designed product is often stimulated by visual information as the vision system provides data at higher speed and rates as compared to other senses [4]. However, consumer responses can be triggered by other senses, namely, taste, smell, touch, and hearing.

Consumer response to designed products has a profound effect on how products are interpreted, approached and used [5]. Based on such response, evoked by product-characteristics, consumers make judgments on the elegance, functionality and social significance of products [3]. The design of products is crucial in determining consumer response and market success [6]. In a styling or aesthetic design process, designers play an important role in formulating intended consumer response. They generate and evaluate concepts to effectively communicate the intended response. The aesthetic design process has a major influence on the consumer response and product success [5].

1.2 Design Research and Aesthetic Design Process

According to Blessing and Chakrabarti [7], the overall aim of design research is to "make design more effective and efficient, in order to enable design practice to develop more successful products". In order to support a more rigorous approach for undertaking design research, they developed the 'Design Research Methodology' (DRM).

As shown in Fig. 1, the DRM consists of the following four stages. (1) *Research clarification*: this stage provides the motivation for undertaking design research.

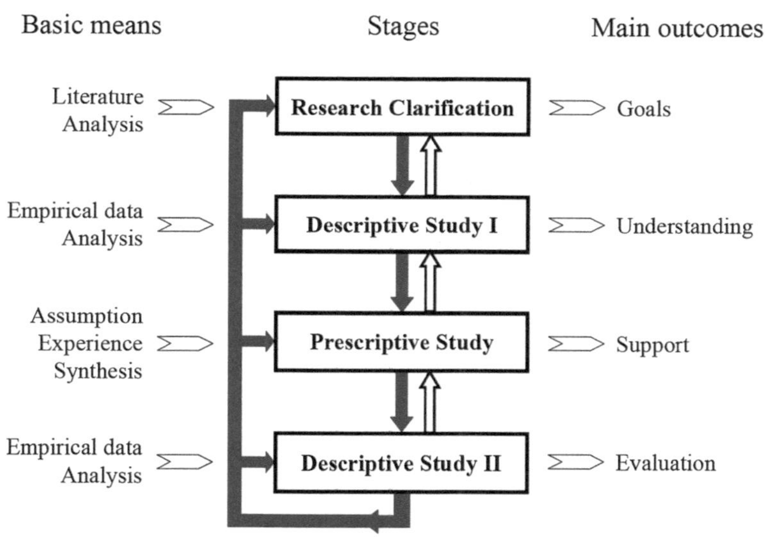

Fig. 1 Design research methodology (adopted from Blessing and Chakrabarti [7])

(2) *Descriptive study* 1 (*DS*-I): this stage is aimed at understating design in the context of the research motivation. (3) *Prescriptive study*: in this stage, based upon the findings of the DS-I study, and including assumptions and experience, a design method is developed. (4) *Descriptive study* 2 (*DS*-II): this stage evaluates the proposed method. While the DS-I aims to understand the design 'as-is', the DS-II aims to understand the impact of a support.

Several research studies, from fields such as aesthetics, product design, psychology, and marketing, have been devoted to studying how consumers respond to products. However, the subject of aesthetic design process has been relatively neglected [5]. In order to improve the aesthetic design process, it is important to understand this process (i.e. DS-I study). This research presents a brief review of DS-I studies (henceforth called descriptive studies) in the area of aesthetic design process. We highlight limitations of these studies, and suggest opportunities for further research in this area.

2 Aesthetic Design Process: Descriptive Design Research

2.1 Aesthetic Design Process: A Broad View

A few studies have explored the general characteristics of a styling design process. Tovey [8] has described the styling design process in the car industry. In the car industry, products generally evolve since there are no radical changes in the design of cars. In this industry, industrial designers, often referred to as stylists, are clearly differentiated from engineers in terms of their role in embodying intended consumer response in the design of cars. The styling process is intuitive and holistic in nature. This intuitive nature of the styling design process has also been reported by Birtley [9]. He also found that stylists use a rich vocabulary of terms for automotive forms. For example, to describe car-forms or connote a feeling, they use words such as 'warp parabolas', 'ellipsoids', 'slippery', 'fluid', 'sheer', 'bath tub', etc.

Several factors influence the aesthetic design process, and thereby the final form of a product. Crilly et al. [5], based on interviews with industrial designers, examined such influencing factors. They developed a framework to represent: the intended consumer response that designers attempt to embody in the design of products; visual sources (e.g. existing products) used by the designers; their activities (e.g. sketching) to represent the product form; and different factors (e.g. manufacturing issues, team characteristics, etc.) influencing the form design. The developed framework is simple in terms of its ability to represent different factors influencing the design of product-forms. Babapour et al. [10] also attempted to investigate designers' activities in the aesthetic design process. They carried out a diary study with Masters students in order to understand how designers generate forms in styling design. However, their study mainly focused on the research method of diary study, rather than investigating designers' activities in the form generation process.

While designers and engineers play a synergistic role in designing a product, differences between their 'general personalities' and roles may pose challenges in the styling design process. Bangle [11], from his experience of heading a design team at the car company BMW, observed several characteristics of the styling process including the communication problems between designers and engineers. He states, "The designers saw perfection as an ephemeral, almost spiritual, quest—a goal to be achieved in stages; for the engineers, perfection was physical and measurable—something to be done right the first time". The designers are emotionally sensitive, and may not respond to "cold, rational arguments". He has also highlighted the importance of using appropriate language (e.g. terms, words) when communicating a concept to different stakeholders so that they can understand the concept easily.

While the above studies have limited scope in providing an in-depth understanding of some specific areas of a styling design process, they do present information on the general nature of the process. For example, these studies point out that a styling design process is intuitive in nature, and that several factors (e.g. manufacturing issues, communication between designers and engineers, etc.) can influence the process. These studies also highlight the artistic and emotional side of designers involved in the process.

2.2 Sketching

Much of the research work on sketching in design has been carried out in the areas of architectural design and engineering design. Research in these areas found that making and perceiving sketches plays an important role in idea generation [12]. There is an interactive dialog between mind and sketch. Ambiguity in sketches facilitates reinterpretation, and can trigger new ideas. Similar findings were observed by Tovey et al. [13], in his study of sketching activity in car styling. The participants in the study were MA Automotive Design students and professional designers from the Ford design studio in Dunton (UK). They found that designers sketch to generate forms, rather than externalising the pre-conceived forms in their heads.

Tovey et al. also found that the professional designers were more fluent in sketching. The professional designers could convey their form ideas mainly by using form lines without using shading. The design students and professional designers mentioned that the form ideas emerged on the paper while sketching. They also found some obvious differences between the students and the professional designers. As compared to the students, the professional designers produced many sketches in several views (see Fig. 2). This can be attributed to their skills developed over time, with practice.

Another study of sketching in the aesthetic design process is by Warell [14]. Based on the examination of sketches from styling design projects and interviews with experienced industrial designers, they investigated the process of form

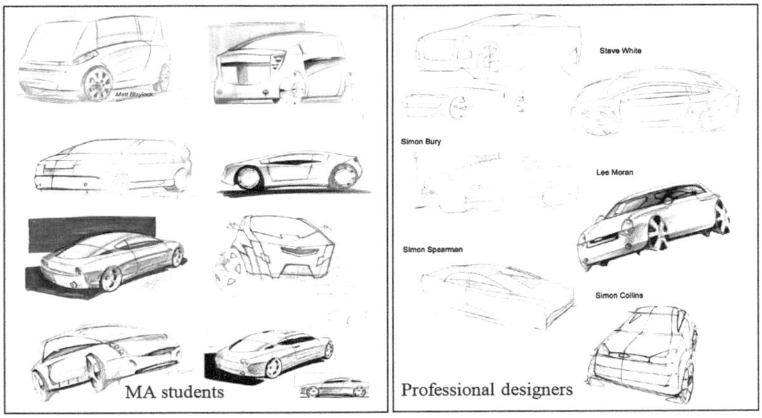

Fig. 2 Sketches produced by MA students and professional designers—adopted from Tovey et al. [13]

development and the function of sketching activity in the process. While their study examined the form development process and sketching activity, the main goal of the study was to verify the ability of their theoretical framework 'design syntactics' to describe and explain the nature of aesthetic form development. The framework 'design syntactics' includes three concepts—form syntactics, form functionality and design format. The concept 'form syntactics' explains the visual structure and content of the form composition, the concept 'form functionality' represents the purpose and function of form aesthetics, and the concept 'design format' takes into account the philosophy and use of form ingredients of product design. Warell found that designers consciously or unconsciously take into account the concepts of the 'design syntactics' framework. Their study mainly focused on styling design projects undertaken by in-house industrial designers (e.g. designers from car industry and large consumer appliances companies).

While there are many studies of sketching in engineering design and architectural design, sketching has been scarcely researched in styling design despite its importance in aesthetic design process.

2.3 Role of Information

In the field of engineering design research, several authors note the significance of the role of information in design activities [15]. The findings of the empirical research conducted by Marsh [16] in an aerospace industry suggest that, on average, 24 % of designers' time is spent in acquiring and providing information. Furthermore, several studies, carried out in laboratory settings with experienced engineering designers or students, note the importance of information in the design process [17, 18]. As compared to these studies, the information behaviour of designers in styling design has been under-researched.

We found one study, which assessed the influence of information gained from consumer research on product aesthetics. Crilly and Clarkson [19], from their interviews with industrial design consultants and consumer researchers, found that information about consumers helps designers: (1) to understand how products are perceived, (2) to establish the direction of future projects, and (3) to gain feedback on completed ones. This information also helps them to demonstrate their clients that their designs satisfy consumer needs, and to increase the clients' confidence in the project. While there are benefits of using information from consumer research, the difficulties in collecting and interpreting this information can reduce its impact on product aesthetics.

2.4 Designers' Perceptions of Product Forms

A few studies have explored how designers perceive product forms. Based on a questionnaire study with practicing car designers and design students, Liem et al. [20] explored designers' perceptions of form characteristics in car styling. The results indicate that designers perceive front emblem, head lamp, radiator grill, tail lamp, and rear bumper as significant components determining the recognition of a car. In their view, the three-quarter front view of a car is important in car-recognition. Balance, scale, and proportion are perceived as the most important principles in car styling. Abidin et al. [21], through a study with professional designer and design students, explored what keywords they assign in expressing their spontaneous reactions to images of cars. While some keywords matched between them, Abidin et al. did not find clear correlations between professional designers and design students. There is large number of studies examining the perception of products by consumers or users. However, the research on the perception of product-forms by designers is sparse.

2.5 Sources of Inspiration

In the aesthetic design process, designers use different sources of inspiration to formulate intentions, and also to generate concepts that will satisfy those intentions. Crilly et al. [5], from their interviews with industrial designers, found that sources of inspiration "may be drawn from almost any sphere", but generally include similar or dissimilar products, historic products and natural objects. The features from these sources can be embodied in the design, and thus the product form is influenced by these sources. Similar observations were made by Eckert and Stacey [22], from their empirical research in the knitwear industry. They found that sources of inspiration help in defining the context for new designs, in the creation of designs, and in communicating designs by reference to those sources.

2.6 Strategic Styling Decisions

In product design, companies make strategic decisions to foster a strong visual identity for their brand since brand recognition is important in competitive market. Decisions to create brand recognition through product design are influenced by a number of factors. Karjalainen and Snelders [23] examined how Nokia (mobile phones) and Volvo (passenger cars) created visual recognition of their brands through product design. By analysing information collected through press releases, annual reports, Internet pages, and a number of in-depth interviews, they found that the design philosophy of creating visual recognition of brands was different in Nokia and Volvo. As compared to Volvo, the application of design features over the product portfolio was flexible in Nokia. In addition, the relation between design features and brand values was more implicit in Nokia. They also observed that factors such as life-cycle stage, brand position, portfolio width, brand heritage, and product history influence the strategy of creating visual recognition of brands through product design. While the above study of Karjalainen and Snelders used data from the 'real life' design projects, Person et al.'s study [24] collected data through controlled experiments to understand how styling decisions are made. By presenting a number of market situations to design professionals, Person et al. examined how styling decisions are made. The results suggest that a number of internal (e.g. company age, company location, company size, etc.) and external factors (e.g. degree of competition, market size, price sensitivity among consumers, etc.) influence styling decisions. In addition, work experience and formal education of designers influence styling decisions.

2.7 Strategies to Elicit Specific Intentions

There are studies that have explored strategies used by designers to elicit specific intentions; in particular to elicit 'surprise' through product design. Based on interviews with designers and analysis of photographs in magazines and websites, Ludden et al. [25] reported the strategies used by industrial designers in eliciting surprise specifically through visual-tactile incongruity. The strategies include using a new material that appears like a familiar material, a new appearance for a familiar product, transparent material, visual illusion, etc. While Ludden et al.'s study focused on strategies to elicit surprise through visual-tactile incongruity, Ramírez's [26] study examined strategies that can be applied beyond visual-tactile incongruity. He interviewed senior designers, design managers or CEOs from influential design organisations. Some of the strategies to elicit surprise through product design are: using archetypes in unexpected contexts, challenging assumptions of appearance, magical interactions, etc.

3 Discussion and Ways Forward

While the number of descriptive studies in the area of aesthetic design process is small, a review of these studies can help to understand the status of descriptive design research in this area, to identify limitations of this research, and to propose further research in this area. In Sect. 2, we presented a brief review of descriptive (DS-I) studies in the area of aesthetic design process. These studies have covered topics such as sketching, role of inspiration, strategic styling decisions, strategies used by designers to elicit a specific intention, etc. In this section, we identify limitations of these studies, and suggest opportunities for further work.

As compared to disciplines such as engineering design, architectural design, systems engineering, and software design, the descriptive design research in the field of styling processes is scarce. The number of topics examined is much smaller in this field. For example, topics such as creativity, novice-expert differences, analogical thinking, design fixation, social processes of design, and studies of design teams have not been explored in the field of styling design processes. Furthermore, the descriptive research in the studied topics (e.g. sketching, role of inspiration, etc.) is highly scarce and fragmented. This can be attributed to the fact that the design research community in the field of product aesthetics has mainly focused on studying the relationship between product features and subjective responses of users and consumers to those features (see Fig. 3). In other words, design research in product aesthetics is centered on answering the question '*what is (or is not) an aesthetic product?*', rather than the question '*how is a successful or an unsuccessful product in terms of eliciting intended consumer response created?*'.

Several researchers in the area of product aesthetics believe that the styling process is intuitive and tacit (e.g. [5, 6, 8]). Real time data collection methods such as protocol studies, ethnographic participation, and observations of ongoing projects can offer promising opportunities to explore the tacit and intuitive aspects of a styling process. However, in this field, research methods such as retrospective interviews and examination of sketches from completed design projects are dominating. While there are some advantages of retrospective interviews, they cannot provide details of the styling process because interviewees may not recall nitty-gritty of the process and they can attempt to rationalize the process for explaining it to the researchers. We believe that future research needs to address these issues by using multiple real-time data collection methods. We also believe that the research community investigating styling processes and the neuroscience

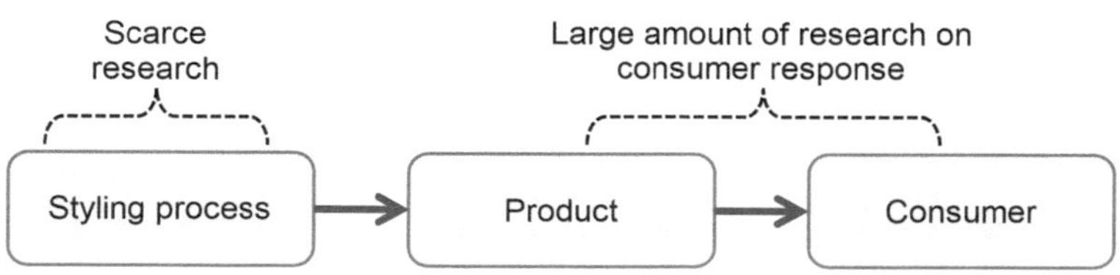

Fig. 3 Scarce research on styling design process

community have much to learn from each other. Intuitive and tacit aspects of a styling process can be studied by employing research methods from the neuroscience area (e.g. fMRI, eye tracking).

In the field of styling design, designers often mention that form-ideas emerge on the paper while sketching. Sketching plays an important role in styling design. However, in this field, the topic of sketching is relatively neglected by design researchers. There is a need of extensive design research to understand sketching activity in styling design. Furthermore, in styling design, designers can use different media such as clay modelling and CAD. Further research needs to investigate designers' activities in dealing with these media and the influence of these media on the styling design process.

Many descriptive studies of styling processes have been focused on the car industry. There can be differences between the styling processes of products such as electronic gadgets, consumer appliances, furniture, kitchen appliances, and vehicles. These differences can be attributed to the difference between companies designing and manufacturing these products. For example, designers in the car industry are, in general, in-house designers and can be responsible for the design and the final manufactured product. On the contrary, designers of consumer appliances may be from industrial design consultancies, and they may be responsible only for the final design and not for final manufactured product. Furthermore, there can be differences between the styling processes of products using established technology and products using entirely new technology. Future research needs to cover a range of products to increase our understanding of styling processes.

We can note that many descriptive studies of styling design processes are limited in terms of the senses considered. These studies have mainly taken into account visual aspects of product design despite the fact that other senses can be important in the design of products. To address this, further research needs to

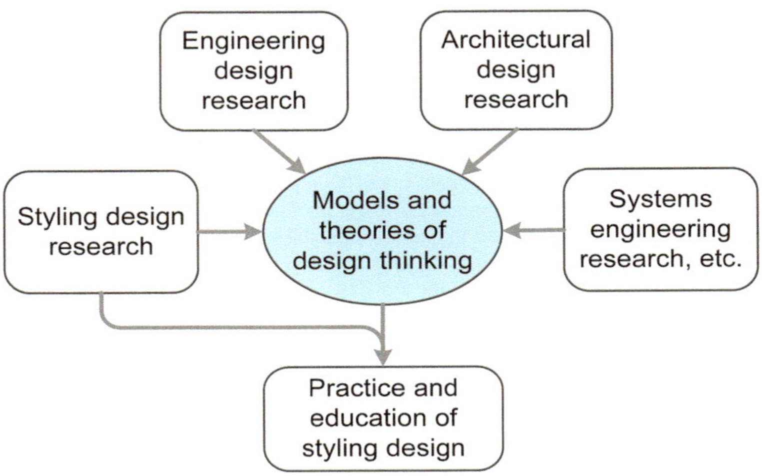

Fig. 4 Descriptive design research in styling design and other disciplines

cover a range of sensory responses (i.e. visual, auditory, tactile, etc.) that designers consider in product design.

Future research investigating styling processes ought to overcome the above limitations. This will help to establish a comprehensive understanding of styling design processes, and will assist in developing tools and methods to improve design practice and education of styling design processes (see Fig. 4). In addition, the understanding of styling design processes will help to develop generic models and theories of design thinking, which in turn can be useful in improving design practice and education of styling processes.

References

1. Dittmar, H.: The Social Psychology of Material Possessions: To Have Is to Be. Harvester Wheatsheaf, NewYork (1992)
2. Hekkert, P.: Design aesthetics: principles of pleasure in design. Psychol. Sci. **48**(2), 157 (2006)
3. Mono, R.: Design for Product Understanding Liber. Stockholm, Sweden (1997)
4. Ulrich, K.T.: Design: Creation of Artifacts in Society, Karl T. Ulrich (2011)
5. Crilly, N., Moultrie, J., Clarkson, P.J.: Shaping things: intended consumer response and the other determinants of product form. Des. Stud. **30**(3), 224–254 (2009)
6. Coates, D.: Watches Tell More Than Time: Product Design, Information, and the Quest for Elegance. McGraw-Hill, NewYork (2002)
7. Blessing, L.T.M., Chakrabarti, A.: DRM, a Design Research Methodology. Springer, London (2009)
8. Tovey, M.: Styling and design: intuition and analysis in industrial design. Des. Stud. **18**(1), 5–31 (1997)
9. Birtley, N.: The conventional automobile styling process. Coventry Polytech. (1990)
10. Babapour, M., Rehammar, B., Rahe, U.: A comparison of diary method variations for enlightening form generation in the design process. Des. Technol. Educ. Int. J. **17**(3) (2012)
11. Bangle, C.: The ultimate creativity machine: how BMW turns art into profit. Harvard Bus. Rev. **79**(1), 47–55 (2001)
12. Schön, D.A.: The Reflective Practitioner: How Professionals Think in Action, vol. 5126. Basic books, NewYork (1983)
13. Tovey, M., Porter, S., Newman, R.: Sketching, concept development and automotive design. Des. Stud. **24**(2), 135–153 (2003)
14. Warell, A.: Design Syntactics: A Functional Approach to Visual Product form Theory, Models, and Methods. Chalmers University of Technology, Sweden (2001)
15. Jagtap, S., Johnson, A.: Requirements and use of in-service information in an engineering redesign task: case studies from the aerospace industry. J. Am. Soc. Inform. Sci. Technol. **61**(12), 2442–2460 (2010)
16. Marsh, J.R.: The Capture and Utilisation of Experience in Engineering Design. University of Cambridge, UK (1997)
17. Eris, O.: Perceiving, Comprehending, And Measuring Design Activity Through The Questions Asked While Designing. Stanford University, Stanford (2002)
18. Jagtap, S., et al.: How design process for the base of the pyramid differs from that for the top of the pyramid. Des. Stud. (2014)
19. Crilly, N., Clarkson, P.: The influence of consumer research on product aesthetics. In: Proceedings of the 9th International Design Conference (DESIGN 2006), Design Society (2006)

20. Liem, A., Abidin S., Warell, A.: Designers' perceptions of typical characteristics of form treatment in automobile styling. In: 5th International Workshop on Design & Semantics of Form and Movement, DesForm (2009)
21. Abidin, S.Z., Warell, A., Liem, A.: The significance of form elements: a study of representational content of design sketches. In: Proceedings of the Second Conference on Creativity and Innovation in Design, ACM (2011)
22. Eckert, C., Stacey, M.: Sources of inspiration: a language of design. Des. Stud. **21**(5), 523–538 (2000)
23. Karjalainen, T.M., Snelders, D.: Designing visual recognition for the brand*. J. Prod. Innov. Manage. **27**(1), 6–22 (2010)
24. Person, O., et al.: Should new products look similar or different? The influence of the market environment on strategic product styling. Des. Stud. **29**(1), 30–48 (2008)
25. Ludden, G.D., Schifferstein, H.N., Hekkert, P.: Surprise as a design strategy. Des. Issues **24**(2), 28–38 (2008)
26. Edgar, R., Ramírez, R.: Industrial design strategies for eliciting surprise. Des. Stud. **35**(3), 273–297 (2014)

Mapping Sentences into Graphics to Communicate Engineering Design Information: Concept Revisited

Zeundjua Tjiparuro

Abstract The use of words in prose form to represent and communicate design ideas is not popular in engineering or at least not to the same extent as graphical representation. Is this because prose is completely defective for communicating design information? A variant question to the above was first investigated in 1987 at Carnegie-Mellon University and revisited nearly 15 years later at Arizona State University. The investigations gave graphics an edge over prose, albeit with the results so close, especially for the latter, that the maxim; the jury is still out there, would be an apt description. It is for this reason that the matter is revisited in the investigation reported in this paper. Ninety-eight (98) freshmen, taking an engineering graphics course (CETG 102: Engineering Graphics with AutoCAD) at the Botswana International University of Science and Technology, were given a 'test' on their first day of the course. In the 'test' the students were given a word description of a system and asked to draw/sketch it. The data gained from the experiment was collated and analyzed for quality and or ambiguity. Major findings from the investigation indicate that a fairly large numbers of students interpreted the word description of the system correctly and the ambiguity was fairly minimal with only 28 different groups of wrong interpretations of the same word description observed.

Keywords Engineering drawings · Engineering graphics · Design representation · Design modelling

Z. Tjiparuro (✉)
Department of Mechanical and Energy Engineering, Botswana International University of Science and Technology, Palapye, Botswana
e-mail: tjiparuroz@biust.ac.bw

A. Chakrabarti (ed.), *ICoRD'15 – Research into Design Across Boundaries Volume 1*, Smart Innovation, Systems and Technologies 34, DOI 10.1007/978-81-322-2232-3_34

1 Introduction

Designers sketch and draw to represent and communicate ideas. When seized with a problem, a designer can draw on anything [1, 2]. It is by this very predisposition that expressions such as; cocktail napkin sketch and back of an envelope sketch, have come to be used in design literature. During the sketching process, it is not uncommon to find notes, in the form of written words or text, accompanying the sketches. Sometimes the notes are scribbled on a note-book page opposite the sketch, sometimes they are written on the sketch itself as labels or attached to a sketch part with an arrow [3]. Thus sketches and notes go together in design communication. That notwithstanding, research, amongst others Bertoline [4], has established that 92 % of the engineering design process is largely graphical (sketches) and only 8 % is divided between mathematics, written and verbal communications. Why? Because, as observed in [5], using another popular metaphor, a diagram is (sometimes) worth ten thousand words. In any case, designers still need to sketch as sketching conveys thoughts that cannot be put in words [6], although it may as well be submitted that the converse is true.

Besides sketching, many other design externalization methods such as: prototypes, solid models, virtual, linguistic and mathematical models exist. A fairly general taxonomy of engineering representation methods is given in [7]. Almost all these representation methods have been investigated and compared for effectiveness. The results are not necessarily in agreement.

On one side of the argument is research work that goes to show that externalization, in the form of sketching, may not necessarily be the be-all-and-end-all of engineering design thinking and representation. The spectrum of suggestions in this regard is wide; from the most abstract of representations to the tangible. For example, an experiment was done in [8] in which designers were denied access to a sketching paper. They were blindfolded and forced to rely on their mental imagery. The experiment showed that imagery alone could act as a design representation medium (called mental representation) on which ideas can be created, revised and analyzed and ultimately externalized by verbalizations. On the tangible side, Faas et al. [9] showed that fast preliminary prototypes are equally as good for design idea representation and exploration as sketches. However, research findings of this type are in the minority.

Hence on the other side of the divide, and which happens to be in the majority of the comparisons, is research that find sketching as a more effective representation method. Research in this regard has also traversed two extreme positions; from pure logic to the empirical. Larkin and Simon [5] compared sketches to verbal linguistic representation. In the study pure logic was used to generate equivalent sentential representations of some engineering devices such as pulley systems. The nub of the study was to demonstrate the complexity of doing such or as aptly put by the author: "Why a Diagram is (Sometimes) Worth Ten Thousand Words." The same comparison between sketching and sentential representation was investigated differently in [1], however with the same verdict. In this particular study, a textual description of a system and sketch representation of another were given to different groups of designers and each group was asked to generate an equivalent representation in the other representational medium (i.e. those given text descriptions were to draw a sketch from the given text and those given sketches were to write text descriptions). Findings from the study gave sketching a slight edge over verbal/textual externalization albeit with the results too close that the matter remained yelling for more work. It had to be revisited and this is what this paper undertook to do.

In this study, the effectiveness of using text in the form of prose to communicate design ideas to novice and amateur subjects was tested. The text was restricted to a simple geometric shape doable by freshmen as the research interest was about enhancing our understanding of how novices/amateurs approach design tasks; itself an active research area (see [10]). Firstly, it was investigated how well each sketch by each subject conformed to the example model. Secondly, the variety of the sketches was investigated to give a sense of how differently the subjects interpreted the same information (this being a measure of ambiguity in the textual information given). The use of novice/amateur subjects is a first as previous studies have tended to use only experienced subjects, especially at sophomore level.

2 Methodology

The methodology (i.e. approach and analysis) used in this study is similar to that normally followed in empirical research studies and is as explained below:

2.1 Approach

Ninety-eight (98) first year students taking an engineering graphics course (CETG 102: Engineering Graphics with AutoCAD) at the Botswana International University of Science and Technology were given a quiz on their first day of the course. To get the best out of students they were instructed that the quiz was an assessable test and the word test was used instead of quiz. The purpose of giving the quiz was mainly to demonstrate to the students the challenges of using prose as a way of describing engineering systems. The research component was an incidental outcome. In the quiz, the students were given a word description of a tool and asked to draw/sketch it in 1 h. The text of the quiz was adapted from an educational blog for the University of Tennessee [11] and rephrased to read:

> A special tool is in the shape of a 140 × 80 × 10 mm rectangular solid. One end rises from zero thickness to the maximum thickness of 10 mm in a length of 40 mm to form a sharp edge. The opposite end is semicircular. A 20 mm diameter hole is positioned so the centre of the hole is 40 mm from the semicircular end and 40 mm from either side of the scraper.

The students were specifically instructed to construct an equivalent visual representation of the above text information as follows:

> Draw and give all necessary illustrations to depict the tool. Show all the step by step working and indicate the final drawing.

How well or badly students reproduced the tool would answer the main research question captured in the abstract of this paper, namely: ***Is prose information completely defective in communicating design information?***

Although the task as given above falls into the ***novel visualisation*** category (sketching from verbal description) as opposed to ***mechanical recall*** (sketching from memory) or live model sketching (see [12]), it was still necessary for the students to divulge whether they have had prior exposure to engineering graphics by answering the following question:

> Have you done any of the following courses at any time during your secondary school educatio (Tick)

Design and technology	Arts
Yes	No
Yes	No

Students who answered "***Yes***" to either question were classified as ***Amateurs*** and those who answered "***No***" were classified as ***Novices***. From these two classes of students it was postulated that: ***Novice students will perform worse off than amateurs***.

2.2 Analysis

The pictorial and multi-view drawings of Fig. 1, as given in [11], were used as the model answer against which the students' sketches were judged.

Firstly the sketches were grouped and collated according to how accurately they depicted the desired geometric shape or what is referred to as representational accuracy [12]. Accordingly, the sketches were assessed for clarity of thought with respect to appropriate use of primitives (i.e. shapes), labels and annotations. Poor quality sketching or messiness, such as overshooting, undershooting and jitters due to hand tremor (see [3]) were not used to disqualify representationally accurate sketches nor were embellishments such as shading. In fact, such attributes

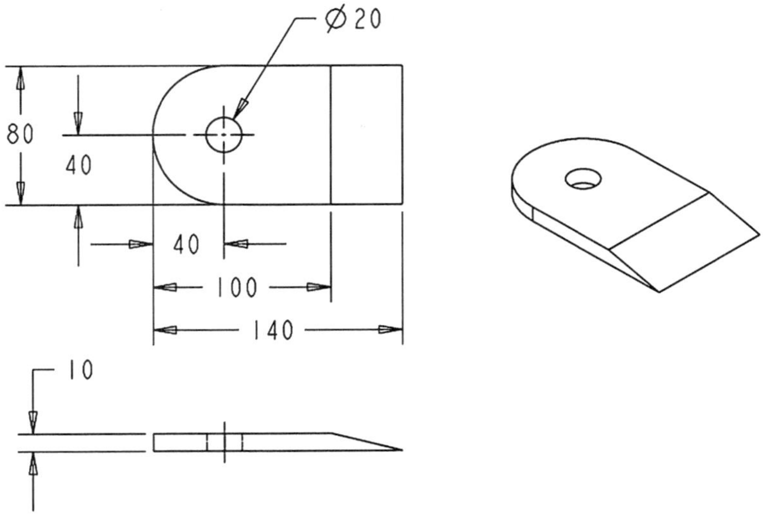

Fig. 1 Model answer

become more relevant when the aim is to measure the skill levels of the sketchers as in the degree of complexity method [13].

From analysing the students' sketches it was established that the sketches could be collated into the following three main classes:

- ***Conforming Sketches***:
  Under this are sketches whose general shape, form and location of features conformed to the pictorial and multi-views drawings of the Model Answer
- ***Deviating Sketches***:
  These are sketches that give the correct form and shape of the tool as per the Model Answer, but misplaced the location of important features (e.g. the sharp edge, the semicircle or the 20 mm hole)
- ***Nonconforming Sketches***:
  In this category are sketches which contain too many mistakes to be accepted as true mappings of the textual information given. This would include invalid objects and what in solid modelling is referred to as nonsense objects.

2.2.1 Ambiguity

To establish the inherent ambiguity of the prose information, deviating and nonconforming sketches were grouped according to their similarities and differences. The number of different groups thus obtained was used as a measure of the extent to which the textual information led to different inferences (i.e. a measure of ambiguity).

3 Results

The major findings of this investigation are as reported below.

3.1 Skill Level: Amateurs versus Novices

Out of the 98 students who took the quiz, 65 (translating to 66 %) were novices without prior knowledge of engineering graphics, while 33 (translating to 34 %) were amateurs with prior knowledge of engineering graphics or arts.

3.2 Sketches

A summary of the sketches produced by both novice and amateur students is as presented in Table 1 under the three categories investigated. Samples from each category are given under the subheadings that follow.

3.3 Conforming Sketches

Notwithstanding differences of scale, neatness and style, Fig. 2 shows samples of sketches that were assessed as conforming.

3.4 Deviating Sketches

Samples of sketches in this category assumed the forms shown in Fig. 3 and explained below.

From A sketches taking this form had the sharp edge terminating into a sharp point like a spear

From B showing sketches whereby the sharp edge is symmetrical or starts from the middle of the basic rectangular shape

Table 1 Number of students per sketch category

	Number (percentages) of students per sketch category			Total
	Conforming	Deviating	Nonconforming	
Amateurs	17 (52)	8 (24)	8 (24)	33
Novices	32 (49)	12 (18)	21 (32)	65
Total	49 (50)	20 (20)	29 (30)	98

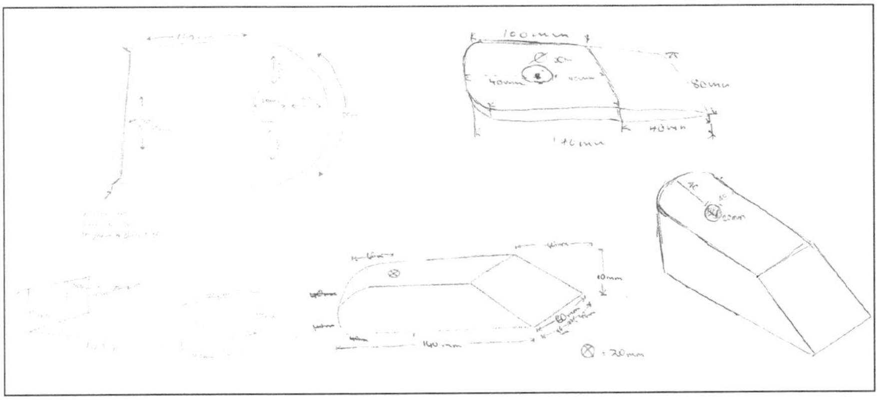

Fig. 2 Conforming sketches

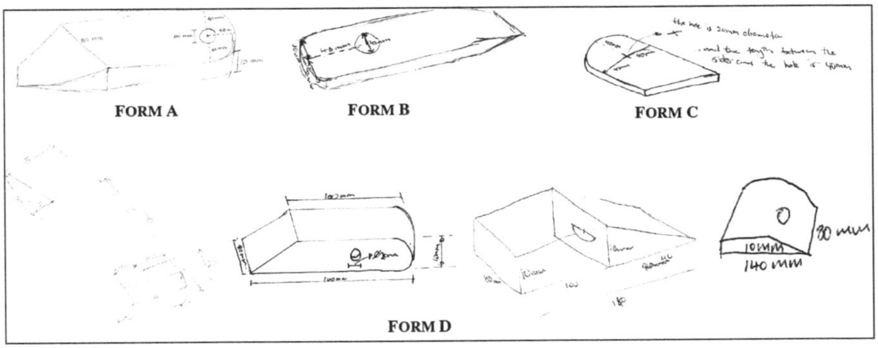

Fig. 3 Deviating sketches

From C whereby the sharp edge is located across the width of the rectangular basic shape as opposed to its length

From D where the sketches contain few mistakes. For instance, the 20 mm hole may be wrongly-located or the semi-circular end omitted as illustrated.

3.5 Nonconforming Sketches

Samples of sketches in this category are shown in Fig. 4 and they were as many as the 29 students who produced them. It was however observed that 12 students (3 amateurs and 9 novices) drew sketches of the Axe-Head form shown in Fig. 4 with the only differences being in the positions of the holes as illustrated.

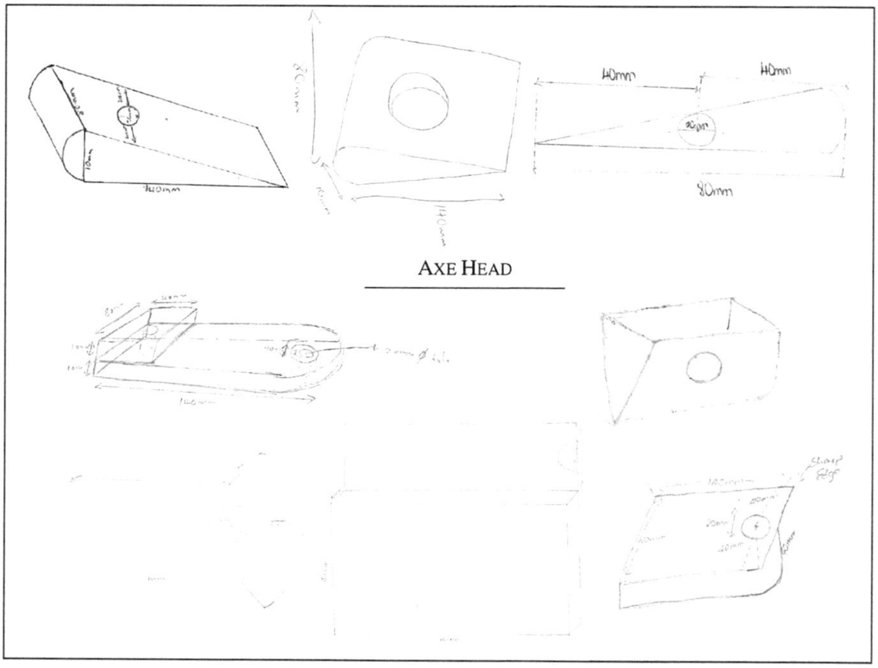

Fig. 4 Nonconforming sketches

3.6 Level of Ambiguity

As noted in Table 1, a total of 49 sketches did not conform to the Model Answer, out of which 12 were of the Axe-Head form mentioned under Sect. 3.5 above. This means that 38 different interpretations of same text information were made in this study.

3.7 Other Observations

Further to the above main investigations, the following salient points were also observed.

Problem decomposition and the use sketching primitives: It was observed that most students were able to use primitives to represent the main features (e.g. semicircle, sharp edge or rectangle) specified in the textual description. These students managed to combine these basic features to generate the final drawing of the tool required. Amateurs' students were especially able to put their knowledge to good use, combining 3-D primitives to give accurate

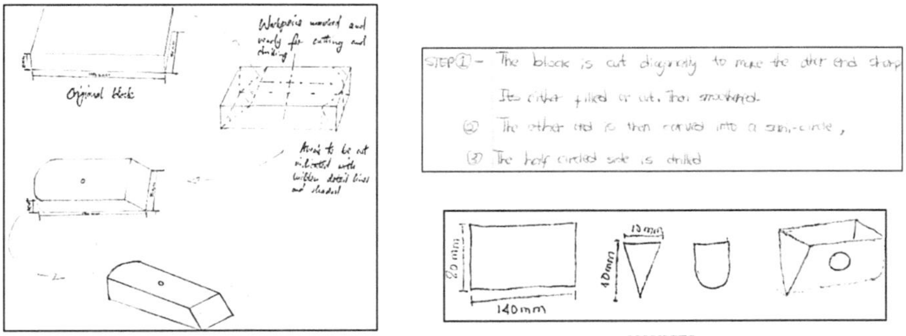

AMATEURS NOVICES

Fig. 5 Students use of drawing primitives

graphical representations. However, this was not so with novices who tended to combine them in an archaic way including resorting to textual descriptions. Figure 5 illustrates these differences between the two groups.

Pictorial representations: It was further observed that in all sketching work, pictorial representations were predominantly used by both novices and amateurs. This was so even without having asked the students to do it. It thus appears that pictorial representation is a preferred way of drawing by most sketchers.

4 Conclusions

This investigation mainly undertook to answer the question: Is prose information completely defective in communicating design information? The results of the study show that 50 % of the subjects correctly interpreted the textual information. The results further show that 20 % more subjects interpreted the information correctly but misplaced some key features. These together makes 70 % subjects who managed to interpret the textual information given fairly well. In fact, notwithstanding the inherent diversity of the sketches including the nonconforming sketches, the basic functions (e.g. the sharp edge, the round and the rectangular basic shape) communicated in the sentential representation were understood and efforts were made to reproduce them. Moreover, the results exhibited a fairly minimal spread of different interpretations of the text information at 28 different groups of wrong interpretations. In total it translates to 29 different interpretations including the single group of correct interpretations.

Finally, this investigation hypothesised that novice students will perform worse off than amateurs. However, results show that 52 % of the amateur students were able to interpret the prose information correctly while 49 % of the novice students managed to do so. The difference is nominal, implying that novices performed as well as amateurs.

5 Future Research Directions

Further investigation of prose for engineering representation needs to be undertaken to validate the results given in this paper. As a corollary to the above, further investigation into the sketching style of the novice students (or people with no prior knowledge of engineering graphics) should be undertaken to help distil knowledge about the basic primitives and symbolism preferred by novices when drawing. This is vital information in the search for helpful computer tools for sketching in the early phases of design as being attempted by various researchers.

References

1. McKoy, F.M. Vargas-Hernandez, N. Summers, J.D. and Shah, J.: Influence of design representation on effectiveness of idea generation. In: Proceedings of ASME Design Theory and Methodology Conference, pp 1–10. (2001)
2. Gross, M.D., Do, E.Y.: Drawing on the back of an envelope: a framework for interacting with application programs by freehand drawing. Comput. Graph. **24,** 835–849 (2000)
3. Fish, J.: How sketches work. Ph.D. thesis, Loughborough University of Technology. (1996)
4. Bertoline, G.R., Wiebe, E.N.: "Fundamentals of graphics communication. (2004)
5. Larkin, J., Simon, H.: Why a diagram is (sometimes) worth ten thousand words. Cogn. Sci. **11**, 65–99 (1987)
6. Sezgin, T.M.: Sketch interpretation using multiscale stochastic models of temporal patterns. Ph.D. thesis, Massachusetts Institute of Technology, (2006)
7. Conole, G.: An overview of design representations. In: Proceedings of the 7th International Conference on Networked Learning, pp. 482–489, (2010)
8. Athavankar, U.A.: Mental imagery as a design tool. Cybernetics Syst. **28**, 25–47 (1997)
9. Faas, D., Bao, Q. and Yang, M.C.: Preliminary sketching and prototyping: comparisons in exploratory design-and-build activities. In: Proceedings of IDETC/CIE. (2014)
10. Lohmeyer, Q., Melboldt, M., Matthiesen, S.: Analysing visual strategies of novice and experienced designers by eye tracking application. In: Proceedings of E&PDE 2013, the 15th International Conference on Engineering and Product Design Education, pp. 1–6. 2013
11. Blog, ef.engr.utk.edu/ef101-2002/as/book/as_chap1.pdf. Accessed April (2014)
12. Yang, M., Cham, J.: An analysis of sketching skill and its role in early stage engineering design. J. Mech. Des. **129**, 476–482 (2007)
13. McGown, A., Green, G., Rodgers, P.A.: Visible ideas: information patterns of conceptual sketch activity. Des. Stud. **19**, 431–453 (1998)

Part III
Human Factors in Design

Competitive Play in Children with Intellectual Disability: Informing Design

Aakash Johry and Ravi Poovaiah

Abstract The purpose of the present multi-disciplinary study was to inform toy and game designers about the presence and degree of competitive play in children with moderate intellectual disability (ID), as a function of their mental age. Toy design industry, especially in developing nations like India, has not looked at children with special needs as a target sector and thus there is a need to address the absence of theoretical understanding about the play behavior of these children and convert it into practical implications in design of playing aids. Considering competitive play in childhood plays a significant role in the social and moral development of the children, it becomes important to understand the nature of competitive play for children with ID. The present study is an attempt to study the emergence and degree of competitive play in children having moderate ID in India through a pilot experimental study involving children (N = 8) divided into four age groups of mental age 2–4, 4–6, 6–8 and 8–10 years as they respond to a situation of building blocks play when controlled competition is introduced explicitly through a facilitator. The results based on a qualitative analysis show that competition emerges after the mental age of around 4 years, however a negative correlation between relative degree of competition with mental age was seen unlike typical children [1]. Also, children in age-group of 2–4 years exhibited self-initiated cooperation. The findings are argued using relevant research literature from a sociological perspective and design recommendations are made like incorporating cooperative strategies and replacing competition in design of play activities. The standard group goals in activities need to be made flexible and customized and process should allow incremental outcomes to account for variation in the abilities of the children. Study also touches on the need for balanced facilitation in play activities and pedagogical interventions that allows children with free play opportunities.

A. Johry (✉) · R. Poovaiah
Industrial Design Centre, IIT Bombay, Mumbai, India
e-mail: aakashj@iitb.ac.in; johry.aakash@gmail.com

A. Chakrabarti (ed.), *ICoRD'15 – Research into Design Across Boundaries Volume 1*,
Smart Innovation, Systems and Technologies 34, DOI 10.1007/978-81-322-2232-3_35

Keywords Competitive play · Intellectual disability · Designing for play · Cooperation · Social interaction

1 Introduction

Competition in children has been a well-researched domain in psychology, sociology, evolutionary science and education for many years. It is found in situations where there is a negative correlation between goal attainments of the participants. The present study uses a more generalized definition by Greenberg [1], who states competition as "a desire to excel, an impulse to do better than our rivals". This may involve rivalrous actions between players (trying to lower other's outcomes) as well as self-reflective actions for enhancing the prospects of goal attainment.

The presence of competition in human behavior has been well established. However it was Greenberg who studied the age-effects as a variable with competition in children during their play session involving building with blocks and concluded that competition emerged as early as around 4 years in children increasing as a function of the age of the children [1]. Studies have further substantiated these results [2, 3]. There are fewer studies that attempt to understand occurrence of competition in children with ID. One such study showed the effect of competition on children with ID in stimulating a better performance judged by enhancement in speed to finish a cognitive task [4]. Madsen and Connor [5] compared the degree of competition and cooperation in a play situation involving marble-pull apparatus among 6–7 years aged and 11–12 years aged children with ID as well as typical children, forming four groups. They concluded that 6–7 years aged children with ID group was actually the most cooperative of all, and competition seemed to increase with age in these children, which confirms to the findings for typical children. However, none of these studies specifically looked competition in play as a function of mental age across all age-groups in children with ID. It is quite possible that this function is different for these children. Social interactions of handicapped children during play have been found to be inhibited and solitary play has been preferred over social play [6]. A recent study showed that to comprehend competition, a child needs capacity for 'instrumental' and 'telic' perspective taking [7]. It becomes important to know how and when competition will emerge for children with moderate intellectual disability with their inherent cognitive limitations. The present study uses a qualitative analytical approach to study this relationship for children with ID by testing two hypotheses derived from the study by Greenberg [1]. Hypothesis 1 states that children with moderate ID will show emergence of competitive play after mental age of 4 years. Hypothesis 2 states that there is a positive correlation between the mental age and degree of competitive play exhibited by children with moderate ID.

Table 1 Demographic data of the subjects taken from school records (assessed 2 years prior)

Group	Name	Sex	IQ	Chronological age (years)	Mental age (years)
1	S1	M	41	7.1	2.9
	A1	F	40	7.4	3
2	A2	M	49	9.3	4.5
	J2	F	46	10.5	4.8
3	A3	M	45	15	6.8
	S3	F	47.5	13.8	6.5
4	S4	M	49	17.3	8.5
	A4	F	48	16.6	8

2 Methods and Procedure

2.1 Subjects

Subjects (N = 8) were selected in pairs from each of the mental age groups namely, 2–4, 4–6, 6–8 and 8–10 years, having moderate intellectual disability (IQ = 35–49), roughly matched upon their mental and chronological age, taken from the school records, updated annually. The subjects attended a special school in Mumbai city, India and all were Indian in ethnicity living in local neighborhood. Each pair included one girl and one boy belonging to different classrooms to keep gender and level of familiarity as a constant during the experiment. S3 and S4 (each subject represented by 1st letter in the name and group no) were the only subjects who had Down's Syndrome besides ID. Table 1 represents the demographic data of the subjects briefly.

2.2 Procedure

The present study derives from the procedure used by Greenberg [1] where he studied competitive play between a pair of typical children using building blocks, while controlling a number of variables unaccounted earlier. The present study involved a pre-experimental phase referred as 'warm-up sessions' lasting fifteen minutes each, conducted for 2 days and in the same week as the experiment to expose the subjects to constructive play using building blocks and balance their level of exposure and comfort with this form of play (constructive play was absent in school prior to the study). The children were brought in pairs into the 'play space' (school recreation room) by the facilitator (recreation teacher) and they sat facing each other, while facilitator helped them play with Jenga brick set (refer Fig. 1a). These were free play sessions however the facilitator modeled the play activity initially for the purpose of explanation. The subjects didn't interact with

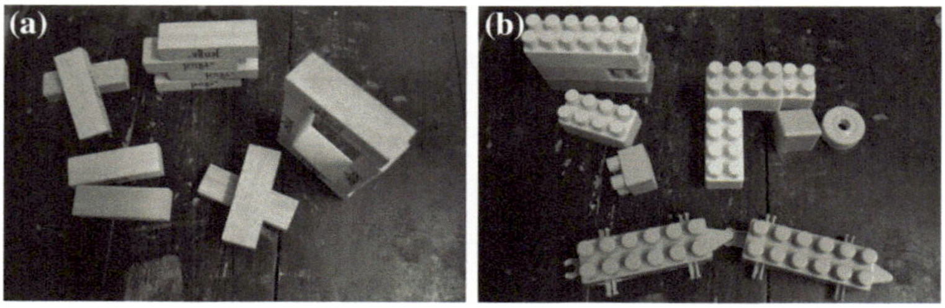

Fig. 1 Constructive toys used for **a** warm up sessions and **b** experimental sessions

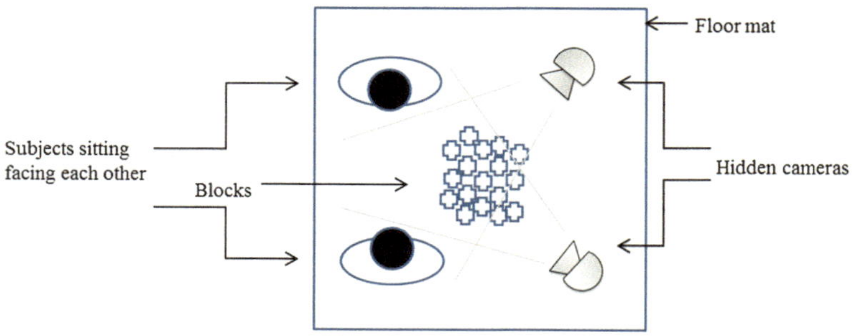

Fig. 2 *Top view* of the experimental setup for the study

their paired experimental partner but a classmate. This was done to avoid experimental pairs developing any bias towards each other.

The experiment was divided into two stages. In the first stage, facilitator escorted the selected pair of children (refer Table 1) to play space. It was checked that subjects were in normal mood and exhibited natural behavior. They were made to sit comfortably, facing each other on a mat and allowed to play, similar to the warm-up sessions (refer Fig. 2). However, they were given a new set of block toys having variable colors and shapes (refer Fig. 1b). This ensured that the play material was novel, interesting and provided more opportunities to elicit competition and at the same time be familiar. Children were not given any instruction on how to interact with each other while sharing the play space.

The facilitator initially checked if the subjects were able to build with new blocks and helped once if needed. Then she left children to play freely and sat at a comfortable distance to make notes of the activity and didn't interfere in play. This session lasted till first 5 min or if all blocks were exhausted before it, or both the children didn't show interest in activity for a minute. The facilitator then approached the subjects to initiate 2nd stage encouraging the children to build again but this time she gave an assignment by saying, "I would like to see who can build prettier this time?" (Translated from Hindi: "Kaun zyada sundar banayega?"), with the objective to induce comparison and introduce competition for the extrinsic reward of teacher's praise. There was no specific target in construction for children and it is possible that the children might interpret notion of making 'prettier' to be different but it would not affect the results since their perception might affect the performance and competition is assessed from the play interactions of children and not the overall performance. She left the play space after pushing blocks back at center. The facilitator ended the 2nd session with a remark that both subjects performed nicely and applauded them. The facilitator, the play-space, the time and the toy remained constant for all the children.

3 Analysis

The activities, remarks and the gestures recorded in video data using hidden cameras were transcribed into series of events on a time line where each event is defined as an action or a set of repetitive actions that can be grouped together to represent a verbal or gestural interaction with the material, competitor or other environmental factors. These codes were then analyzed along with facilitator's notes to draw insights. The aim was to determine presence and to some extent relative degree of competition.

The subject indicates the degree of competition which he experiences through three parameters—his interaction with the competitor, interaction with the material and his interest in play activity. The parameters were divided into sub-categories (refer Table 2) taken from the categorization used by Greenberg [1] with introduction of new sub-categories like 'smiled', 'imitated', 'cooperated' and 'played without building' emerging from the data. Interaction was compared along the sub-categories between the two stages of experiment to support the qualitative insights and inter-rater reliability was checked. It is important to realize that performance of subject including the shape and form of their final construction was not included in the analysis to identify competition as it might vary with child's ability, and competition had to be studied independently.

4 Results

Each subject was individually analyzed in detail to establish relative degree of competition as summarized in Table 3. Children were always seen making their own construction rather than collaborating on one. Hypothesis 1 was found to be true as competitive play emerged after mental age of 4 years (age group 4–6 years) in children with moderate ID. Hypothesis 2 was nullified as a positive correlation between the mental age and degree of competitive play exhibited by the children could not be established. Competitive play was most convincingly seen in age group 4–6 years and seemed to decrease as the mental age increased. It is to be noted that these results do not represent conclusive remarks but give substantial

Table 2 Operational definitions of sub-categories of parameters for classifying events

Parameter	Sub-category	Operational definition
Interaction with the competitor	Looked at competitor	Looks or gives a glance to competitor or his construction, but doesn't smile or speak anything or imitate
	Smiled at competitor	Smiles looking at competitor or his construction, but doesn't speak anything or imitate
	Imitated competitor	Copies one or more of competitor's action while building
	Verbal communication	All the verbal interaction directed clearly towards competitor excluding the subsequent statements made in conversation during cooperation
	Cooperated with competitor	Helps competitor by guiding him on how to build or helping him in imitating
Interaction with the material	Blocks used as needed	Builds using blocks as needed, reaching out for one or two at a time, either taking those nearest at hand or which seem best fitted to the construction
	Grabbed	Picks up a block rising from their position deliberately from near the competitor, or from the other side, often disregarding the competitor's activity or cautiously trying to disturb the competitor
	Given	Offers a block to the competitor to help him or guide him in building
	Played without building	Holds, fiddles and plays, exploring the block itself but not indulging in the building activity
Interest in play activity	Distracted	Gets distracted from building without any external influence (self-initiated) and excluding instances of imitation and cooperation since they are a part of building
	Physical signs	Calls facilitator or competitor to show his construction, shows interest at start, and/or shows hesitation at end to submit and continues building

Table 3 Summarized results showing relative degree of competition across age-groups

Age group (mental) (years)	Relative degree of competition	Signifiers
2–4	Absent	Cooperation shown, relaxed state in 2nd session
4–6	Clearly present	Competitive verbal comments, grabbing the blocks, increased interest in building in 2nd session
6–8	Might be present	Disliking being imitated, using side-glance to imitate, reluctance to end and submit in 2nd session
8–10	Absent	Negligible interaction, *inability to build of one player may have led to error in sustaining competition*

directions for further enquiry and need to be validated by increasing the number of participants. However, inter-rater reliability was assessed for 50 % of randomly selected video data by an independent play researcher and 95 % of agreement in code assignment was found. Pearson's r was calculated and found to be significantly correlated for sub-categories having higher disagreement, namely looked at competitor ($r = 0.89$, $p < 0.01$), smiled at competitor ($r = 0.86$, $p < 0.01$) and cooperated with competitor ($r = 0.92$, $p < 0.01$). Results of each age-group are briefly discussed below derived from rich qualitative insights and comparing parameters quantified from sessions (refer Table 4).

4.1 Group 1: 2–4 Years Mental Age

Subjects in this group didn't show any competition. On the contrary, there was a display of self-initiated cooperation in 2nd session when A1 who had been imitating S1 was guided and helped by S1 as seen from following transcript:

{64}[1] Talks to A1 and points at his construction saying, "A1, make like this". Looks into her face for the first time and gives a big smile. {72} Replies "Like this", smiling and pointing to his construction. {78} Calls A1 by saying "Do it. This way". He is very happy and smiling as he shifts his construction (train) towards A1 to explain her the design. {82} Reaches A1's construction and points at locations and speaks, "It has to be placed there". Smiles as she places it.

This result supports the finding of study conducted by Madsen and Connor [5] showing that younger children (chronological age 6–7 years) with ID were more cooperative than their elder counterparts. The clear increase in interaction among competitors in 2nd session (refer Table 4) might again be attributed to the cooperative behavior.

Increase in physical signs for S1 was because he repeatedly called facilitator and competitor to show that he constructed a train from blocks using wheels and felt happy about his achievement. However, A1 showed a decline in her interest in building in 2nd session and looked more anxious as she was not constantly smiling anymore. This further supports that she was not looking to compete with S1.

4.2 Group 2: 4–6 Years Mental Age

Competition was clearly visible in both the subjects as a remarkable difference was seen in behaviour between the two sessions. A2 was the only subject in the experiment who communicated competition verbally as well as through his body language in 2nd session, as seen in following excerpts.

[1] The numbers in {} represent the event number as transcribed during analysis for ease of reference.

Table 4 Frequency and duration of various parameters as observed during both sessions for all subjects to estimate relative degree of competition

Subject name	S1		A1		A2		J2		A3		S3		S4		A4	
Parameter	Frequency (duration in seconds-s)															
Session	1st	2nd	1st	2nd	1st	2nd	1st	2nd	1st	2nd	1st	2nd	1st	2nd	1st	2nd
Interaction with competitor																
Looked	1 (1s)	0	1(4s)	3(32s)	6(11s)	6(8s)	3(3s)	6(6s)	8(8s)	8(9s)	6(10s)	9(11s)	2(6s)	0	1(1s)	1(1s)
Smiled	2(13s)	3(10s)	2(5s)	3(17s)	0	0	1(1s)	0	0	0	0	0	0	0	0	0
Imitated	0	0	7(64s)	5(141s)	0	0	0	0	1(37s)	1(56s)	0	0	0	0	0	0
Verbal communication	0	4	2	2	0	4	0	1	0	0	0	0	0	0	0	0
Cooperated	0	3(36s)	0	0	0	0	0	0	0	0	0	0	0	0	0	0
Interaction with material																
Used as needed	Each time	Each time	Each time	Each time	Each time	Each time	Each time	15 (179s)	Each time	Each time	Each time	Each time	Each time	Each time	Each time	Each time
Grabbed	0	0	0	0	0	0	0	5(25s)	0	0	0	0	0	0	0	0
Given	0	0	0	0	0	0	0	0	0	0	0	0	0	0	0	0
Played without building	0	0	0	0	0	0	0	0	0	0	0	0	0	0	0	0
Interest in play-activity																
Distracted	3(14s)	5(12s)	10(20s)	9(38s)	6(21s)	6(12s)	7(20s)	8(8s)	4(4s)	3(4s)	7(13s)	10(12s)	4(63s)	3(167s)	3(4s)	0
Physical signs	2	7	5	2	1	3	0	3	0	0	1	1	0	1	0	1

{274} Raises both his hands, smiles and exclaims, "I will win". {276} Seems excited. Sits upright smiling and claps couple of times raising his shoulders and says, "Yes". {277} Stops smiling. Gives a side glance to ***J2***. {288} Exclaims loudly, "I will win it" while putting blocks. {319} "Mine is better. 11 floors".

J2 showed competitiveness though her interaction with material as seen in grabbing and body language including instances of babbling, tightening of muscles and restlessness, seen as follows.

{400} Rises up and picks a pink block from other side, although it was available near her. {401} Turns and looks at facilitator. {402} Again rises up to pick up a yellow block from near A2 and babbles as she places it over. {403} Rises up and picks a set of block from other side. {404} Gives a quick side-glance to *A2*. {405} Holding the previous block in one hand, rises and picks up a yellow block from very near *A2*. {408} Puts a block on construction moving restlessly. {412} Beats a block on construction for the first time to fit it over.

Both subjects showed a clear increase in interest towards play-activity in 2nd session as distractions got reduced (refer Table 4).

4.3 Group 3: 6–8 Years Mental Age

Subjects didn't show competition as apparently as group 2 showing similar behaviour in both sessions but there were some signs of competitive spirit (refer Table 4). However, A3 was seen using side-glance while imitating S3, more often in 2nd session. A3 also showed disappointment on his face when asked by facilitator if he had finished at the end of 2nd session, which might be due to dissatisfaction, shown in the following excerpt.

{508} Looks at facilitator and tightens his teeth, drops forward and looks down.

While A3 imitated, S3 was seen showing displeasure when she noticed being imitated in both sessions, as shown below.

{535} Looks at A3's construction. Scratches her head and twists her mouth side-ways. {572} Looks at A3 and his construction as he is imitating. Twists her mouth sideways. Gets distracted.

While these instances do not clearly establish presence of competition, further experimentation in this age-group should give a clearer picture.

4.4 Group 4: 8–10 Years Mental Age

S4 was largely, unable to build, as later reported to be generally lethargic by teachers. It would be unfair to derive conclusive insights from this group, however some interesting patterns emerged. Though S4 was seen disinterested almost throughout the experiment, it became much more prominent in 2nd session as his duration of being distracted increased almost three times (refer Table 4).

On the other hand, A4 displayed good building ability and was not at all a match to S4. She also didn't show any sign of competition and hardly any interaction with S4. There was however a period at the start of 2nd session when she was seen struggling with building activity and shivered, which appears to be an outcome of nervousness, as seen below. A1 has also been found to show anxiety due to competition.

{724} Tries putting a block over other but fails 4 times as her hands are shaking.

5 Discussion

Hypothesis 1 showed that competitive play in children with moderate ID emerges at the same age (around 4 years) as seen in typical children when matched on mental age, however hypothesis 2 was nullified and the results didn't match the progressive pattern in degree of competitive play shown by typical children [1]. In fact, results suggest that competitive play decreased with growing mental age. It can be argued by looking at the sociological perspective of competition which emphasizes that it is the social and cultural environment in which an individual is brought up that determines the character of this competitive urge [8, 9]. Vaughn and Diserens [10] sum this up well by stating, "The particular form, intensity, and objects of competition are largely dependent on the nature of the social environment. They vary considerably among individuals and groups, and seem to be dependent on the degree of socialization which the individual and the group have achieved." Children with intellectual disability are subjected to a very different social environment and exposure as they often fail to compete with their typical peers. This vulnerability in social and peer acceptance, and failures in peer acceptance substantiate the negative self-concept among special children [11]. This could have led the children to avoid a competitive play situation as their age increased.

Presence of significant amount of social associative play in children with ID has been identified by Johry and Poovaiah [12] in their ethnographic study at the same facility. In such situations, competitive strategies should be replaced with cooperative strategies during play activities and pedagogy interventions. Present study reported emergence of anxiety due to competition in two children. Using cooperative learning techniques has been linked with reduction in tension and conflict among typical children [13], which becomes more significant considering the existing low self-esteem in these children [11]. Focusing on group goals and cooperatively co-constructing solutions also aids in the moral development of the children [14]. While the study did not analyze the performance of children, it is fairly evident that performance varied significantly within groups. The design of play activities and toys involving challenge can do away with having a standard target to compete against and allow facilitators to set individual specific goals which can stretch children beyond their abilities while still being achievable.

Furthermore, design should involve incremental outcomes, so the child can self-regulate his play and set his own target. Block play is a good example of a toy with such flexibility. The design of toys should also be adaptable and flexible so that it can address wider audience by being applicable at different, interrelated stages of development, from the instructor giving maximum help, to the user operating the toy unaided [15]. The present study controlled facilitation where the child was observed but not led into play. Such a balanced facilitation will provide the opportunities for free play found lacking with these children [12]. Exposure to social open and permissive conditions will also benefit in creating a positive self-concept among children thus reducing their avoidance tendencies.

The present study didn't address a number of factors like effect of gender or socio-economic background of children with ID on competitive play exhibited by them. While the present study used inter-rater reliability, there is still a need to confirm the validity of the findings with increase in the number of participants.

References

1. Greenberg, P.J.: Competition in children: an experimental study. Am. J. Psychol. **44**, 221–248 (1932)
2. Leuba, C.J.: An experimental study of rivalry in young children. J. Comp. Psychol. **16**, 367–378 (1933)
3. Owens, L.L.: Competition in children as a function of age, race, sex and socioeconomic status. Ph.D. thesis, Texas Tech University (1969)
4. Stoneman, Z., Keilman, P.A.: Competition and social stimulation effects on simple motor performance of EMR children. Am. J. Mental Defic. **78**, 98–100 (1973)
5. Madsen, M.C., Connor, C.: Cooperative and competitive behavior of retarded and nonretarded children at two ages. Child Dev. **44**, 174–178 (1973)
6. Guralnick, M.J., Groom, J.M.: The peer relations of mildly delayed and non-handicapped preschool children in mainstreamed playgroups. Child Dev. **58**, 1556–1572 (1987)
7. Priewasser, B., Roessler, J., Perner, J.: Competition as rational action: why young children cannot appreciate competitive games. J. Exp. Child Psychol. **116**(2), 545–559 (2013)
8. Booth, A., Nolen, P.: Choosing to compete: how different are girls and boys? J. Econ. Behav. Organ. **81**(2), 542–555 (2012)
9. Gneezy, U., Leonard, K.L., List, J.A.: Gender differences in competition: evidence from a matrilineal and a patriarchal society. Econometrica **77**(5), 1637–1664 (2009)
10. Vaughn, J., Diserens, C.M.: The experimental psychology of competition. J. Exp. Educ. **7**(1), 76–97 (1938)
11. Pijl, S.J., Frostad, P.: Peer acceptance and self-concept of students with disabilities in regular education. Eur. J. Spec. Needs Educ. **25**, 93–105 (2010)
12. Johry, A., Poovaiah, R.: Paradigms of play in a special school setting in India. In: 19th International Play Association World Conference, Istanbul, May 2014
13. Hertz-Lazarowitz, R., Sharan, S.: Enhancing prosocial behavior through cooperative learning in the classroom. In: Development and Maintenance of Prosocial Behavior, pp. 423–443. Springer, New York (1984)
14. Kruger, A.C.: Peer collaboration: conflict, cooperation, or both? Soc. Dev. **2**(3), 165–182 (1993)
15. Airey, B., Hirst, R.: Toys for Mentally Handicapped People. Leads and District Active Toy Project, Leads, UK (1982)

Redesign and Ergonomic Analysis of Scoop Stretcher for Full Body Immobilization During Casualties

Mohammed Rajik Khan, Preeti Giri and Pavan Kumar Palarapu

Abstract In the context of pre-hospital care, incapacitated persons need to be transferred from places of casualties with proper precautionary measures. Immobilization of body especially spine, is of prime concern during transfer of patients for averting further consequences. The aim of this research work is to redesign the existing scoop stretcher for full body immobilization during casualty transfer. An attempt has been made to provide head, vertebral column and pelvis immobilization and to provide comfort to both the patient as well as the person carrying the stretcher. The proposed design is simulated and analyzed for ergonomic consideration in a virtual CAD environment. The designed scoop stretcher overcomes the loop holes in the existing one in terms of comfort, usability and motion restriction. For physical visualization and design verification, a prototype has been fabricated. The intuitive design may bring a radical change in providing pre-hospital services.

Keywords Pre-hospital care · Scoop stretcher · Full body immobilization · Human factors

1 Introduction

Proper pre-hospital care is the most essential step in saving the patient's life in cases of accidents. This period is very crucial and has to be handled with utmost care. Any negligence with the casualty would result into further injuries or may even cause death of injured. Under such traumatic conditions, patients need to be transferred to the hospital with the proper casualty transfer device and patient's comfort is of prime concern. This can be effectuated by manual lifting methods,

M.R. Khan (✉) · P. Giri · P.K. Palarapu
Department of Industrial Design, National Institute of Technology, Rourkela, India
e-mail: khanmr@nitrkl.ac.in

A. Chakrabarti (ed.), *ICoRD'15 – Research into Design Across Boundaries Volume 1*, Smart Innovation, Systems and Technologies 34, DOI 10.1007/978-81-322-2232-3_36

ambulance stretchers, vacuum mattress, spine boards, scoop stretchers, etc. depending upon the severity of the injuries.

In the area of emergency medical services, work is being done to develop various assistive devices to enhance the patient's safety and comfort. Lopes et al. [1] and Peixoto et al. [2] have devised brain actuated and voice controlled robotic wheelchair. Hsiang et al. [3] have examined the influence of lifting technique on low back pain and potentially related mechanical and physiological factors. Misra and Kalita [4] have discussed the effect of collar therapy in cervical spondylitis. An affordable lifting device to reduce musculo-skeletal injuries has been developed and evaluated by Heacock et al. [5]. Stretchers have been in use since ages with its most simple variant as a sheet of cloth which was used to carry the incapacitated person. With time, stretchers are being improvised so as to increase patient's comfort during transfer and rationalize the standards of emergency services. Various types of stretchers like wheeled stretcher, long spine board, water filled mattresses, foldable stretcher, multipurpose stretcher, etc. are being used for lifting casualty during accidents. Bon and Forrester [6] has performed the ergonomic evaluation on patient handling devices. Cooper and Ghassemieh [7] have used biomechanical failure criteria to assess the risk involved in patient handling with ambulance stretchers. Shelton and Wright [8] have conducted and interpreted interface pressure evaluations of clinical support services. The major loophole in all such stretchers is that, the patient has to be transferred onto the stretcher following the traditional lift and drop technique even if the patient is suffering from traumatic spinal injury. Krell et al. [9] have conducted experiments on scoop stretcher and compared it with long back board for spinal immobilization in trauma patients. They showed that such scoop stretchers can restrict motion to a higher degree during casualty transfer. Even then, complete full body immobilization has not been achieved till date [5–7, 9].

The work aims at proposing a novel design of scoop stretcher for casualty transfer. The new design is a modification of the existing scoop stretcher in terms of scooping the trauma patient, comfort and completely restricting the motion during emergency transfer. The designed stretcher is broadly divided into three main parts: the first scoops the head and the neck portion, the second scoops the trunk portion of the human body and the pelvis portion gets scooped by the third. Inserting any part of the stretcher between the body and the ground won't induce any motion in the other thus ensuring full body immobilization and the patient's comfort during transfer. The desired shape of the stretcher is developed in accordance with the anthropometric data of Indian population [10]. The proposed design is virtually developed and ergonomically verified for 5th, 50th and 95th percentile persons in a CAD environment. A prototype is also fabricated using rapid prototyping technique.

The composition of this research work is further organized as follows. Section 2 presented the design of various segments of scoop stretcher. Implementation of the proposed design in a commercial CAD package is shown in Sect. 3. Section 4 discusses the design verification and ergonomic analysis of the novel scoop stretcher. Finally, conclusions and future recommendations are suggested in the last section.

2 Design of Scoop Stretcher

The scoop stretcher is an casualty transfer device, essential to lift a person from the ground, either due to unconsciousness or in order to maintain stability in trauma, especially spinal injuries. There exists a variety of stretchers in the market; one such is the Ferno scoop stretcher [9] which provides maximum immobility during casualty transfer. In this work, we have re-designed the existing scoop stretcher so as to achieve full body immobilization during emergency transfer.

The geometry of the proposed novel scoop stretcher projected on two-dimensional (2D) orthographic planes is shown in Fig. 1. l_x, l_1, l_2, l_3, l_4 are the length of the scoop stretcher and lengths for accommodating head, trunk, thigh and pelvis parts respectively. w_1, w_2, w_n, w_s and w_p are the width of the stretcher at head end, lower pelvis end, half width of head part, trunk part and lower pelvis part respectively.

The proposed design has been segmented into three parts as per the critical sections of the human body forming:

- Geometry of head and neck part
- Geometry of trunk part
- Geometry of thighs and lower pelvis
- Geometry of supporting structures

Each part of the stretcher is again divided into left and right halves with a small clearance between them so as to avoid pinching of skin, clothes or hair between the two halves while scooping. The left and right halves of the scoop stretcher are joined with the help of a clip joint at both the extreme ends of the stretcher. After attachment, this joint also enables turning movement which helps to apply scissor technique to scoop the body of the patient.

The two halves of the head and neck part scoop the head and the neck portion of the casualty. The two halves of the trunk segment scoop the trunk portion. The trunk and the head part of the stretcher are connected with the help of turning joints which makes the two parts independent of each other's motion. This restricts unwanted motion in the vertebral portion, when head of the injured is being scooped and vice versa. The thighs and lower pelvis parts are integral to

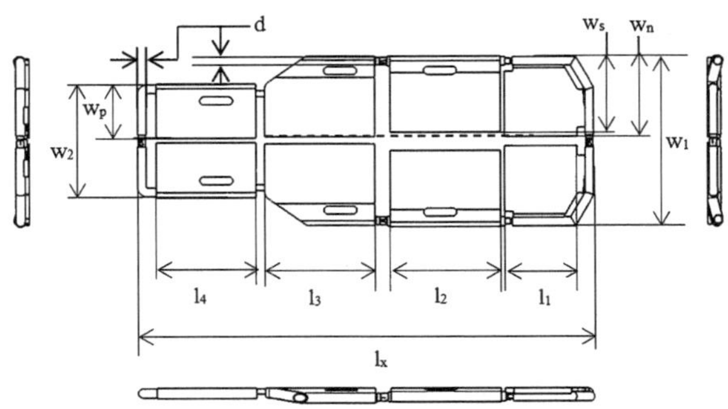

Fig. 1 Two-dimensional projected geometry of a scoop stretcher

each other and act as a single unit while scooping the pelvis portion of the patient. Here, scooping is performed with the turning (clip) joint provided at the bottom-most end of the stretcher. After scooping the left and the right halves of the thighs and pelvis region of the patient, this segment of the stretcher is connected with the trunk one through a detachable joint. The proposed design of the stretcher enables us to increase or decrease its length with the extension feature provided at the lower pelvis part. When not in use, the stretcher can also be detached in two parts and can be kept compact occupying less space. Various sections of the scoop stretcher are supported with the hollow metal rods casted within the frame at the outer region.

2.1 Geometry of Head and Neck Part

Section $1-1$ to $1'-1'$ in Fig. 2 represents the head segment of the scoop stretcher. The geometry of the head part as shown in Fig. 2 is having a tapered cross-section in the longitudinal direction. This enables ease in scooping during casualty transfer. The geometry of the head part has been designed keeping in view the anthropometric dimensions [10] of the 95th percentile male. The boundary manikin approach is followed so that persons of all build can be fitted comfortably. Table 1 shows the standard anthropometric dimensions of the Indian population as mentioned by Chakrabarty [10].

2.2 Geometry of Trunk Part

Section $2-2$ to $2'-2'$ in Fig. 2 represents the trunk segment of the scoop stretcher. While designing the width of the trunk portion, body breadth of 95th percentile is taken into consideration as the stretcher needs to occupy the maximum population [10]. For determining the length of trunk part, the difference between

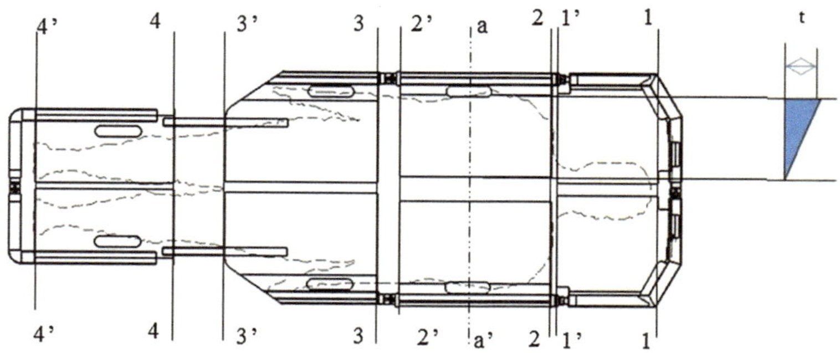

Fig. 2 2D projected geometry of various segments of a scoop stretcher

Table 1 Anthropometric data of Indian population [10]

S. No.	Anthropometric parameter	Percentile of data used (male)	Values (cm)
1	Stature	95	178.1
2	Cervical height	95	151.1
3	Waist height	50	97
4	Mid patella height	5	41.9
5	Maximum body breadth	95	61.9
6	Upper thigh breadth	95	55.4
7	Ear to ear distance	95	19.2
8	Head length, maximum	95	22.6
9	Vertex to chin distance	95	25.7

Source Chakrabarti [10]

waist and cervical height of average population's (50th percentile) is considered so that the trunk and pelvis portions of the patients can be accommodated in the respective segments of the stretcher [10]. The gap between the trunk and the pelvis portion of the stretcher will accommodate the extra trunk length of the 95th percentile person. The left and right portions of the trunk part are separated by a small gap which is same as the maximum width of the lumbar vertebral column of man [11]. While scooping patients with vertebral fractures, this gap restricts the contact between the edges of the flaps and the vertebral column. This may ensure increased comfort of the patient.

The trunk part is attached with the head segment of stretcher with a permanent turning joint allowing rotation between the two about an axis perpendicular to the plane of the stretcher. This helps in scooping the trunk/head section of the body independently and thus provides complete immobilization, when either of the section is scooped.

The cross-sectional geometry of the trunk part (as shown in Fig. 2) is tapered in the longitudinal direction. In order to calculate the thickness 't' of the tapered cross-section, either half of the trunk segment is considered as a cantilever beam. Pressure mapping data for anthropometric dimension of 95th percentile male lying on a hospital bed as worked out by Shelton and Wright [8] is shown in Fig. 3a. Using the pressure distribution (Fig. 3a), the forces acting on various segments of the stretcher can be computed. The load distributions at any cross-section along the trunk half are shown in Fig. 3b.

Thickness 't' of the tapered cross-section of the trunk part (as shown in Fig. 2) has been calculated using the governing equation referred from Timoshenko and Young [12].

$$\begin{aligned} M = L \times w_S + W_x \times x \times (w_S - 0.5 \times x) + W_y \times y \times (w_S - x - 0.5 \times y) \\ + W_z \times z \times (w_S - x - y - 0.5 \times z) \end{aligned} \tag{1}$$

$$\frac{M}{I} = \frac{\sigma}{Y_n} \tag{2}$$

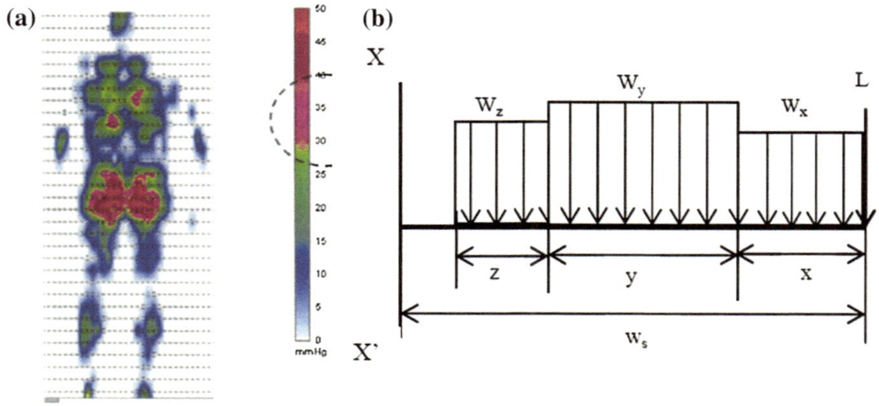

Fig. 3 **a** Full body interface pressure distribution on a hospital bed for anthropometric dimension of 95th percentile male *Source* Shelton and Wright [8]. **b** Load distributions at any cross-section along the spinal half for anthropometric dimension of 95th percentile male

$$t = \sqrt{\left.\begin{matrix}(L \times w_S + W_X \times x \times (w_S - 0.5 \times x) + W_y \times y \times (w_S - x - 0.5 \times y) \\ +W_Z \times z \times (w_S - x - y - 0.5 \times z)) \times 1{,}000 \times 12 \times F.S.\end{matrix}\right/ 2 \times \sigma_{ut}} \tag{3}$$

where, σ, σ_{ut} = tensile strength and ultimate tensile strength of the high density polyethylene material respectively. F.S. = factor of safety, I = area moment of inertia of the cross-section and Y_n = distance between the neutral axis and the edge.

2.3 Geometry of Thighs and Lower Pelvis Part

Section 3−3 to 4′−4′ in Fig. 2 represents the entire pelvis portion of the stretcher. This portion can be completely separated from the rest of the stretcher which promotes compact storage when not in use. It also enables scissor technique to be applied from the bottom most ends. As the patient has been scooped in completely from the pelvis portion, this segment gets attached to the trunk part of the stretcher through a detachable joint. The bottom part of the stretcher is divided into two parts; one of which accommodates thighs and the other to the lower pelvis part. The part bearing knee portion can be elongated so as to accommodate varying heights of patients. The length of the bottom part of the stretcher has been given the dimensions according to the 50th percentile male [10]. The gaps given in between the left and right halves of the pelvis portion of the stretcher enables us to tie the Velcro straps during casualty transfer, so that complete immobilization can be achieved. Table 2 gives the required anthropometric data needed to construct

Table 2 Dimensional parameters of a scoop stretcher

S. No.	Part name	Length (cm)	Breadth (cm)	Half breadth (cm)	Thickness (cm)	External diameter of rod (cm)	Internal diameter of rod (cm)
1	Head and neck	$l_1 = 30$	$w_1 = 64$	$w_n = 30$	$t = 6$	$d = 2.5$	$d_i = 2.1$
2	Trunk	$l_2 = 44$	$w_1 = 64$	$w_s = 29$	$t = 6$	$d = 2.5$	$d_i = 2.1$
3	Pelvis (thigh)	$l_3 = 44$	$w_1 = 64$	$w_n = 30$	$t = 6$	$d = 2.5$	$d_i = 2.1$
4	Pelvis (knee)	$l_4 = 40$	$w_2 = 42$	$w_p = 20$	$t = 6$	$d = 2.5$	$d_i = 2.1$

the pelvis part of the stretcher. The thickness of the stretcher at entire pelvis portion is taken same as the thickness of the trunk segment in order to achieve more robust design.

2.4 *Geometry of Supporting Structures*

The complete frame of the scoop stretcher is integrated with the hollow metal rods at the boundaries for proper strength and support to the structure. Here, uniform load distribution of Wr Newton/cm is considered on the rod which behaves as a simply supported beam. The outer diameter of the rod 'd' is calculated using the governing equations [12].

It's clearly inferred that bending moment is maximum at the mid of the rod that is at a distance of l2/2. So bending moment 'Mr' is calculated as

$$M_r = W_r \times l_2^2 / 8 \tag{4}$$

Using Eq. 2, the diameter 'd' is calculated using the following expression,

$$d = \sqrt{(M_r \times F.S.)/(0.032 \times \pi \times \sigma_r)} \tag{5}$$

where, σ_r = tensile strength of the rod material.

3 Implementation

The proposed design of the scoop stretcher is implemented in CATIA V6 R2013x modeling environment, which is a commercial CAD package. This allows users to create three-dimensional models in a Graphical User Interface (GUI). Based on the boundary manikin approach, Indian anthropometric standards of the 5th, 50th

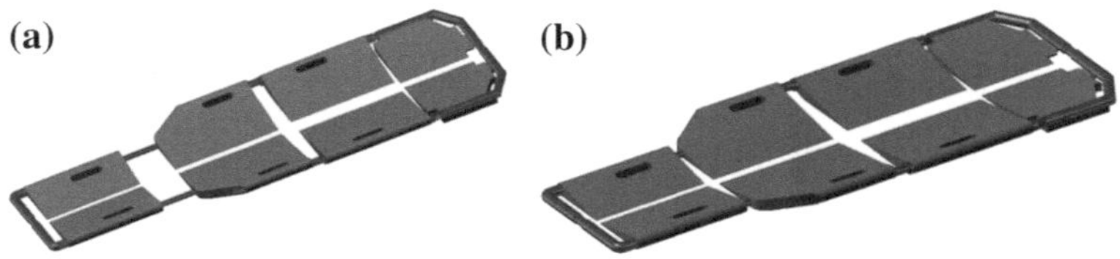

Fig. 4 CAD model of scoop stretcher **a** with elongation. **b** without elongation

and 95th percentile male have been taken. The dimensional parameters needed to completely model the novel scoop stretcher are shown in Table 2. The output is the 3D virtual model of the scoop stretcher in CAD environment as shown in Fig. 4. The scoop stretcher demonstrated in Fig. 4a can accommodate patients ranging from 50th to 95th percentile male height whereas the one shown in Fig. 4b accommodates patients' height ranging from 5th to 50th percentile.

4 Demonstration and Validation

For the purpose of physical visualization and the design verification, the scaled model of the scoop stretcher is manufactured using rapid prototyping (RP) machine. The RP model of the scoop stretcher is shown in Fig. 5a. It is manufactured from Fused Deposition Modeling (FDM) RP technique in Stratasys, Dimension 1200es 3D printer. The RP model is manufactured using Acrylonitrile Butadiene Styrene (ABS-Plus) with soluble support (SST 1200es). The actual model is to be fabricated using high density polyethylene (HDPE) for the various segments of the stretcher and aluminium for rods and joints. Further, the demarcation in terms of operation between the existing and the designed stretcher can be visualized from Fig. 5b–d.

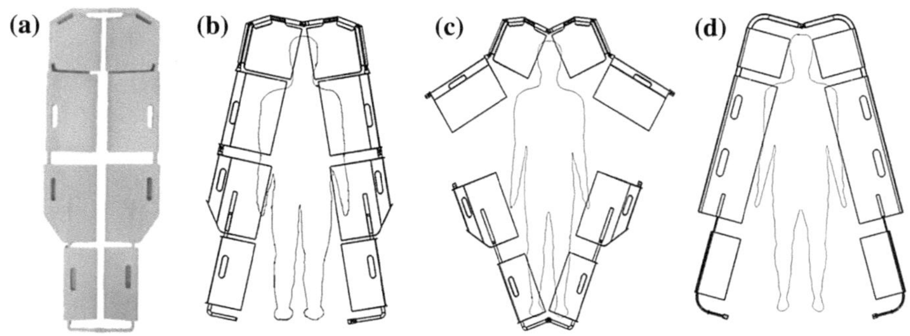

Fig. 5 **a** RP model of the scoop stretcher. **b**, **c** Designed stretcher showing multiple orientations for scooping operation. **d** Existing stretcher in operation

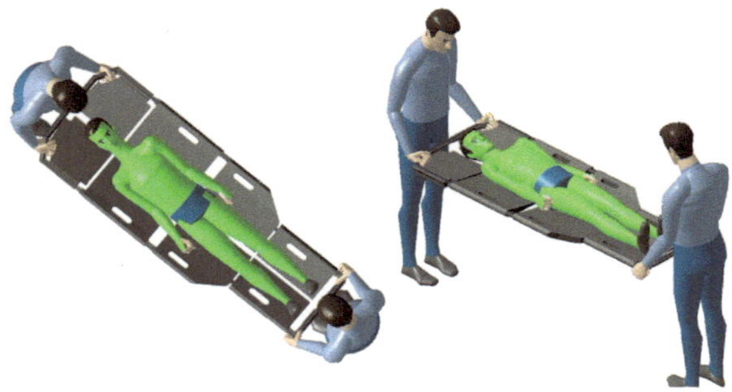

Fig. 6 Scoop stretcher bearing persons with 5th percentile male

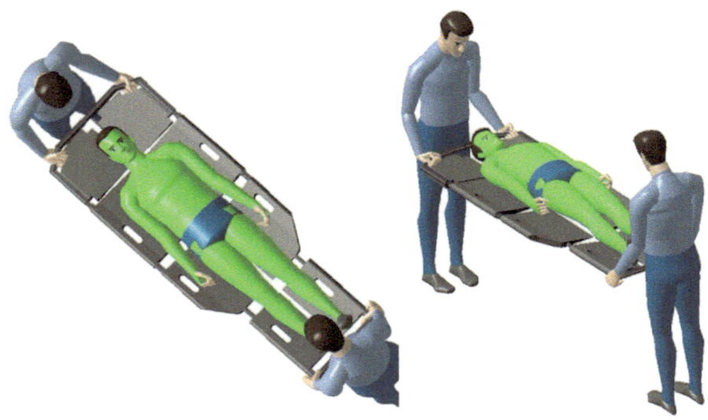

Fig. 7 Scoop stretcher bearing persons with 50th percentile male

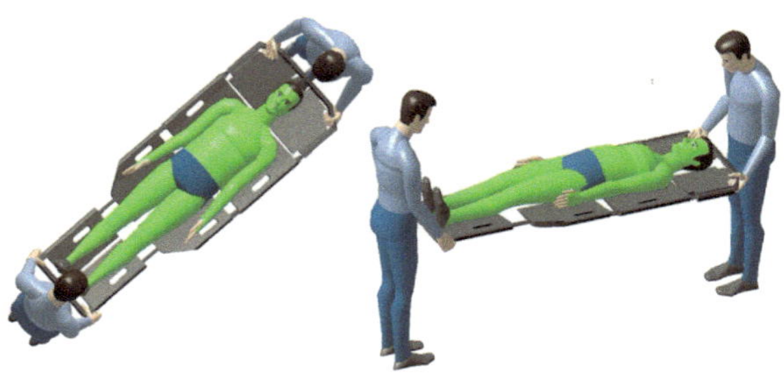

Fig. 8 Scoop stretcher bearing persons with 95th percentile male

Virtual ergonomic analysis has been performed on the 3D CAD model of the scoop stretcher in ergonomic simulation module of CATIA software. As per the boundary manikin approach, the analysis is conducted to verify that various physiques can be accommodated comfortably or not. Figures 6, 7 and 8 shows the lifted scoop stretchers accompanying persons' anatomy ranging from 5th, 50th and 95th percentile male respectively. Results clearly depicts that various sizes of manikin are comfortably accommodated into the stretcher.

5 Conclusions

This work presents re-designing of existing scoop stretcher for casualty transfer during emergency medical services. The proposed design modifications ensure higher degree of full body immobility during emergency transfer of patients, especially persons with traumatic spinal/multiple injuries. The design, development and ergonomic assessment of scoop stretcher done in a virtual CAD environment clearly demarcates it from the existing one in terms of features, restriction of unwanted mobility and patient's comfort. To physically visualize the novel scoop stretcher, a prototype is also fabricated through RP process. In future, real time testing and ergonomic analysis of the designed scoop stretcher can be done, so as to make it a robust one. The proposed design of scoop stretcher may prove a unique lifting/transfer device in emergency medicine to noticeably reduce low back pain, pains during musculoskeletal disorders, other injury claims, etc.

References

1. Lopes, A.C., Pires, G., Nunes, U.: Assisted navigation for a brain-actuated intelligent wheelchair. Robot. Auton. Syst. **61**, 245–258 (2013)
2. Peixoto, N., Nik, H.G., Charkhkar, H.: Voice controlled wheelchairs: fine control by humming. Comput. Meth. Programs Biomed. **112**, 156–165 (2013)
3. Hsiang, S.M., Brogmus, G.E., Courtney, T.K.: Low back pain (LBP) and lifting technique—a review. Int. J. Ind. Ergon. **19**, 59–74 (1997)
4. Misra, U.K., Kalita, J.: Motor evoked potential is useful for monitoring the effect of collar therapy in cervical spondylitic myelopathy. J. Neurol. Sci. **154**, 222–228 (1998)
5. Heacock, H., Seeleya, N.P., Tokunoa, C., Frederkinga, S., Brian, K., Mattieb, J., Kanigana, R., Watzke, J.: Development and evaluation of an affordable lift device to reduce musculoskeletal injuries among home support workers. Appl. Ergon. **35**, 393–399 (2004)
6. Bon, C.L., Forrester, C.: An ergonomic evaluation of a patient handling device: the elevate and transfer vehicle. Appl. Ergon. **28**, 365–374 (1997)
7. Cooper, G., Ghassemieh, E.: Risk assessment of patient handling with ambulance stretcher systems (ramp/winch, easy-loader, tail-lift) using biomechanical failure criteria. Med. Eng. Phys. **29**, 775–787 (2007)
8. Shelton, F., Wright, J.: Conducting and interpreting interface pressure evaluations of clinical support surfaces. Geriatric Nur. **24**, 222–227 (2003)
9. Krell, J.M., McCoy, M.S., Patrick, J.S., Fisher, G.L., Stoy, W.A., Hostler, D.P.: Comparison of ferno scoop stretcher with long back board for spinal immobilization. PreHospital Emerg. Care **10**, 46–51 (2006)
10. Chakrabarti, D.C.: Indian anthropometric dimensions for ergonomic design practice. National Institute of Design (1997)
11. Gocmen-Mas, N., Karabekir, H., Ertekin, T., Edizer, M., Canan, Y., Duyar, I.: Evaluation of lumbar vertebral body and disc: a stereological morphometric study. Int. J. Morphol. **28**(3), 841–847 (2010)
12. Timoshenko, S., Young, D.H.: Engineering Mechanics. Schaum Outline Series, International Second Revised edition (1964)

Design and Ergonomic Evaluation of Multipurpose Student's Bed

P. Pavan Kumar and B.B.V.L. Deepak

Abstract A bed is a piece of furniture used as a place to sleep or relax. Students generally need a wide range of accessories to meet their daily requirements like beds for sleeping, tables for studying, shelves for keeping things etc. which demands a lot of space. So, here a multipurpose bed is designed targeting especially on college students living in hostels. Different customer needs have been identified, existing models have been studied and bed has been ergonomically designed. Concept of ergonomics and its applications in product design along with anthropometric data has been reviewed. The additional features of the bed are attached reading table, shelf for storing books, hanger for hanging clothes, and place for keeping foot wear. Bed and all other components are designed and mechanisms have been developed using CAD software (CATIA V5R17) and simulated. The model has been also evaluated using digital human modeling.

Keywords Multipurpose student bed · CAD · Digital human modeling · Anthropometry

1 Introduction

Beds, tables, chairs and shelves are the basic requirements of students. All these things occupy most of the space of their rooms leaving no further space for their free movement. This motivated to the design a multipurpose bed which satisfies most of the needs and more comfortable, economical than the traditional ones. Products generally achieve overall success in terms of generating sufficient sales revenue, user compatibility and user satisfaction only if ergonomic design

P. Pavan Kumar · B.B.V.L. Deepak (✉)
Department of Industrial Design, National Institute of Technology Rourkela, Rourkela, India
e-mail: bbv@nitrkl.ac.in

A. Chakrabarti (ed.), *ICoRD'15 – Research into Design Across Boundaries Volume 1*,
Smart Innovation, Systems and Technologies 34, DOI 10.1007/978-81-322-2232-3_37

considerations are engineered into it [1, 2]. The purpose of ergonomic design is to improve the performance of systems by improving human machine interaction. This can be done by 'designing-in' a better interface or by 'designing-out' factors in the work environment, in the task or in the organization of work that degrade human–machine performance [3]. Investigations for studying human-machine/ product compatibility from a traditional physical ergonomic perspective, necessarily involved building real physical mockups and subsequently trials with real human beings which is time consuming and quite expensive [4]. Computer aided digital human modeling and simulation technology has emerged as the state of art expertise for human centric ergonomic evaluations and is highly useful because of many associated benefits [5]. Digital human model (DHM) is the computer generated 2D or 3D structure of a human used to represent the complex physical and cognitive aspects of human beings. It can also be considered as a digital representation of the human inserted into a virtual environment to facilitate prediction of safety and performance [6–8]. DHM has many applications in design such as vehicle design [9], vehicle interior design [10], physically-based grasp posture generation [11], automatic head and facial feature extraction based on geometry variations [12] and human body shape modeling [13] etc.

In design of bed, ergonomics is considered in width of the bed, its length and height from the ground, reading table dimensions. Anthropometric data which deals with human body measurements are used in ergonomics to specify the physical dimensions of workspaces, equipment, furniture and clothing to ensure that physical mismatches between the dimensions of equipment, products and the corresponding user dimensions are avoided [14]. The anthropometric variables used in the design of bed are Stature (full body length in erected stretched posture), Maximum body breadth when relaxed, Upper lumbar (height from seat to first lumbar vertebra), Elbow rest (height from seat to lower most point of elbow) and Foot length (distance parallel to the long axis of the foot, from the back of the heel to the tip of the toe).

Researchers are working on various kinds of beds so as to improvise and rationalize the design of beds. They came with many innovative multifunctional beds but cost, comfort and space are still the key challenges. Wang et al. [15] developed a multifunctional student bed including bed board, bed frame, wardrobe, drawer with oblique surface, computer desk, writing board, keyboard supporting frame etc. Similar work has also been done by Yang et al. [16]. Chen [17] developed a bed with writing board, comprising a bed body and a guard bar provided with an eversible writing board, on which a locking pin is fixed. Yezhi et al. [18] designed a simple bed table which comprises carrying rails and a table. Wu et al. [19] designed a combined student bed for military schools which can be combined into a bed and a study table, and belongs to the field of tools used in student dormitories. Xue [20] designed a simple student's bed with more concentration on appearance and structure.

The proposed design consists of additional features like adjustable reading table, shelf, hanger, inclined head board etc. The proposed design is virtually developed and ergonomic evaluation was done in CAD environment.

2 Concept Exploration

This phase consists of:

- *Identification of target users*: College students (Primary target users) and school students (Secondary target users) living in hostels.
- *Identification of need*: A small user survey on students of NIT Rourkela has been done which led to some useful need statements. Statements include comfort in studying, ease of shifting, space for keeping books and hanging clothes and occupying reasonable space etc.
- *Concept development*: Based on the need statements, some rough concepts have been generated and the concept satisfying all the needs has been selected and modelled.
- *Study of Anthropometric data*: Before starting the design, anthropometric variables are taken into consideration as shown in Table 1. This data is collected from [14].

Stature and maximum body breadth of 95th percentile male is considered to accommodate maximum population on the bed. For determining comfortable height of the head board, upper lumbar height of 50th percentile male is considered so that 5th percentile and 95th percentile persons may also accommodate comfortably. If the 5th percentile person's lumbar height is considered, 95th percentile may face problem and if 95th percentile person's lumbar height is considered 5th percentile may face problem. So, average of both (50th) is considered. Similarly other anthropometric variables are taken into consideration in order to accommodate maximum population comfortably.

Considering all these variables, CAD model of bed has been made.

2.1 Assembly of Model

First, the whole bed consisting of book shelf and provision for foot wear as shown in Fig. 1a–c and then reading table and hanger are assembled.

Table 1 Anthropometric data used in the design of bed

S. No.	Anthropometric variable	Percentile of data used	Value of percentile data (cm)	Purpose
1	Stature	95th (male)	175.1	Length of the bed
2	Maximum body breadth when relaxed	95th (male)	61.9	Width of the bed
3	Upper lumbar	50th (male)	30	Height of head board
4	Elbow rest	95th (male)	27	Maximum height the table can be reached above the bed
5	Foot length	95th (male)	27.4	Height of hanger from the bed

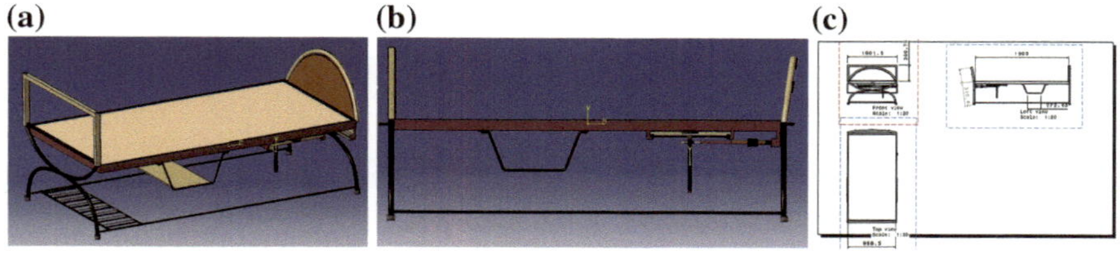

Fig..1 **a** Isometric view of bed, **b** side view of bed, **c** 2D model of bed

2.1.1 Assembly of Reading Table

A provision is given beneath the bed to adjust the height of the reading table. Its mechanism can be explained as follows:

- Worm gear mechanism is used here. The knob beneath the reading table is connected to the worm rod which can be rotated. The rotational motion of worm rod along its longitudinal axis can be transferred to the gear in conjugation.
- The rotation of gear causes vertical motion of a screw that runs through its axis.
- A provision is given at the top of the screw where the protruded part of table gets inserted and the table slides. With the help of prismatic and revolute joint given here, table can rotate and slide in the horizontal plane.
- Ball bearings are also added between the screw and rotating bar (at the top of the screw) so that when one part is stationary, other rotates and also to reduce frictional force.
- As the table is drawn out from the bed, the whole worm gear mechanism also comes out with it. Thereafter, on rotating the knob, table is lifted up. The table assembly is shown in Fig. 2c.

2.1.2 Assembly of Hanger

Two holes are made in the hanger and these holes are inserted into the slot present in the bed through which revolute joint is created. Whenever necessary, the hanger can be drawn out from the bed and two provisions can be obtained for hanging, one is hanger itself and the other is slot for hanger as shown in Fig. 3a–c.

2.2 Merits

- It consists of reading table which is adjustable. Height of the table can be adjusted due to worm gear mechanism and it can be rotated in horizontal plane due to revolute joint.

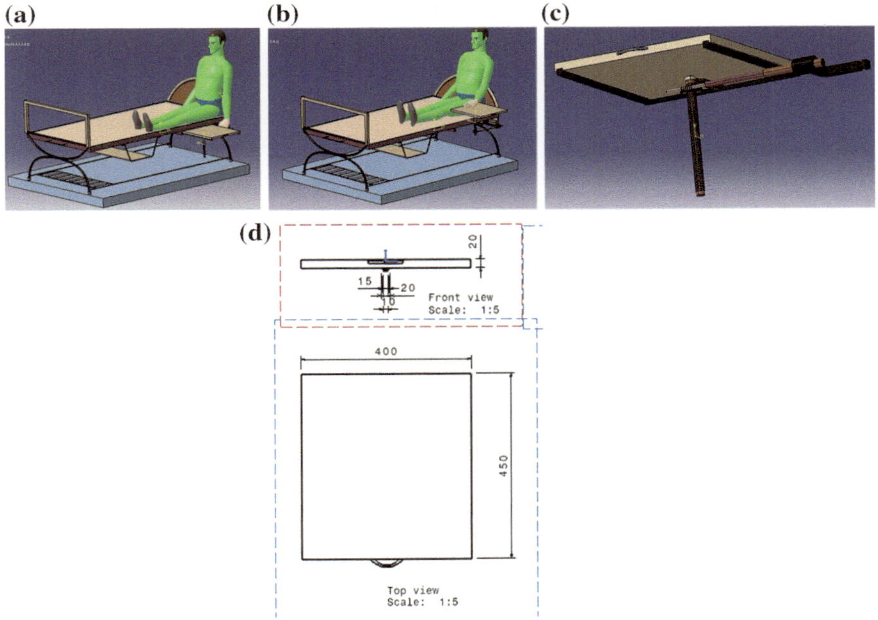

Fig..2 **a**, **b** Person using reading table. **c**, **d** Assembled table and its dimensions

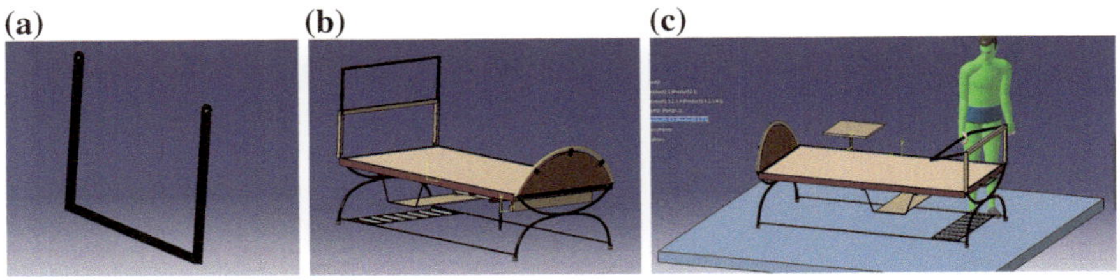

Fig. 3 Representation of **a** hanger, **b** bed with hanger and **(c)** usage of hanger

- It has a provision for storage of books and laptops.
- It has a provision for keeping footwear.
- The back rest of bed has been given an angle of 8°(app.) so that it is comfortable.
- It has provision for hanging clothes.
- Holding rods are provided for easy shifting of the bed.
- The space under the bed can be utilized for keeping bags and suitcases.
- The floor beneath the bed can be cleaned easily.

2.3 Comparison with Some Existing Beds

S. No.	Reference No.	Limitations of existing designs	Limitations overcame by proposed design
1	[21]	Though this design has features like inclined head board, under storage and back shelves, this can't be used for reading or studying by students	Proposed design consists of attached reading table which is adjustable. Also consists of hanger for hanging clothes
2	[22]	Consists of fixed reading table and has no shelf or storage space	Consists of adjustable reading table and storage space
3	[23]	These beds (murphy beds) are specially used by students living in dorm rooms. They can be folded for saving space but doesn't have head boards, foot boards or bed rails. Folding a big bed every day can be a tiresome job and these are generally costly	Already occupies less space and has many additional features
4	[24]	Though completely and easily foldable, it can't accommodate reading table, shelf and other features	It consists of reading table, shelf, and hanger

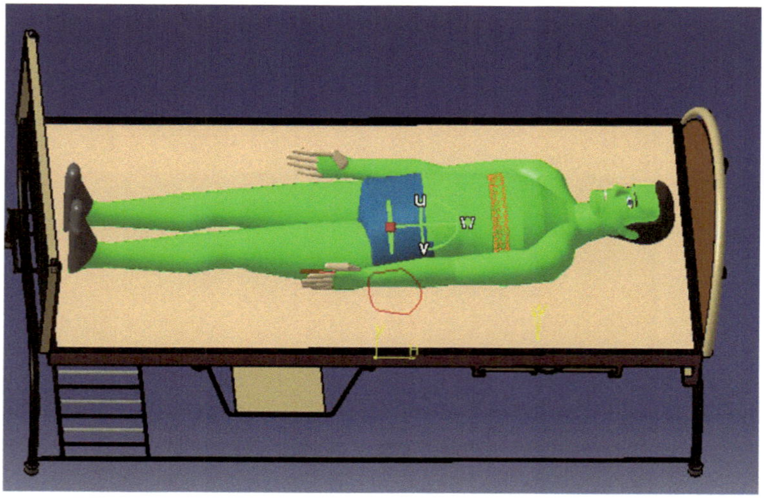

Fig. 4 95th percentile male manikin sleeping on bed

3 Ergonomic Evaluation Using Digital Human Modeling

After preparing the model of bed, ergonomic evaluation is done using digital human modelling. For this purpose, manikins are prepared according to Indian population dimensions. A 95th percentile male manikin is taken and made him to sleep on the bed as shown in Fig. 4.

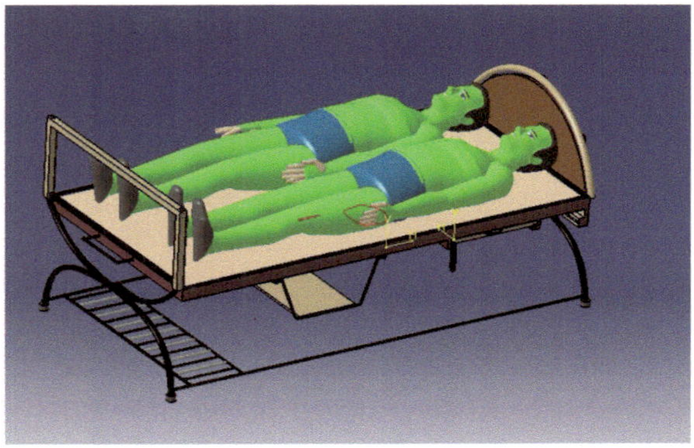

Fig. 5 Two 95th percentile male manikins sleeping

Table 2 Joint angle data of manikin in sleeping posture

Segment	Degree of freedom (DOF)	Right side comfort angles (in deg.)	Left side comfort angles (in deg.)
Arm	1 (flexion/extension)	0 (flexion)	0 (flexion)
	2 (abduction/adduction)	0 (abduction)	0 (abduction)
	3 (medial/lateral rotation)	0 (lateral rotation)	0 (lateral rotation)
Clavicular	1 (flexion/extension)	0 (extension)	0 (extension)
	2 (elevation/depression)	0 (elevation)	0 (elevation)
Forearm	1 (flexion/extension)	0 (flexion)	0 (flexion)
	2 (pronation/supination)	160 (pronation)	160 (pronation)
Head	1 (flexion/extension)	0 (flexion)	
	2 (lateral left/right)	0 (lateral left)	
Leg	1 (flexion/extension)	0 (Flexion)	0 (Flexion)
	3 (medial/lateral rotation)	0 (medial rotation)	0 (medial rotation)
Lumbar	1 (flexion/extension)	−8.297 (extension)	
Thigh	1 (flexion/extension)	7.03 (flexion)	7.03 (flexion)
	2 (medial/lateral rotation)	0 (abduction)	0 (abduction)
Thoracic	1 (flexion/extension)	−4.269 (extension)	
Hand	1 (flexion/extension)	−20.39 (extension)	−11.52 (extension)
	2 (radial/ulnar deviation)	20 (radial deviation)	20 (radial deviation)

Here, it is clear that the bed can accommodate the whole length of the person.

There may be cases in hostels where due to the lack of availability of beds, two persons are provided with only single bed. So, bed is designed for two 95th percentile persons taking their body widths into consideration. So, two 95th percentile male manikins are considered for ergonomic evaluation as shown in Fig. 5.

Here, it is clear that two persons can also fit in the bed and the angles of manikins are shown in Table 2.

Next, 95th percentile manikin is again taken and some comfort ranges are given to segments for evaluating his reading posture as shown in Table 3. Green color is given

Table 3 Comfort ranges of manikin in reading posture

Segment	Degree of freedom (DOF)	Comfort range (in deg.)
Arm	1 (flexion/extension)	−15 to 35 [25]
	2 (abduction/adduction)	0 to 30 [25]
	3 (medial/lateral rotation)	−27.758 to 45 [25]
Clavicular	1 (flexion/extension)	−3 to 5 [25]
Forearm	1 (flexion/extension)	15 to 100 [25]
Lumbar	1 (flexion/extension)	−10 to 25 [26]
Thoracic	1 (flexion/extension)	−10 to 25 [26]
Head	1 (flexion/extension)	−5 to 20 [26]
	2 (lateral left/right)	−20 to 20 [26]

(a) **(b)**

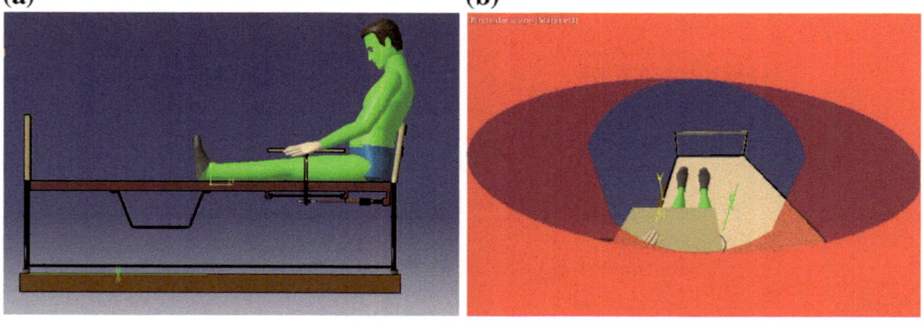

Fig. 6 **a** 95th percentile in reading posture and **b** eye reach in reading posture

Table 4 Joint angle data in reading posture of 95th percentile male manikin

Segment	DOF	Right side comfort angles (in deg.)	Left side comfort angles (in deg.)
Arm	1 (flexion/extension)	0 (flexion)	22.708 (flexion)
	2 (abduction/adduction)	7.323 (abduction)	7.323 (abduction)
	3 (medial/lateral rotation)	9 (medial rotation)	9 (medial rotation)
Clavicular	1 (flexion/extension)	0 (extension)	0 (extension)
	2 (elevation/depression)	0 (elevation)	0 (elevation)
Forearm	1 (flexion/extension)	28.25 (flexion)	47.194 (flexion)
	2 (pronation/supination)	160 (pronation)	160 (pronation)
Head	1 (flexion/extension)	17.792 (flexion)	
	2 (lateral left/right)	0 (lateral left)	
Lumbar	1 (flexion/extension)	4.83 (flexion)	
Thoracic	1 (flexion/extension)	1.089 (flexion)	
Hand	1 (flexion/extension)	−20.39 (extension)	−11.523 (extension)
	2 (lateral left/right)	20 (radial deviation)	20 (radial deviation)

to the manikin in the given comfort range and red color is given to range outside the comfort and manikin is made to sit in reading posture as shown in Fig. 6.

Here, total manikin color is green which indicates that he is comfortable and it is clear that the maximum height reached by table is above the thigh height and it is also in comfortable height when a mattress of about 3 cm is used and table is in the eye reach and the angles in reading posture are shown in Table 4. The angles are within the range of the standards and hence it is comfortable.

4 Conclusion

This work presents design of multi-purpose student's bed which ensures less space usage. The proposed model is compared with some existing ones and it is different from them in terms of features, space usage and user's comfort. The model is developed in virtual CAD environment and ergonomic evaluation is also done using digital human modeling. In the future, the work can be extended for structural analysis so as to make it a robust one.

References

1. Hanson, L., Högberg, D., Nåbo, A.: Digital human modeling in automotive product applications. In: Duffy, V.G. (ed.) Handbook of Digital Human Modeling, pp. 40.1–40.16. Taylor and Francis, London (2009)
2. IEA: Definition of ergonomics. Available at: http://www.iea.cc/01_what/What%20is%20Ergonomics.html (2000). Accessed 15 June 2013
3. Bridger, R.S.: Introduction to Ergonomics, 3rd edn, Chap. 1 (Introduction). McGraw-Hill, Inc., New York City (1995)
4. Karmakar, S., Majumdar, D., Pal, M.S., Majumdar, D.: Ergonomic study of ingress-egress of an army vehicle in simulated environment. In: Proceedings of International Conference on Trends in Product Lifecycle, Modeling, Simulation and Synthesis (PLMSS), Bangalore, India, pp. 69–74 (2011)
5. Karmakar, S., Pal, M.S., Majumdar, D., Majumdar, D.: Application of digital human modeling and simulation for vision analysis of pilots in a jet aircraft: a case study. Work **41**, 3412–3418 (2012)
6. Demirel, H.O., Dufy, V.G.: Applications of human digital modeling in industry. In: Duffy, V.G. (ed.) Digital Human Modeling, pp. 824–832, LCNS 4561. Springer, Berlin (2007)
7. Sanjog, J., Karmakar, S., Agarwal, H., Patil, C.D.: Designing and ergonomic evaluation of a shoe-rack in CAD environment. Int. J. Comput. Appl. (0975–8887) **49**(20), 38–41 (2012)
8. Chaffin, D.B.: Invited plenary paper: improving digital human modeling for proactive ergonomics in design. Ergonomics **48**(5), 478–491 (2005)
9. Meulen, P.V.D., Seidl, A.: Ramsis—the leading cad tool for ergonomic analysis of vehicles. In: Duffy, V.G. (ed.) Digital Human Modeling, pp. 1008–1017. HCII, LNCS 4561. Springer, Berlin (2007)
10. Yang, J., Jim, J.H., Abdel-Malek, K., Marler, T., Beck, S., Kopp, G.R.: A new digital human environment and assessment of vehicle interior design. Comput. Aided Des. **39**, 548–558 (2007)

11. Endo, Y., Miyata, N., Kouchi, M., Mochimaru, M., Kanai, S.: Physically-based grasp posture generation for virtual ergonomic assessment using digital hand. In: Duffy, V.G. (ed.) Advances in Applied Digital Human Modeling, pp. 166–175. Taylor & Francis, CRC press, New York (2011)
12. Fang, S.Y., Fang, J.J.: Automatic head and facial feature extraction based on geometry variations. Comput. Aided Des. **43**, 1729–1739 (2011)
13. Baek, S.Y., Lee, K.: Parametric human body shape modeling framework for human-centered product design. Comput. Aided Des. **44**, 56–67 (2012)
14. Chakrabarti, D.: Indian Anthropometric Dimensions. National Institute of Design. Chapter: Statistical treatment of anthropometric data (1997)
15. Wan, M.Y., Wang, D., Wang, S.: Multi-function bed for student. Patent no. CN 1483371 A, 24 March 2004
16. Wan, M.Y., Yang, X.L., Lung: Multifunction bed for school students. Patent no. CN2666279 Y, 29 Dec 2004
17. Chen, Q.: Bed with writing board. Patent no. CN2341481 Y, 6 Oct 1999
18. Qiao, D., Yezhi, B., Li, Z., Gan, Y., Caisheng, Z., Zhao, X., Zhao, P., Zou, K.: Simple bed table. Patent no. CN102342663 A, 8 Feb 2012
19. Wu, X.L., Linbao, F., Yuan, X.: Combined student bed for military schools. Patent no. CN201727134 U, 2 Feb 2010
20. Xue, S.: Student bed. Patent no. CN301630357 S, 3 Aug 2011
21. http://www.godrejinterio.com/godrej/Godrejinterio/products.aspx?id=29&menuid=310&catid=41&subcatid=43&sec=det&prodid=57. Accessed 21 Oct 2013
22. http://www.tradeindia.com/fp221545/HOSTEL-BED.html. Accessed 21 Oct 2013
23. http://www.furniturefashion.com/10_cool_murphy_beds_for_decorating_smaller_rooms/. Accessed 21 Oct 2013
24. http://www.sz-wholesaler.com/p/1304/1334-2/metal-folding-bed-hl-002-710184.html. Accessed 21 Oct 2013
25. Henry Dreyfuss Associates: The Measure of Man and Woman, Human Factors in Design. Wiley, New York (2002)
26. Macleod, D.: The Office Ergonomics Kit. CRC Press Publishers, Boca Raton (1999)

Ergonomic Risk Assessment on Women Handloom Weavers in Assam with the Introduction of Jacquard

Sangeeta Pandit and Debkumar Chakrabarti

Abstract Weaving was cultural and traditional activity, for every woman of Assam, which has now been transferred to commercialization. Earlier, weavers spend their leisure time in weaving as passion or for domestic needs. Now with commercialization, both time and competition plays an important role, giving rise to a new set of occupational issues. Introduction of Jacquard is the main difference between domestic and commercial frame-loom structure. The present study was conducted to find, effect of Jacquard on pedalling force with work-shift duration for commercial production. Weavers participated were from the Kamrup district of Assam, known for commercial production. Two way ANOVA was performed to see effect of two independent variables (pedal force and work shift) on a single dependent variable (lower limb pain). Significant increase of lower limb pain (LLP) was observed for both pedalling force ($p < 0.001$) and work shift duration ($p < 0.001$), but no relation was found between increased pedalling force and work-shift with LLP ($p = 0.658$).

Keywords Women weavers · Handloom industry of assam · Jacquard · Lower limb pain

1 Introduction

Handloom is the ancient cottage industry of India and second largest industry after agriculture [1]. India produces 85 % handloom products of the world. Other countries having handloom industries are Sri Lanka, Nepal, Bangladesh, Norway,

---

S. Pandit (✉)
Visiting Faculty, Design Discipline, IIITDM Jabalpur, Jabalpur 482005, Madhya Pradesh, India

D. Chakrabarti
Department of Design, IIT Guwahati, Guwahati 781039, Assam, India

A. Chakrabarti (ed.), *ICoRD'15 – Research into Design Across Boundaries Volume 1*, Smart Innovation, Systems and Technologies 34, DOI 10.1007/978-81-322-2232-3_38

West Indies, Indonesia, Afghanistan, Pakistan, Iran etc. [2]. The vast population of the country, directly or indirectly depends upon handloom industry for their livelihood. Handloom forms part of the heritage of India and exemplifies affluence and diversity of our country and artistic quality of the weavers. North East part of India has the highest number of handlooms and among North Eastern States, Assam consists maximum handlooms where 99 % weavers are women [3]. Almost every household in rural areas of Assam are connected with weaving. It is the oldest and largest industry of the state and major contributor to the state revenue [4].

Ergonomic study in textile industry of India is very rare and particularly with reference to women workforce [5]. Literature reviews in handloom industries in different parts of India proved the existence of occupational health problems among weavers. Study on male handloom weavers of West Bengal, were found to suffer from repetitive strain injuries of upper limbs due to the nature of the repetitive weaving task [6]. Another study at Ahmedabad on handloom weavers was found to suffer from back and knee pain. The authors of this study have mentioned poor workstation design aggravates the effect of musculoskeletal disorders among the weavers [7]. Study on handloom weavers of Uttarakhand was found to suffer from lower limbs and back pain due to improper seat design and fixed posture [8]. Studies on handloom weavers of Assam have also revealed, high prevalence of ergonomic risk factors in weaving workstation causes development of musculoskeletal disorders (MSDs) among women weavers [9]. In another study in Assam it was found ergonomic intervention in seat modification following ergonomic principles, was found to be highly effective by handloom weavers [10]. The attainments of awkward posture to undergo weaving have emerged in the development of muscular pain and discomfort reported by women weavers of Assam [11].

Previously, weaving was a traditional household activity for Assamese women. Now in the context of various limitations to the development of organized industries in this area, this household practice is gradually transferring to commercial industry by Government initiatives. Many small handloom units have now taken weaving at the commercial level under different co-operative organizations. This results in more women weavers getting associated with weaving as part time or full time profession. Activities that women performed for their domestic needs during leisure time, has now been transformed into professional work Fig. 1. Duration of loom used for commercial weaving is quite long, even extending beyond 8 h of work schedule [12]. Handloom workstation for commercial purposes are different from domestic requirements. In this paper, a comprehensive ergonomics study was carried out in the existing commercial handlooms to find out major risk factors associated with the development of musculoskeletal problems.

(a) (b)

Fig. 1 Weaving in **a** domestic loom without Jacquard and **b** commercial loom with Jacquard

2 Materials and Methods

The present study was carried out in three phases.

2.1 Phase 1

Kamrup district of Assam was selected purposively for its contribution towards commercial production [13]. From Kamrup 118 handloom clusters were identified from both organized and unorganized sectors. Discussing with administrative bodies of both sectors, number of weavers working were identified and from them 150 women weavers were sampled randomly. Handloom clusters were visited and meetings were conducted with loom owners, managers and weavers in order to identify the problems of commercial work practice. Subjective assessment was carried out on handloom weavers to quantify discomfort or pains in different body parts with the help of Body Map [14] Fig. 2 and Standard Nordic questionnaire [15] along with frequency and severity of pain assessment. Weaving consists of different sub-tasks, weaving sub-tasks were analyzed and weavers were asked to identify body regions, with the help of the body map responsible for pain in executing a particular sub-task and asked to rate in the Borg's CR-10 Pain scale [16].

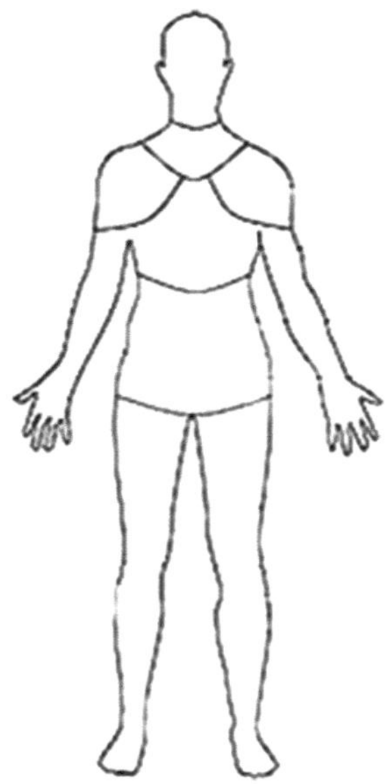

Fig. 2 Body map

2.2 Phase 2

Direct observations of weaving workstation and video recording of weaving activity was further analyzed to verify the findings of subjective assessments which helped to further investigate the problems in details. Video recording of weaving activity was conducted in different frame looms and analyzed in slow motion in the laboratory. The purpose of weaving task analysis is to make a step-by-step investigation, to find out the causative factors for the development of muscle pains in those body regions, which the subjects reported while, subjective pain assessment.

2.3 Phase 3

Subjective assessments and direct observations from field study was further experimentally tested and data analysis was performed using SPSS statistical software, version 20.0. The responses of the weavers on commercial work practices were compared by ANOVA.

3 Results

3.1 Phase 1

Region wise mapping of pain revealed that 95 % weavers reported pain in the neck and back. Of the upper limbs, 78 % reported pain in the shoulder, 85 % in the upper and lower arms and 77 % in the wrist area. For lower limbs, 89 % weavers reported pain in the ankle and foot, 83 % in the knees and 87 % in the thighs shown in the Fig. 3.

From the frequency of pain reporting it was found neck, back, hand, ankle and foot were the most affected body areas where more than 80 % weavers reported of having pain "always" followed by the shoulder, wrist and knees where 70 % and more reported of having "always" pain Fig. 4. To further investigate, the body was categorized into four segments—neck, back, upper limbs and lower limbs which helped to understand which segment of the body was mostly affected. Reporting from weavers revealed that maximum weavers were having acute pain in the back and lower limbs for handloom weaving Fig. 5.

By analyzing weaving sub-task Table 1, it was observed that, back muscle has the highest mean pain rating of 8 followed by lower limb and neck, having a pain rating of 7. The second highest pain rating was observed for both neck and lower limbs, but noticeable reduction in the range was observed in case of neck pain. From observation, it was clear that the highest pain rating of the back muscle was due to constant flexion of the trunk for inspection of thread breakage and for performing beating. Incidence of neck pain was due to neck flexion for inspection of thread breakage. Pain in the lower limb was due to both pedalling and poor seat design [5]. In the right hand, shoulder and wrist, pain rating was found to be 6, which was due to repetitive forceful exertion to carry picking operation. Roughness of the sley race with age of the loom and improper loom maintenance increases the frictional force which resulted more force for picking operation. Pain rating for the left hand and wrist was found to be 5, this was due to the repetitive

Fig. 3 Percentage of pain in different body parts

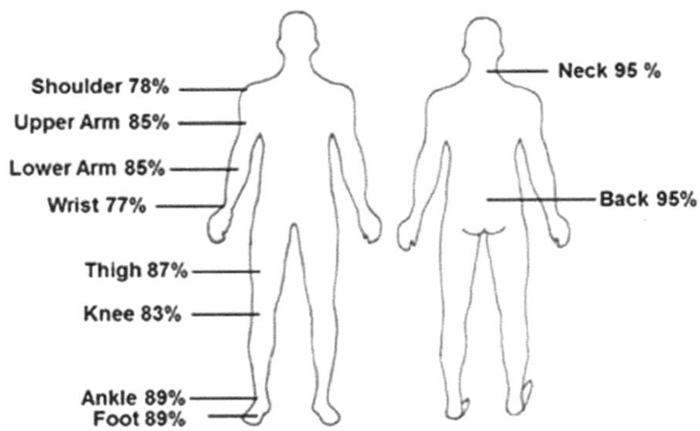

Fig. 4 Frequency of muscular pain

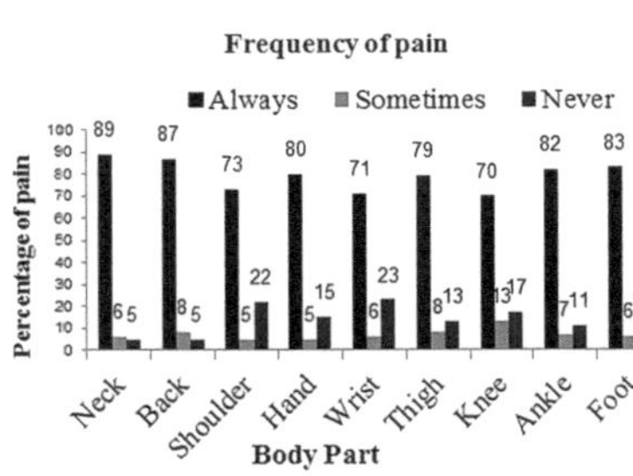

Fig. 5 Severity of muscular pain

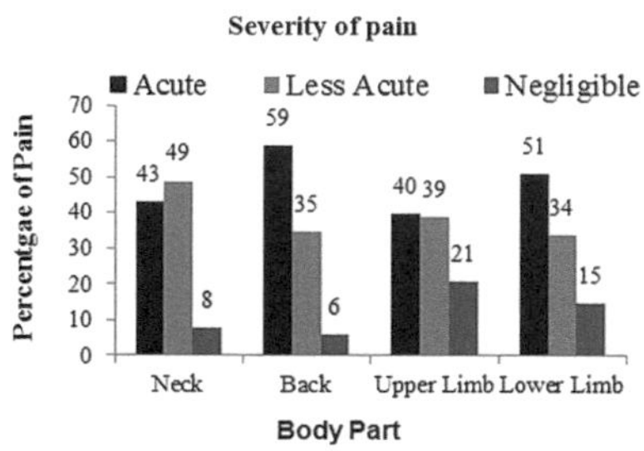

Table 1 Analysis of pain rating with CR-10 scale for performing different weaving sub-task

Sub-task	Body region	Rating(mean) (range)
Picking	Rt shoulder	6 (4–8)
	Rt hand	6 (4–8)
	Wrist (Rt)	6 (3–7)
Beating	Lt shoulder	4 (2–6)
	Lt hand	5 (3–7)
	Wrist (Rt)	5 (3–6)
	Back	8 (5–9)
Inspection task by seating	Neck	7 (3–7)
	Back	8 (5–9)
Pedalling	Lower limb	7 (4–9)

oscillatory movement of sley frame. The movement was sudden, quick with considerable force for a proper beating of the weft thread to give rise to the fell of the cloth.

Subjective assessments and weaving sub-task analysis revealed that back and lower limbs were the most affected body segments reported by the weavers undergoing weaving. Causative factors for highest pain rating were further studied with the help of video-event analysis of weaving activity.

3.2 Phase 2

An important finding of an observational study of ergonomic assessments of commercial weaving workstation surfaced out, that Jacquard is the main difference in loom structure between domestic and commercial horizontal frame loom of Assam Fig. 1. Initiatives from different organizations to increase production rate, Jacquards are provided to the weavers at a very low interest rate by government organizations as well as train weavers with its operational use.

Jacquard for producing design, controls, warp threads separately by means of harness cord/yarn. In order to form shed on warp threads of loom with jacquard, application of a large force is required on the pedals against the reverse action of a

Fig. 6 Jacquard frame loom. **a** Top view of the loom with jacquard and harness yarn. **b** View of the pedal connected with heald yarn and harness yarn

harness cord of Jacquard and heald yarns of healds Fig. 6. The operation is highly repetitive, forceful and carried throughout the working hours. This downward pushing force of the pedal increases consequently with:

1. Width of the cloth: Cloth width is directly related to the number of heald yarns/ thread. With an increased cloth width, number of warp threads passing through the "eye" of the heald yarns rises. This enhances tension on the warp threads against which more force is required on pedals for shedding.
2. No of motifs: With increased number and size of motifs amount of harness thread of Jacquard increases. Harness threads are connected with "ends" (a single thread of the warp) based on schedule patterns of the design cards. Harness yarns keep the warp threads in tension from the direction of Jacquard. With shedding motion, Jacquard shedding pulls the warp threads or ends on the reverse direction of pedal movement. This develops more force on the pedals for shedding.

Cumulative effects of both the factors increased tension on the warp threads against which the weavers push the pedals downward for shedding.

Domestic looms are without any jacquard Fig. 1 and motifs are created manually by the weavers. For domestic weaving, weavers generally used to weave "Gamocha" (towel) or "Meghla Chadar" (traditional dress attire of Assam), which are shorter by width and consist of smaller and lesser numbers of motifs. For commercial production in the Jacquard loom, Assamese Sarees with heavy traditional motifs, bed-sheets with motifs were produced which were wider in width resulting increased numbers of both heald yarns and harness yarns.

Importance of Jacquard for commercial production cannot be ignored. Previously weavers used to create motifs manually, which resulted constant neck flexion and eye strain. With Jacquard, incidence of neck pain and eye strain was reducing which was reported by both weavers and loom owners but on the other hand LLP was increasing. Same observation surfaced out, during weaving sub-task analysis, though neck pain was found to be the second highest pain, but there was significant difference was observed in case of range of pain reporting Table 1. Due to the use of larger leg muscle group, the weavers were unaware of the amount of force they were executed to operate the pedal.

3.3 Phase 3

In order to confirm the above findings field experiment was conducted where 'Two way ANOVA' was run to determine, whether increased pedalling force due to Jacquard and longer duration of work shift for commercial production has any effect on LLP. The effects of two independent variables (weight for pedal depression and work shift) were tested on a single dependent variable (LLP). For this experiment 150 weavers were questioned regarding LLP in executing pedalling and asked to rate, in Borg's CR-10 Pain scales [16]. 150 frame looms with jacquard was chosen and depression of the pedal was measured by adding weights on top of the pedals in place of foot pressure. 118 handloom clusters were selected for this experiment and working hours of these looms were noted from loom owners of unorganized sectors and management bodies of organized sectors.

Normality test was performed graphically with Q-Q plot, Fig. 7. The data points were observed to be normally distributed from a close situation of data points from diagonal line. Homogeneity of variance was tested with Levene's Test Table 2. The resulting $p = 0.036$ thus, the null hypothesis of equal variances was

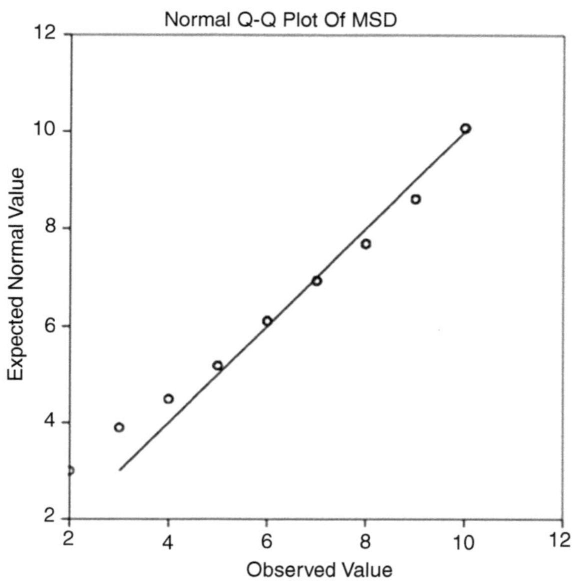

Fig. 7 Graphical representation of normality

Table 2 Levene's test of equality of error variances

F	df1	df2	Sig
1.557	92	57	0.036

Table 3 Tests of between-subjects effects

Source	Type III sum of squares	df	Mean square	F	Sig.
Corrected model	512.423	92	5.570	5.726	0.000
Intercept	4,278.443	1	4,278.443	4,398.039	0.000
Wt	256.626	22	11.665	11.991	0.000
Work_shift	76.008	8	9.501	9.767	0.000
Wt * Work_shift	54.280	62	0.875	0.900	0.658
Error	55.450	57	0.973		
Total	9,585.000	150			
Corrected total	567.873	149			

rejected and it was concluded that there was a difference between the variance in the population.

From ANOVA analysis Table 3, it was found with increased force of pedalling there was a significant increase in LLP, where $F(22,57) = 11.99$, $p < 0.001$. A similar observation was also been noticed in case of duration of work shift with LLP. As the duration of work shift increases in case of commercial weaving so does the LLP as observed by $F(8,57) = 9.77$, $p < 0.001$. But within group, there was no significant relationship was observed with increased pedalling force and work shift with LLP, $F(62,57) = 0.90$, $p = 0.66$.

From above findings, it was clear that increased pedalling force against Jacquard for longer duration of loom usage for commercial production significantly increased incidence of LLP.

4 Discussion

Observations from the present study revealed, as the duration of loom usage increases for commercial production, it had a positive impact on the occurrence of MSDs Table 3 which is similar to the previous study [17]. The findings of the present study revealed that the reported percentage of weavers having back pain (95 %) and knee pain (83 %) Fig. 3 was much higher than reported by the studies on handloom weavers of Ahmedabad and Nagpur [7, 18]. The populations of weavers in the present study were all women, so one of the possibilities of higher incidence of MSDs, might be due to the physiological demands to perform household activities. This reduces their physical recovery throughout the day, which gets adverse to commercial weaving practice. Consideration of non-work related social factors for MSDs was not carried out in the present research study, but similar observations revealed that women weavers were more affected with MSDs in the handloom sectors due to household and workplace activities [7, 19].

Reporting of previous studies on handloom weavers related to leg pain was due to maximum leg stretching for pedaling [7] and fixed leg posture [8], but present research findings revealed some other causative factors for leg pain. Leg pain in the existing condition, was due to commercial production on Jacquard operated handloom and for improper seat design [10]. Pedalling was found to be highly repetitive task involving 74 cycles/min $\pm$ 13.59 with implication of high force against Jacquard shedding.

From the ANOVA analysis, it was found significant relations exist between pedal force and work shift duration with LLP Table 3. As the pedalling force increase there was a significant increase on LLP, similarly as work shift duration increases so do LLP. But no relation was observed within group for increased pedalling force and work shift on LLP. This might be due to two types of weaving practices, continuous weaving and weaving for motif making. Weaving heavy motifs increases pedalling force due to increase number of harness yarn, but at the same time, micro pauses were there during warp stop motion to move the extra weft. During these pauses, short leg muscle recovery takes place. For continuous weaving for plain weave, pedalling force was less, but constant pedalling takes place relatively for longer duration. ANOVA indicated that an increase in pedaling force and work shift duration increases incidence of the LLP, among handloom weavers in commercial production. But due to two types of pedalling, no significant relation was observed within group for pedalling force and work shift on LLP.

5 Conclusion

Comprehensive study showed the addition of Jacquard on frame loom has increased incidence of LLP but the importance of Jacquard in increasing production cannot be ignored. Ergonomic intervention in reducing the pedaling force is the need of the hour for this sector so that early leg muscle fatigue could be overcome.

Acknowledgments Whole hearted cooperation of State Institute of Rural Management (SIRD), Khadi and Village Industries Commission (KVIC), Institute of Rural Management (IRM), Artisans Cooperative Federation Limited (ARTFED), Fabric Plus (FP) and weavers and loom owners of Sualkuchi, Bijoynagar, Self-help group Dolgovindo (SHG) and Hajo is gratefully acknowledged.

References

1. A brief report on textile industry in India (2013). Retrieved from http://www.cci.in/pdfs/surveys-reports/Textile-Industry-in-India.pdf on 26 March 2014
2. Garg, M.S., Jain, M., Paul, B.B., Ulaganathan, S.: Handloom, a rich heritage of India-needs protection and promotion (2012). Retrieved from http://handlooms.nic.in/hl_heritage_india.pdf on 16 July 2013

3. Handloom Census of India 2009–2010. Retrieved from http://handlooms.nic.in/handloom_census_2009-10.pdf on 4 Jan 2012
4. Goswami, K.: Impact of Globalisation of Silk industry in North East India: An assessment From Gender Perspectives (2005). Retrieved from http://faculty.washington.edu/karyiu/confer/beijing06/papers/goswami.pdf on 25 April 2014
5. Metgud, D.C., Khatri, S., Mokashi, M.G., Saha, P.N.: An ergonomic study of women workers in a woolen textile factory for identification of health-related problems. Int. J. Occup. Environ. Med. **12**(1), 14–19 (2008)
6. Banerjee, P., Gangopadhyay, S.: A study on the prevalence of upper extremity repetitive strain injuries among handloom weavers of West Bengal. J. Hum. Ergol. **32**, 17–22 (2003)
7. Nag, A., Vyas, H., Nag, P.K.: Gender differences, work stressors and musculoskeletal disorders in weaving industries. Ind. Health **48**, 339–348 (2010)
8. Goel, A., Tyagi, I.: Occupational health hazards related to weaving. Int. J. Appl. Math. Stat. Sci. **1**(1), 22–28 (2012)
9. Pandit, S., Kumar, P., Chakrabarti, D.C.: Ergonomic problems prevalent in handloom units of North East India. Int. J. Sci. Res. Publ. **3**(1), 1–7 (2013)
10. Pandit, S., Chakrabarti, D.: Ergonomic approach of seat design in the handloom industry of Assam. In: Proceeding HWWE, pp. 133–145 (2013)
11. Pandit, S., Chakrabarti, D. Investigation of work-related muscular pain among women handloom weavers of Assam a state of North- East India using REBA method. ACED, 21–24 May, South Korea (2014)
12. India. The Factories Act, 1948 (Act No. 63 of 1948), Chapter VI section51 Weekly hours. Retrieved from http://www.ilo.org/dyn/natlex/docs/WEBTEXT/32063/64873/E87IND01.htm#a051 on 1 April 2014
13. Chakravorty, R., Dutta, P., Ghose, J.: Sericulture and traditional craft of silk weaving in Assam. Indian. J. Tradit. Knowl. **9**(2), 378–385 (2010)
14. Corlett, E.N., Bishop, R.P.: A technique for assessing postural discomfort. Ergonomics **19**(2), 175–182 (1976)
15. Kuorinka, I., Jonsson, B., Kilbom, A., Vinterberg, H., Biering, S.F., Andersson, G., Jorgensen, K.: Standard nordic questionnaires for the analysis of musculoskeletal symptoms. Appl. Ergonomics **18**(3), 233–237 (1987)
16. Borg, G.: Perceived Exertion and Pain Scales. Human Kinetics, Champaign (1998)
17. Costa, G., Samantha, S., Torbjorn, K.: Influence of flexibility and variability of working hours on health and well-being. Chronobiol. Int. **23**(6), 1125–1137 (2006)
18. Tiwari, R.R., Pathak, M.C., Zodpey, S.P.: Low back pain among textile workers. Indian. J. Occup. Environ. Med. **7**(1), 27–29 (2003)
19. Motamedzade, M., Moghimbeigi, A.: Musculoskeletal disorders among female carpet weavers. Ergonomics **55**(2), 229–236 (2012)

Evolving Process of Application of Methodology for Visual Perception of Urban Place: Case Study of Kolkata

Mainak Ghosh, Sanjib Nag and Satyaki Roy

Abstract Vision is the most dominant sense for human beings. Visual perception has a substantial bearing on cognition, impression and importance of an urban place. Along with infrastructural needs, appropriate perceptual quality of a place is of considerable significance. Especially in developing countries the short fall has been alarming as reported by McKinsey. In this context, the paper brings forth a methodological approach, integrating relevant disciplines. The disciplines of visual communication design and urban design focussing on the core areas of visual perception and urban place respectively, have been analyzed. Though there have been overlaps in literary context, a holistic methodology has not been hitherto considered in unison. The various inter-relationships have been unearthed and a logical interpretation of its applicability has been adjudged in this ongoing research. Assorted urban places of city of Kolkata has been considered for analysis, however the findings of the paper is an endeavour to evolve the holistic methodology, generic in nature.

Keywords Visual perception · Urban place · Visual communication design · Urban design · Place making · Urban development · Environmental design · Imageability

M. Ghosh (✉)
Department of Architecture and Regional Planning, Indian Institute of Technology, Kharagpur, India
e-mail: mg@arp.iitkgp.ernet.in; mainak.ghost@gmail.com

S. Nag
Department of Architecture, Jadavpur University, Kolkata, India

S. Roy
Design Programme/HSS, Indian Institute of Technology, Kanpur, India

A. Chakrabarti (ed.), *ICoRD'15 – Research into Design Across Boundaries Volume 1*,
Smart Innovation, Systems and Technologies 34, DOI 10.1007/978-81-322-2232-3_39

1 Introduction

Vision being the most dominant sense for human beings, it provides more information than the other senses combined. Orientation in space is achieved visually. **Visual Perception** is the process of registering visual sensory stimuli as meaningful experience. "A perceptual approach to visual communication" delves into the mechanism of processes of vision and how those forms attitudes and ideas [1]. Visual Communication design (VCD) is both the process and product, predominantly in visual media, which conveys intended information comprehensively; often it is produced through industrial means. It has targeted audience to cater to effectively.

Space is the physical manifestation of a location, whereas a place contains higher level aspects. Montgomery's research sheds light on formation of **Urban Place** amongst space. According to him urban places are constructed by the physical form, activity and meaning [2]. Meaning generate perception and that is associated with individual's internal psychological and social processes [3, 4] that generate perception [2]. Urban Design (UD) is the art of place making, associated with its arrangement, appearance and function. Like VCD, it is both the process and outcome. People create urban places, and eventually they themselves get influenced by those places [5]. In contemporary perspective, visual perception of urban place has substantial importance due to the following five aspects:

- It acts as a strong "image building" parameter; often serves as a continuum in restoring "historical, cultural and community background" of the place [2].
- It creates a specific character or trait of the place; forms a recognizable feature of the place, makes it "attractive or unattractive in comparison to another" [6].
- It is a proponent of commercial developments; "higher rental levels, enhanced regeneration, increased public support for development" [7].
- It creates aesthetically "pleasing ambience" and positive stress-free psychological effect, invites "community interaction", participation and commerce [8].
- It directly or indirectly affects "developments, investments, occupancy levels, planning and design" [7].

Kolkata, like many other cities in developing countries, lacks a holistic methodology of design for visual perception of urban places. They often develop in a sporadic and subjective manner, eventually taking up a gruesome shape. McKinsey reports in 2010, major deficiency in Indian metropolitan and urban developments, with no mention of urban form or design to create a unique visual feel and to build the city's character [9]. Kolkata is no exception. Numerous urban places face the same fate in India.

In present context, Visual Perception of Urban Places of Kolkata is getting importance in the light of urban transformation and changing political scenario. Kolkata, is thriving to get a new perceptual quality to establish itself at par with other developed metropolitan cities in the world.

For good place making, the knowledge of visual perception and urban place should be considered simultaneously. In this paper, an attempt has been made to understand the gap and establish the inter-relationship of these two subject matters pertaining to visual perception of urban place, hitherto practiced independently, i.e. VCD and UD. A holistic **methodology for visual perception of urban place** has been reputed embracing these two disciplines and its appropriateness for application.

It is important to understand that though UD has certain principles pertaining to perceptual parameters, the core subject area which deals with visual perception and communication of human beings is VCD. Hence a wholesome place making would rely on appropriate synthesis of both UD and VCD. Both of them are relatively nascent subjects areas, coined only in late twentieth century and little has been done to integrate them till date.

In present context of rapid urbanization and mindless interplay of market forces leading to chaotic environment, research is necessary to bridge the gap between these two disciplines. For wholesome and articulated urban environment they should act in unison. A holistic methodology has been evolved which hitherto ceased to exist or not egalitarian in nature.

2 A Methodological Approach

2.1 Understanding the Gaps of VCD and UD: Literature Study

In continuum to the previous discussion it is important to deep dive into understanding the methodology encompassing VCD and UD for visual perception of urban place. And there after indicate the process of its application relevant to a context. Kolkata metropolitan has been selected for case study of the research.

2.1.1 Visual Communication Design and Its Aspects

VCD could be defined as "the process of conceiving, programming, projecting and realizing visual communications that are usually produced through industrial means and are aimed at broadcasting specific messages to specific sectors of the public. This is done with a view towards creating an impact on the public's knowledge, attitudes, or behaviour in an intended direction" [10].

It has evolved in the last 50 years only, as a professional discipline. Though graphic communication with specific objectives had formations since 25,000 years, it pedagogy vastly changed only recently with advancements ranging from cognitive psychology to marketing [10]. Until beginning of World War I, the conception of visual communication was entrusted to artists or printers. Often these professionals evolved from schools of art and crafts and indulged in usage of artistic ornamentation, playing with various forms of typefaces, styles and sizes, where typography took a back seat and ornamentation was of priority. However things started to change in nineteenth century. With the advent of Art Novaeu, along with strong complexity of design, stylistic blend with higher visual order started evolving. VCD developed its essential components in the 1920s (Art Novaeu, Bauhaus, De Stilj). It then changed from artistic notions of creation to effective communication in the 1950s when new developments in psychology, sociology, linguistics, and marketing attracted the attention of designers. "Important areas connected to communication design are experimental psychology (studies of perception, learning, and behavior); social psychology (studies in cultural trends, statistics, and behavior); research on perception, communications, learning, and behavior for the armed forces; marketing research for advertising agencies and consumer products manufacturers; and research in linguistics, rhetoric, and semiotics" [10]. This clearly indicates an essential feature of VCD considering inter-disciplinary, evolutionary nature of the same.

Now it is witnessing the next stage, based on developments in technology, which enhances interaction between the public and information. Also with the rise of consumerism and globalization, it is becoming more versatile and appreciated. People are becoming aware of and interactive to this easily discernible visual form into almost everything they see or visually encounter. VCD covers four basic areas based on nature of expertise, function and objectivity: namely, Design for information, Design for persuasion, Design for education, and Design for administration.

Design for information primarily aims at categories of design where dissipation of information is of prime importance [10]. While information design consists of two fold steps, organization of the information, then planning of its visual manifestation, design for persuasion is directed towards affecting behavior. It could be further subdivided into three areas namely, "advertising (commercial and non-commercial); propaganda (political and ideological); and social interest communications" [10]. These designs strongly get influenced by trends of globalization, consumerism unlike design for education which also has an impact on behavior but in a subtle manner. It is somewhere between information and persuasion. The active participation of the users of educational materials is central in this case. "Design for administration is a category all its own. It does not involve information, persuasion, or education, as the previous cases do, but it contributes to the organization of certain communications inside administrative systems" [10]. It is noteworthy, that essentially VCD relies on two elements: typography and image. Again, images can be representational or abstract in nature.

2.1.2 Urban Design and Its Aspects

City Beautiful Movement had a bearing on the then limited approaches of civic design, which was primarily concerned with orientation and design major civic buildings and their relationship with open spaces. Urban design evolved from this initial concept of building masses and space in between buildings to more expansive approach. It imbibed the qualities of physical and socio-cultural attribute of places affecting people. "It includes the way places work and matters such as community safety, as well as how they look. It concerns the connections between people and places, movement and urban form, nature and the built fabric, and the processes for ensuring successful villages, towns and cities" [11].

"Urban design is concerned with the arrangement, appearance and function of our suburbs, towns and cities. It is both a process and an outcome of creating localities in which people live, engage with each other, and the physical place around them. Urban design involves many different disciplines Urban design operates from the macro scale of the urban structure (planning, zoning, transport and infrastructure networks) to the micro scale of street furniture and lighting" [12].

Though the term Urban Design as it is referred today ceased to exist before 1965, urbanization and urban design has been an intrinsic part of human civilization. Man started taking his first step towards urbanization when the Paleolithic man started making first shelters forsaking caves; followed by Neolithic man envisaging on agrarian population, protecting surplus resources and villages. With civilizations passing over ages, we have witnessed growth of buildings and cities. Urban developments have been formed as a functional core to lavish ostentations. From picturesque Barouque city to congested industrial ghettos the tale of the cities has been similar, until George Pakins Marsh introduced the concept of ecology. Olmsted took a leading role to bring out the new thinking process into city design. It has embraced newer agenda in twenty first century, aligning to inclusive design, neo-urbanism, sustainability and technology, along with integrative professional activity, interdisciplinary in nature.

"Kevin Lynch (1918–1984) was the first author who focused his work on visual elements and cognitive concepts of the urban environment" [13]. Lynch's notion of mental image of the city was a macro level classification based on following five aspects: District, Edge, Pathway, Node and Landmark.

In a similar perspective, Godon Cullen developed another concept of the mental image of a city, through observation of various urban places and elements, these are: Serial Vision, Place and Content. Speriregen constructed the skeletal framework at a greater granular level comprising of: Landform and nature, Local climate, Size, Shape, Pattern, Density, Grain and Texture, Districts, Routes, Activity Structure, Urban Spaces and Open Spaces, Built Form, Vista and Skyline, Orientation, Details and Non-physical Aspects [14]. While Lynch, Cullen and other authors have talked about visual concepts and imageability at a generic level, Speriregen, mentioned the micro-level elements.

During discussion of these aspects of UD, he has referred to elements several times, predominantly visual communicational in nature. These have special connotation with regards to visual perception of urban place. He pointed about signs, advertisements, paving, night lighting, materials, greenery, colors along with building form and density during discussing on appearance of District. He has further indicated in his findings, "a visual survey of urban details should, therefore, include sign studies. More broadly, it includes quality and conditions of park benches, wastebaskets, streetlamps, pavements, curbs, trees, fences, doorways, shop windows, etc.—the street furniture and hardware of the city" [14]. In another place he has aptly pointed out the usage of trees as an effective visual treatment and sidewalks being adorned with displays, cafes, kiosks or simply a place to seat as a part of the ‘scene’. While describing programs he has mentioned about illumination as ‘untapped reservoirs of modern urban possibility’, street furniture, street hardware emphasizing on advertising, urban sculpture and art and landscape [14].

2.1.3 Aspects of VCD and Aspects of UD and Kolkata

One of the core objectives of UD is concerned with the appearance of a place. Also from the preceding literature study it is understood various forms of VCD are prevalent in visual regime of urban places affecting its appearance. Hence it is established that there exists relationship between VCD and UD, so far as visual perception of urban place is concerned, which directly or indirectly affects urban development and good place making. This paper unravels a remarkable aspect in this context of the inter-relationship of VCD and UD which has not been documented before. The Rhizomatic map of visual communication with various disciplines are shown in Fig. 1 [15]. The relationship of VCD with various other disciplines are conspicuously discernible, however, it is interesting to note that though it clearly connects up with Architecture and Engineering, Art alike, there is no existence of urban design in the diagram.

Again strongly inspecting the interdisciplinary nature of urban design, quoting Carmona “Urban design is not, however, simply an interface. It encompasses and sometimes subsumes a number of disciplines and activities” [7]. Hence it is expected that the relatively nascent discipline of UD should conjugate with various relevant domains and disciplines, including that of visual perception of place which directly correlates to appearance of a place. On scrutiny of Fig. 1, it reveals VCD has not yet been specified into the fabric of inter-disciplinary genre of UD, though ‘Art’ exists [16]. Today due to vast number of disciplinary specializations, urban design, needs to be egalitarian and holistic embracing the discipline related to appearance of urban place.

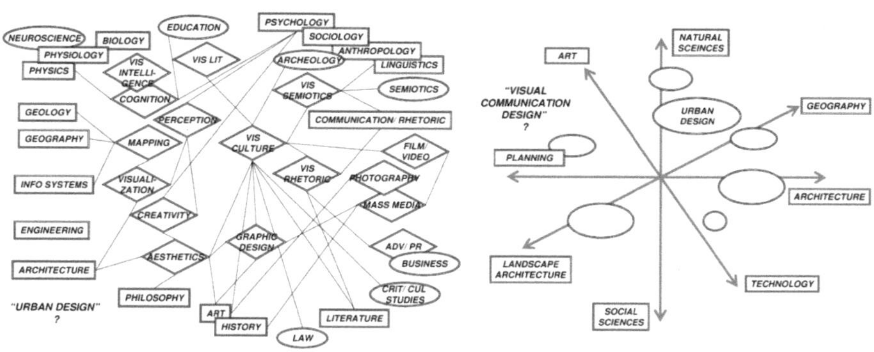

Fig. 1 A comparison of Rhizomatic map of visual communication and urban design with various disciplines. *Source* Smith et al. 2005 and Arida 2002 respectively [15, 16]

In context to urban development, VCD plays a pivotal role for visual communication of information, persuasion, administration and to some extent education, with direct usage such as Signages, Maps, Outdoor Advertisements, Banners, Posters, Graphics, Graphitti-Murals, Displays (print/digital/physical), Kiosks, Propaganda, Social Interest, Branding of a place or city etc. These are influential features so far as visual perception of urban place is concerned. However, these two disciplines lack integrated approach.

Paper published by Ghosh, Nag and Roy declares this context with Kolkata as case example study and case application study. It reveals substantial lacuna in some important places of Kolkata metropolitan, at the urban nodes aligned north-south, namely, Dumdum, Shyambazar, Esplanade, Hazra and Tollygunge. Though the city has propagated aspects of UD to some extent, it lacks cohesion with aspects of VCD. The pitfalls have been visually identified as against aspects of VCD such as lack of proper signage and its hierarchy, chaotic advertisement banners, lack of identity or congruent visual features of places, cluttered and non-coherent street furniture or hardware including hutments, slums and temporary shelters etc. Though some beatification efforts have been made but they are devoid of visual rationale and specification [17].

2.2 Establishing Inter-Relationships of VCD and UD: Evolving the Methodology

From the preceding discussions, it is clearly discernible that there exists a strong relationship between VCD and UD hitherto unattended. Now to objectively scrutinize the inter-relationship and derive at a methodology based on the literature study, a pilot survey has been conducted. The survey was crafted to validate the hypothetical literature findings. Moreover this hints towards various parameters which could be utilized by the methodology of visual perception of urban place. An online survey was conducted with 8 questions, each rated on a Likert scale, randomly posed to audience unrestricted to any geography. The questions were based on literature study, and Kolkata being considered as a case example. 111 responses were received out of which 6 were disqualified. The respondents were mostly of middle or high income group, with varied age groups, 33 % being female respondents. The respondents include both layman and people with prior education on various forms of design and aesthetics, also including people with experience of visiting or residing in one or more cities of developed nations. Such a mixed group has been chosen to analyze the variability of data based on variability of perception, experience and culture. Interestingly the data hints similar patterns irrespective of respondents' variability. It is evident from the survey, as shown in Fig. 2, that apart from buildings, various other elements of UD (U_E) such as signage, advertising, street furniture etc. play a key role for visual perception of a place.

Again, 84 % respondents relate the four areas of VCD (V_A) to appearance of urban place, followed by its function and arrangement. Hence, the appearance of

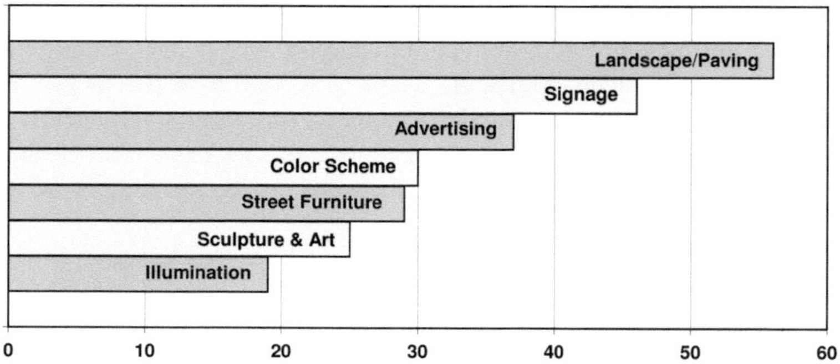

Fig. 2 Priority of U_E for visually perceiving urban place (x: %, y: U_E)

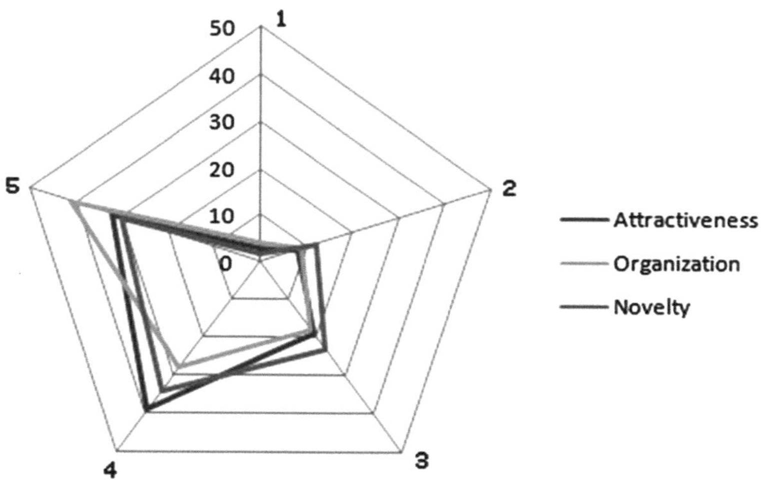

Fig. 3 Rating of P_Q of visually perceived urban place (% values mentioned)

the place is intrinsically affected by various artifacts from various areas of VCD which again has an overlap with U_E. Putting a parametric dimension to the findings, the authors propose the dimensions of people's impression of a place in terms of attractiveness, organization and novelty (P_Q). Though Ernawati's research on these attributes (P_Q) have been targeted towards tourist-historic districts, it could be well adopted for urban places [18]. Survey indicates people relate to P_Q strongly so far as visual perception of urban place is concerned as shown in Fig. 3 where 1 is Strongly Disagree and 5 is Strongly Agree.

Based on these findings and validations, the inter-relationship between VCD and UD for visual perception of urban place has been proposed as depicted in Fig. 4, which eventually leads to the formulation of the methodology addressing the parameters identified therein.

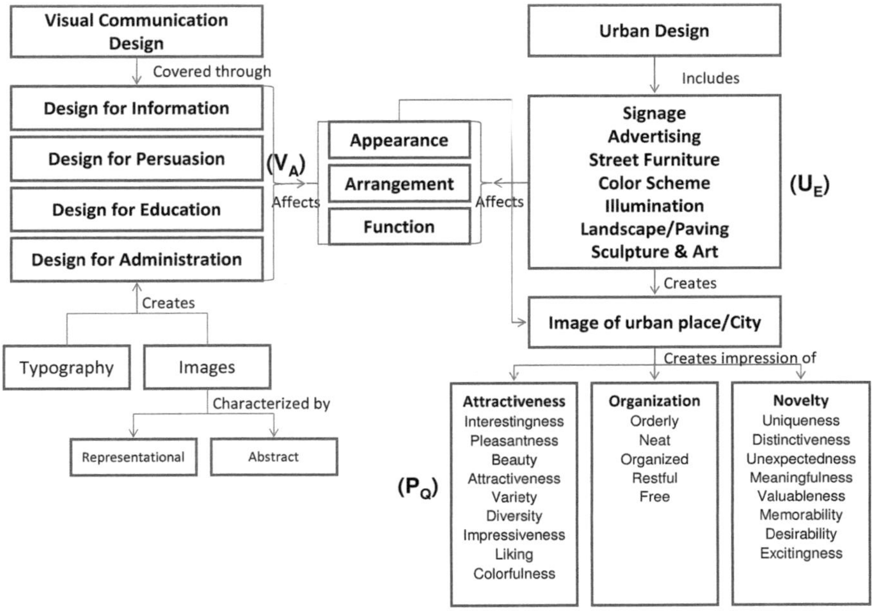

Fig. 4 Relationship of aspects of VCD and UD used for evolving the methodology for visual perception of urban place

Based on the above inter-relationship a methodology of visual perception of urban place has been derived as shown in Fig. 5. It is important to note, that based on the discussion earlier such a methodology or tool would come handy generating non-conventional approaches of assessing, designing, regenerating urban places. It may serve as an effective instrument of visual perception of urban place related to environmental design, aesthetics, identity building, imageability etc. which is a holistic blend of VCD (V_A) and UD (U_E) on one hand and parametric on measurable grounds (P_Q) on the other.

2.3 Evolving the Process of Application of the Methodology

To apply the thus evolved methodology, the relationship of U_E and P_Q needs to be objectively analyzed, though in previous Figs. (2 and 3) individual characteristics of U_E and P_Q could be understood. Graph in Fig. 6 shows importance of P_Q against each U_E based on the Likert scores (1 = Very unimportant, 5 = Very important). The gross pattern is similar agnostic of Likert scores. Now, the weight of attractiveness, organization and novelty (constituents of P_Q) have been found to be close, 0.99, 1 and 0.96 respectively. Hence to apply this methodology based on the case application, equal importance of these factors with respect to U_E to

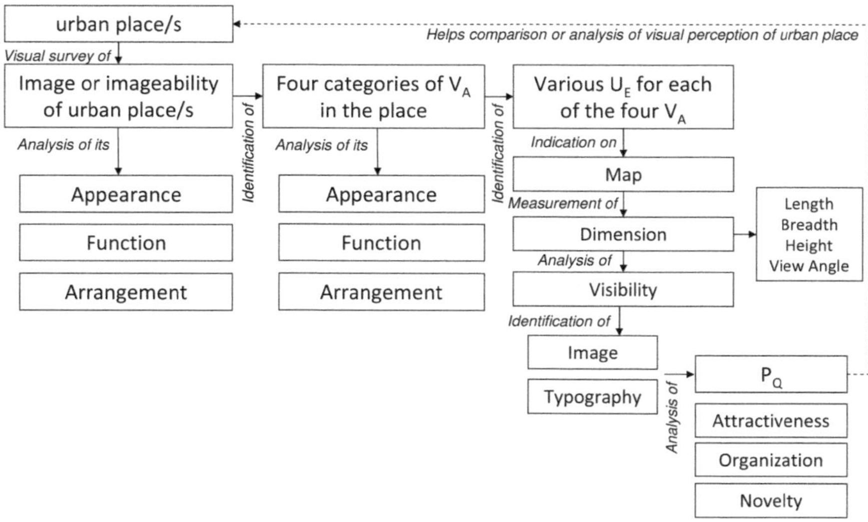

Fig. 5 Evolving a methodology for visual perception of urban place

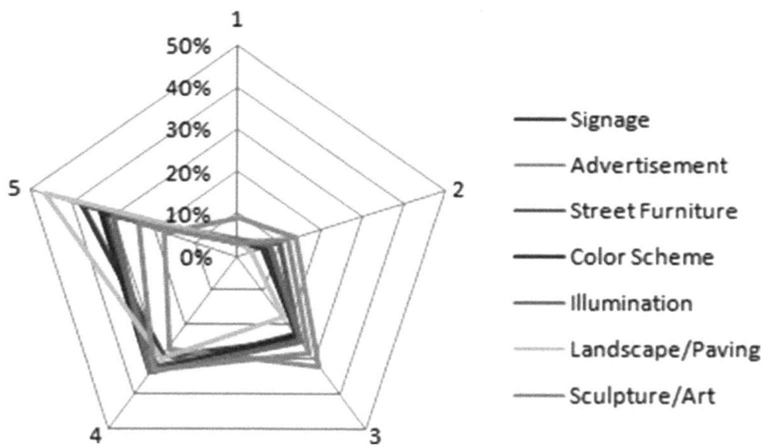

Fig. 6 U_E rated against P_Q

be considered. However precise ordering indicates 'organization' to be of higher weight followed by attractiveness and novelty. As shown in Fig. 4 P_Q is parameterized into 22 factors, under three broad heads. These factors need to be taken into account for particular case application.

Again, for the application of the methodology, the parameters of each U_E pertaining to various V_A needs to be investigated. The substantial overlap of U_E and V_A establishes the very conception of this research. Literature study indicates VCD

V_A	U_E	P_Q →	Attractiveness (P_{QA})					Organization (P_{QO})					Novelty (P_{QN})	
			Interestingness	Pleasantness	Beauty	Attractiveness	Variety …	Orderly	Neat	Organized	Restful	Free	Unique-ness	Distinctive-ness …
Design for Information	Signage	Image	1-2-3-4-5	1-2-3-4-5	1-2-3-4-5	1-2-3-4-5	1-2-3-4-5	1-2-3-4-5	1-2-3-4-5	1-2-3-4-5	1-2-3-4-5	1-2-3-4-5	1-2-3-4-5	1-2-3-4-5
		Typo	1-2-3-4-5	1-2-3-4-5	1-2-3-4-5	1-2-3-4-5	1-2-3-4-5	1-2-3-4-5	1-2-3-4-5	1-2-3-4-5	1-2-3-4-5	1-2-3-4-5	1-2-3-4-5	1-2-3-4-5
	Maps	Image	1-2-3-4-5	1-2-3-4-5	1-2-3-4-5	1-2-3-4-5	1-2-3-4-5	1-2-3-4-5	1-2-3-4-5	1-2-3-4-5	1-2-3-4-5	1-2-3-4-5	1-2-3-4-5	1-2-3-4-5
		Typo	1-2-3-4-5	1-2-3-4-5	1-2-3-4-5	1-2-3-4-5	1-2-3-4-5	1-2-3-4-5	1-2-3-4-5	1-2-3-4-5	1-2-3-4-5	1-2-3-4-5	1-2-3-4-5	1-2-3-4-5
Design for Persuasion	Advertisement	Image	1-2-3-4-5	1-2-3-4-5	1-2-3-4-5	1-2-3-4-5	1-2-3-4-5	1-2-3-4-5	1-2-3-4-5	1-2-3-4-5	1-2-3-4-5	1-2-3-4-5	1-2-3-4-5	1-2-3-4-5
		Typo	1-2-3-4-5	1-2-3-4-5	1-2-3-4-5	1-2-3-4-5	1-2-3-4-5	1-2-3-4-5	1-2-3-4-5	1-2-3-4-5	1-2-3-4-5	1-2-3-4-5	1-2-3-4-5	1-2-3-4-5
	…	Image	1-2-3-4-5	1-2-3-4-5	1-2-3-4-5	1-2-3-4-5	1-2-3-4-5	1-2-3-4-5	1-2-3-4-5	1-2-3-4-5	1-2-3-4-5	1-2-3-4-5	1-2-3-4-5	1-2-3-4-5
		Typo	1-2-3-4-5	1-2-3-4-5	1-2-3-4-5	1-2-3-4-5	1-2-3-4-5	1-2-3-4-5	1-2-3-4-5	1-2-3-4-5	1-2-3-4-5	1-2-3-4-5	1-2-3-4-5	1-2-3-4-5
Design for Education	…													
Design for Admin	…													

Fig. 7 Survey format derived from the evolved methodology as a process of application of methodology for visual perception of urban place

relies on images and typography as its constituents. It is interesting to note, this research brings forth, so far as visual perception of urban place is concerned, visual communication is possible through elements, which compositionally do not constitute of typography and image, i.e. Landscape/Paving (76 %) and color shceme (72 %) followed by street furniture (60 %), illumination (56 %) and art/sculpture (56 %) as shown in Fig. 2. These elements constitute majorly to the image of a place or scene rather than constitution of a VCD elements in singularity. Based on this discussion, a process of application of the methodology has been evolved. A matrix as shown in Fig. 7 has been crafted to examine the applicability of the methodology, which would be applied at some of the urban places of Kolkata.

Kolkata has seen its rise from conglomeration of villages into a colonial town. It served as the glorified capital of the nation giving way to various anomalies post independence. It has seen a diverse interplay of culture and socio-economic variations. Along with British colonial squares there exists slums, vernacular mansions, congested shopping areas alike. Similar to most of the cities in developing nations, places have evolved with no visual perception kept in mind. Morphologically Kolkata is a linear city stretching north-south. Major arterial roads are aligned accordingly. Important urban places evolved over course of time at various nodes of these roads. Considering the diverse visual and cultural characteristics of Kolkata, along with fundamentally potential place make-over tendency, several important nodes (places) has been identified for application of this methodology. As an ongoing research, an initial visual survey as indicated in the methodology (Fig. 5) has been conducted which brings out crucial observation in support to the applicability of the methodology. Figure 8 indicates V_A and its issues for four different urban places of Kolkata. □, ○, △ and + markers are used to indicate issues pertaining to V_A, namely, design for information, persuasion, administration and education, respectively. The process of application of methodology in context to Kolkata as both case example and case application study, has been used as a precursor for

Fig. 8 Adjudging the process of application of methodology in context to visual perception urban places of Kolkata

further data gathering and its analysis, as part of an ongoing research endeavour. However a holistic attempt has been made bridging VCD and UD in a parametric way, deriving a methodology and herewith adjudge it process of applicability.

3 Conclusion

Conventional place making approach often lacks disciplinary cohesion of VCD with UD. The deficiency has already been discussed in detail as pointed out by McKinsey's report. Along with lack of a holistic, inter-disciplinary and integrated approach, many cities of the developing nations are often subjected to urban development, environmental design and regeneration models without proper tools and course of action. Based on the ongoing research endeavor, the paper evolved a process of application of methodology for visual perception of urban place. We visually perceive the environment around us, and it affects our actions, reactions and feelings. Urban places are juxtaposed, with innumerable visual elements and treatments, often developed without any holistic planning over course of time. It often has perceptual inappropriateness and complexity. It is at par with basic infrastructural qualities of a city. Also it has been found that visual perception has a strong influence on urban development and commercial aspects. Hence, application of the methodology in urban paradigm would lead to promising findings and implementations towards better place making. The approach of the research is twofold to evolve the methodology and its process of application and thereafter adjudge the data gathered through them. Hence this paper based on research work in progress, is directed towards the theoretical base or the process of the empirical data gathering. Future studies in direction holds immense possibilities of knitting place perception and environmental design with enhanced interaction, interface, robotics, sustainability etc.

References

1. Barry, A.M.: Perception and visual communication theory. J. Vis. Lit. **22**, 91–106 (2002)
2. Montgomery, J.: Making a city: urbanity, vitality and urban design. J. Urban Des. **3**(1), 93–116 (1998)
3. Stokols, D., Shumaker, S.A.: People in places: A transactional view of settings. Cognition Social Behaviour and the Environment, Hillsdale, NJ. Lawrence Erlbaum Assoc., 441–488 (1981)
4. Stedman, C.R.: Is it really just a social construction?: The contribution of the physical environment to sense of place, Society and Natural Resources, **16**, 671–685 (2003)
5. Weeks. J. R.: Defining urban areas. Remote Sens. Urban Suburban Areas **10**, 37 (2010)
6. Hidalgo, M.C., Berto, R., Galindo, M.P., Getrevi, A.: Identifying attractive and unattractive urban places: categories, restorativeness and aesthetic attributes. Medio Ambientey Comportamiento Humano **7**(2), 115–133 (2006)
7. Carmona, M., de Magalhaes, C., Edwards, M., Awuor, B., Aminossehe, S.: The Value of Urban Design: a research report commissioned by CABE and DETR to examine the value added by good urban design. Thomas Telford Publishing, CABE (2001)
8. Gehl, J.: Life Between Buildings: Using Public Space. The Danish Architectural Press, Copenhagen (1987)
9. Sankhe, S. et al.: India's Urban Awakening: Building Inclusive Cities, Sustaining Economic Growth, p. 112. McKinsey and Company, New Delhi (2010)
10. Frascara, J.: Communication Design: Principles, Methods, and Practice. Allworth Press, New York (2004)
11. Carmona, M., Heath, T., Oc, T., Tiesdell, S.: Public Places Urban Spaces: The Dimensions of Urban Design. Architectural Press, Oxford (2003)
12. INFRA1219: Places for People: An Urban Design Protocol for Australian Cities, p. 10. Australian Green Infrastructure Council, Australia (2011)
13. Šiđanin, P.: On lynch's and post-lynchians theories. Archit. Civ. Eng. **5**(1), 61 (2007)
14. Spreiregen, P.D.: Urban Design: The Architecture of Towns and Cities. McGraw-Hill, New York (1965)
15. Smith, K., Moriarty, S., Barbatsis, G., Kenney, K. (eds.): Handbook of Visual Communication Design. Theory, Methods and Media. Lawrence Erlbaum Associates, Mahwah (2005)
16. Arida, A.: Urban Design and the Quantum Worldview. Architectural Press, Quantum City (2002)
17. Ghosh, M., Nag, S., Roy, S.: Visual perception of urban place: issues need attention in kolkata. J. Indian Inst. Archit. **78**(7), 49–51 (2013)
18. Ernawati, J.: People's impressions of a tourist-historic district. In: International Conference on Architectural Research, pp. 144–151. Architectural Research Centers Consortium (ARCC), USA (2012)

A Study on Entrances and Foyers in Shopping Malls and Their Role in Influencing Perceptions

Himanshu Bansal, Pradeep Yammiyavar and P.Y. Anita

Abstract Growing numbers of shopping malls in India have compelled shopping mall managers and retailers to find new ways to attract customers to their enterprises. This paper presents a study of four shopping malls in India aiming at identifying important factors contributing to attitude and perception formation of the visitors entering into the premises. The factors posited are categorized under Aesthetic Appearance, Qualitative edge, Way-finding, De-Stressing facilities and Pre-shopping services. The study was conducted in three stages: (i) Questionnaire on importance of factors in general; (ii) Eye-tracking sessions; (iii) Comparative questionnaire on four malls. We analyzed the data obtained from these sessions both quantitatively and qualitatively. We also propose design guidelines which designers and architectures can make use of while designing entrance space in Malls.

Keywords Shopping mall · Entrances and foyers · Human factors · Eye tracking

1 Introduction

India is already a big market for unorganized retail sector in the form of kirana stores. It is now emerging as an investment destination in organized retail sector. Indian retail trade has been seen to be the fastest growing in the world [1]. India has been ranked fifth of the thirty emerging retail markets of the world by global real estate consulting group Knight Frank [2]. Initially, these organized retailing formats were mainly established in metropolitan cities and now are springing up in smaller cities and towns [3].

H. Bansal (✉) · P. Yammiyavar
Department of Design, Indian Institute of Technology, Guwahati, India
e-mail: himanshubansal99@gmail.com

P.Y. Anita
Royal School of Architecture, Guwahati, India

A. Chakrabarti (ed.), *ICoRD'15 – Research into Design Across Boundaries Volume 1*,
Smart Innovation, Systems and Technologies 34, DOI 10.1007/978-81-322-2232-3_40

Increasing numbers of shopping malls and its consumers in India has made it necessary for retailers to be aware of shoppers' motivations and value so that they can attract more customers [4]. Shopping mall image has been defined by Houston and Nevin [5] as the total of consumers' perceptions of a shopping mall based on functional and emotional attributes. Shopping mall image plays an important role for customers while selecting choosing between different shopping malls. Research finding e.g. [6] had shown that store image is correlated to store choice. Bearden [7] indicated price, quality of the merchandise, assortment, atmosphere, location, parking facilities and friendly personnel as the characteristics affecting the store image. Other factors which can affect shopper's decision of entering into the mall are efficient access to products i.e. wayfinding [8], attractiveness [9], variety of tenant stores [10].

Place-making is a movement which rethinks public spaces from the perspective of people, their communities and their connections within the community instead of just buildings or cars. It involves the planning, design, management and programming of public spaces [11]. This approach has been used in multiple case studies like Shangai [12] and Malaysia [13]. Being a public space, it is important to consider the concerns of the visitors while designing shopping mall. While many published papers mention design factors, very few highlight the role of shopping mall entrance (facade and foyer) and their contribution to overall perception in the consumer. One particular study related to mall entrance was done in Surabaya aiming at identifying problems of the application of the entrance and circulation facilities [14].

In contrast to previous works, we used empirical and qualitative methods to identify factors to be considered while designing for entrances and foyers of the shopping mall and propose design guidelines for these. Category labels formulated and mentioned under 'Aim of the study section' below have been made partly based on published literature and partly based on a part of the initial field study undertaken.

2 Initial Field Study and Description of Malls

In order to observe the mall environment especially area near the entrance, four malls in the Indian city of Guwahati were chosen for observational study. Shopping Malls for field study were chosen randomly based on how different they are to each other as well as their popularity amongst different segments of consumers. These four malls are labeled H-Mall; D-Mall; S-Mall and B-Mall in this paper.

In terms of their internal layouts each of the four Malls had their own distinct schema. In H-Mall, different brands were present in common open spaces without barriers/physical boundaries. In D-Mall, different brands had their allotted individual space separated by wall/partition boundaries. S-Mall has each floor dedicated to a product category such as Clothes, Food items, Interior decor items etc. B-Mall was the biggest in terms of space and had floor wise product categorization.

On observation of the crowds entering these malls two distinct categories emerged which we label as (i) Hedonist—those who visit malls more as an entertaining past time and not necessarily with pre-planned buying purpose. (ii)

Utilitarian—those who entered the mall with a specific product buying purpose. These terms 'Hedonists' and 'Utilitarian' were adopted from the work of Babin [15]. Therefore, mall is playing multiple roles here: place to purchase goods for consumers, gathering space for hedonists as well as source of profit for tycoons. Further observations have been listed below:

(a) People have to sit on the stairs due to unavailability of benches.
(b) Three of the malls did not display information as to which brands are available inside.
(c) New, first time visitors were observed to spend more time standing still just inside at the entrance door deciding on the direction to proceed. As a result the small entrance area becomes crowded during rush hours and blocks the outgoing crowd.
(d) Each Mall had its front facade (building face) crowded with multiple visual displays and messages often obstructing the name of the Mall. There is multiplicity in types of colours, forms and material used in front wall design of the building.
(e) There are some facilities present in some of the Malls but not in others like ATMs, signage and food-joints etc.
(f) In one of the Mall baggage submission room was not near entrance. Irritation of having to go back and forth to deposit their purses and baggage was evident on the faces of the shoppers. Such disruptions tend to interfere with the expectation' and anticipation built up by the shopper just about to enter a mall.

3 Aim of the Study

Based on the above heuristic observations during our initial field visits, we planned a detailed study. The aim of this study was to propose a set of design guidelines for the Shopping Mall Entrances in the Indian context. There are multiple factors associated with the design of a Mall entrance which affects user's perception of the Mall. We divided these factors into five major categories as shown below, on the basis of our observations from field study and literature review:

(a) *Aesthetic appearance of the mall*: This category includes factors relating to the aesthetics of front of the mall which are responsible for first impression in visitor's mind e.g. colour, architectural layout and attractiveness etc.
(b) *Qualitative edge of the mall*: This category includes visibility of important qualities (e.g. available brands, tenant variety and offers) of the shopping mall from the entrance itself which persuades visitors to enter in the mall.
(c) *Way-finding*: This category includes factors like ease in finding the entrance, ease in finding the way to desired product after entering into the mall. These factors save visitors time and make their experience in the mall smoother.
(d) *De-stressing facilities*: This category includes factors or facilities available near the entrance of the mall which help in maintaining the stress level of the visitors low while or after shopping e.g. sitting area and parks/greenery etc.

(e) *Pre-shopping services*: This category includes services provided by shopping mall management which are accessed by visitors before entering into the mall e.g. Baggage submission, ATMs and vehicle parking etc.

4 Methodology

In order to gather insights from visitors keeping five categories in mind, we conducted three different user study sessions. In the first study, 34 visitors were asked to rate the importance of listed 18 factors based on which they took a decision to 'enter' into the particular Mall. The second study eye-tracking sessions using Eye Tracking Recorder were conducted with 16 participants. Digital images of shopping Mall entrances were used for tracking eyes. In third study, 25 participants were asked to rank the four shopping malls on the basis of each of the five categories namely: (i) Aesthetic appearance of the mall; (ii) Qualitative edge of the mall; (iii) Way-finding rating; (iv) De-stressing facilities; (v) Pre-shopping services.

4.1 Questionnaire About Importance of Eighteen Listed Factors

While designing the layout of the shopping mall, an architect/interior designer needs to consider constraints such as economics of space, availability of space; orientation of the entrance to the street amongst others. In the first part of the study, we aimed at identifying 'most' and 'least' important factors which contribute to forming first impression resulting in a perception value about the shopping mall at its entrance. Responses regarding importance of factors were collected from 34 visitors by asking them to fill questionnaire when they were visiting malls (Fig. 2a).

4.1.1 Randomized Sample Selection

Questionnaires were filled by visitors of 3 different shopping malls. Diversity in the demographics was also maintained to collect unbiased responses. The respondents were also categorized into hedonists and utilitarian, based on a demographic question (Table 1).

4.1.2 Instrument Used

A questionnaire was prepared having 18 likert items on 5 scale importance (from 'Not important at all' to 'Very important'). Each item was representing one factor associated with mall perception. Few items were derived from observations and literature

Table 1 Sample distribution

	Gender wise	Age wise	Shopping value wise
No. of participants	Male: 22	<=25:16	Hedonist: 16
	Female: 12	>25:18	Utilitarian: 18

research [16]. The items were prepared in such a way that these can be grouped in five main categories, mentioned previously (Table 2). Visitors were instructed to rate each of the eighteen factors according to their importance in contribution towards creating a positive experience at the shopping Mall's entrance. One sample item was: 'To what extent, you feel *Adequate open space* near the entrance important'.

4.1.3 Analysis

We analyzed the data obtained from questionnaire responses in three steps. First, we computed the Pearson correlation coefficient between each pair of factors to check whether the any responses of any two factors are correlated so that we could club or omit any of the two similarly perceived factors (highly correlated). Secondly, 95 % confidence intervals for responses of each of the eighteen factors were calculated and plotted (Fig. 1). Using this, for each individual factor we calculated number of rest of the factors which were significantly less and significantly more important than this particular factors. High number of rest of the factors signifies that this particular factor is less important and vice versa. Third, we applied Mann-Whitney U test (as the data type is ordinal) between hedonist and utilitarian groups for each of the 18 factors to identify the factors which are significantly more important for one group.

4.2 Eye-tracking Sessions

Aim of this session was to identify primary user attraction areas in different mall facades and check whether users look at important information displayed on

Table 2 Items/factors grouped into five categories

Categories	Factors/Likert scale items
Aesthetic appearance	Attractiveness, Architectural Layout and Sighting name of the mall
Qualitative edge of the mall	Tenant variety, available brands and offers
Way-finding	Ease in finding, knowing sections, visibility of elevators and prominence of entrance
De-stressing facilities	Sitting area, adequate space and greenery
Pre-shopping services	Baggage, promotional stalls, cineplex counters, ATMs visibility and sufficient parking

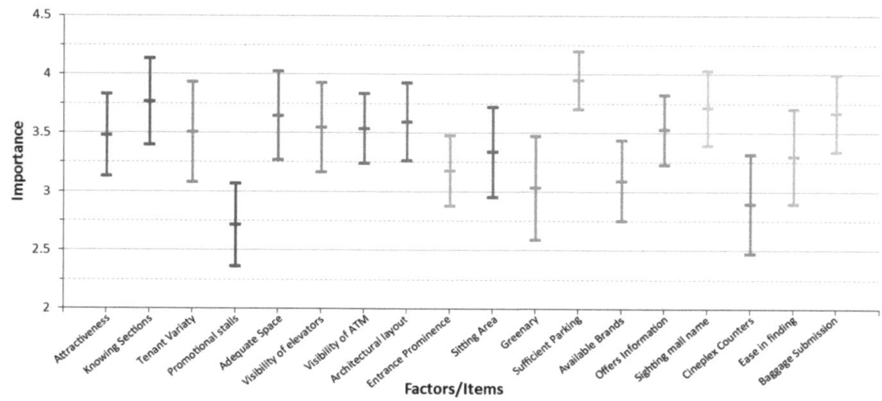

Fig. 1 95 % confidence interval plot of 18 factors according to their importance

(a) (b)

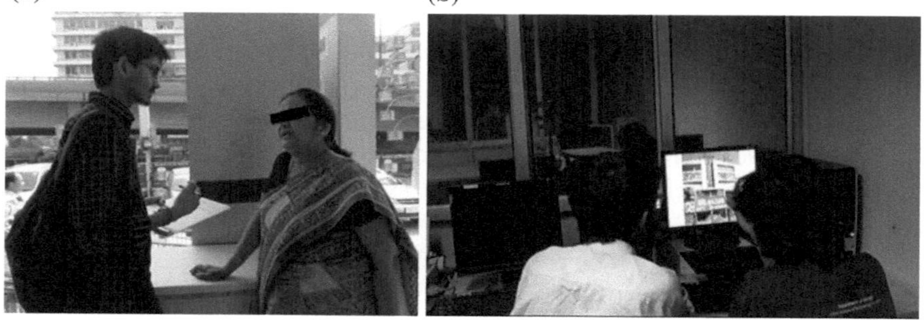

Fig. 2 (**a** and **b**) Respondent during questionnaire about importance of factors (*left*); Eye tracking session (*right*)

facades. Sessions were conducted with 16 participants by showing images of facades of four shopping malls on computers (Fig. 2b). Due to static nature of the eye-tracker, it was not possible to track users' gaze on actual mall. While they were looking at the images, their gaze behaviour which includes position and time-duration of fixations and saccades were recorded using an eye-tracker. The aim of this study was to identify the elements in the Mall facades which visitors look and notice more than others.

4.2.1 Instruments Used

Tobii Eye Tracker, ClearView Analysis Software, Self-clicked images of the facades of four malls in Guwahati, India, Video Camera for documentation.

4.2.2 Experiment Design

Total 16 participants were divided into the four groups of four participants each. Latin square technique was used to assign four different orders of images to these four groups so that results didn't get affected by the order of images shown to participants. Each image was shown for 20 s. Participants were initially briefed about aim of the study and its protocol. Eye-tracking session for each participant started with tracker calibration according to current participant's profile. After calibration, mall façade images were shown to them. A white blank image was also shown for 3 s before each mall façade image for in-between rest purpose and also not to let initial fixation of the mall image affected by the last fixation of previous image. After completing the eye-tracking session, we showed participants computer generated scanpaths of their eye tracking sessions and followed retrospective method to inquire about the reasons for looking at some particular regions in the image more than the others.

4.2.3 Analysis

Participants were divided into 4 groups to remove any bias due to sequence of façade images. Aggregated heat maps for all four groups were analyzed mall wise and each mall image was labeled through areas of Interests (Fig. 3a, AOEs: where looked more). After that, frequency for having a particular spot to be an area of interest in each group were noted down for all four groups and compiled into tables (Fig. 3b).

These tables were helpful for us to analyze which element of a particular Mall façade was noticed with what relative frequency. We drew qualitative inferences about their gazing behaviour using this analysis.

(a)

(b)

AOE	G1	G2	G3	G4	TOTAL
1	1	1	1	1	4
2	0	1	0	0	1
3	1	1	0	0	2
4	1	0	1	1	3
5	1	0	0	0	1
6	1	1	1	1	4

Fig. 3 **a** and **b** Areas of interest (AOEs) labeled on H-mall front (*left*); *Right figure* contains the frequency of occurrence of an AOE in all four groups for H-mall

Fig. 4 Image-collage of H-Mall shown to respondents

4.3 Comparing Four Malls on the Basis of Five Categories

In this study, we showed image-collages (Fig. 4) of the four malls of Guwahati, India digitally to randomly selected 25 respondents and asked them rank the malls relatively on the basis of each of the five categories mentioned before. In retrospective manner, they were asked to give views on their feedback. The aim of this study was to identify most frequent parameters which visitors consider while preferring one mall over the others. Collage of a shopping mall was created such that it could provide maximum visual information. While giving ranking, respondents were allowed to view any of four images as many times as they wanted.

4.3.1 Analysis

Mean of the rankings provided by all respondents were computed for each category individually. We subtracted these mean values from 5 in order to make the numbers and perceived performance of the mall in that category proportional (larger the value, better the mall in the category). Radar graphs for each of the mall were generated where axis are the five categories (Fig. 5).

5 Results

5.1 Questionnaire About Importance of Factors

After computing the Pearson's correlation coefficient for each pair of the factors, we figured out that all the correlation values lay inside the range of -0.6 to 0.6

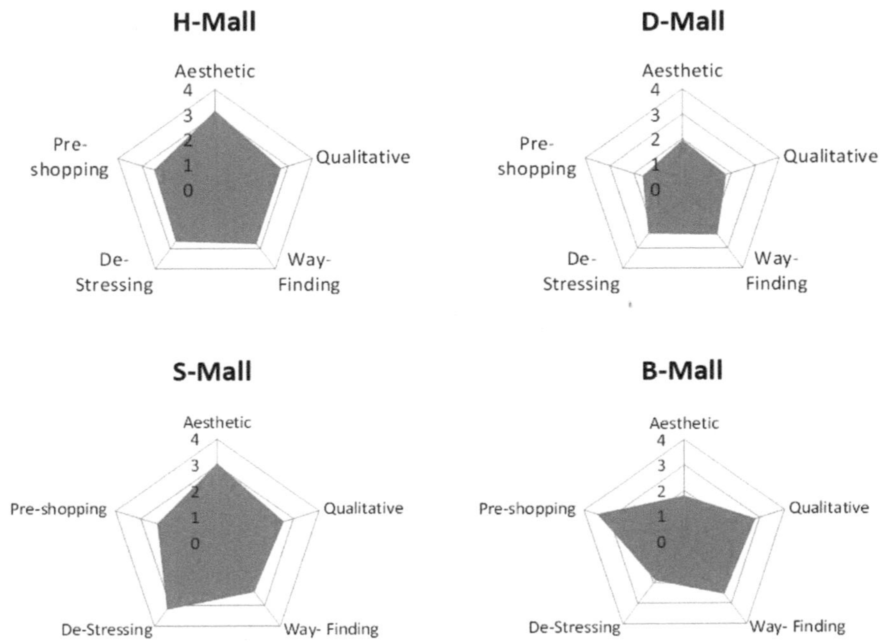

Fig. 5 Radar graphs showing aggregate response on five categories for four malls

Table 3 Most and least important factors

Most important factors	Parking
	Knowing different sections
	Sighting name of the mall
	Baggage submission
Least important factors	Ease in finding way to required location
	Greenery/parks
	Prominence of entrance
	Cineplex counters
	Promotional product stalls

which means no two factors are strongly correlated. Therefore, it can be inferred that all the factors are independent to each other and there is no need to club or omit any of the 18 factors. We plotted 95 % confidence interval charts of all 18 factors (Fig. 1) and compared them pairwise to find out the factors which are significantly more important that other factors. These factors were then arranged according to their importance using method mentioned in Sect. 4.1.3 from which few most and least important factors are compiled in Table 3.

In order to identify the factors significantly more important for utilitarian than hedonist and vice versa also, we compared the responses of the independent groups for each of the factor using Mann-Whitney U test. Adequate open space near entrance

(U(31) = 83.50, Z = −2.24, $p = 0.03$) and ease in finding ways (U(31) = 83.00, Z = −2.18, $p = 0.03$) were found out to be significantly more important for Hedonists than Utilitarian. For other factors, there was no significant difference.

5.2 Eye-tracking Sessions

(a) Mostly crowded or highly dense (with information, visual elements or people) areas became area of initial interest (AoE).
(b) For all four malls, their main name displays on façade wall were areas of interest.
(c) Immediate areas around the entrance were more attention gathering than rest of the building.
(d) Lower part of the malls contained more AoEs as compared to upper part.
(e) Front walls without any information display were not liked.

5.3 Comparing 4 Malls on the Basis of Five Factors

H- Mall was found to be good in all five categories resulting in most balanced among four malls (Fig. 5). Likewise, it can be inferred from radar graph of D-Mall was rated lower than other 3 malls in our study. Particularly, S-Mall was ranked best due adequate space near entrance and good lightings whereas B-Mall was ranked last to lack of space to roam around and cluttering in organization of space. On the other hand, B-Mall was ranked best in pre-shopping services due to availability of ATM inside, sufficient parking space, baggage submission room and promotional stalls outside the entrance.

6 Design Guidelines and Conclusion

Creating a positive perception by using design elements, much before the shopper enters the Mall, is as crucial as after the shopper makes his/her entry. Here, we propose design guidelines which designers and architectures can refer while designing entrance space in malls to create positive perception in visitor's mind beforehand:

(a) *Aesthetic appearance*: Warm and vibrant colors (like red and yellow) for front facade are suggested with the aim to attract visitors. Cool colors are suggested to make visitors feel comfortable as soon as they enter the foyer. Attention in terms of size and placement should be given to main logo display board of the mall as it adds to the attractiveness of the mall.
(b) *Qualitative Edge*: Eye tracking results suggest that banners and billboards of available brands should be placed in the lower half of the facades as they get

ignored in upper half. For banners placement both façade plane and sideways vertical planes should be used. Only single image, few words on big font size and with color contrast should be preferred for design of banners.

(c) *Way-finding*: Entrance can be emphasized by adding symmetric front layout or use of converging forms towards entrance gate. Usage of signage at the entrance, foyer and labeling of products/section should be visible and maximized. Clear sight of view as soon as the shopper enters the foyer leads to spatially position oneself in terms of wayfinding, once inside the mall.

(d) *De-stressing facilities*: Abundant greenery at the entrance along with resting street furniture and presence of water leads to de-stressing, both before and after shopping experience. In comparative study, S-mall had best mean in this category due to spacious foyer and good lighting inside the mall.

(e) *Pre-Shopping services*: Promotional product stalls should either be avoided near the entrance or positioned such that they don't affect visitors' primary actions. More the pre shopping felicities (parking at street level, ATMs, Depositaries) higher is the feeling of being welcomed by the Mall. Positive perception starts much before entering the Mall thereby resulting in increased anticipation and excitement.

This paper presents most and least important factors related to mall entrance according to visitors. Also, here we purpose an approach to compare multiple malls on the basis of five categories (Sect. 4.3). Using similar methodology, this study can be extended for other countries as well in future.

Acknowledgments We would like to thank managers of the Shopping Malls Hub, Dona Planet, Sohum Emporium and Big Bazaar, Guwahati for granting us permission to photograph and make observations on the premises. We also acknowledge our survey respondents for their informed consent to use their pictures and videos for academic research purposes.

References

1. Technopak: Impending Economic Impact of the Resurgent Indian Retail Sector, Images Retail Forum, New Delhi, 9 Sept 2006
2. Real trends: The boom continues. Indian Real Estate Blog: http://www.indiarealestateblog.com/?p=491 (2006)
3. Organized Retail Industry in India: ICFAI Centre for Management Research: http://www.icmr.icfai.org/casestudies/catalogue/Marketing/MKTG114.htm (2006)
4. Sinha, P.K., Banerjee, A.: Store choice behaviour in an evolving market. Int. J. Retail Distrib. Manage. **32**(10), 482–494 (2004)
5. Houston, M.J. and Nevin, J.R.: Retail shopping area image: structure and congruency between downtown areas and shopping centers. In: Monroe, K.B. (ed.) Advances in Consumer Research, vol. 8. Association for Consumer Research, Ann Abor (1980)
6. Doyle, P., Fenwick, I.: How store image affects shopping habits in grocery chains. J. Retail. **50**, 39 (1974)
7. Bearden, W.O.: Determinant attributes of store patronage: downtown versus outlying shopping areas. J. Retail. **53**, 15–22 (1977)
8. Passini, R.E.: Wayfinding design: logic, application and some thoughts on universality. Des. Stud. **17**(3), 319–331 (1996)

9. Taneja, K.: Mall Mania in India—Changing Consumer Shopping Habits. Business School, University of Nottingham, Nottingham (2007)
10. Baker, J., Haytko, D.: The Mall as entertainment: exploring Teen Girls' total shopping experiences. J. Shopping Center Res. **7**(1), 29–58 (2000)
11. What is Placemaking, Project for Public Spaces (extracted on Aug. 2014): http://www.pps.org/reference/what_is_placemaking/
12. Wu, F.: The global and local dimensions of place-making: remaking Shanghai as a world city. Urban Stud. **37**(8), 1359–1377 (2000)
13. Zakariya, K.: Refining tourist's place experience through placemaking: a case study on Middle East tourists in Kuala Lumpur City Centre, Doctoral dissertation, Universiti Teknologi Malaysia, Faculty of Built Environment (2006)
14. Kusumarini, Y., de Yong, S., Thamrin, D.: Entrance and circulation facilities of Malls in Surabaya: auniversal interior design application. Procedia Soc. Behav. Sci. **68**, 526–536 (2012)
15. Babin, B.J., Darden, W.R., Griffin, M.: Work and or fun: measuring hedonic and utilitarian shopping value. J. Consum. Res. **20**(4), 663–673 (1994)
16. Wakefield, K.L., Baker, J.: Excitement at the mall: determinants and effects on shopping response. J. Retail. **74**(4), 515–539 (1998)

PSS for Healthcare Service Engineering, a User-Centered Approach Using Social Network

Tu Anh Duong, Romain Farel, Julie Stal-Le-Cardinal and Jean-Claude Boquet

Abstract User centered design methods appear to integrate user preferences, expectations, use (product or service), and feedback in the design process. Social networks bring to PPS new capacities to increase the access to user rapid feedbacks and to new gathering and analyses methods, mainly for products. By integrating users and their utilities, social networks may answer to one of the main stakes of complex system design and user-centered approach. Telemedicine i.e. clinical healthcare using IT-technologies is an innovative service suffering from long development processes and difficulties to merge and implement in the medical landscape aside the conventional care process. In this context, user centered design approach is challenging to target new perspectives. Considering an application of telemedicine for skin disorders, we used collaborative tools and social-networks to gather data about users (patients) expectations with their skin problems in both conventional and innovative services.

Keywords User centered design · Social networks · Innovative service · Healthcare

1 Introduction

In the highly marketing-time sensitive context of innovative products or services, there is challenge for designers to be able to match the new design to the users' needs. Product-service-system (PSS) approaches enable the emergence of innovative services or products to increase the value for customers [1]. It also integrates

T.A. Duong (✉) · R. Farel · J. Stal-Le-Cardinal · J.-C. Boquet
Laboratoire de Génie Industriel, Ecole Centrale Paris, Chatenay Malabry, France
e-mail: tu-anh.duong@ecp.fr

R. Farel
Paris Saclay Energy Efficiency Institute (PS2E), Loges-en-Josas, France

A. Chakrabarti (ed.), *ICoRD'15 – Research into Design Across Boundaries Volume 1*,
Smart Innovation, Systems and Technologies 34, DOI 10.1007/978-81-322-2232-3_41

business models throughout the product or service lifecycle stages creating innovative value. In the perspective to design the right product or service answering users or customers needs or expectations, PPS design integrates users preferences and stakeholders expectations at the first stages of the design process.

1.1 User-Centered Design

In today very competitive market, user-centered design (UCD) is a relevant approach to meet customer needs satisfactions and to improve the quality of service or products [2–7]. Product or service successes stand on a strong match between users and services [2]. Product design methods embodied users description or preferences using either a top/bottom approach from context to product or a bottom up approach i.e. from product or service to context. At the conceptual stage of the design process it enables the description of users profile, role, need and field of interest [3, 5, 8]. It also enables users needs or use to highly match the product or service functional requirements [2]. UCD emphasizes the knowledge of users psychological ethnological, sociological, organizational, economical aspects [9–12].

For highly competitive complex products such as high-tech products or service, the trend is to integrate users in the product/service design process, at early stage of the design process for ideas generation via co-creation or throughout the design process via co-design [2, 7]. This past decade several tools or methods were developed to determine user experiences, preferences, sensorial perception, daily use, ergonomic, satisfactory and to integrate them in the product or service design [2, 11, 13]. The use of social networks i.e. online communities, networks, groups and supports provided to designers a cheap and fast tool to gather users and customers preferences [14].

UCD through a bottom up design approach enables either to improve preexisting product or service by the integration of users needs or prototype experiences and for non-existent product or service to create them. Users or "clients" influence are set as giving the brief and the constraints, influencing the choice of design methods, giving information, participating in user research and fixing the budget of their involvement. For product service design, social networks enable the caption of end-users feedback about a concept or idea at the conceptual stage and about a product after its release in the market. They contribute to integrate human dynamics behaviors in the co-production process, opening the field of collaborative product development methods research [14].

1.2 Design in Medical Domain

In medical domain, design of devices or services is considered as a specific product/service market segment because of the regulatory requirements, obligations, post marketing engagement, safety assessment and so on. Thus the design process faces business constraints strongly influencing the design process [15–17]. In result designing medical products or services becomes highly complex, risky and costly comparing the same product/service. The strong integration of regulatory constraints and the comprehension of expert users experience are mandatory from the beginning of the design process.

Good design in healthcare requires an appropriate risk analysis or evaluation with a real difficulty to define verification factors to validate the design requirements [16]. Thus, the success of product or service in healthcare relies on the ability for a solution to meet expert users expectations i.e. facility, convenience, easy to understand and daily life improvement [15].

For a standard product, design process includes a step of needs assessment or problem definition, a step of conceptualization, a step of preliminary design and evaluation, a step of detailed design and a step of production [11]. In their process for medical device development, Lourdes et al. added a step of clinical need and team formation, a step of feasibility, and underlined the necessity of a robust step of verification and validation, production planning and qualification, then market introduction and post-launch [15]. The authors introduced in addition to customers preferences or needs assessment the necessity to integrate information about the market environment, the technology, the competition, the regulations, the legal/IP landscape, the reimbursement strategy and finances. At the feasibility stage they also integrate as design requirements: performance characteristics, quality characteristics, product characteristics, market characteristics and regulatory and quality requirements.

1.3 Telemedicine: An Innovative Service

Telemedicine (TM) is the use of IT-technologies to provide medical care or medical advice [18]. TM differs from E-Health services business, as it is considered as a medical, technological and organizational innovation, a medical care process that requires healthcare professionals involvement [19]. E-Health services covers a larger field including information services on Internet, connected medical devices or applications linked to health, it business model mainly depends on the number of users or customers.

This past decade, the technology maturity enabled TM experimentation, and various financial investors gave interest in its development. For public investors such as State, it was considered in a context of resource scarcity as a potential answer to provide an equal medical access to care and to optimize the healthcare system (costs or medical time) [20]. For IT-companies, TM was considered as new field in the huge healthcare market and a promising source of profit.

Unlike E-health business, TM financial model differs from a country to another [19]. In France, the French national Health Insurance Fund for employees (Sécurité Sociale) and complementary private insurances reimbursed medical acts, but none of TM medical acts. In 2010, public investment for TM reached

Table 1 TM and E-health financial investors

			Telemedicine[a]	E-health[b]	Fix the law	Ratio yes/no
Financial Investors	Public	State	Yes	No	Yes	2/1
		Local state	Yes	No	No	1/2
		FNHISW	No	No	No	0/3
		Hospitals	Yes	No	No	1/2
	Private	Hospitals/ clinics	Yes	No	No	1/2
		CHI	Yes	No	No	1/2
		IT-companies	Yes	Yes	No	2/1

FNHISW French National Health Insurance for Salaried Workers
CHI Complementary Health Insurance
[a]IT technologies to provide medical care
[b]Medical devices Medical services using IT technologies Social networks

77 millions euros with a total investment of 180 millions euros [19, 21]. Its turnover was expected to account for 80–140 millions euros, while E-health business was 200–300 millions euros [19, 21]. Table 1 summarizes the role of financial investors in both TM and E-health and identifies at a macro stage level two major financial investors (State and IT-companies).

In France, the public authority is the first investor and the only decision maker able to change the law, for restrictions and new constraints. However, TM suffers from a lack of sustainable funds or developing usage. To date although experiment or solutions exist, there is no participation of national health insurance while public or private sources are major funders without clearly identified financial benefit (Table 1).

Considering TM as a complex system with multiple stakeholders and finalities, we aimed at searching sustainable financial model using a design engineering approach to design a user-centered service. Focus was given in a TM service to provide medical dermatological care, teledermatology (TD). Hypothesizing that social networks users were tomorrow healthcare 2.0 services customers, we conducted a web survey to gather one of the main TD users (the patients) expectations for an user-centered approach.

2 Materials and Methods

TD is the TM application for skin disorders with two described modalities: non real time store-and-forward (clinical images and data transfer, can be performed without the patient) and real time live interactive using video to give a medical consultation at distance (requires the patient) [18].

Figure 1 shows our PPS design approach considering TD as a complex system i.e. service to design. In a previous work, all stakeholders involved in TD service were

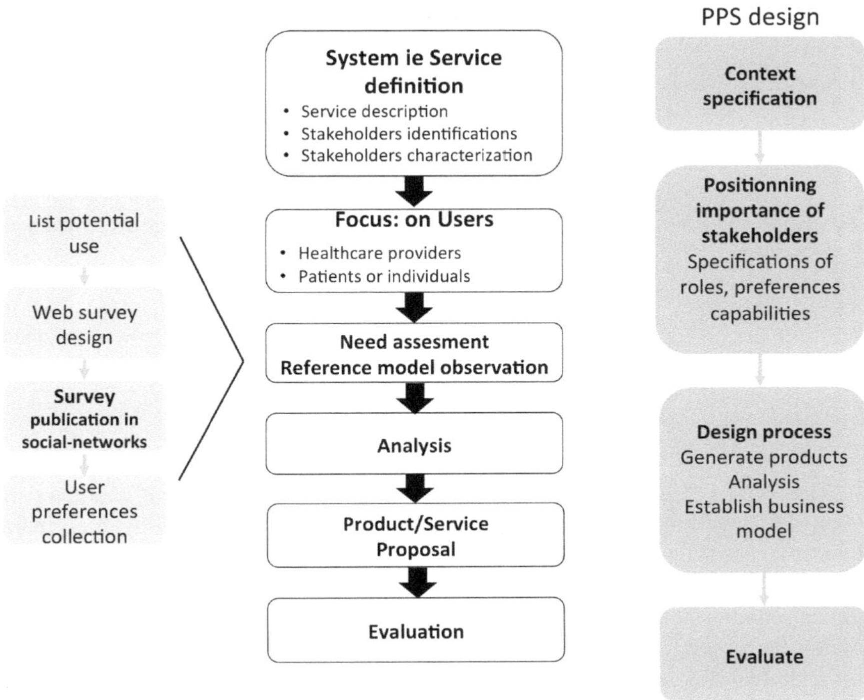

Fig. 1 Methodology to design a teledermatology service regarding a PPS approach

identified and their expectations described. They were classified as financial investors, assessors, requesting physicians, requested physicians, IT professionals. Considering the French law and regulation, focus was given on the system end-users listed in Table 2. Users classification differentiated expert-users i.e. requesting/requested physicians, healthcare professionals and users i.e. patients. Interestingly non-expert users cannot directly interact with TM or TD service that provide medical care, while they can directly use any e-health service for advice or to gather additional information.

This paper describes the characteristics and expectations of one product end user i.e. the patient. To identify their individual route (medical history) or attitude when skin disorders, we conducted a web-survey using social networks. Individual medical routes for a patient with skin disorders were identified using experts' knowledge. First after noticing their disorders, patients or users could check themselves (internet or ask a friend), ask a non-practitioner, go to the pharmacy, go to the primary care physician (private practice or hospital), go to the dermatologist. Then after visiting their primary care physician skin disorder may be solved or sent to a specialist.

With the strong hypothesis that social network users could be potential E-health services or medical care users or more interested in TD deployment or development, we built a questionnaire and sent it using a specific social network link. Only adults ($\geq$18 year-old) were included in the study.

Table 2 Teledermatology, store and forward and live interactive users

Users	Store and forward[a]	Live interactive[b]
Requesting physician	Provide clinical data Take photo, integrate the data in the application	Ask for medical consultation, assist or not consultation
Patient	Is taken in picture, his data are sent	Assist to at distance consultation
Requested physician dermatologist	Give a medical advice, write a report	Give a medical consultation at distance
Healthcare professionals: nurses, assistant	Integrate or not the data in the application	Assist the patient and the dermatologist during the consultation and provide additional information

[a]Non real time transfer of clinical images and data
[b]Real time video-consultation with a physician at distance

The aim of this 20 questions survey was to describe individual characteristics (age, sex, location, socio-professional category, mean salary), individual characteristics regarding dermatology (mean expanses, have a dermatologist, number of consultation for skin disorders within the past 5 years, have chronic dermatological conditions), individuals attitude regarding skin disorders (what did you do within the past 5 years when you had a skin disorders? mean waiting time to ask for a dermatologist if skin disorders, mean lead time to get a consultation), individuals knowledge of TD (do you know what is TD? interest for TD, mean acceptable cost for this service without reimbursement, interest in getting medical advice via E-health application, mean acceptable cost for this service).

Qualitative variable were described as percentages (%), and compared using χ^2 test, $p < 0.05$ indicated statistical significance. Data were analyzed using SPSS™ statistics.

3 Results

3.1 Main Web Survey Results

Social networks enabled to collect 319 answers within 10 days. Individuals mean age was 38 year-old (17–85) and sex ratio (F/M) was 2.3. Higher managerial or professional occupations represented 56.7 % ($n = 181$) of the investigated individuals, students 13.2 % ($n = 42$). Main results are summarized in Table 3.

In case of skin disorders, most of the patient did not visit a dermatologist. Individuals were interested in E-Health services to identify whether they needed a medical visit or to get additional information. The percentage of individuals without a regular dermatologist interested in directly getting an advice via E-health application 46 % was significantly higher than those with a regular dermatologist 35 %, $p = 0.028$. In univariate analysis, individuals interested in TD had

Table 3 Questionnaire main answers

		Percentages % (n)
Individuals characteristics regarding dermatology	Mean expanses for dermatology	50–100 = 31 % (99) 100–500€ = 36 % (120)
	Have a dermatologist	Yes: 42.6 % (136)
	Number of consultation for skin disorders within the past 5 years	Never: 30.4 % (97)
Individuals attitude regarding skin disorders	What did you do within the past 5 years when you had a skin disorders? (Missing data = 99)	Nothing 8.7 % (21) self medication 5.8 % (14)
		Ask a friend 3.7 % (9) Check on internet 1.7 (4)
		Go to the pharmacy 7.5 (18)
		Go to the general practitioner 28.6 % (69)
		Go to the dermatologist 44 % (106)
Individuals knowledge of TD	Mean waiting time to ask for a dermatologist if skin disorders	>15 days = 38.6 % (80)
	Mean lead time to get a consultation	1–2 weeks = 38.7 % (87)
	Do you know what is TD?	No: 78.6 (220)
	Interest for TD	No: 11.3 % (36)
	Mean acceptable cost for this service without reimbursement	No: 35.1 % (112) Maximum 25 € = 21.6 % (68)
	Interest in getting medical advice via E-health application	No = 18.8 % (60)
		Yes to see if I need a medical visit 58.6 % (187)
		Yes to get information 36.7 % (117)
		Yes to check my prescription 19.1 % (61)
	Mean acceptable cost for this service	No: 32.9 % (105) Maximum 25€ 39.8 % (127)

higher expanses for their skin $p = 0.028$ and were significantly more interested in E-health application or paying extra fees for E-health services or TD medical care ($p < 0.001$). In multivariate analysis, sole the interest for E-health or paying for E-health services or TD medical care remained significant ($p = 0.045$, $p = 0.03$, $p = 0.009$).

Considering our results, one of TM or TD potential financial model especially for IT-companies could be to develop E-health services that are not submitted to high regulation constraints in comparison with TM services. For public financial stakeholders investment in a medical innovative service could favor the development of non-medical services providing advices and not medical care.

4 Discussion

Social networks enabled us to rapidly gather user expectations and potential financial implication. Our study limitations mainly rely on the non-representative sample. Most of the individuals belong to the higher managerial or professional occupations category, 67 % spending between 50 and 500€ for their skin products. Nevertheless this survey confirms the great interest of this category in new technologies or new services. Most of the interviewed individuals were interested in TD application for medical consultation, but also in an E-service giving them the possibility to directly solicit the dermatologist only for an advice without medical prescription. Most of them accepted to pay for this new service without reimbursement for both medical consultations and advices. To date, regarding TD pilot experimentations, individuals as patients do not pay for this new service. Looking for a sustainable financial model, our study highlights that TM should not only rely on public finance or national health insurance reimbursement. Our study displays in a selected population category their readiness to pay for a new medical service but also for medical advice (Fig. 2).

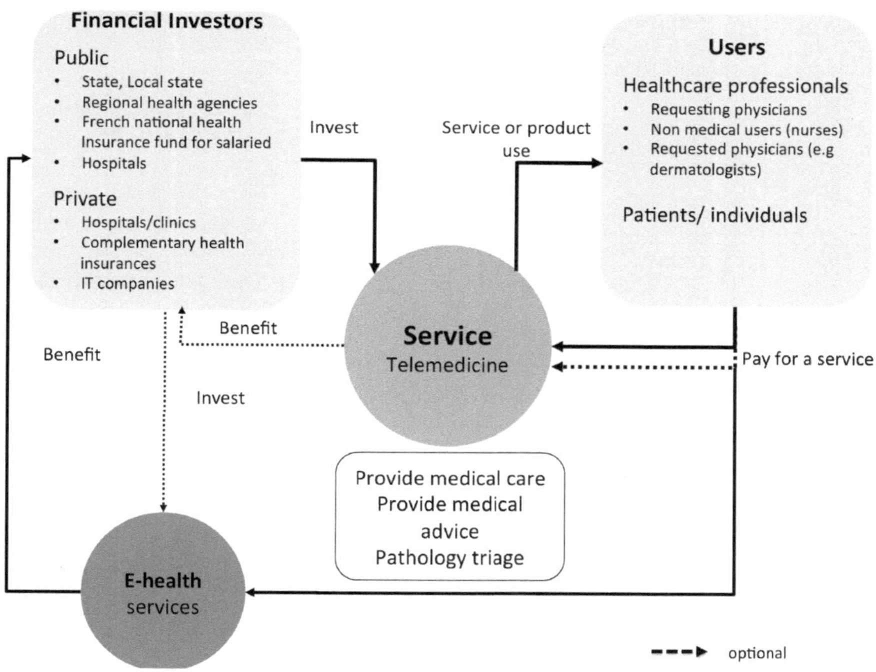

Fig. 2 TM and E-health services financial investors and users role

5 Conclusion

Our PPS approach enabled us to identify users/patients interest in TM new service and their potential ability to pay for this new service. This user-centered approach also highlighted users interest in connected services as medical advice. Then paying for E-health services may contribute for industrials to consolidate their investment in TM and provide a sustainable financial system. Medical care at distance owing to TM, seems in the particular field of skin diseases to find it path in the users needs. For a dermatological medical service, one option could be to offer a scalable service at first providing advice and information then if necessary specialist consultation at distance. This provides new perspectives in the field of medical service. Our paper only focused on one users whom role is not until now considered as crucial in the financial landscape of medical care, the financial role of expert users i.e. healthcare professional should also be investigated. For a non existing large-scale service, these investigations highlight other financial source to strengthen the model of medical care at distance aside any classical reimbursement. To confirm our social network survey or unbiased the results, this questionnaire will be send to a representative sample of the French population and compared to these results.

Acknowledgments The authors acknowledge Pr Pierre Wolkenstein from the department of dermatology, Hôpital Henri Mondor.

References

1. Vasantha, G.V.A., Roy, R., Lelah, A., Brissaud, D.: A review of product-service systems design methodologies. J. Eng. Des. **23**(9), 635–659 (2012)
2. Stappers, P.J., van Rijn, H., Kistemaker, S.C., Hennink, A.E., Sleeswijk Visser, F.: Designing for other people's strengths and motivations: three cases using context, visions, and experiential prototypes. Adv. Eng. Inform. **23**(2), 174–183 (2009)
3. Cardoso, C., Clarkson, P.J.: Simulation in user-centred design: helping designers to empathise with atypical users. J. Eng. Des. **23**(1), 1–22 (2012)
4. Ward, J., Buckle, P., John Clarkson, P.: Designing packaging to support the safe use of medicines at home. Appl. Ergon. **41**(5), 682–694 (2010)
5. Goodman-Deane, J., Langdon, P., Clarkson, J.: Key influences on the user-centred design process. J. Eng. Des. **21**(2–3), 345–373 (2010)
6. Santoro, F.M., Borges, M.R.S., Pino, J.A.: Acquiring knowledge on business processes from stakeholders' stories. Adv. Eng. Inform. **24**(2), 138–148 (2010)
7. Sanders, E.B.N., Stappers, P.J.: Co-creation and the new landscapes of design. Co-Des. **4**(1), 5–18 (2008)
8. Ki Moon, S., McAdams, D.A.: A market-based design strategy for a universal product family. J. Mech. Des. **134**(11), 111007 (2012)
9. Kouprie, M., Visser, F.S.: A framework for empathy in design: stepping into and out of the user's life. J. Eng. Des. **20**(5), 437–448 (2009)
10. Smith, G.C., Smith, S.: Latent semantic engineering—A new conceptual user-centered design approach. Adv. Eng. Inform. **26**(2), 456–473 (2012)
11. Yazdani, B.: Four models of design definition: sequential, design centered, concurrent and dynamic. J. Eng. Des. **10**(1), 25–37 (1999)

12. Rezgui, Y., Hopfe, C.J., Vorakulpipat, C.: Generations of knowledge management in the architecture, engineering and construction industry: An evolutionary perspective. Adv. Eng. Inform. **24**(2), 219–228 (2010)
13. Red, E., French, D., Jensen, G., Walker, S.S., Madsen, P.: Emerging design methods and tools in collaborative product development. J. Comput. Inf. Sci. Eng. **13**(3), 031001 (2013)
14. Ahram, T., Karwowski, W., Amaba, B.: Collaborative systems engineering and social-networking approach to design and modelling of smarter products. Behav. Amp. Inf. Technol. **30**(1), 13–26 (2011)
15. Medina, L.A., Kremer, G.E.O., Wysk, R.A.: Supporting medical device development: a standard product design process model. J. Eng. Des. **24**(2), 83–119 (2013)
16. Alexander, K., Clarkson, P.J.: A validation model for the medical devices industry. J. Eng. Des. **13**(3), 197–204 (2002)
17. Clarkson, P.J., Buckle, P., Coleman, R., Stubbs, D., Ward, J., Jarrett, J., Lane, R., Bound, J.: Design for patient safety: a review of the effectiveness of design in the UK health service. J. Eng. Des. **15**(2), 123–140 (2004)
18. Warshaw, E.M., Hillman, Y.J., Greer, N.L., Hagel, E.M., MacDonald, R., Rutks, I.R., Wilt, T.J.: Teledermatology for diagnosis and management of skin conditions: a systematic review. J. Am. Acad. Dermatol. **64**(4), 759–772 (2011)
19. livre_blanc_telemedecine_avril2013.pdf »
20. Loi Hôpital Patients Santé Territoires, *Loi Hôpital Patients Santé Territoires*. [En ligne]. Disponible sur: http://www.sante.gouv.fr/la-loi-hopital-patient-sante-et-territoires.html
21. La télémédecine, un secteur d'avenir, *Le Figaro*, 04-juin-2012. [En ligne]. Disponible sur: http://www.lefigaro.fr/societes/2012/06/04/20005-20120604ARTFIG00693-la-telemedecine-un-secteur-d-avenir.php. [Consulté le: 29-mai-2014]

Skimming and Scrutinizing: Quantifying Two Basic Patterns of Visual Behavior in Design

Quentin Lohmeyer, Moritz Mussgnug and Mirko Meboldt

Abstract In design research recently an increasing number of eye tracking experiments have been conducted. The evaluation of the recorded data usually bases on analyzing a set of individual scan paths. In order to gain more value from the data and thus, to better explain human behavior in design, the authors proposed a differentiation of two basic scan path patterns. The first, skimming, is applied when a person wants to get an overview of a visual stimulus, while the second, scrutinizing, indicates that a person tries to understand its details. Although these patterns of visual behavior describe basically different cognitive processes, their differentiation still bases on personal judgment. This paper presents the results of an investigation that aimed to quantify the patterns of skimming and scrutinizing. It introduces an algorithm that, based on numerical criteria, allows an automated detection of both skimming and scrutinizing sequences in eye tracking data.

Keywords Human behavior · Eye tracking · Scan path pattern · Design representation

1 Introduction

Applying eye tracking technologies in the research of human behavior in design recently has gained in importance. Due to advances in the usability of modern eye tracking systems, the reduction of their costs and improvements in their accuracy, an increasing number of eye tracking experiments has been conducted to investigate the visual behavior of designers and users during their specific processes of thinking and acting.

Q. Lohmeyer (✉) · M. Mussgnug · M. Meboldt
Product Development Group, ETH Zurich, Zurich, Switzerland
e-mail: qlohmeyer@ethz.ch

A. Chakrabarti (ed.), *ICoRD'15 – Research into Design Across Boundaries Volume 1*,
Smart Innovation, Systems and Technologies 34, DOI 10.1007/978-81-322-2232-3_42

A basic concept in analyzing eye tracking data is the 'scan path'. A scan path represents a person's visual attention as a sequence of visually focused locations (fixations) and eye movements from one focused location to another (saccades) [1]. A fixation is especially characterized by the location coordinates (Where did the person looked?) and the fixation duration (How long did the person looked at this location?). Consequently, a saccade is also determined by the location coordinates (From where to where did the gaze jump?). A common measurement that results from this data is the saccade length, which can be specified relative to the stimulus (in pixel) or relative to the eye (in degree). The latter is also called saccade amplitude.

Most research results gained on basis of scan path data refer to measured fixation durations and saccade amplitudes. However, these central measurements are often evaluated separately and without an adequate consideration of interdependencies. This paper presents two basic scan path patterns that are defined based on a combined evaluation of a scan path's fixations and their subsequent saccades.

The definition of the two scan path patterns goes back to an observation made in a previous eye tracking experiment that investigated the scan paths of engineering designers trying to understand a mechanical system represented in a 2D sectional drawing [2]. As one key result of this study, it has been shown that while analyzing the drawings, the test persons alternately used two contrary scan path patterns that are called 'skimming' and 'scrutinizing' (cf. Fig. 1).

Skimming is defined by short fixations and long saccades, which indicates the visual exploration of a larger area. It is assumed that skimming is applied, when a person aims to get an overview of a system and tries to identify relations between single elements from a high-level perspective. Scrutinizing is characterized by long fixations and short saccades. Scrutinizing indicates an accurate analysis of a small area. Consequently, it is applied when a person tries to understand certain details. Although skimming and scrutinizing describe basically different cognitive processes, their differentiation still bases on personal judgment.

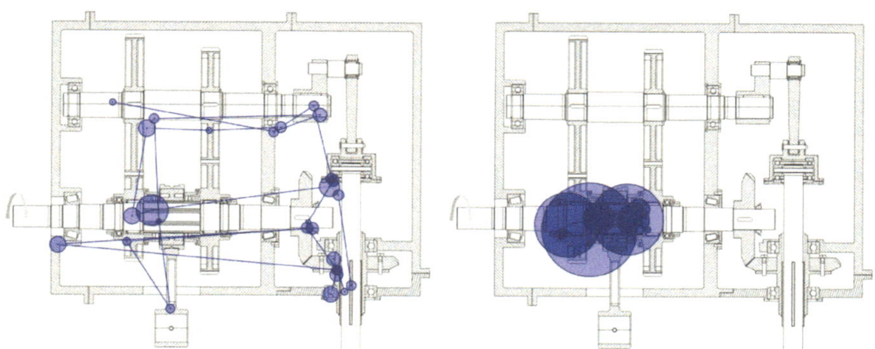

Fig. 1 Basic types of visual behavior: skimming (*left*) and scrutinizing (*right*) in 7 s scan paths of one test person trying to understand an engineering drawing [2]

This paper investigates the basic questions of how the patterns of skimming and scrutinizing can be quantified and how skimming and scrutinizing can be automatically detected from regular scan path data.

2 Basic Patterns of Visual Behavior

2.1 Stimulus-Driven Examination of Pictures

In the context of researching visual behavior during free examination of pictures, the two scan path patterns presented above have already been known for decades. In fact, there was a famous eye tracking experiment conducted in 1935 by the American psychologist Guy Thomas Buswell, who recorded and evaluated the eye movements of 200 test persons looking at pictures of several works of art.

Buswell [3] reports that two general patterns of perception became apparent in his records. He states that "one of these patterns consists of a general survey, in which the eye moves with a series of relatively short pause over the main portions of the picture; and a second type of pattern was observed, in which a series of fixations, usually longer in duration, are concentrated over small areas of the picture, evidencing detailed examination of those sections".

In the 1980s researchers proposed different terminologies to describe these basic patterns. Groner et al. [4], for instance, distinguish global and local aspects of scan paths, whereas Leibowitz et al. [5] differentiate between ambient and focal visual processing. In contrast to these examples, recent research tends toward emphasizing the cognitive processes rather than the eye movements and thus, the terms of skimming and scrutinizing have become increasingly accepted [6, 7].

Although these patterns can be found in almost every eye tracking data set and a systematic data evaluation regarding phases of skimming and scrutinizing provides the opportunity to better understand human cognition, only few research has aimed to quantify the patterns' characteristics. In this context, valuable findings were made by Unema et al. [8], who showed that in the first 20 s of free pictures examination, fixations durations increase exponentially and saccade sizes decrease correspondingly (cf. Fig. 2). The curves indicate that there is just one single changeover from skimming to scrutinizing at about 3 s of viewing time. Important threshold values seem to be the fixation duration of 240 ms and the saccade amplitude of 5.5°.

2.2 Task-Driven Examination of Engineering Drawings

In engineering design the examination of visual representation is usually not only stimulus-driven, but determined by the specific design task. Here, eye tracking experiments (e.g. Fuge and Kara [9], Cölln et al. [10]) have clearly shown strong dependencies of human visual behavior and the task assigned to the observer.

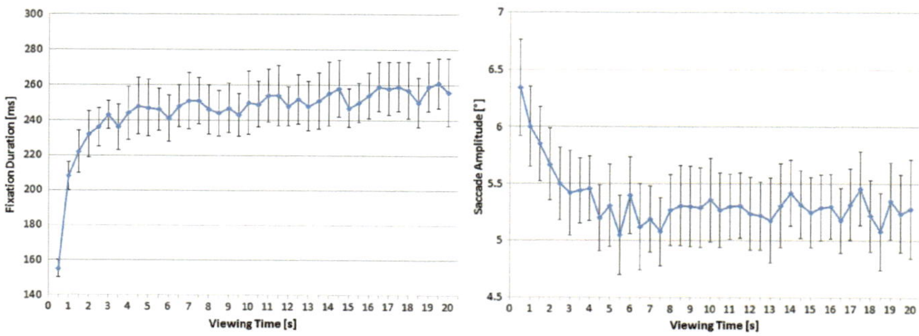

Fig. 2 Fixation duration (*left*) and saccade amplitude (*right*) as a function of viewing time (20 s) in free picture examination (n = 15) according to Unema et al. [8]

In the context of examining engineering drawings, Lohmeyer et al. [11] investigated task-dependencies in the visual behavior of engineering designers. Based on the data gained in their eye tracking experiment, they showed similarities in the visual behavior of test persons (n = 26) solving the same task as well as differences in the visual behavior across different tasks. In contrast to free picture examination, understanding an engineering drawing requires considerably more time than 20 s. Based on these results, Ognjanovic et al. [12] showed significant differences in fixation time proportion by subsystems depending on the specific task assigned.

Figure 3 depicts a typical course of fixation duration and saccade amplitude across the first 120 s of a single test person reading an engineering drawing. In contrast to Fig. 2, the measurements are not averaged. Thus, the curves represent the actual visual behavior of this individual person. Based on these results the following observations can be made: (1) The maximum values of fixation duration are partly much higher than 240 ms. In fact, there are peeks up to 1,740 ms. (2) Most saccade amplitudes are smaller than 5° and there are also a few very long saccades with amplitudes above 30°. However, the most important difference is that (3) both curves show no overall trend of increase or decrease. Instead, fixation

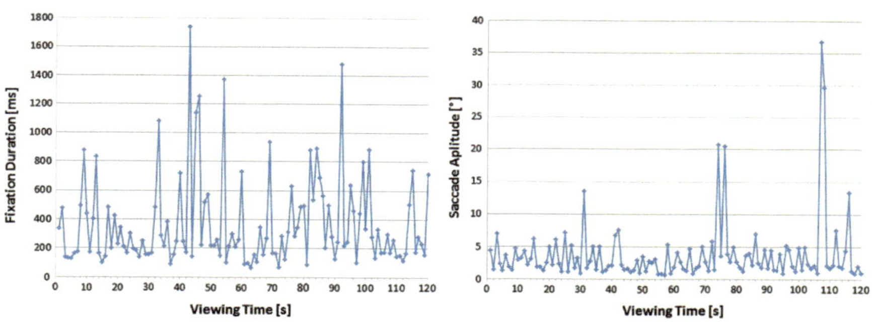

Fig. 3 Fixation duration (*left*) and saccade amplitude (*right*) as a function of the viewing time (120 s) in task-depending examination of the sectional drawing shown in Fig. 1 (n = 1)

duration as well as saccade amplitude are continuously alternating between a rather high and a rather low value. In the scan paths this alternation is recognized as skimming and scrutinizing patterns.

2.3 Research Question

The comparison of the results from stimulus-driven and task-driven visual behavior shows that, although the scan path patterns of skimming and scrutinizing appear in both evaluations, only in task-driven scan paths the patterns alternate continuously. While it is relatively easy to qualitatively identify skimming and scrutinizing by just watching the scan paths visualizations, the definition of adequate numerical differentiation criteria is more challenging. Thus, the basic research question addressed in this paper is: How can the patterns of skimming and scrutinizing be automatically detected in a scan path and what are the numerical criteria required for this purpose?

3 Methods

In order to address the research question, a three-step course of action has been chosen. Its basic idea is to first select one specific reference scan path to be the central object of investigation. By application of a qualitative evaluation method, in a second step, this scan path is systematically split into skimming and scrutinizing sequences. Due to this, the numerical characteristics of the change points can be analyzed and described. Based on these results finally an algorithm has been developed that is able to reproduce the scan path pattern differentiation by using numerical criteria.

3.1 Reference Selection

The reference scan path was recorded in a previous eye tracking study, which investigated the visual behavior of engineering designers (cf. [2]). Here, the stimulus was the 2D sectional drawing of an advanced gear drive that is also shown in Fig. 1. The test person was a male Ph.D. student graduated in mechanical engineering. In the experiment he was assigned to analyze the drawing in detail and to find and mark an essential functional design error.

The experiment was conducted by application of the remote eye tracking system SMI RED 250 (sampling rate: 250 Hz, position accuracy: 0.4°, tracking resolution: 0.03°) with a 22″ flat screen (solution: 1,680 × 1,050 pixels, operating distance: 70 cm). Fixations were defined after duration criterion ($\geq$80 ms) and dispersion criterion ($\leq$100 px).

3.2 Change Point Detection

In the study, three additional test persons were asked to subdivide the whole 5 min of the reference scan path in sequences of skimming and scrutinizing as exactly as possible and to note the corresponding points in time, when the pattern changed. All test persons were familiar with scan path visualizations in general and with the concept of the skimming and scrutinizing in particular. The data was made accessible to the test persons by using a video showing the scan path in front of the engineering drawing. The test persons were asked separately and they were allowed to watch the video several times. After their evaluation the test persons were asked about how they had come to their decisions. In this context special attention was drawn on the numerical criteria they used to detect a change point.

3.3 Algorithm Development

The change points detected by the test persons were visualized in a sequence diagram to represent and compare the skimming and scrutinizing patterns in this specific reference scan path. This diagram represented the benchmark for the algorithm able to detect the same change points and sequences by a quantitative evaluation. Insights from the interviews regarding the single test persons' decision process were used to develop the logical sequence of the algorithm.

4 Results

4.1 Results of Change Point Detection

Within the reference scan path the test persons detected at least 22 points in time, where the scan path pattern changed from skimming to scrutinizing or the other way round. Figure 4 depicts the evaluation details and highlights the resulting scrutinizing sequences.

The figure also shows that the test persons have come to similar results, although there are some variations in the temporal recognition of the change points. In addition to this, it became apparent that some sequences have not even been recognized by all test persons. For example, in the period from 03:50 to 03:55 only one single test person detected a scrutinizing sequence.

Furthermore, the interviews revealed the following key insights about how the test persons made their decisions: (1) The test persons rather looked for the starting and ending points of scrutinizing than of skimming sequences to detect change points. (2) They further stated that they had always tried to simultaneously consider fixation duration and saccade length. Thus, the test persons especially looked at pairs consisting of a fixation and its subsequent saccade.

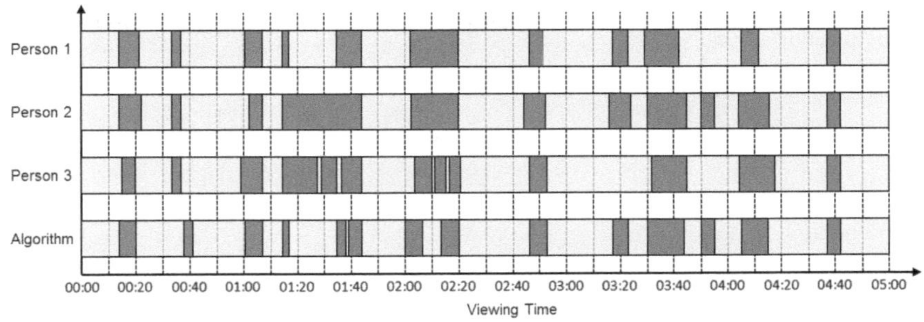

Fig. 4 Skimming sequences (*light gray*) and scrutinizing sequences (*dark gray*) detected in the reference scan path by three tested test persons and the developed algorithm

In those cases where two sequent pairs were characterized by both a noticeable long fixation (>240 ms) and a short saccade (<0.5°), the test persons supposed the start of a scrutinizing sequence. But to reach the final decision, they usually waited until they had seen 3–5 following pairs. Only if all these following pairs had almost similar fixation duration and saccade length, the test persons classified a sequence as scrutinizing. The ending points of scrutinizing sequences were detected by recognition of at least two misfit pairs characterized by too short fixations (<180 ms) or too long saccades (>1.2°).

Based on the results presented in this section, an algorithm was developed, which is able to calculate the detected change points directly from an eye tracking data file.

4.2 Results of Algorithm Development

The basic idea of the algorithm is to emulate the evaluation process applied by the test persons. In contrast to human beings, the algorithm is able to directly process numerical eye tracking data without an intermediate step of a graphical scan path representation. This allows an exact calculation of the change points, along with an objectified differentiation of skimming and scrutinizing sequences and thus, an improved comparability of the patterns across several recorded scan paths.

Figure 5 shows a flow chart describing the algorithm's structure in detail. It starts by reading in the specific measurements of the first pair (i = 1) of a fixation and its subsequent saccade. The fixation duration (t_i) is tested to be larger than criterion a (=240 ms) and the saccade amplitude (α_i) to be smaller than criterion b (=0.5°). If none or just one criterion is met, this pair definitely is part of a skimming sequence, but if both criteria are met, the pair might be the starting point of a scrutinizing sequence. To check this, the second pair (i = 2) is tested for the same criteria a and b.

As known from the qualitative evaluation, two pairs successfully passing this test are enough to suppose the start of a scrutinizing sequence (i.e. n = 2). It is

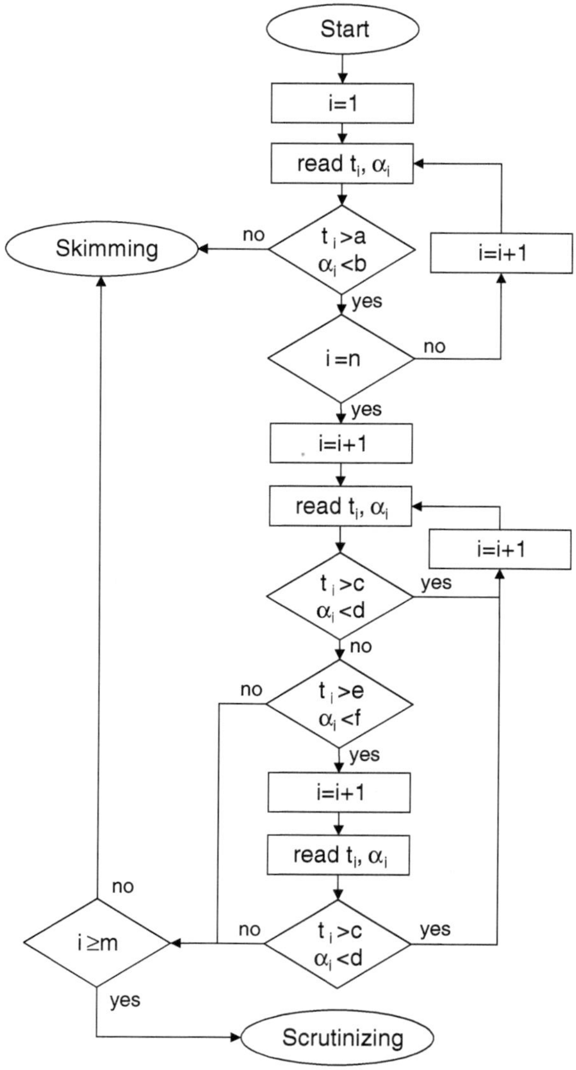

Fig. 5 Flowchart of an algorithm allowing an automated detection of skimming and scrutinizing sequences in eye tracking data based on numerical criteria

also known that the criteria for the following pairs are less hard. Thus, the fixation durations (t_i) are now tested to be larger than criterion c (=200 ms) and the saccade amplitudes (α_i) to be smaller than criterion d (=0.8°). After a certain number ($m = 6$) of passed tests in series, the decision is made that the tested pairs are part of a scrutinizing sequence.

If a pair fails in the test on criteria c and d, it is checked whether or not this pair is an acceptable outlier. In case the fixation duration is larger than criterion e (=180 ms) and the saccade amplitudes smaller than criterion f (=1.2°), it is acceptable, which means the series continues, but the next pair again has to meet criteria c and d. If a pair fails the final test, the algorithm stops. In this case the number of pairs passing the tests are counted. If the number is equal or higher than m, the pairs build a full scrutinizing sequence and its starting and ending point are displayed. If the number is lower than m all pairs are added to the previous skimming sequence. Then the algorithm starts again.

The algorithm was exemplary applied to automatically subdivide the reference scan path into skimming and scrutinizing sequences. The input values (a–f, n and m) were set as presented above. The input file contained eye tracking data from the 5 min of the reference scan path including all fixation durations and saccade amplitudes. Based on this setup the algorithm detected a total of 24 change points. The resulting skimming and scrutinizing sequences are also depicted in Fig. 4.

It can be seen that the sequences calculated by the algorithm basically correspond to the results from the test persons' evaluations. However, there are two differences that need to be highlighted. Some scrutinizing sequences have been detected correctly regarding their duration, but are displayed slightly time shifted. Another point became apparent is that the algorithm detected more change points than the single test persons. This indicates that the algorithm is able to graduate sequences more finely.

5 Conclusion and Outlook

5.1 Conclusion

This paper introduced the two basic scan path patterns skimming and scrutinizing, which especially in task-driven examinations allow to distinguish between sequences of visual detail investigation and visual overview exploration. It was shown that skimming and scrutinizing can be qualitatively detected by persons analyzing a video that displays the course of a scan path. Since this form of evaluation strongly depends on personal judgment, the exact results slightly vary from one person to another.

The algorithm presented in this paper allows an objective detection of a scan path's skimming and scrutinizing sequences via an evaluation based on clearly defined numerical criteria. By using the example of a reference scan path recorded in an eye tracking study from engineering design, skimming and scrutinizing sequences were evaluated by both (1) test persons, who are familiar with the analysis of eye tracking data, and (2) the algorithm, whose logical structure is oriented to the human evaluation process. The comparison of the results show that the algorithm, in this specific case, is able to detect the change points of the sequences with an adequate accuracy. When long scrutinizing phases were interrupted by very short skimming sequences the algorithm's differentiation was even more precise than the one detected by the test persons.

However, there have also been single discrepancies in the results. As an example, the algorithm detected one of the scrutinizing sequences more than 4 s too late.

Effects like this need to be further investigated in order to better understand the scan path patterns' characteristics and to improve the performance of the algorithm.

5.2 Outlook

In this paper the presented algorithm was only tested regarding a single reference scan path. In a next step it is important to investigate its applicability within the evaluation of a full eye tracking data set including the scan paths of several persons.

It can be assumed that across different eye tracking experiments the input values need to be adjusted to the specific characteristics of each study. On one hand the variety of the stimuli shown and the tasks assigned might influence the threshold values. On the other hand the chosen criteria for event detection strongly determine the range of fixation durations and saccade amplitudes, which has to be considered in the definition of the algorithm's input values. Consequently, future research work will address the analysis of relevant interdependencies in order to support the adjustment of these input values.

In the context of design, the application of mobile eye tracking seems to be most promising to investigate the behavior of designers and users in their natural environment. In contrast to remote systems, eye tracking glasses measure the scan path relative to the scene video showing the test person's field of view, which is recorded by an integrated camera. Future experiments will show if the patterns of skimming and scrutinizing are also suitable to evaluate the data gained from mobile eye tracking studies.

References

1. Holmqvist, K., Nyström, M., Andersson, R., Dewhurst, R., Jarodzka, H., van de Weijer, J.: Eye tracking: a comprehensive guide to methods and measures. Oxford University Press, New York (2011)
2. Lohmeyer, Q., Mussgnug, M., Matthiesen, S., Meboldt, M.: Analysing visual behaviour in engineering design by eye tracking experiments. In: International Symposium on Tools and Methods of Competitive Engineering (TMCE 2014), Budapest, Hungary (2014)
3. Buswell, G.T.: How People Look at Pictures. Chicago Press, Chicago (1935)
4. Groner, R., Walder, F., Groner, M.: Looking at faces: local and global aspects of scanpaths. Theor. Appl. Aspects Eye Mov. Res. **22**(9), 523–533 (1984)
5. Leibowitz, H.W., Shupert, C.L. Post, R.B.: The two modes of visual processing: implications for spatial orientation. NASA PVHD Conference 1983. Edwards, CA (1983)
6. Duggan, G.B., Payne, S.J.: Skim reading by satisfying: evidence from eye tracking. In: ASM Conference on Human Factors in Computing Systems (CHI 2011), Vancouver (2011)
7. Sharma, S.Y., Chakravarthy, B.K.: How people view abstract art: an eye movement study to assess information processing and viewing strategy. In: International Conference on Research into Design (ICoRD'13), Bangalore (2013)
8. Unema, P.J.A., Pannasch, S., Joos, M., Velichkovsky, B.M.: Time course of information processing during scene perception: the relation between saccade amplitude and fixation duration. Vis. Cogn. **12**(3), 473–494 (2005)

9. Fuge, M.D., Kara, L.B.: A testing method and cognitive model of human diagram understanding for automated design sketch recognition. In: ASME International Design Engineering Technical Conferences (IDETC 2009), San Diego (2009)
10. Cölln, M.C., Kusch, K., Helmert, J.R., Kohler, P., Velichkovsky, B.M., Pannasch, S.: Comparing two types of engineering visualizations: task-related manipulations matter. Appl. Ergon. **43**(1), 48–56 (2012)
11. Lohmeyer, Q., Matthiesen, S., Meboldt, M.: Task-dependent visual behaviour of engineering designers: an eye tracking experiment. In: International Design Conference (DESIGN 2014), Dubrovnik (2014)
12. Ognjanovic, S., Lohmeyer, Q., Hölscher, C., Meboldt, M.:Visual behavior in task-driven reading of engineering drawings. In: International Conference on Human Behavior in Design (HBiD 2014), Ascona (2014)

Sound Association for Product-Sound Design Using Semiotics

Kumari Moothedath Chandran, Prajakta Prabhune and Dibakar Sen

Abstract Sounds from products which carry information about its performance can be called as product sounds. Product sounds can aid visual display and controls if designed effectively. In this work, the authors collected 6,190 sound samples. Out of this, 670 sounds were coded, based on three categories of abstractions, namely *index* (on causality), *icon* (on similarity) and *symbol* (arbitrary). The coded sounds were used to associate them to selected examples of two product interfaces. Design exercises and user surveys were done to confirm the coding of individual sounds, as well as the method of association. Three different methods were used for the surveys. One—to associate sounds based on a given theme; two—to match the sounds provided to the product interface given; three—the subjects were made to listen to selected sounds and asked to describe the meaning they perceive of it. The results from the surveys tallied well with the method of categorization of sounds and their associations, and it provides a structured way of product-sound design.

Keywords Product sound · Semiotics · Sound association

1 Product Sounds

Sounds emanated from products which carry information about its functioning can be called as product sounds. A suitable and supportive product sound contribute positively to a product's perceived quality by the user [1]. Consumers perceive auditory cues to be as important as visual cues in many product categories such as vehicles, domestic durables, and high-tech products [2, 3]. Hence, product sounds

K.M. Chandran (✉) · P. Prabhune · D. Sen
Centre for Product Design and Manufacturing, Indian Institute of Science, Bangalore, India
e-mail: kumari@cpdm.iisc.ernet.in

A. Chakrabarti (ed.), *ICoRD'15 – Research into Design Across Boundaries Volume 1*, Smart Innovation, Systems and Technologies 34, DOI 10.1007/978-81-322-2232-3_43

can support user interaction and communication, and aid visual display and controls, if designed effectively.

There are two different kinds of sounds emanating from products, based on its source of origin. The first kind of sounds are the structure-borne sounds, which are the audible sounds produced from the mechanical, electrical or manual functioning of the product or its parts. For example, sound from a vacuum cleaner when it is switched 'on' and performing the cleaning action, or the sound produced while turning a mechanical switch 'on' or 'off. The second kind of sounds are those used in product interfaces and controls which are intentionally added to a product. Examples are, the ring sound produced by a telephone as an incoming call alert, and beep sounds produced by a washing machine to alert user when a washing cycle is over. The first type of sounds are named as consequential sounds and the second as intentional sounds by van Egmond [4–6]. In this paper, the authors focus on intentional sounds in a product which are digitally synthesized or recorded and refer to them as product sounds. The objective of this study is to support the use and design of intentional sounds in product-user interfaces by taking examples of user-interfaces from two domestic appliances. While consequential sound design has been primarily a domain of acoustic engineers, and not entirely that of product designers, it is often a matter of production quality of the physical parts, materials involved, and the physical phenomena involved in the sound producing function. However, intentional sound design has been a domain of product designers. In this work, the authors aim to support product designers to systematically design sounds on a product's user interface.

2 Product-User Interface

A product interface is a medium between the core of a product and the user, to establish communication between the user and the product, by providing and receiving appropriate inputs and outputs needed for its efficient intended functioning and use. The user-friendliness of a product relies on its interface design, as it is on the overall ergonomic design of the product.

Communication through a product's user-interface can happen through five media—visual, tactile, audio, olfactory, gestures and brainwaves. The first three media are widely used in user-interfaces of consumer products and machines worldwide. While, olfactory signs are yet to be explored as a communication medium in products, scented ads are used in print-magazines of cosmetic and beauty industry. Gestural control is in early stages of commercialization, and it is becoming common in commercial computer games. Brainwaves are being explored in actuating machines in its nascent stage. The input-output controls on the user-interfaces (Fig. 1) attract user attention and action using these media either alone or together.

Considering the first three media for its design parameters—in visual interaction, use of texts, graphical icons and symbols, colour, shape and composition

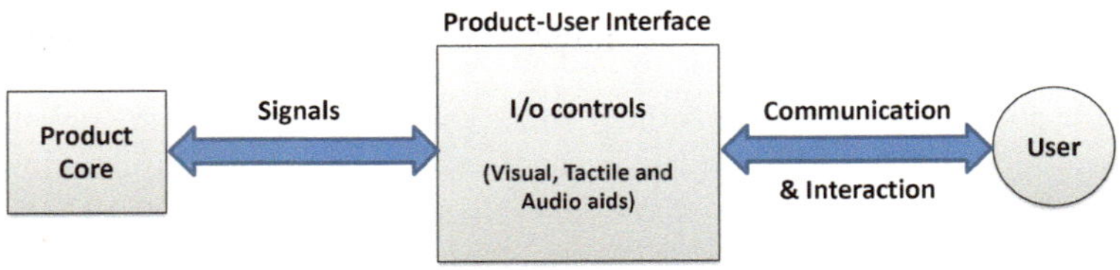

Fig. 1 Role of product-user interface

are the parameters. In tactile interaction, shape and form of the controls, material, texture, effort and movement of operating controls are the design parameters. In audio interaction, the mechanical sounds associated with the controls, the digital sounds of the interface, sound volume and quality, operation by user voices or recording of it are the parameters.

For product designers, the scope and use of digital sounds (intentional sounds) in product user-interfaces for product innovation are to be explored. The area of visual design is well explored in interface design, but not sound design. It has been shown that dual-coding, i.e., adding semantic (i.e., verbal) information congruent with perceptual (i.e., visual) information, has an additive effect on memory [7, 8]. The studies of Tindall-Ford, Chandler, and Sweller have shown that participants studying instructional materials in multiple modalities (i.e., audio-text and visual diagrams/tables) perform better than those studying in a single modality (visual-only format) [7, 9]. Hence, interface design may benefit from adding multiple modalities for better user-interaction. The work presented here, aims to support product designers in assigning sounds to user-interfaces which are meaningful associations to the user.

3 Semiotics and Semiosis in the Context of Product Sounds

Semiotics is a theory of signs [1, 10, 11]. In semiotics, communication is based on the processing of perceptual events as sign events [1]. Jekosch [1] explains sign as a mental phenomenon and it denotes the processing of a perceptual object. Constituents of signs are the sign carrier, the object of prior experience to which it refers and the meaning which evolves [1]. The process of selection, organisation, coordination and structuring, items of perception and objects of experience is termed as *semiosis* [1].

Consider an example of cooking food in a microwave oven. A bowl of milk was kept for reheating for 10 s by pressing a button, and immediately a visual countdown starts in the LCD display. The user can take out the bowl by seeing the visual feedback, or by hearing the consequential sound produced by the microwave stopping. In case, the bowl is still not taken, the microwave oven gives beep sounds at regular intervals to catch user-attention. The user then understands that the reheating process is over, and the bowl is ready to pick up, or in other words *semiosis* takes place.

In the above example, there is the product, the user, the sound, the user-interface involving controls and visual display involved in completing the communication process (or for achieving semiosis) while reheating a bowl of milk. While assigning sounds to a product-user interface, there are four elements involved in a product-user communication process to be kept in mind viz., function of the control for which the sound is assigned, the feedback to be communicated to the user through the user-interface, the sound itself, and perception of the user. To achieve communication or semiosis, the perception of the user is important. For a successful product-sound design, Jekosch explains the requirement of 'no incongruity' between user expectation and the perceived auditory image of the product [1]. However, 'no incongruity' can be a necessary bare minimum condition, and once it is achieved, the need for increasing degree of congruity begins. Therefore, the authors propose that need to achieve 'congruity' is a more natural direction to go in, for a successful product-sound design. Hence, the meaning of the function of a control, the meaning of the visual/tactile or other aids in an interface, the meaning of the sound assigned, as perceived by the user, should be congruent with each other for the user to be able to associate it with the information expected to be conveyed (see Fig. 2).

According to Jekosch, semiotic theory can be applied to product-sound design, as it is a basic theory for sciences which deal with phenomena that are perceived as sign carriers [1]. Jekosch [1] explains, for the sign carriers, here the product sound, there can be three categories of abstractions to its content: (a) Index—the sound refers the listener directly to its physical source, e.g., sound of a chisel on stone; (b) Icon—the relation between sound and source is based on similarity,

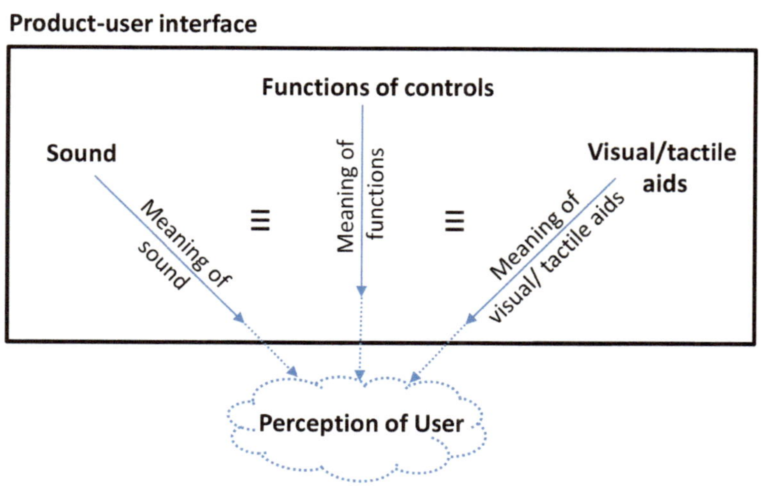

Fig. 2 Congruency of meaning to achieve semiosis

e.g., an artificial (digital) clack sound which sounds like the clack sound of a door being locked; (c) Symbol—the relation between sound and what it signifies is based on conventions, e.g., a sound of an applause, where a person should know what it signifies as (see Fig. 4). The next section explains how these categories are used in our work to support product-sound design.

4 Collection and Coding of Sound Samples

For sound related communication in design teams, product sound designers need to reproduce sound information from memory [7]. Several methods have been proposed to reduce the load on cognitive activities, which involves, verbalizing, audiolising, imitating and using pictograms, to reproduce the perceived aspects of the sound [7, 12–14]. For product-sound designers, it is difficult to explain graphically or lexically the particular sound they are looking for exactly. A database of sounds will be of help here. The authors felt the need for having a database of sounds, for product designers to find sounds suitable for the product they are designing. Three set of sound samples were used as base for this study: (a) 125 sound samples were recorded from the environment around (both natural and man-made), (b) 68 sound samples were downloaded from *Soundsnap* website [15], (c) a free library called *6000 Sound Effects* [16] was downloaded from their internet site. The samples from this library were primarily used for our study along with the samples the authors recorded.

1,000 different sound samples from various categories like natural (environmental) sounds (birds, animals, wind etc.), household sounds, speech sounds (expressions, disappointment, cheering etc.), machine sounds and cartoon sounds (synthesized sounds) were selected almost equal in number from each category from the above three sets. Sounds are given onomatopoeic names like cuckoo clock, buzz, clack sound, etc., to give the reader an understanding of the sample sounds being referred to. Out of the 1,000 sounds, 670 sounds were shortlisted based on sound quality and duration (3–5 s) of the recordings, for coding using the principles of semiotics. The abstractions of *index*, *icon* and *symbol* discussed in the previous section were mapped to each of these sounds. Each of these sounds were coded based on the dominating semiotic relation it carries. Figure 3 shows an example of how the coding was followed by authors with 3 sound samples given. If more than one relation was found, those too were coded to the sounds. For example, for the sound named 'windup alarm clock', it referred to its physical source—an alarm clock, which is *index*, and also it referred to conventions of raising an alarm that it is time to perform some action, which is *symbol*. Table 1 shows a sample list of coded sounds by authors. This categorization and coding helps in the subsequent phase of sound selection and association, i.e., if one needs *icon* type sounds, one can pick that and use it to associate to a product user-interface control.

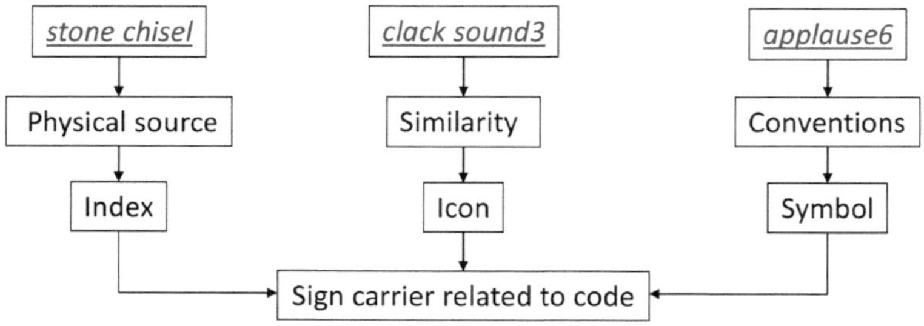

Fig. 3 Abstraction of product sounds based on its content

Table 1 Categorisation of sound samples

Sound samples	Index	Icon	Symbol
Chipmunk chatter		✓	✓
Cannon fires	✓		✓
Water gushing	✓		
Winding down theme		✓	
Cola burp	✓		✓
Cash register rings 2	✓		
Child toy duck quacks		✓	✓
Office phone ring	✓		✓
Chop vegetables	✓		
Brushing teeth			✓

5 Sound Association to Products

With the database of categorized and coded sounds, experiments were conducted to see whether assigning sounds to product user-interface controls based on the semiotic relations will help in making meaningful associations to the user. Three approaches were followed for associating sounds to products: (1) sound association based on theme by *index, icon* or *symbol*; (2) sound association for individual function or control by *index, icon* or *symbol*; (3) single sound checked for prominent component out of *index, icon*, or *symbol*. For the cases 1 and 2, user interfaces of a washing machine and a microwave were chosen. Interfaces from these house-hold appliances were chosen because the subjects were familiar with these products, and the interface was complex enough in terms of number of operations, and variability in input controls. For example, interfaces of an electric kettle or a vacuum cleaner has only very few input-controls such as POWER ON and OFF, whereas the interfaces of washing machine and microwave oven have at least more than four different controls.

5.1 Case 1—Theme Based Sound Association

5.1.1 Method

Fifteen postgraduate students of design and research and of age group of 20–30 years participated in the experiments. All subjects reported normal hearing and 6/6 corrected vision. An interface of a washing machine was used for this experiment (Fig. 4). The authors chose two themes for assigning sounds—first theme 'water' based on 'icon' or similarity (as washing involved water), and second theme 'orchestra' based on 'symbol' or conventions (as washing involved an ensemble of actions and sounds like that of an orchestra). Four sound samples each for the two themes were chosen from the database by the authors. The subjects were given a PowerPoint slide with sounds' links and a digital image of the washing machine interface. Names of the sounds were not shown to the subject and were merely numbered as sound 1, 2, etc. with the volume set at a comfortable listening level. Function of four controls, WASH, RINSE, SPIN and DOOR LOCK, were explained. The interface was shown for a minute and then sounds were played. Sounds were present throughout the experiment for the subjects to playback if needed. After the sounds were heard, the participants were asked to do a matching task. Subjects were asked to match the sound of their choice to the control in the interface by drawing a line in the PowerPoint slide. The results of the matching task were checked to see if the associations by subjects have a pattern. One such outcome of a subject matching sounds to the interface given, is shown in Fig. 4.

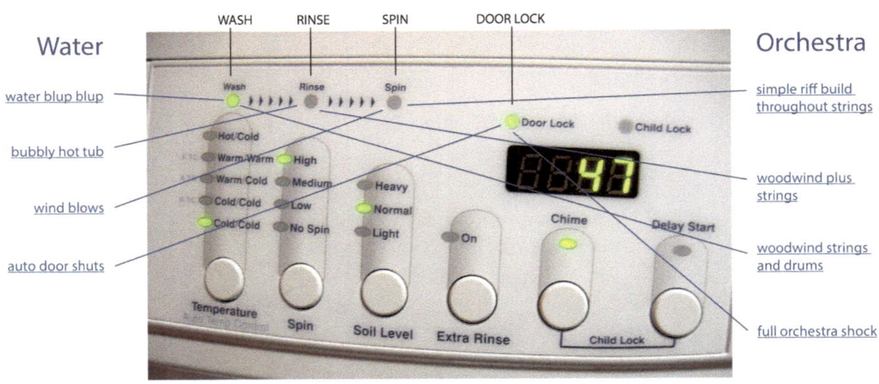

Fig. 4 Sound association diagram of a washing machine interface

5.1.2 Results

The tabulation of the matching task is shown in Figs. 5 and 6. From Fig. 5, it can be seen that for SPIN and DOOR LOCK, subjects favored two of the provided sounds in particular. The controls, WASH and RINSE, were matched with two different sounds equally by the subjects. This reflects that the interpretation of the meanings of the controls by the subjects were not distinct enough for WASH and RINSE, and the words also have a closely related meaning. Also, the two sounds matched for WASH and RINSE, were different from those matched for the other two controls, showing clear distinction of the perception of meanings of these controls and the sounds. In the symbol based orchestra theme, there was ambiguity in matching sounds to the first three controls. A clear choice was made for the control DOOR LOCK to one particular sound. This demonstrates the need for users to have prior associations to these sounds to find meanings in symbol or convention based relations.

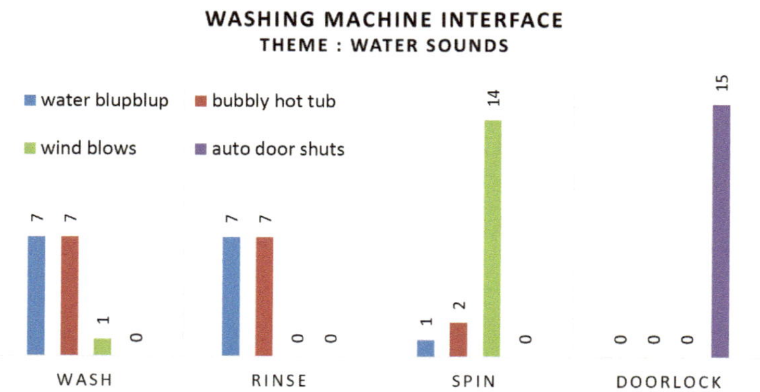

Fig. 5 Case 1: Experiment 1—Washing Machine. Theme: Water (Icon based)

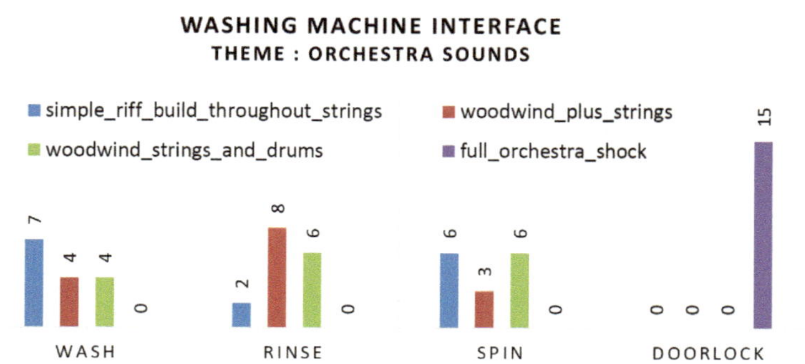

Fig. 6 Case 1: Experiment 2—Washing Machine. Theme: Orchestra (Symbol based)

5.2 Case 2—Function Based Sound Association

5.2.1 Method

The same subjects as in Case 1 were involved in this experiment. An interface of a microwave oven was chosen for this experiment. Depending on the function of the individual controls, sounds with either 'index',' icon' or 'symbol' relations were chosen from the database by the authors. The subjects were given a digital image of the interface and functions of eleven controls were explained. The interface was shown for some time and then eleven sounds were given. After the sounds were heard, the participants were asked to do a matching task in a PowerPoint slide, similar to Case 1. Figure 7 shows the outcome of one such matching task by a subject. The results from this experiment were checked to see whether the sound association to the controls have a pattern.

5.2.2 Results

The tabulation of the outcomes of the matching task is shown in Fig. 8. From Fig. 8, one could see that out of the 11 sounds given—3 controls (CLOCK, BABYFOOD, KIDS MEAL) were associated with only 2 sounds; 5 controls (POWER LEVEL, AUTO REHEAT, AUTO DEFROST, MORE, KITCHEN TIMER) were associated with only 3 sounds; and the remaining 3 controls were associated with 4 or 5 sounds. In 10 out of the 11 controls, the subjects were predominantly matching a particular sound to a particular control. One control, KITCHEN TIMER, had 2 sounds preferred equally, which was a symbol based

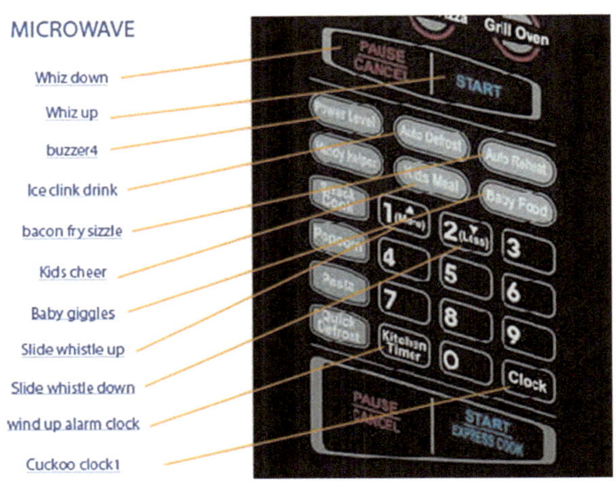

Fig. 7 Case 2—Microwave Oven

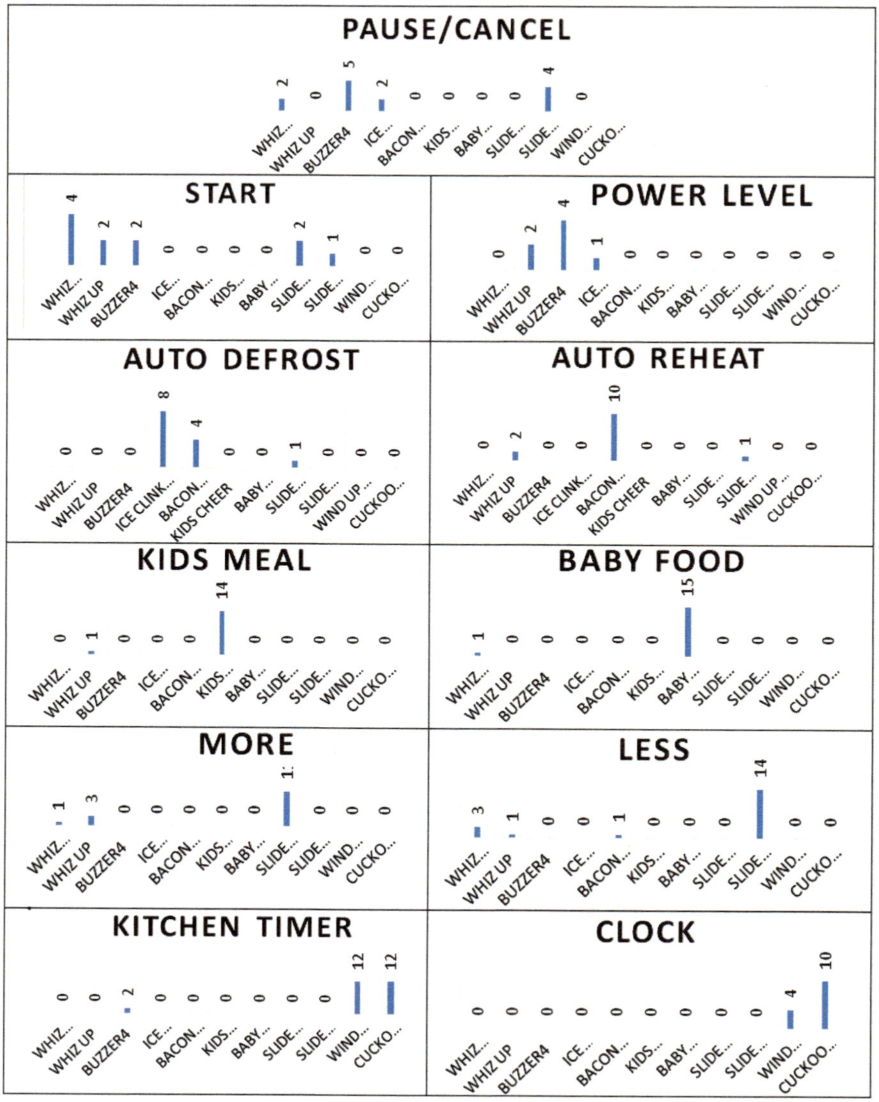

Fig. 8 Case 2—Microwave Oven

association. The results from Experiment 2 show that if the sounds chosen are based on the semiotic relation the functions of individual controls are demanding, the associations are more meaningful to the listeners, and are easily understood and interpreted by them, to detect and discriminate the controls.

5.3 Case 3

5.3.1 Method

Ten sound samples, five each from natural sounds and product sounds categories, were taken from the database by the authors. A survey was conducted to find out whether a single sound has a dominating semiotic association (index, icon or symbol) as perceived by the subjects. Same subjects from the previous 2 studies participated in this study. The subjects were made to listen to the selected sounds and were asked to speak about what they felt, infer or identify, upon hearing the sound. These psychoacoustic descriptions were then categorized based on the semiotic relations of index, icon, or symbol.

5.3.2 Results

A sample of descriptions of sounds by subjects and categorisations are given below.

Thunder cracks: *crackers*(icon); *bomb explosion and drum*(icon); *lightening* (index); *thunderstorm lightening*(index); *land mines blasting sound* (icon); *something falling down*(icon); *lightening*(index); *crackers*(icon); *explosion*(icon);*bomb explosion while a racial riot*(symbol);*computer game sounds*(symbol); *bullet firing* (icon); *bomb explosion destroyed the house*(icon), *thunder light rain*(index); *drum*(icon).

Egyptian goose: *horse driven carriage*(icon); *laughing man or turning old wheel*(icon); *ghost or cruel laugh* (icon); *laugh with coughing of older people*(icon);

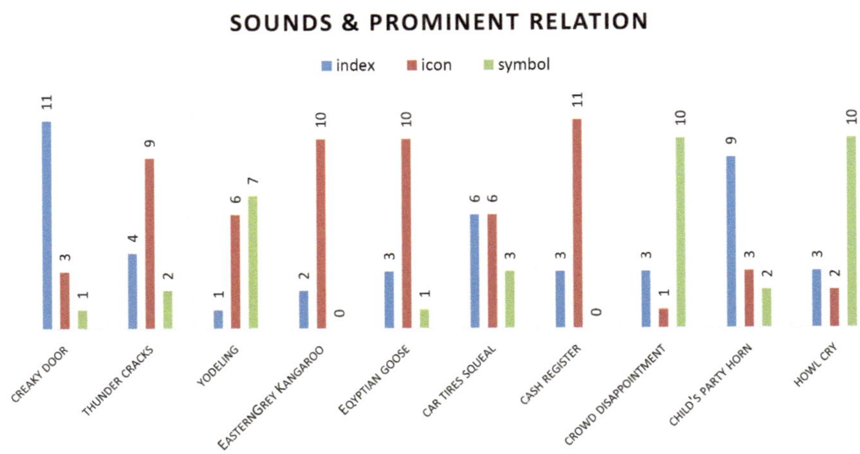

Fig. 9 Case 3—Sounds checked for prominent relation

bird crying(index); *a jungle bird*(icon); *mechanic testing horn sound*(icon); *pipani fair noise*(symbol); *a man mocking or a bird or animal*(icon); *a sound human being can't make*(icon); *some bird sound in background or animal sound in focus*(index); *thought of squirrel but too loud for it, a big mammal's sound*(icon); *dog barking, cycle horns, blow horns or duck like bird*(icon); *ducks*(index).

The results of Case 3 is as shown in Fig. 9. The results show that 8 out of the 10 sound samples have all the three semiotic relations, viz., index, icon and symbol. Nine out of the 10 sounds have one relation dominating substantially when compared to the remaining two. This shows that sounds have a prominent semiotic relation, and subjects can distinguish, and perceive, the dominating relation of a sound. This categorization can hence be applied to associate sounds to product interfaces in an effective manner.

6 Conclusions

Understanding of the semiotic relations of sounds and product interface controls, can support product designers in assigning sounds to products in a systematic and innovative manner. However, there is a dichotomy that a product sound be 'novel' and yet 'known' to the user at the same time [1]. By use of themes for products, user surveys, and a database of sounds with semiotic categories, it is possible to design novel sound associations to product interfaces which the users can relate to. The need for a semiotic relations based database of sounds for product designers, is emphasized in this study. As seen from the experiments, user surveys can help in identifying ambiguous sounds and hence in refining the sound design of the interface. User surveys can also help in choosing sounds to assign, or in modifying and synthesizing new sounds by giving more prominence to the identified semiotic relation by the user. It has been shown that synthesized sounds can also elicit semantic associations [5, 6, 17–21]. The approach presented in this work can bring in more structure to product-sound design process which has been done in an intuitive manner by product designers at present.

References

1. Jekosch, U.: Assigning meaning to sounds—semiotics in the context of product-sound design. Communication Acoustics, pp 193–221, 2005
2. Knöferle, K.: Using customer insights to improve product sound design. Mark. Rev. St. Gallen. **29**(2), 47–53 (2012)
3. Schifferstein, H.N.J.: The perceived importance of sensory modalities in product usage: a study of self-reports. Acta Psychol. **121**(1), 41–64 (2006)
4. van Egmond, R.: The experience of product sounds. In: Schifferstein, H.N.J., Hekkert, P. (eds.) Product Experience, pp. 69–89. Elsevier, Amsterdam (2008)
5. Özcan, E., van Egmond, R.: Basic semantics of product sounds. Int. J. Des. **6**(2), 41–54 (2012)

6. Langeveld, L., van Egmond, R., Jansen, R., Özcan E.: Product sound design: intentional and consequential sounds. In: Coelho D. (Ed.) Advances in Industrial Design Engineering. ISBN: 978-953-51-1016-3, InTech, doi:10.5772/55274. Available from: http://www.intechopen.com/books/advances-in-industrial-design-engineering/product-sound-design-intentional-and-consequential-sounds. (2013)
7. Özcan, E., van Egmond, R.: Memory for product sounds: the effect of sound and label type. Acta Psychol. **126**(3), 196–215 (2007)
8. Paivio, A.: Dual coding theory—retrospect and current status. Can. J. Psychol. **45**(3), 255–287 (1991)
9. Tindall-Ford, S., Chandler, P., Sweller, J.: When two sensory modes are better than one. J. Exp. Psychol. Appl. **3**(4), 257–287 (1997)
10. Eschbach, A., Trabant, J. (eds.): History of Semiotics. Benjamins, Amsterdam (1983)
11. Eco, U.: Einfu¨hrung in die Semiotik. 38. Authorized German ed by J Trabant. Fink. München (1994)
12. Özcan, E., van Egmond, R.: Pictograms for sound design: a language for the communication of product sounds. In: Kurtgözü, A. (ed.) Proceedings of the 4th International Conference on Design and Emotion. Middle East Technical University, Ankara (2004)
13. Özcan, E., van Egmond, R.: Product sound design and application: an overview. In: Desmet, P.M.A., Karlsson, M.A., van Erp, J. (eds.) Proceedings of the 5th International Conference on Design and Emotion. Chalmers University of Technology, Gothenburg (2006)
14. van Egmond, R.: Designing an emotional experience for product sounds. In: Desmet, P.M.A., Karlsson, M.A., van Erp, J. (eds.) Proceedings of the 5th International Conference on Design and Emotion. Chalmers University of Technology, Gothenburg (2006)
15. Soundsnap website. http://www.soundsnap.com/
16. 6000 sound effects. http://www.sound-ideas.com/sound-effects/series-6000-sound-effects-library.html
17. von Bismarck, G.: Timbre of steady sounds: a factorial investigation of its verbal attributes. Acta Acustica United Acustica **30**, 146–159 (1974)
18. Kendall, R.A., Carterette, E.C.: Verbal attributes of simultaneous wind instrument timbres: I. von Bismarck's adjectives. Music Percept. **10**(4), 445–468 (1993)
19. Björk, E.A.: The perceived quality of natural sounds. Acustica **57**(3), 185–188 (1985)
20. Solomon, L.N.: Semantic approach to the perception of complex sounds. J. Acoust. Soc. Am. **30**(5), 421–425 (1958)
21. Edworthy, J., Hellier, E., Hards, R.: The semantic associations of acoustic parameter commonly used in the design of auditory information and warning signals. Ergonomics **38**(11), 2341–2361 (1995)

Experience Tradeoff with Technological Advancement

Bharat Sarkar and Shatarupa Thakurta Roy

Abstract In the course of making life convenient, human activities tend to be losing the richness in interaction, i.e. much lesser engagement and involvement of senses to address all the different human skills like perceptual motor, emotional and cognitive, but rather address mostly cognitive skills. Hence, it not only becomes cognitive load but people don't get to practice the skills. Instead of decrying over technology, the author tries to understand the different reasons that contribute to the above. The author proposes that technology if used properly can bring back the lost experiences and also enrich the interaction. An effort has been made to help designers realize the importance of emotion, experience and rich interaction, then exemplify it with a case study of coffeemaker, and design of a tea maker and communicating the message to users while outlining the different guidelines that can help designers fulfill their role.

Keywords Experience design · Emotional design · Rich interaction · Tangible interaction · Technology · Human skills · Cognitive load · Senses · Electronics

1 Introduction

There are many reasons which contribute to the current scenario in the forefronts of technological advancement where it is common to see a new technology finding way into human life. And the factors contribute on part of different stakeholders.

B. Sarkar (✉) · S.T. Roy
Design Programme, Indian Institute of Technology, Kanpur, Kanpur, India
e-mail: bsarkar@iitk.ac.in; bharatsarkarw@gmail.com

A. Chakrabarti (ed.), *ICoRD'15 – Research into Design Across Boundaries Volume 1*, Smart Innovation, Systems and Technologies 34, DOI 10.1007/978-81-322-2232-3_44

1.1 What are the Causes for it?

On the part of the users: Indulgence in new technology and urge to play with it blinds the user to other rich interactions possible. It could be the new wave to rush to get the latest technological wizardry provoked by jealousy.

On the part of the designers: Every time with a new invention in technology, designers get fascinated and make a new feature out of it. "They look forward, not behind, so they repeat the same problems over and over again" [1].

Why does it happen?

1. The reward structure of the design community and demands of competitive market.
2. The designers are not typical users and they become so over obsessed with what they are doing that they tend not to be looking backwardbased paradigm to a cognitive [2]
3. A sense of overwhelming that comes with the new and fascinating tech. that blocks the view of what else can be done.

On the part of Product and Service providers: The world has grown highly competitive, more than ever before. Even if the designers restraint themselves, the need to stand out calls for added features. Forced deviation from the normal results in alienating ways of using products demand high cognitive abilities. "All this brings up an important lesson in design. Once a satisfactory product has been achieved, further change may be counterproductive, especially if the product is successful. You have to know when to stop" [3].

As a result of technological development: During the rise of interactive products, information of use gradually became more and more abstract. For example a person can directly perceive what a drinking glass is for, and how it can be used, from its form. In contrast, a person needs to read and interpret the labels on the controls of a digital camera to understand what he or she can do with it. In this case the information-for-use is much more abstract. There is a difference in how people's skills are addressed. Overbeek et al. identify three types of human skills that are relevant in this context: perceptual-motor, emotional and cognitive skills. Where earlier products use all of people's skills, the new electronic interactive products mainly address people's cognitive skills. The focus of interaction with products has shifted from an action-based paradigm to a cognitive-based one [4].

1.2 Why Should we be Concerned?

"It is the fact that we have moved from a utility centered approach to a lifestyle where technology has significance, it's how we associate emotions, expressions and values with it" [5].

Whether technology should be a part of daily life and how it should be integrated with life is for people to decide for themselves. It occupies a major part of daily interaction and hence it's very hard to completely isolate the technical interaction and hence we allow a large part of our daily interaction to be boring and dull.

This all calls for a new paradigm of design where emotions are valued, experiences are preserved and interactions with products are richer.

2 The Constituents of the Paradigm

2.1 Introducing the Factors

2.1.1 Communication/Perception

The importance of communication
Good design is also an act of communication between the designer and the user, except that all the communication has to come about by the appearance of the device itself. And that is where the interface plays an important role. Also, it is totally in the hands of the designer, what he wants to convey to the user about a product [6].

What is optimal communication?

The designer has to find the correct balance of what the user should know and what not. Many a times it is even important to hide certain things from the users, whereas at other times the user should know exactly everything about how exactly certain products work or it may instill fear.

Understanding the factors
Making things visible
Too less and it adds to the complexity of the task, too much and it causes information overload, or makes things just too easy and boring. When things are visible they tend to be easier than when they are not. In addition, there must be a close, natural relationship between the control and its function: a natural mapping [7]. Make it possible to determine the system state readily, easily and accurately, and in a form consistent with the person's goals, intentions, and expectations.

Feel/Haptic/Touch, Sound and Other senses
Haptic communication can provide us with so many rich interactions. Touch can convey a lot and can communicate more precisely like love, anger, etc.

Earlier products had visual or tactile qualities that communicate directly with great effect [5]. —John Hesket, Toothpicks and Logos. e.g., MacBook's clickable track-pad.

We associate a number of experiences just by the sound. And many a time sound is used as an auditory skeuomorph. Like the mechanical camera shutter sound in mobile phones. Other senses enable perception of taste, body balance, acceleration, gravity, position of body parts, temperature, pain, time, and perception of internal senses such as suffocation, gag reflex, etc. [8].

Multimodal Interaction
It is a principle of providing multiple modes of interfacing with a system, which helps in forming "interaction with the virtual and physical environment through natural modes of communication" [9], i.e., the modes involving the five human senses [10].

2.1.2 Skills and Loads

Whenever the number of possible actions exceed the number of senses or skill that can address it, there is most likely to be load concentrating, somewhere.

Recreational activities should be wide and deep, for we do them when we have the time and wish to expend the effort. We don't want to engage in deep thought while doing the daily life activities. It should always be optional and there should always be an escape if someone doesn't want to engage to that extent [11].

2.1.3 Mental Model

Designers should provide users with appropriate models, i.e. try to create the mental models which are correct and relevant because not only will it help the designers in building upon the same model to work forward and help the user interact with the product for further actions but also because then the fear of the unknown can be reduced and as when they are not supplied, people are likely to make up inappropriate ones [12], which can rather pose difficulty while they interact with other devices. This soon relates to the concept of making things visible as in if the system image transparently represents how the system actually works then the job gets easier.

2.1.4 Mapping

The common problems have simple solutions, which properly exploit affordances and natural constraint. Thus most of the common products have simple mappings and usually doesn't need special/arbitrarily created conceptual models or system image.

2.1.5 Levels of Specifications

Actions can be specified at different levels; High level specifications and low level specifications. People likely communicate in terms of high level specifications/ global description. Assumption: We usually care about the higher level of specifications, but by working at lower levels mostly subconsciously. Thus in many cases the high level experience can be provided without wondering about the low level specification which makes it work. A clear example can be how the power steering works and is able to provide feedback to compensate for the feedback lost in the process of assisting it with technology [13].

2.1.6 Sustainability and Well Being

Sustainability is not just about the continuation of life but knowing when to stop, find the balance and continue to stay in.

Much human behavior is done subconsciously, without conscious awareness and not available to inspection. It is good at detecting general trends. And it is good at generalizing, at making predictions about the general trend based on few examples. Conscious thought is quite different. It is slow and labored. Here is where we slowly ponder decisions, think through alternatives, and compare different choices [11].

Designers can try to bring the actions into conscious thinking from the sub conscious thinking. Hence be able to ponder upon, think through alternatives and compare different choices, to be more conscious of the environment and the surrounding.

2.1.7 Emotions and Associations

Humans have emotions and they like to express them. And our immediate environment and the activities associated become ways of achieving the same.

Tableware, cutlery, and glassware serve specific purposes while dining, but again can be manifested in a huge variety of forms, often with complex decorative patterns. Perhaps the classic example of our age is the automobile, which has from early years been an early extension of ego and personal lifestyle [5].

2.1.8 Being Less Human with Time

There are numerous such evidences which point out to the over reliance of humans on technology rather than tacit knowledge and where humans are not being able to practice certain skills and senses because, someone with the use of technology has made the job just too easy. For example:

Sensory dynamism
The concept has to do with our perception. When you look out of a window, you perceive millions of variances - color, perspective, etc. But when you gaze at an iPad, you're sensing just a few variables - and with email and SMS, you may barely be using your senses.

Neema Moraveji, director of the Calming Technology Lab at Stanford University says sensory dynamism can be a problem when it comes to an over-reliance on computer technology [14].

Similar other examples are, Implantable electronics, Search dependence, Social networking, Texting and email, Geolocation, etc.

2.2 The Unfortunate Compromise

There is no need to compromise rich experiences for the merits that technology bestows upon us.

Whether their meaning stems from the social taste of a particular fashion or age, or an intensely personal evocation of relationship and meaning, their significance is intrinsic and not dependent upon any specific affordance. In addition between the poles where utility and significance can be clearly identified as the dominant characteristic, there are innumerable products that unite efficiency and expression in an astonishing range of combinations [5]. Many products can express both at different levels and time.

3 The Framework

Field Study:
A preliminary session of interviews was taken of users from all age groups to verify the problem statement, empathize with the users and understand their take on the situation.

The results conveyed their understanding of the role of technology, different examples of their recollection of lost experiences, and willingness to engage more.

A solution to the above problem could not be a formula to apply, but a set of simple rules and design cues which if followed in a certain way can ensure the optimum design of the system.

3.1 The Decision Maker

It is important to understand that no one person can decide, whether the people should move on with the technology wave or should they care about the emotions and experiences. It is but for the individuals to decide for themselves, the level of engagement they would like to have.

Is it same as decrying over technology? No, it is just a reminder of the fact that technology has great potential and hence has to be put in the best use of people, which can be ensured with the proper use of design. And it is the role of the designers to first of all define the 'best" and then achieve the same.

For e.g., the save icon is used in different software and applications since a long time, mostly represented as a floppy disk. Now, there is a big debate over whether to move onto new icons representing the current state of the art technology like cloud.

Likewise is the case with technology usage. It is upon people to decide whether to move on or not. But, it is better to stay with the well-established mental model than to blindly follow a trend. Again, a balance has to be sought which works for the people.

The history of technology shows that we are not very good at prediction, but that does not diminish the need for sensitivity to possible changes. New concepts will transform society, for better or worse [15]. For example, the concept of paperless office, where the concept of paper vanishing in near future due to the emergence of new technology was very much predicted, but it never saw the light of day.

3.2 The Capabilities of Technology

Computer as chameleon

Technology has become very powerful it can even disappear (behind the system image), and take any form as desired. This property can be highly utilized while designing for the experience [16].

3.3 Interface

The interface is the most crucial element where the system image is the point of interaction between the users and the service/product. It is at this point where the user exchanges information, connects and interacts by different means.

All the activities and products have their inherent properties and attributes, which contributes to the richness of possibilities of interaction. From here it is the designer's influence that comes into play.

Mr. Norman speaks about two different ways of getting a task done. One way is to issue commands to someone else who does the actual work: call this "command mode" or "third-person" interaction. The other way is to do the operations yourself: call this "direct manipulation mode" or "first-person" interaction. Both forms of interactions are needed. Third-person interaction is well suited for situations in which the job is laborious or repetitive, as well as those in which you can trust the system (or other person) to do the job for you properly [17].

This is where the designers can choose between the activities to engage user in and the ones to be assigned to technology.

If the designer wants then the activity can stay in its purest form or it can be presented to the user in a totally changed form. On one hand it could be simplified and robbed of its genuine properties to be represented as mundane, whereas on the other hand not only the inherent properties be retained and the unwanted attributes be selectively removed or modified, but other different interactions added and experience enriched.

3.4 Cognitive Load

These interactions, if well balanced ensure practicing of skills and engagement, but when their concentration starts towards the cognitive skills, they become cognitive load. Then there are following consequences like information overload, information loss, split attention, etc.

Mr. Norman has given a guideline for designing which can help prevent cognitive load. According to him, "A task can be restructured through technology, or technology might provide aids to reduce the mental load. Technological aids can show the alternative courses of action; help evaluate implications; and portray outcomes in a more complete, more easily interpretable manner. These aids can make the mappings more natural. Four major technological approaches can be followed":

1. Keep the task much the same, but provide mental aids.
2. Use technology to "make visible" thus improving feedback and control.
3. Automate but keep the task much the same.
4. Change the nature of the task [18].

3.5 Layers of Control

All tasks have several layers of control. The lowest level is the details of the operation, the nimble finger work of sewing or playing the piano. Higher levels of control affect the overall task, the direction in which the work is going. Here we determine, supervise, and control the overall structure and goals. Automation can work at any level. Sometimes we really want to maintain control at the lower level. For some of us, it is the nimble execution of the finger or mind that matters. In cases like these, we would not want automation to interfere. At other times we want to concentrate on higher level things. Perhaps our goal is to listen to music, and we find the radio more effective for us than the piano [19]. So, people who are inclined towards the low level of controls will anyways not like automation, but rich interaction is a pleasure to be enjoyed by all.

The Guidelines:
It works in a process, where all the above mentioned design solutions, methodologies and factors come into play but in general the sequence has to be like this;

Step 1: Try to retain the properties and communicate it with transparency:
So, first we try to design the product/interface in such a way that the mappings are most natural and are directly connected to the functions all the while including the various senses and excluding the undesirables like the effort and "conversion of skills and enjoyment into loads".

Next we try to "Bring transparency in the system image" if the image represents the correct mapping at the back. For example to show the events inside the coffee machine and know that the user will enjoy richer experience in that case.

Thus most of the common products have simple mappings and usually doesn't need special/arbitrarily created conceptual models or system image.

With modern computers and their powerful graphic displays, we now have the power to show what is really happening, to provide a good, complete image that matches the person's mental model of the task-thereby simplifying both understanding and performance. Today, computer graphics are used more for show than for legitimate purposes. Their powers are wasted. But there exists great potential to make visible what should be visible (and to keep hidden what is irrelevant) [20].

It is important to make things visible. Otherwise people form explanations for the things they can see, explanations that can easily be false/wrong.

What not to do: Make things invisible, make things arbitrary and nonobvious, make operations unintelligible, induce forceful behavior upon people, create random mental models, present with random system images, and mappings.

Step 2: Redefine the interaction
Make the users see and experience what they want to, irrespective of how it actually works on the inside of the device. Create system image and interactions that mimic the rich interactions that have been lost or try to create new other interactions irrespective of how it works inside. Create mappings after restructuring the activities/interactions with the product.

Again, People are most likely to communicate in terms of high level specifications/global description. And hence if the system becomes too complex to directly map the functions with image, we can care less for the low level specifications provided the interactions are correct on higher level.

Skeuomorphism:
Designers can recreate certain experience artificially through the use of different technologies using mimicry, trying to simulate feelings.

One such example is power steering; power steering provides some feedback of forces acting on the front wheels to give an ongoing sense of how the wheels are interacting with the road; this is typically called "road feel". This is one very good example of a selective use of technology, where the experience associated with the conventional way of driving was retained and the efforts were reduced to make it more accessible and enhance its functionality. Similar other examples:

- The shutter-click-sound, auditory skeuomorph
- Visual Skeuomorph- like simulated wood grains on automobiles, etc.

Similarly, not only can the existing experiences be preserved but other new ways of interaction can be incorporated, making it much richer.

Step 3: Design with options
Sometimes redefining the interactions is also not sufficient, because of the following:

Extrinsic and Intrinsic Experience
The extrinsic part depends on the context and hence depends on different factors like, the time available for getting the work done, how hospitable the environment is, etc. And hence this is where it becomes difficult. Since, a design has to account for different such situations; it's very hard to standardize or present one solution and fixed ways of interaction. Similarly, intrinsic part of experience depends on the personal and subjective traits, and hence while someone may dislike the vibrations in force feedback and find no sense of "road feel" in steering, for a different class of people it may be indispensable.

And here's where options become important. In the best of the worlds, users would be able to choose automation or full control and there will be option of personalization [21]. And that's how the designers can ensure that they continue to meet the demands of the people and fulfill their responsibility as designers too.

References

1. Norman, D.A.: Design of Everyday things, p. xv, Basic Books, New York (2002)
2. Norman, D.A.: Design of Everyday things, p. 151, Basic Books, New York (2002)
3. Norman, D.A.: Design of Everyday things, p. 150, Basic Books, New York (2002)
4. Michel, R., Design Research Now, Birkhäuser Verlag AG, Germany, pp. 139 (2007)
5. Heskett, J.: Tooth picks and Logos, Oxford University Press, UK, 21 Mar 2002
6. Norman, D.A.: Design of Everyday things, p. xi, Basic Books, New York (2002)
7. Norman, D.A.: Design of Everyday things, p. 27, Basic Books, New York (2002)
8. Perception, Wikipedia.org, http://en.wikipedia.org/wiki/Perception
9. Bourguet, M.L.: Designing and Prototyping Multimodal Commands. In: Proceedings of Human-Computer Interaction (INTERACT'03), pp. 717–720 (2003)
10. Ferri, F., Paolozzi, S.: Analyzing multimodal interaction. In: P. Grifoni (Ed.), Multimodal Human Computer Interaction and Pervasive Services. Hershey, PA: Information Science Reference. pp. 19–33 (2009) doi:10.4018/978-1-60566-386-9.ch002
11. Norman, D.A.: Design of Everyday things, p. 125, Basic Books, New York (2002)
12. Norman, D.A.: Design of Everyday things, p. 70, Basic Books, New York (2002)
13. Norman, D.A.: Design of Everyday things, p. 111, Basic Books, New York (2002)
14. Brandon, J.: Is Technology Making us Less Human, 6 Aug 2013 http://www.techradar.com/news/world-of-tech/future-tech/is-technology-making-us-less-human–1171002/2
15. Norman, D.A.: Design of Everyday things, p. 210, Basic Books, New York (2002)
16. Norman, D.A.: Design of Everyday things, p. 180, Basic Books, New York (2002)
17. Norman, D.A.: Design of Everyday things, p. 184, Basic Books, New York (2002)
18. Norman, D.A.: Design of Everyday things, p. 191, Basic Books, New York (2002)
19. Norman, D.A.: Design of Everyday things, p. 197, Basic Books, New York (2002)
20. Norman, D.A.: Design of Everyday things, p. 192, Basic Books, New York (2002)
21. Norman, D.A.; Design of Everyday things, pp. 194, Basic Books, New York (2002)

Cross-Cultural Issues in Working with Users in the Design of Interactive Systems

Helen Petrie, Tanja Walsh, Olufunmilayo Odutola and Lei Ang

Abstract Globalization of markets means that interactive systems need to be usable and provide positive user experience (UX) to users in many different cultures. There is an increasing realization that systems may need to be localized to different cultures, but less realization that methods used to work with users also need to be localized. We present two case studies which investigate cultural differences in users' interpretation of and reactions to methods used to elicit usability and UX information. The first case study investigated the use of photos and sketches with Nigerian and Anglo-Celtic participants, the second investigated the use of three different question types, Likert items, sentence completion questions and open-ended questions with Chinese and British participants. Conclusions are drawn for conducting cross-cultural user research.

Keywords Cross-cultural differences · Usability · User experience · Design methods

1 Introduction

There is currently considerable globalization of markets for interactive systems, from smartphones to washing machines. This creates a need for good usability and user experience (UX) with these systems for users from many different cultures.

H. Petrie (✉) · O. Odutola · L. Ang
Human Computer Interaction Research Group, Department of Computer Science, University of York, York, UK
e-mail: helen.petrie@york.ac.uk

T. Walsh
Unit of Human-Centered Technology, Tampere University of Technology, Tampere, Finland

O. Odutola
Simple Usability, Leeds, UK

A. Chakrabarti (ed.), *ICoRD'15 – Research into Design Across Boundaries Volume 1*, Smart Innovation, Systems and Technologies 34, DOI 10.1007/978-81-322-2232-3_45

There is growing awareness of the need for consideration of cross-cultural issues in the design of interactive systems, but as yet not nearly as much consideration that the methods used in eliciting user requirements and evaluating usability and UX are also potentially subject to cultural differences. The ways people in India or China typically interpret and react to particular kinds of questions and materials used in usability and UX work may be very different from the ways people in North America or northern Europe typically interpret and react to them [1].

Our research objective was to study how cultural issues affect methods used for eliciting usability and UX information with a web survey. We present preliminary findings from two case studies both using the remote web survey. First we will review background relevant research on this topic, on internationalization and localization of products, on cross-cultural differences in usability and UX methods and on frameworks for understanding cultural differences between users in their interpretation of the methods. Our first case study investigated the reaction of Nigerian and Anglo-Celtic (AC) participants to two types of visual material, photos and sketches, in the elicitation of their experiences with smartphones. The second case study investigated the reactions of Chinese and British participants to three different question types (Likert rating items, sentence completion questions and open ended questions), again in the elicitation of their experiences with smartphones. Finally we will draw some conclusions about cross-cultural in working with users in the design of interactive systems.

2 Background

There is a growing interest in the relationship between people's culture and their use of interactive systems [2, 3], largely due to the globalization of markets for interactive systems. In addition, users around the world expect more from their interactive systems than only utility and usability: they are looking for positive user experiences [4]. Thus the design of UX needs to take into account users' cultural contexts as well as the focus on content, brand and emotion emphasized by UX [5]. Therefore understanding the concept of culture and how it impacts on the UX of an interactive system is becoming an important part of the design of the system. As Ito and Nakakoji [6] note, much of cognitive reasoning depends on social norms and background culture. This cultural diversity makes it unrealistic for developers to rely only on intuition or personal experience when designing for good UX in cross-cultural contexts [7]. Ouygi et al. [8] suggest that cultural differences in signs, meanings, actions, conventions, norms and values raise challenging issues in the design of usable interactive systems for different cultures.

Two major strategies are adopted in the attempt to meet the demands of globalization: internationalization and localization [2]. Internationalization is the designing of systems in such a way that they do not need to undergo changes to their core form and functionality in order to be adopted for various target markets [e.g. 2, 9]. Conversely, localization means that interactive systems are adapted to suit the specific cultures and languages of target markets thereby making the systems usable and acceptable by members of the target cultures [2, 9].

To create a basis for localization of interactive systems, the key factors differentiating cultures from one another need to be clearly identified [10]. Two broad types of issues related to cross-cultural design have been identified. Firstly, there are objective issues, such as language and format conventions of time of day, dates and number, text directionality in writing systems and so on [2, 7]. Secondly, there are subjective issues such as value systems, behavioural and intellectual systems of the cultural groups and the ways in which people in different cultures interact with interactive systems [7]. Young [9] argues that localization requires authentication of the design through methods such as ethnographic research to make sure that the design specifications are truly representative of the target cultures. Chavan et al. [1] also support the idea of authenticating localization to ensure that designs are not biased to the designers' own culture.

Furthermore, Oyugi et al. [8] argue that cultural aspects need to be considered also in the selection and use of methods used in UX research and design. Cultural differences potentially affect the manner in which users interpret and react to user requirements studies, design exercises and evaluation studies. A majority of the methods used to conduct user research have originated in the "western" world. These "western" methods may not always be appropriate for research with users from other cultures or users from other cultures may interpret aspects of these methods differently. These views are also supported by research by Vatrapu and Pérez-Quiñones [11] on structured interviews with Anglo-American and Indian participants which found that interviewees found more usability problems and made more suggestions when the interviewer was from the same cultural background.

The most well known framework for thinking about cross-cultural differences in value systems was developed by Hofstede [12], although it has also been strongly criticized. Hofstede studied over 116,000 people from over 50 countries. It was later discovered that all the participants were employees of IBM, and this discovery along with other observations lead to considerable criticism of Hofstede's work. Critics believed that IBM may have a culture of its own that might have influenced Hofstede's findings in ways that his dimensions do not apply to people outside the IBM world [13]. Another prominent criticism is the fact that Hofstede refers to culture in the national sense [14]. His study did not seem to take into account the fact that a country could have several distinct cultures within it.

Despite these criticisms, Hofstede's findings have served as a useful foundation on which to conduct cross-cultural studies. Hofstede's five cultural dimensions [12] provide a way to understand the influence of cultural differences on human-computer interaction. Therefore, these cultural dimensions have been one of the ways used to characterize the target cultures investigated in the current case studies. Hofstede's cultural dimensions are as follows and examples are mostly from China, Nigeria and the UK, as an example of the Anglo-Celtic cultures used in this study.

Power Distance: the degree to which the less powerful members of a society accept that power is unequally distributed. Hofstede's results found that both

China and Nigeria have relatively high Power Distances while the UK has a relatively low Power Distance.
Individualism versus Collectivism: the high end of this dimension represents cultures in which individuals are more concerned with their individual needs. The low end represents cultures where the community acts as a whole and the collective needs of its members are more important than the needs of any one individual. Hofstede's results found that the UK has relatively high Individualism when compared to high Collectivism in China and Nigeria.
Masculinity versus Femininity: is perhaps the most controversial of Hofstede's dimensions, it is concerned with the expected social and emotional roles of women and men in a culture. The higher end of this dimension represents a masculine culture where there is a preference for heroism, assertiveness, achievement and material reward for success. The low end represents a feminine culture where there is a preference for modesty, caring for the weak, cooperation and quality of life. China, Nigeria and the UK all have similar values on Masculinity-Femininity according to Hofstede's findings.
Uncertainty Avoidance: is concerned with the degree to which a culture tries to deal with the unpredictability of the future. The high end of this dimension represents cultures that are intolerant of unorthodox ways and use strict laws and rules of conduct to maintain some kind of predictability. The low end represents those cultures that are more flexible and welcoming of change. According to Hofstede's findings, Nigeria has higher uncertainty avoidance than China or the UK.
Long-term versus Short-term Orientation: is representative of the degree to which a culture considers the future in its present actions. The high end of this dimension represents cultures with long-term orientations, they are more interested in the long run outcome of present situations and individuals in these cultures are more likely to save and invest. The low end represents cultures that have short-term orientation, they have great respect for tradition and have little consideration for long-term outcomes and individuals in these cultures do not have a strong habit of saving and investing and are more likely to buckle under social pressure to maintain an appearance of high social standing. According to Hofstede's findings, the UK has a higher long term orientation than Nigeria and China has the highest long term orientation of the three.

Bearing this research and particularly Hofstede's dimensions in mind, we now turn to the two case studies.

3 Case Study I: Cross-Cultural Differences in Reactions to Visual Materials

The first study investigated the use of two types of visual materials, photographs and sketches, in the elicitation of users' experiences with smartphones. The latter are often perceived by designers to be less susceptible to cultural influences [15].

We worked with respondents from two very different cultures, Nigeria and the Anglo-Celtic (AC) cultures, being the UK and its ex-colonies that are populated largely from the UK and Ireland: Australia, Canada and New Zealand. Obviously, both Nigeria and the AC countries have a variety of cultures within them, but each have an identifiable dominant culture.

3.1 Method

3.1.1 Participants

Participants were 92 people who responded to an online survey. 50 were Nigerian and 42 were British or from other Anglo-Celtic countries.

The Nigerian participants comprised 25 women and 23 men. Ages ranged from 18 to over 60, with a median age group of 21–30 years. Respondents rated their proficiency in English (on a scale from 1 = beginner to 5 = native speaker level) and gave a mean rating of 4.43 (SD = 0.93). 48 (96.0 %) reported having at least one mobile phone (two respondents did not answer this question, so we cannot say whether they had a mobile or not) and 32 (64.0 %) reported having at least one tablet computer.

The AC participants comprised 24 women and 18 men and included 36 respondents from the UK, three from Canada, two from Australia and one from Ireland. Ages ranged from 18 to over 60, with a median age group of 31–40 years. All the AC participants reported having at least one mobile phone and 31 (73.8 %) reported having at least one tablet computer.

3.1.2 Materials

Three scenarios were created, with both photo and sketch versions, starting from the internationalized storyboards developed by Walsh et al. [15]. The first task was to critique these materials from a Nigerian perspective. A critique group of 7 Nigerians was used for this purpose. Members were given the four Walsh et al. storyboards and asked to work through them and note all the things that would seem out of place in a Nigerian setting, those things that would make it difficult for them to identify with the characters depicted in the storyboards and the activities being shown. The group produced 16 points on which the storyboards would not be appropriate for a Nigerian context, even though they were attempting to be international. Using these critiques as a basis, three of the scenarios were localized for the Nigerian cultural context, each with a photo and a sketch version. For example, Scenario 1 was about taking a photo on holiday, the main sketch showing a woman taking a photo of herself in front of the Eiffel Tower. The critique revealed that Paris is not a popular holiday destination with Nigerians, so the scene was changed to Dubai (with the Burj Al Arab in the background). In addition, the critique revealed that Nigerians would be unlikely to take a photo of themselves, a “selfie”, so the scenario was changed to a woman posing for a photo being taken by someone else. The four images used for this scenario in the current case study are shown in Fig. 1. The other two scenarios were checking information on the Web while travelling (on a bus or in a car) (Scenario 2) and having a casual social get together with friends and sending a photos to other friends (Scenario 3). Although the original sketches had been considered international, in light of the critique by the Nigerian group, we now considered them localized to the AC culture.

An online questionnaire presented the three scenarios, each with a short textual introduction. After questions about the three scenarios, respondents were asked to rate the following two 5 point Likert items: *Did you feel that the photos/storyboards helped or hindered you in answering the questions about the situations? Were the photos appropriate to the kind of person you are?* The questionnaire also included a number of questions to collect demographic data and information about mobile phone use.

Fig. 1 Photos and sketches for the Holiday Photo scenario (*top panels* Nigerian photo and sketch, *bottom panel* AC photo and sketch)

3.1.3 Procedure

The study was publicized widely amongst professional and personal contacts of the authors in the AC countries and Nigeria, via personal emails and messages to online discussion groups. The survey took 20–30 min to complete. All participants were entered into a prize draw for one of four prizes of gadgets worth £15 (approx USD 24).

3.2 Results

To analyse the Likert scale rating items, we first investigated the relationship between the responses to the questions. The two Likert scale questions that asked respondents to reflect on all three scenarios they had seen correlated significantly ($p < 0.001$), so we created an Overall Reaction score, being the mean response for each respondent on these two questions.

A three way ANOVA on the Overall Reaction scores found a significant main effect for Material Culture ($F = 17.27$, $df = 1, 82$, $p < 0.001$), that is the culture depicted in the scenarios. Nigerian participants answered significantly more positively overall, with a mean score of 3.80 (SD: 0.82) compared to a mean score of 3.11 (SD: 0.78) for AC participants.

We were surprised to find that the Nigerian participants were more positive about both the Nigerian and the AC visual materials, we had expected that there would be a cultural matching, with Nigerian participants favouring Nigerian materials and AC participants favouring AC materials. However, this result may be explained using Hofstede's cultural value framework. Nigeria is a high Power Distance culture which promotes reverence for authority and an unwillingness to offend those in authority. This unwillingness to offend may have hindered Nigerian participants in responding negatively. In this study, the researcher who asked the Nigerian participants to undertake the study may have been seen as an authority figure and expert. The Nigerian participants may not have wanted to offend the researcher by providing negative ratings that could be seen as criticism of the researcher's work.

There was also a significant interaction between Materials Culture and Material Type ($F = 3.53$, $df = 1, 82$, $p < 0.05$). Figure 2 shows that all participants (both Nigerian and AC) were equally positive about the photo materials, whether they depicted Nigerian or AC culture, whereas for the sketches, participants were more positive about AC materials than about Nigerian materials. This result is more difficult to interpret, but may relate to the fact that Nigerian participants are exposed to much AC visual material through the mass media, whether AC participants are not generally exposed to Nigerian visual material.

We are continuing analysis of other results from this case study, in particular the open-ended comments that participants made to explain their ratings and their ratings of the individual scenarios.

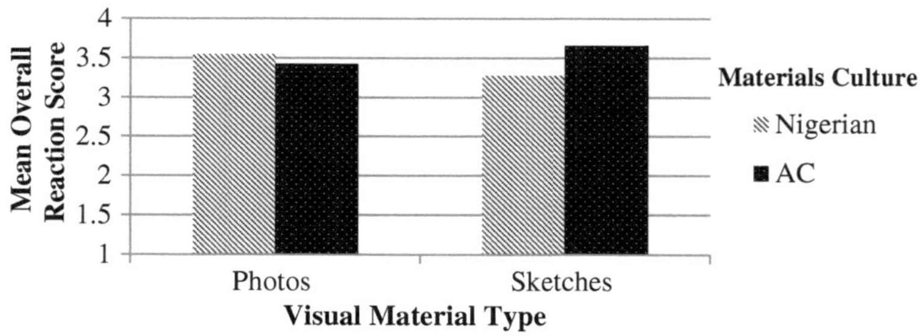

Fig. 2 Interaction in overall reactions to visual materials (photos and sketches) in the two cultures depicted in the scenarios (Nigerian and AC)

4 Case Study II: Cross-Cultural Differences in Reactions to Different Question Types

The second study investigated the use of three different question types with British and Chinese respondents: Likert rating items, sentence completion questions and open-ended questions, again in assessing UX with smartphones. We wished to investigate both how participants from the two cultures react to providing information in different formats and when answering in their native language or a second language (so half the Chinese participants responded in Chinese and half in English).

4.1 Method

4.1.1 Participants

Participants were 96 people who responded to an online survey, 56 women and 40 men, average age 26.8 years (range 18–60 years). 36 British respondents answered in English, 30 Chinese respondents answered in Chinese and 30 English speaking Chinese respondents answered in English. The Chinese participants who answered in English considered themselves proficient in English, they had studied English for an average of 11.0 years (range 1–20 years) and rated their proficiency on average 4.74 on a 7 point scale (1 = beginner to 7 = near native speaker).

4.1.2 Materials

A series of questions was developed about UX with smartphones that could be answered in three different formats: 7 point Likert rating items, sentence completion questions and open-ended questions. It was not possible to make questions

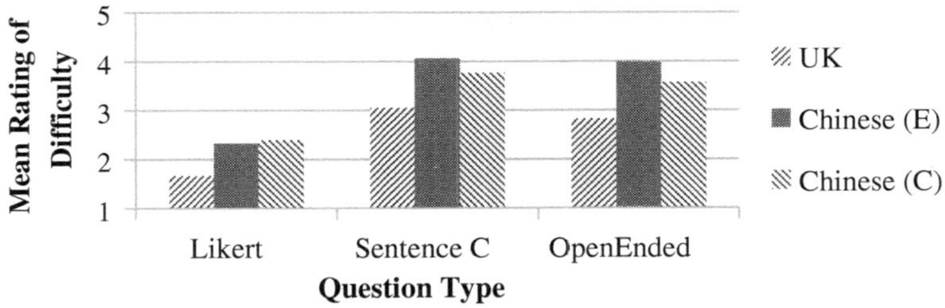

Fig. 3 Mean ratings of difficulty of answering questions of different types for UK participants and Chinese participants (answering in English and Chinese)

across the three formats exactly equivalent, but they addressed similar issues. For example, an open-ended question asked "How does your smartphone make you feel?", a sentence completion question asked "When I use my smartphone, I feel …" and Likert items asked participants to rate how much they agreed with the statements "I feel attached to my smartphone", "My smartphone feels good in my hand" and "I find my smartphone inspiring". In all there were 14 Likert items, 12 sentence completion questions and 7 open-ended questions.

These questions were embedded into a web survey that included questions about the participants experience with the survey itself and in particular with the different question types and demographic information. The survey was translated into Chinese by the native Chinese-speaking co-author and back-translated by another native Chinese speaker to ensure an accurate translation.

4.1.3 Procedure

The study was publicized widely amongst professional and personal contacts of the authors in China and the UK, via personal emails and messages to online discussion groups. The survey took 15–20 min to complete. All respondents were entered into a prize draw for one of four Amazon gift vouchers worth £10 (approx USD 17). The web survey was available for approximately 3 weeks to gather data from a sufficient number of participants.

4.1.4 Results

Participants were asked to rate the difficulty of answering each question type. The mean ratings for the three question types and the three groups of participants are shown in Fig. 3. A two way Analysis of Variance showed there was a significant difference between the question types ($F = 40.44$, $df = 2, 186$, $p < 0.001$) and a significant difference between the three groups of participants ($F = 8.17$, $df = 2, 93$, $p < 0.001$), but no significant interaction between the two variables. As can be seen in Fig. 3, all participants found the Likert items the easiest (mean rating: 2.10) and there was little overall difference between the sentence completion and open-ended questions (Sentence Completion mean: 3.59; Open-Ended mean: 3.43). The UK participants found all question types the easiest (mean: 2.52), and although the Chinese participants found all the question types harder, there was surprising little difference from them answering in English and Chinese (mean Chinese participants answering in English: 3.46; overall mean Chinese answering in Chinese: 3.24).

We are continuing analysis of the data from this case study, to investigate both the relationship between the different question types and differences between the three groups of participants.

5 Discussion and Conclusions

The results of our two studies show that designers need to think carefully about the cultural biases in the methods they use with potential users of systems that they are developing. The first case study showed that storyboard which we thought were internationalized actually contained many cultural references that may be interpreted differently by participants from different cultures. In addition, it showed that participants with different cultural values may react differently to questions and that levels of agreement cannot be considered comparable across different cultural groups. The second case study showed a similar pattern of perceived levels of difficulty across the three groups. Contrary to our expectations, Chinese participants did not report greater difficulty or a different pattern of difficulty when answering in English compared to Chinese. However, further analyses may reveal differences in the quantity and quality of their answers. Our results will contribute to recommendations for conducting cross-cultural user research.

References

1. Chavan, A.L., Gorney, D., Prabhu, B., Arora, S.: The washing machine that ate my sari—mistakes in cross-cultural design. Interactions **16**(1): 26–31 (2009)
2. Aykin, N.: Overview: where to start and what to consider? In: Aykin, N. (ed) Usability and Internationalization of Information Technology. Lawrence Erlbaum (2005)
3. Clemmensen, T., Röse, K.: An overview of a decade of journal publications about culture and human-computer interaction. In: IFIP Advances in Information and Communication Technology, vol. 316, pp. 98–112 (2010)
4. Karat, J., Karat, C.-M., Vergo, J.: Experiences people value: the new frontier of task analysis. In: Stanton, A., Diaper, D. (eds.) Handbook of Task Analysis for Human-Computer Interaction. Lawrence Erlblaum (2004)
5. Marcus, A.: Cross-cultural user experience design. In: Barker-Plummer, O., et al. (eds.) Proceedings of Diagrams 2006 (LNAI 4045). Springer, Berlin (2006)
6. Ito, M., Nakakoji, K.: Impact of culture on user interface design. In: del Galdo, E.M., Nielsen, J. (eds.) International Users Interfaces. Wiley, New York (1996)
7. Smith, A., Gulliksen, J., Bannon, L.: Building usability in India: reflections from the Indo European systems usability partnership. In: Proceedings of HCI. Springer, Berlin (2005)

8. Oyugi, C., Dunckley, L., Smith, A.: Evaluation methods and cultural differences: studies across three continents. In: Proc. NordiCHI. pp. 318–325 (2008)
9. Young, P.A.: Integrating culture in the design of ICTs. Br. J. Educ. Technol. **39**(1), 6–17 (2008)
10. Honold, P.: Learning how to use a cellular phone: comparison between German and Chinese users. Tech. Commun. **46**(2), 196–205 (1999)
11. Vatrapu, R., Pérez-Quiñones, M.: Culture and usability evaluation: the effects of culture in structured interviews. J. Usability Stud. **1**(4), 156–170 (2006)
12. Hofstede, G.H.: Culture's consequences: comparing values, behaviors, institutions and organizations across nations. Sage, Thousand Oaks (2001)
13. Kirkman, B.L., Lowe, K.B., Gibson, C.B.: A quarter century of culture's consequences: a review of empirical research incorporating Hofstede's cultural values framework. J. Int. Bus. Stud. **37**, 295–320 (2006)
14. McSweeney, B.: Hofstede's model of national cultural differences and their consequences: a triumph of faith—a failure of analysis. Hum. Relat. **55**(1), 89–118 (2002)
15. Walsh, T., Nurkka, P., Koponen, T., Varsaluoma, J., Kujala, S., Belt, S.: Collecting cross-cultural user data with internationalized storyboard survey. In: Proceedings of OzCHI 2011. ACM Press, New York (2011)

Design of UI Component Model for Evaluation of Medical UI of Ventilator System in Intensive Care Unit

Ganesh Bhutkar, G.G. Ray, Dinesh Katre and Shahaji Deshmukh

Abstract A ventilator system provides a respiratory support to critically-ill patients. Its touch screen-based onscreen User Interface (UI) helps medical users in treatment of the patient. The usability problems in medical UIs may contribute to medical errors, affecting the patient condition. Thus, the usability evaluation of medical UIs is extremely important for patient safety. The study of usability evaluation models has revealed that there is a need of domain-specific evaluation model for medical UIs, focusing on medical context; especially critical aspects in Intensive Care Unit (ICU). The proposed UI Component Model is based on UI components, Norman's action model and identified usability problems.

Keywords Usability evaluation model · UI component model · Ventilator system · Intensive care unit

1 Introduction

There are several medical devices such as patient monitoring system, ventilator system, electrocardiogram (ECG) machine and defibrillator in Intensive Care Unit

G. Bhutkar (✉) · G.G. Ray
Industrial Design Centre (IDC), IIT Bombay, Mumbai, India
e-mail: ganesh.bhutkar@vit.edu

G.G. Ray
e-mail: gaurgray@gmail.com

G. Bhutkar
Department of Computer Engineering, VIT, Pune, India

D. Katre
Human Centered Design and Computing (HCDC) Group, C-DAC, Pune, India
e-mail: dineshkatre@yahoo.co.in

S. Deshmukh
Department of Surgery, BVU Medical College, Pune, India
e-mail: surgeonshahaji@yahoo.co.in

A. Chakrabarti (ed.), *ICoRD'15 – Research into Design Across Boundaries Volume 1*,
Smart Innovation, Systems and Technologies 34, DOI 10.1007/978-81-322-2232-3_46

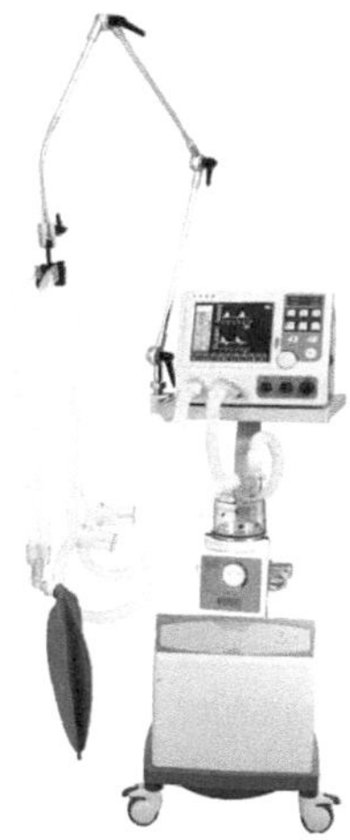

Fig. 1 Ventilator system with its medical UI

(ICU). A ventilator system is vital therapeutic device, providing a respiratory support to critically-ill patients [1]. A Fig. 1 depicts a ventilator system with its UI. Its User Interface (UI) helps medical users to input and control diversified system parameters during patient care.

There are various commonly observed usability problems with medical UIs such as poor legibility, lack of support in local language, poorly distinguished alarms and poor system feedback [2–4]. Such problems may contribute to medical errors. Too many patients have been injured or died because someone pressed a wrong button, misread a number or skipped a step [5] when using medical devices. In fact, medical error is a leading cause of death along with motor vehicle accidents, breast cancer and AIDS [6]. Thus, usability evaluation of medical devices is extremely important for patient safety.

There are many usability evaluation models applied in healthcare domain; but most of them are general-purpose models. They can be used for evaluation of medical devices by professionals who are trained for usability as well as domain knowledge of medical devices. Also, these models miss out medical context; especially criticality aspect. In this paper, **a comprehensive usability model is proposed for evaluation of medical UIs of ventilator systems in ICU.** For developing this model, the evaluation models applied in critical domains such as aviation, nuclear power plant and software, are analyzed. A comprehensive UI Component Model is designed through adaptation of selected evaluation model and with consideration of related medical context and usability problems observed in ICU.

2 Need of a Comprehensive Evaluation Model in Healthcare Domain

There is a need of comprehensive evaluation model for medical UIs in ICU. Related study of UIs and evaluation models used in various domains is discussed in this section.

2.1 Study of Usability Problems with UIs of Ventilator Systems in ICU

There are several usability problems, which are related with UIs of ventilator systems. And hence, medical users such as physicians and nurses face difficulties in operating them during the patient care. In this section, such problems are identified and categorized through Cognitive Walkthrough (CW) method.

There are two major types of users of ventilator systems—physicians and nurses in ICU. They are primary users and actually operate the ventilator systems. Four models of ventilator systems along with user interaction are studied and related videos are captured. Each video involves one user. Three videos depict interaction of physicians and fourth video has interaction of a nurse. The names of manufacturers, device models and user identities are not disclosed to maintain confidentiality. In patient care, several tasks are performed by medical users with ventilator systems, such as setting a system, pretesting of a system, changing parameter values, setting up alarms, updating patient record and generating reports [5, 6]. It is observed that four tasks—setting up a ventilator system, changing parameter values, entering patient data and setting up alarms or alerts, are most frequently employed tasks by medical users with ventilator systems. Therefore, these tasks are selected for video-recording and in turn, for CW.

A cognitive walkthrough is one of the most popular and widely used task-oriented Usability Evaluation Methods (UEM) in healthcare domain [7]. It helps in identifying reasons for errors during user interaction and is selected for usability evaluation of ventilator systems. It is a technique that attempts to evaluate system usability with involvement of expert evaluators usually usability experts or software developers with a focus on learning through exploration [8]. The evaluators specify the sequence of actions required to perform a certain task and then step through that sequence to identify potential usability problems using four major questions [9]. During CW, related videos involving four different medical users are evaluated to identify usability problems, which also reflect several critical aspects. These usability problems are categorized based on their mean severity rating (out of 10.0) perceived by a group of 10 medical users into types such as catastrophic ($\geq$8.0), critical ($\geq$6.0) and serious ($\geq$4.0) [3]. Such identified problems along with related categorization [10] based on problem severity are depicted in descending order of mean severity rating in Table 1.

2.2 Study of Evaluation Models Applied in Healthcare Domain

Three evaluation models are identified during literature survey related with healthcare domain. These models are proposed by Drews et al., Zhang et al. and Chapman et al. respectively. The study of evaluation models in healthcare domain

Table 1 Type of usability problems based on problem severity

Problem no.	Usability problem	Mean problem severity rating	Problem type
1	Inappropriate data entry in profile	8.1	Catastrophic
2	Missing valid ranges for parameter values	8.0	Catastrophic
3	No or unreadable feedback/message	7.8	Critical
4	No onscreen help or tip provided	7.7	Critical
5	High waiting time	7.6	Critical
6	No confirmation about important therapeutic action	7.5	Critical
7	No provision of screen lock	7.5	Critical
8	No use check	7.4	Critical
9	No storage of multiple patient records	7.0	Critical
10	Unnoticed LEDs/Alerts	7.0	Critical
11	Reflections on the screen	6.8	Critical
12	Long-distance visibility of visual alerts	6.5	Critical
13	Options hidden inside the menu	6.4	Critical
14	Lack of security to patient record	6.4	Critical
15	No electronic communication of patient record	6.2	Critical
16	Invalid field values or dates	6.0	Critical
17	No use of icons or selection of poorly designed icons	5.9	Serious

Table 2 Comparison among evaluation models applied in healthcare domain

Item	Evaluation Models in Healthcare Domain		
	Drews et al.	Zhang et al.	Chapman et al.
Type	General-purpose	General-purpose	Domain-specific
Purpose of evaluation	Error producing condition (EPC) identification	Error categorization	Error categorization
Input	EPCs	Medical problems observed in stages of Norman's model	Reported issues related with stages of Norman's model
Output	Failures and their severity levels	Categories of medical errors	Categories of errors and related recommendations
Categories	Condition-based and error-oriented	Action-based and error-oriented	Action-based and error-oriented

[11–13] has some useful findings as observed in Table 2. First, most of the models are general-purpose models. Two, all evaluation models are error-oriented and deal with errors or problems and, help in their categorization. Three, the errors are categorized with respect to stages in Norman's action model in an iterative process [14]. Four, these models may miss out many usability problems as they have different focus and purposes. Hence, there is a need for a comprehensive evaluation model, which helps in identification of missed out medical context in terms of errors or problems with respect to UIs of ventilator systems.

2.3 Study of Evaluation Models Applied in Other Domains

During literature survey, four evaluation models have been identified that are applied in critical domains such as aviation and nuclear power plant. These models are proposed by Kornecki et al., Latorella et al., Yun et al. and Norros et al. respectively. The evaluation models applied in critical domains [15–18] are studied and they are either general-purpose or domain-specific models. Most of them are either error-oriented or quality-oriented. The problems identified are categorized into groups such as 'technical', 'engineering' and 'management' or prioritized based on their severity. Most of the methods have used 'inspection' approach involving experts for detection of problems.

Four evaluation models are also identified during literature survey related with software domain. These models are proposed by Abran et al., Diniz et al., Kainda et al. and Winter et al. respectively. These evaluation models in software domain [19–22] are general-purpose and attribute-based models, in which a focus of evaluation is on identifying major attributes and related assessment. Most of the models have hierarchical structure with two-three levels of attributes and they consider 'security' as vital attribute of evaluation. Two evaluation models also consider quality standards (ISO) and Nielsen's Model [23] in evaluation process.

3 Design of Usability Evaluation Model for Ventilator Systems

Initially, work analysis of ventilator systems is performed for more comprehensive understanding of related tasks as well as identified usability problems. Then, UI components of ventilator systems are identified based on study of related tasks. A reference model is selected on the basis of the study of evaluation models in several domains. Finally, required changes are adopted in selected reference model to propose a design of usability evaluation model for ventilator systems in ICU.

3.1 Work Analysis of Ventilator Systems

The medical users such as physicians and nurses operate the ventilator systems, and related workflow diagram for ventilator systems is depicted in Fig. 2. It shows several important tasks and related action sequence.

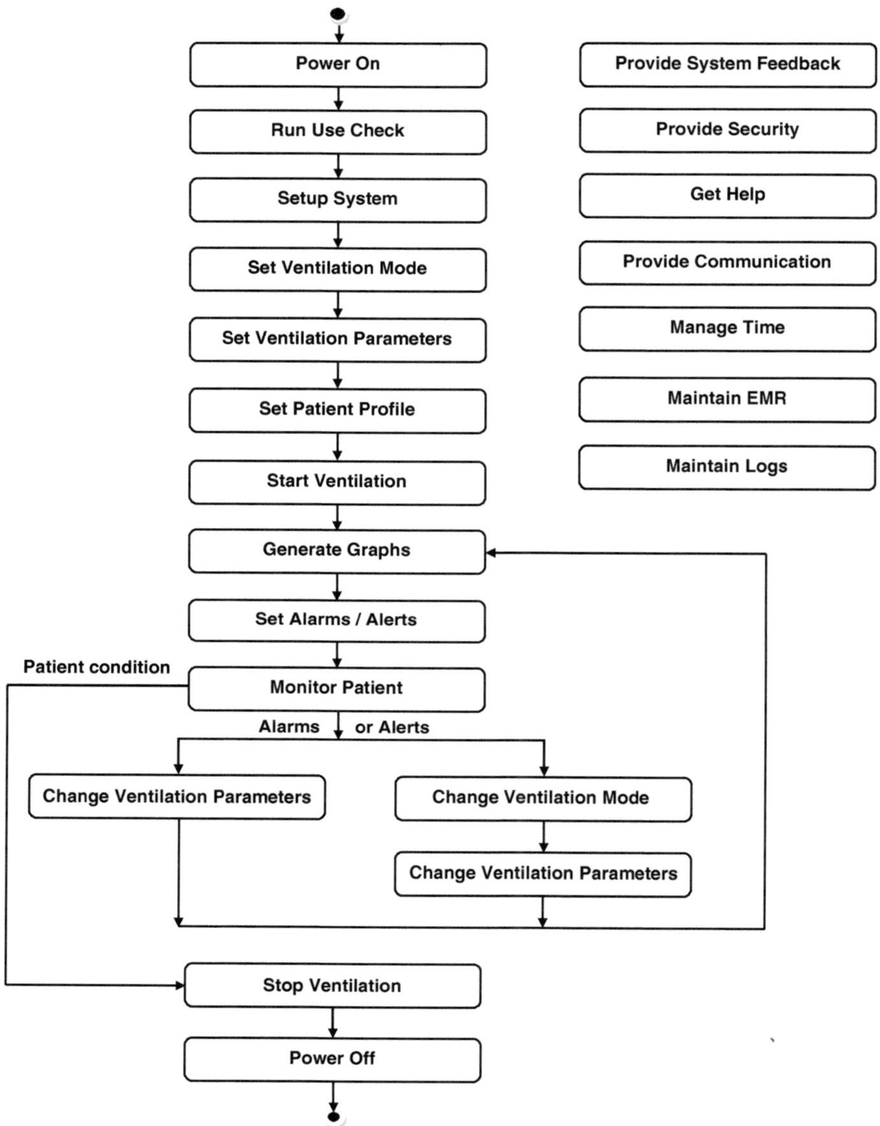

Fig. 2 Workflow in ventilator systems

3.2 Identification of UI Components in Ventilator Systems

The work flow in ventilator systems is further analyzed to identify UI components associated. Each task in workflow and its association with UI component is studied. For example, the task—'Power ON/OFF' is related with a button in a control panel.

Similarly, use check is identified as another essential component, which verifies the working of several functional constituents of ventilator system. The identified UI components are mainly of two types—'Physical components' and 'On-screen components'. The control panel and display are major items in 'Physical components'. Other components, which are displayed on the screen are 'On-screen components'.

3.3 Selection of a Reference Model

One way of proposing a model is selecting a reference model and then, extending it to a full-fledged model. The major criterion for selection of reference model is that model should accommodate recognized UI components. Another criteria is that such model should have a space for medical context; mainly the identified usability problems. The error-oriented model can predominantly fulfill this criterion, supporting even a categorization of problems. A use of Norman's action model is also desirable aspect among evaluation models in healthcare domain. The evaluation model may also accommodate a security aspect; vital take from models in software domain.

The most suitable evaluation model fulfilling these criteria is two-dimensional model by Winter et al. as shown in Fig. 3. This model associates system properties (rows) with the interaction stages [22]. The UI components can replace row-wise items related with system properties, which are further categorized into sub-components. The relationship between UI components and interaction stages is marked in terms of usability problems along with their categorization in proposed two-dimensional model.

3.4 Adaption of a Reference Model

The selected two-dimensional model by Winter et al. [22] is adopted with inclusion of categorized usability problems. These problems have categories such as 'catastrophic', 'critical' and 'serious' problems, and are highlighted in Table 1. Likewise,

Fig. 3 Two-dimensional model by Winter et al.

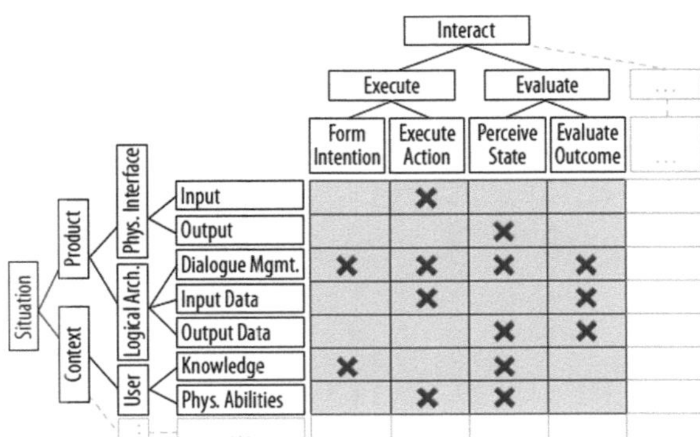

the UI components are of two types viz. physical components and onscreen components. Similarly, related UI Component Model is divided into two parts: Model for Physical Components and Model for Onscreen Components. There are two major physical components—Control Panel and Display. A control panel can be further divided into Buttons and Circular Knob. A display can be further divided into Touch Screen, Screen Lock, LEDs and Indicator Lamp. The major problems with Physical Components are indicated by 'Onscreen Reflection', 'Icon Design' and 'Long-Distance Visibility'. Similarly, there are about 15 onscreen components in UI Component Model as observed in Fig. 4. Some components have few more subcomponents. For example, Patient Record can be further divided into Patient Profile, Patient Data, Security and Communication. The major problems with Onscreen

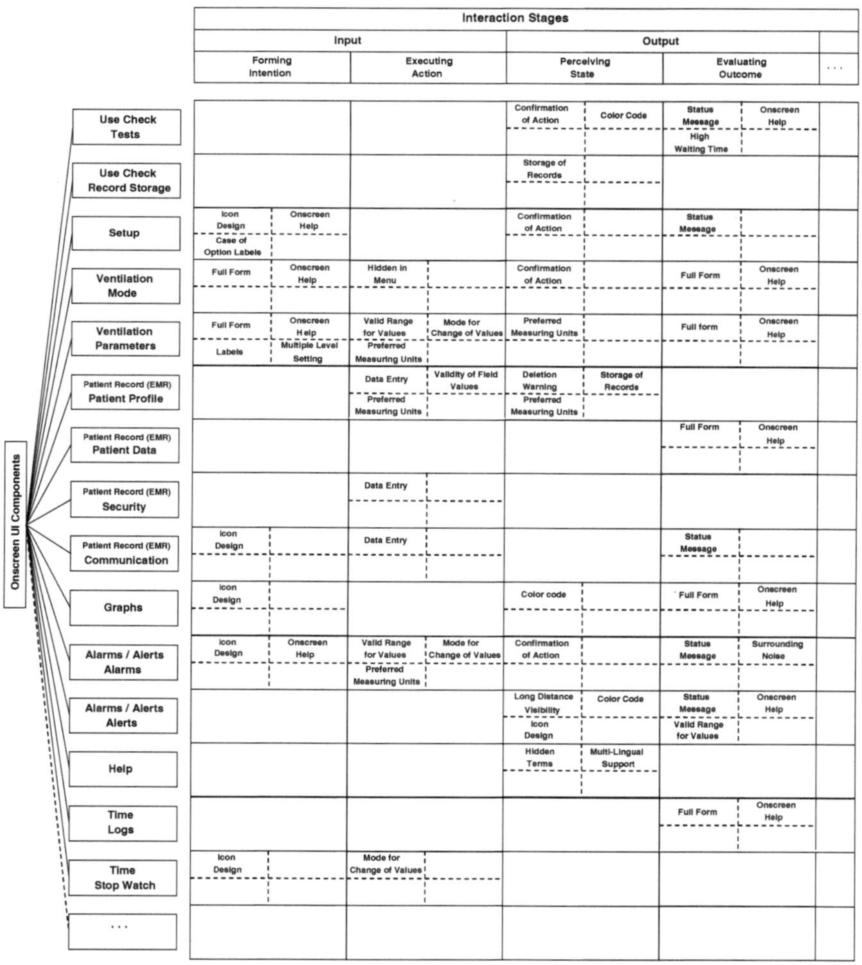

Fig. 4 UI component model for onscreen components

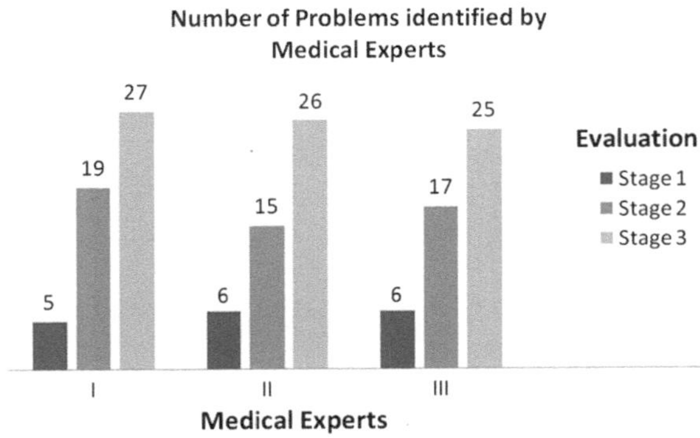

Fig. 5 Graph depicting number of problems identified by medical experts

Components are indicated by 'Onscreen Help', 'Valid Range of Values', 'Data Entry', 'Icon Design' and 'Confirmation of Action'. An observation table is prepared for each model which has entries depicting the presence or extent of related usability problem. For example, for a problem 'icon design', an entry in observation table will be Appropriate/Inappropriate/Not applicable, which help medical expert to identify usability problem in medical UI more effectively.

3.5 Validation of UI Component Model

For model validation, three medical experts are provided with a recorded video of medical UI of ventilator system. The validation is performed sequentially by each expert in three stages of evaluation viz. using a preferred UEM, using problem list (depicted in Table 1) and even, using UI Component Model. The experts have identified number of problems in all three evaluation stages. The graph depicting these problems can be observed in Fig. 5. There are total 27 problems in the list of identified problems. The average number of problems identified in stages—1, 2 and 3 are 6, 17 and 26 problems respectively. Thus, the well-defined structure and options in the proposed model help experts to identify more than 95 % of the usability problems in medical UI of ventilator system. Therefore, UI Component Model is the most suitable option for evaluation of medical UIs.

3.6 Impact of UI Component Model

A UI Component Model can be useful to device designers and usability experts for identifying most of usability problems and improving the related medical UI design. An expert-based usability evaluation using proposed UI Component Model is likely to avoid user errors, prevent patient injuries and deaths, reduce the demand for customer support, reduce the chance of liability claims and increase device sales [5].

4 Conclusion

All general-purpose evaluation models used in healthcare domain miss out important medical context; especially criticality aspect. Thus, there is a need for a more comprehensive usability model for evaluating medical UI of ventilator system in ICU. Based on the study of evaluation models in critical domains, an evaluation model—UI Component Model is proposed with a two-dimensional model by Winter et al. as a base model, which can accommodate both—major UI components and categorized usability problems. This model can be useful to medical device designers and usability experts for identifying most of usability problems and improving the medical UI design as shown during model validation.

Acknowledgments We are grateful to **Prof. Uday Athavankar** and **Prof. Azizuddin Khan** for their critical comments and suggestions during Annual Progress Seminar (APS) presentations. We appreciate a cooperation extended by physicians and healthcare usability experts. At last, we thank **Prof. Dhiraj Jadhav** and Mahadev Karad for their assistive support in documentation during the research work.

References

1. Gould, T., de Beer, J.: Principles of artificial ventilation. Anesth. Intensive Care Med. **8**(3), 91–101 (2007)
2. Bhutkar, G., Katre, D., Rajhans, N., Deshmukh, S.: Scope of ergonomic and usability issues with intensive care unit (ICU): an indian perspective. HFESA J.—Ergon. Aust. **22**(1), 26–32 (2008)
3. Wiklund, M., Wilcox, S.: Designing usability into medical devices. CRC Press, Taylor & Francis (2005)
4. Bhutkar, G., Katre, D., Rajhans, N.: Usability survey of medical devices used in ICU. J. HCI Vistas **IV** (2008)
5. Wiklund, M., Kendler, J., Strochlic, A.: Usability testing of medical devices, pp. 30–33. CRC Press, Taylor & Francis (2011)
6. Zhang, J., Johnson, T., Patel, V., Paige, D., Kubose, T.: Using usability heuristics to evaluate patient safety of medical devices. J. Biomed. Inform. **36**, 23–30 (2003)
7. Bhutkar, G., Konkani, A., Katre, D., Ray, G.: A review: healthcare usability evaluation methods. Biomed. Instrum. Technol.: Hum. Factors IT **47**(s2), 45–53 (2013)
8. Lewis, C., Wharton, C.: Cognitive walkthrough. In: Helander, M., Landauer, T., Prabhu, P. (eds.) Handbook of human-computer interaction, pp. 717–730. Elsevier Science, North-Holland (1997)
9. Wharton, C., Rieman, J., Lewis, C., Polson, P.: The cognitive walkthrough method: a practitioner's guide. In: Nielsen, J., Mack, R. (eds.) Usability inspection methods, pp. 105–140. Wiley, New York (1994)
10. Bhutkar, G., Katre, D., Jadhav, D., Ray, G.: Cognitive walkthrough of medical user interface of ventilator system in intensive care unit. In: Sixth International Conference on Design Computing and Cognition (DCC). London, UK (2014)
11. Drews, F., Musters, A., Samore, M.: Error producing conditions in intensive care unit, pp. 1–13 (2008)
12. Zhang, J., Patel, V., Johnson, T., Shortliffe E.: Toward cognitive taxonomy of medical errors. AMIA 934–938 (2002)
13. Chapman, R., Taylor, L., Wood, S.: Cataloging errors from reported informatics patient safety adverse events. In: Proceedings of Human Factors and Ergonomics in Health Care Symposium, Baltimore, Maryland, pp. 87–94 (2012)

14. Norman, D.: The design of everyday things, pp. 45–53. Basic Books (2002)
15. Kornecki, A., Zalewski, J.: Experimental evaluation of software development tools for safety-critical real-time systems. Innovations Syst. Softw. Eng. **1**, 176–188 (2005)
16. Latorella, K., Prabhu, P.: Review of human error in aviation maintenance and inspection. Int. J. Ind. Ergon. **26**, 133–161 (2000)
17. Yun, M., Han, S., Hong, S., Kwahk, J., Lee, Y.: Development of a systematic checklist for human factor evaluation of the operator aiding system in a nuclear plant. Int. J. Ind. Ergon. **25**, 597–609 (2000)
18. Norros, L., Savioja, P.: Usability evaluation of complex systems: a literature review. Research report: STUK-YTO-TR-204, pp. 1–44. Helsinki (2004)
19. Abran, A., Khelifi, A., Suryn, W., Seffah, A.: Consolidating ISO usability models. In: 11th International Software Quality Management Conference (2003)
20. Diniz, E., Porto, R., Adachi, T.: Internet banking in Brazil: evaluation of functionality, reliability and usability. Electron. J. Inf. Syst. Eval. **8**(1), 41–50 (2005)
21. Kainda, R., Flechais, I., Roscoe, A.: Security and usability: analysis and evaluation. In: 5th International Conference on Availability, Reliability and Security (2010)
22. Winter, S., Wagner, S.. Deissenboeck, F.: A comprehensive model of usability. In: Engineering Interactive Systems Conference. LNCS Springer (2007)
23. Nielsen, J.: Usability engineering. Academic Press, Boston (1994)

Designing of Mould for Brickfield Workers

Bijetri Bandyopadhyay, Amar Kundu and G.G. Ray

Abstract Brick manufacturing is a traditional, unorganized and very booming industry in India. More than 150,000 brick units are engaged in brick making with employment to more than 8 million workers. The main drawback of existing brick moulding method is that a large amount of load is handled by fingers and wrists of both hands in squatting posture. There is wrist bending of (32.78° ± 12.84) which mostly consists of ulnar deviation, which in turn causes pain in wrist movement. Thus, an ergonomic study is done for improving the hand, wrist-palm and finger movement. Three different new concepts of moulds were designed which will not only increase the productivity by delaying the physiological fatigue, but also will prevent the expense of the human cost. So that the workers can continue working for a longer period under the conducive and safe work condition which in turn will influence their social security, health and safety.

Keywords Moulder · Brick · Wrist · Mould

1 Introduction

Brick manufacturing is a traditional and unorganized industry. Off late, due to rapid urbanization, concept of high rise metropolises and spurt in real estate development, brick making has become one of the flourishing and very booming industries. In order to meet this demand, over 150,000 brick units in India are engaged

B. Bandyopadhyay
Ergonomics and Work Physiology Laboratory, Department of Physiology, Presidency University, Kolkata, India
e-mail: bijetri.banerjee@gmail.com

A. Kundu · G.G. Ray (✉)
IDC, IIT Bombay, Mumbai, India

A. Chakrabarti (ed.), *ICoRD'15 – Research into Design Across Boundaries Volume 1*, Smart Innovation, Systems and Technologies 34, DOI 10.1007/978-81-322-2232-3_47

in brick making with possible direct employment to more than 8 million workers. According to NCEUS 2007 report, 422.6 million workers are part of the unorganized sector. Previous studies in India showed that workers in brick making industries suffer from assorted health problems due to awkward postures and carrying heavy load [1, 2]. Poorly designed work station, frequent bending and twisting of the trunk, repetitive and sustained handling of are also reported by other study [3].

2 Description of Activity

The process of brick making involves six major activities like mud preparation, moulding, carrying raw bricks to brick kiln, arranging raw bricks inside kiln for firing, firing of raw bricks and carrying out fired red bricks from brick kiln. Out of these, moulding is an essential part of brick making process, which includes six sub processes like, (1) making a sizable portion of mud (good mixture of clay, water, coal dust, rice husk, and salt) for the mould. (2) Rolled in sand and thrown to the mould. (3) Excess mud is discarded and sand is spread above the mould. (4) The mould with mud is taken to the site. (5) The mould is turned upside down by the help of fingers of both hands so that the soft brick comes out of the mould and (6) The mould is again used for a new brick making. This molding activity is performed by groups of workers designated as moulders. Thus, the objectives of the present study is to identify and evaluate the physiological and biomechanical stresses accumulated by the molders due to load handing in awkward work posture, and to formulate probable remedial measures to reduce these stresses so that the musculoskeletal disorders are prevented and health status of the workers are improved.

3 Methodology

3.1 Selection of Site and Subjects

The brickfields under study were situated around the city Ahmedabad in Gujrat. 10 brick moulders (5 male and 5 female) with mean age of 26.0 ± 4.7 years and 22.6 ± 4.0 years respectively, having minimum 3 years of working experience were randomly selected for the study.

3.2 Assessment of Physical Characteristics of the Subjects

The height of the subjects was measured by the Martin type anthropometric rod (Mfg by Seiber and Heigner, Switzerland) and weight was measured by a portable, calibrated bathroom weighing scale. Hand grip strength was determined by using a hydraulic hand dynamometer (Mfg by Baseline fabrication enterprises Inc., USA) to test the maximum voluntary contraction. Blood pressure was measured with the help of a Sphygmomanometer and stethoscope.

3.3 Biomechanics

Work posture analysis: Working postures were assessed by video capturing the operation by using Sony Handicam camera, followed by using the following protocols:

OWAS method (Ovako Working Posture Analysis System) [4].
RULA method (Rapid Upper Limb Assessment) [5].
REBA method (Rapid Entire Body Assessment) [6].

4 Problem Identification

For casting bricks, a MOULD is essential. The mould used by the workers is a hollow rectangular cube (23.8 cm × 11.5 cm × 8.5 cm) made of aluminum, weighing 1.34 kg as shown in Fig. 1.

Moulding is basically a repetitive and continuous work performed in awkward squatting posture in which a load of around 6 kg (mud plus mould) is handled each time by finger and wrist of both hands with a frequency of 3 bricks per minute. This leads to a total load handling equivalent to 18 kg per minute at a stretch for minimum half hour to forty five minutes. A person in average makes 600–900 bricks a day. Taking it on minimum side, a worker handles minimum 3,600 kg of cumulative load by both hands a day. In every brick making cycle they use thumb for generating the force, and remaining fingers as pivot to topple the mould so that the brick comes easily as shown in Fig. 2.

Fig. 1 Brick moulding tool known as mould

Fig. 2 Steps of brick moulding

Initially when the workers began their life in this job three years ago, they had pain in wrists, fingers, entire arm and shoulders during work, but now they have taken it as a part of life.

During moulding a high range of wrist movement at the rate of three movements per minute is seen among the workers as depicted in Fig. 3. This unnatural degree of wrist bending ($32.78^{\circ} \pm 12.84$) as shown in Table 1 mostly consists of ulnar deviation, which in turn causes more adverse Musculo-skeletal-disorder (MSD) effect on worker's health especially to wrist joint movement.

The main drawback of existing brick moulding method is that a large amount of load is handled by fingers and wrists of both hands demanding a high range of wrist bending in squatting posture.

It is, therefore, absolute to make our workers aware about the negative impact of the work condition and better way to perform the work. This will help them to continue their working life for a longer period under the conducive and safe work condition which in turn will influence not only on the working conditions, but also on the social security, health and safety of the workers.

Fig. 3 Degree of wrist bending during moulding activity of different workers

Table 1 Deviation of wrist during moulding activity of different workers

Parameters	Different Subjects					
Type of deviation	Radial (N = 3)			Ulnar (N = 3)		
Deviation angle	25°	27°	36°	24°	20°	27°
Mean ± SD	29.33° ± 5.86			34.5° ± 12.84		
Mean ± SD	32.78° ± 12.84					

5 Result

The physical characteristics of the brick kiln workers are presented in Table 2. It shows the Mean (±SD) values of age and other physical parameters like height, weight, BMI, Hand Grip strength, and Blood pressure, of the subjects.

From the result of BMI it is seen that the values for both genders were just at the lower level of the normal range (18–22.9) [7] and all the workers fall below ideal body weight based on Brocas's index [8].

5.1 Biomechanical Assessment

From the view point of biomechanics, two approaches were made: (1) work posture assessment and (2) assessment of localized hand, wrist, palm, and fingers fatigue during moulding.

Table 2 Physical characteristics of the brickfield workers

Variables	Mean ± SD	
	Male	Female
Age (yrs)	26.00 ± 4.69	22.60 ± 3.97
Body weight (kg)	48.50 ± 5.72	43.40 ± 6.54
Height (cm)	160.28 ± 7.13	151.28 ± 6.07
BMI	18.84 ± 1.33	18.96 ± 2.53
Maximum grip strength in Left hand (kg)	35.00 ± 7.28	25.80 ± 2.68
Maximum grip strength in Right hand (kg)	37.20 ± 6.26	27.00 ± 1.73
Systolic pressure (mmHg)	125.00 ± 10.95	110.40 ± 7.80
Diastolic pressure (mmHg)	78.80 ± 6.10	74.40 ± 5.37

Table 3 Postural assessment of the brick moulders

Postures	OWAS score	Action level	RULA	Action level	REBA score	Action level
Cutting mud	2, 1, 6, 1 (2)	Corrective measures in near future	7	Investigate and change immediately	12	Very high risk, implement change
Inserting mud on mould	2, 1, 6, 1 (2)	Corrective measures in near future	7	Investigate and change immediately	10	High risk, investigate and implement change
Take out brick from mould	4, 1, 6, 1 (4)	Corrective measures in near future	7	Investigate and change immediately	12	Very high risk, implement change

5.1.1 Work Posture Assessment

The entire process of manual brick moulding was divided into certain components for postural assessment. Different standardized methods were applied for each component to identify the risk factors associated with a particular task. Cutting mud, inserting mud into the mould and take out brick from the mould, were the main task components. Assessment by OWAS method revealed that all the postures were highly risky (Table 3) and corrective measures were required as soon as possible. In every stages of brick moulding RULA score was seven suggesting action level 4 where intervention and changes are immediately required. REBA score was 10 or more than 10 in all the postures indicating corrective action including further assessment is immediately required. Based on posture evaluation scores (Table 3) it was clear that adoption of sustained squatting posture and moulding the bricks by forward bending, for hours after hours throughout the day, in the field is very detrimental for the workforce. It was evident that prolonged squatting posture caused numbness in the lower leg resulting from lack of blood supply due to sustained muscle compression. Changes in work posture and associated tools need immediate attention.

5.1.2 Assessment of Hand, Wrist, Palm and Finger Movement

With reference to Fig. 3, it was reported that all moulders were suffering from pain in the fingers, wrists, hand and shoulder which are primarily related to moulding job because to the mould design. To perform this job works had to adopt the squatting posture which demanded burden on upper back, lower back, both knees, both thighs and both ankles. Brick moulders were having more pain in the wrists and lower part of the body compared to upper part, because most of the time they sit continuously in the same awkward squatting posture for long hours to mould the bricks without taking proper rest interval. Their task is also repetitive in nature contributing to their major discomfort level.

6 Design Intervention

Based on anthropometric data 3 different conceptual model of the mould were developed by using solid works CAD software. The CAD drawing was then converted to a full scale prototype for stimulation study. Three different new concepts were designed and fabricated with proper gripping for moulding activity. It will increase the productivity by delaying the physiological fatigue, by preventing the expense of the human cost. Workers comfort was also taken into consideration while designing the mould.

6.1 Concept 1

To shift the whole load during moulding from fingers to both hands, two additional handle on both sides of mould were added as shown in Fig. 4, so that the workers can transfer the inside raw brick by turning the handle. But during testing the design was rejected as it required too much of wrist bending. Moreover addition of fixed handles also demanded extra space in the row.

6.2 Concept 2

In this case four additional grips on two sides and at the bottom of the mould were provided by extending the surface of the mould on four sides as shown in Fig. 5. The grips on the sides were for turning the mould on opposite sides and the grips at the bottom were given so that it will be easier to take out the mould after turning it upside down on the floor. A high fillet (round shape) was made on the bottom edges of the mould which helped in turning it. In this concept 2, force generation was shifted from fingers to palm. Lateral movement of hand was required to turn the mould which was convenient to execute. This minimized the wrist bending and so

Fig. 4 Prototype and solidworks (3D) model of concept 1 with additional handle and anti-sliding pins/nails

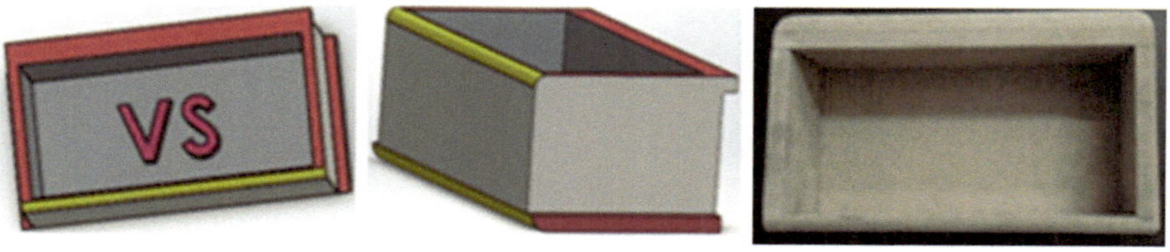

Fig. 5 Prototype and solidworks (3D) model of concept 2 with additional grip and high fillet

Fig. 6 Actual field testing of prototype

force generation by fingers was reduced to a greater extent. The prototype was tested Fig. 6 and the workers appreciated the design. Multiple numbers of prototypes on concept 2 is now under fabrication for study on a large number of moulders.

6.3 Concept 3

In existing mould, there is no sufficient space for applying force by thumb. In this design we increase the space for thumb. In existing mould due to insufficient

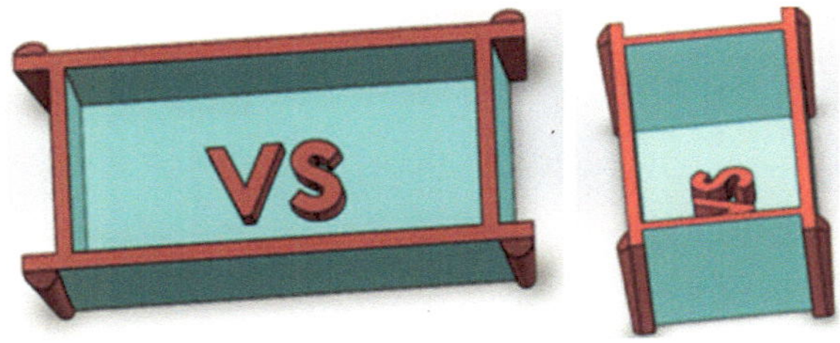

Fig. 7 Solidworks (3D) model of concept 3 with additional larger grip

handle, the workers feel more pain in their finger. In this concept we introduce bigger and rounded smooth handle to decrease the finger pain as shown in Fig. 7. This prototype was not yet tested in field.

7 Conclusion

Biomechanical analysis indicates that the workers are continuously adapting awkward squatting postures, and lifting of load on their hands which results in whole segmental pain and discomfort. It is needless to mention that the importance lies in a systematic intervention to relook at the better work posture and redesigning of the mould accordingly. In parallel steps must be taken to create awareness for both the workers and management to protect this limited and valuable work force from depletion.

Acknowledgments Authors convey sincere thanks to the brick field workers for their active participation throughout the study. Heartfelt thanks to Prayas for sponsoring the project.

References

1. Mukhopadhyay, P.: Risk factors in manual brick manufacturing in India. HFESA J. Ergon. Aust. **22**(1), 16–25 (2008)
2. Sett, M., Sahu, S.: Ergonomic study on female workers in manual brick manufacturing units in West Bengal, India. Asian-Pacific Newsl. Occup. Health Saf. **15**(3), 59–60 (2008)
3. Basra, G., Crawford, J.O.: Assessing work-related upper limb disorders in a brick making factory. In: Robertson, S.A. (ed.) Contemporary Ergonomics, pp. 480–485. Taylor & Francis, London (1995)
4. Karhu, O., Kansi, P., Kuornika, I.: Correcting working postures in industry: a practical method for analyses. Appl. Ergon **8**(4), 199–201 (1977)
5. McAtamney, L., Corlett, E.N.: RULA: a survey method for the investigation of work related upper limb disorders. Appl. Ergon **24**(2), 91–99 (1993)
6. Hignett, S., McAtamney, L.: Rapid entire body assessment (REBA). Appl. Ergon **31**(2), 201–205 (2000)
7. http://www.igovernment.in/site/India-reworks-obesity-guidelines-BMI-lowered sited on 1st March (2014)
8. Visweswara, R., Balakrishna, N.: Feasibility of Broca's index for the nutritional status of adults. J. Med. Res. **102**, 173–178 (1995)

Objectifying Usability in Product Design

Ranjit Konkar

Abstract Even though the interactions of any product with its target users are predictable, every so often products are designed with a missing usability feature that on hindsight should have been easily anticipated and provided for, like a desktop CPU cabinet that has no easy way of lifting it off the table. Can the process of usability design be objectified and standardized to the extent that it avoids lack of attention to known interactions of user with product? Is it possible to articulate, list, codify, and thus capture predictable aspects of product design and thus provide designers a check-list or an algorithm to not forget some aspect of the product? This paper proposes that the ad hoc part of some of the designers' work is replaced by a check-list approach of attending to known classes of usability thereby removing some uncertainty in the process.

Keywords Usability · User-interaction · Industrial design

1 Motivation and Approach

The motivation for this paper is the author's frequent encounter with products that seem to completely miss certain usability features as if the designer forgot to check for the need for that feature or did not run through the various ways in which users would interact with the product. For example: the cuboid-shaped desktop computer (Fig. 1)—from the world's most reputed PC manufacturer, no less—has no facility to be lifted off the table easily without tilting it sideways to

R. Konkar (✉)
National Institute of Design, Paldi, Ahmedabad, India
e-mail: ranjit_k@nid.edu

A. Chakrabarti (ed.), *ICoRD'15 – Research into Design Across Boundaries Volume 1*,
Smart Innovation, Systems and Technologies 34, DOI 10.1007/978-81-322-2232-3_48

Fig. 1 Most desktop computers have no facility to be lifted. They have to be tilted to get a purchase underneath

reach under it. A thermos that is too large to hold conveniently in one hand. A glue-stick that rolls off the table. How did the designers of these products overlook aspects of the products' usability whose importance simple usage for a few minutes makes evident?

It is the thesis of this paper that leaving aside the creative, inventive side of design, the usability side of it can be systematised rather than be left knee-jerk and ad hoc. The industrial design process could be steered along a path of known and predictable behaviours and interactions with the user. Certainty of interaction should reduce uncertainty of design outcome.

The approach of the paper is to conduct a task analysis of a user's interactions with a generic consumer product over its life-cycle and to derive the design needs of products from the analysis. We track the interactions of a generic product with its users from the point that it is picked up from the retail shelf to its disposal after a lifetime of handling, usage, maintenance, transport, and repairs. An interactions-to-features mapping is attempted and a list of generic features that service the given interactions is derived.

2 Previous Related Work on Usability

In Engineering Design, there are principles such as Design for Manufacturability and Design for Assembly developed by Boothroyd et al. [1] that deal with how to design a part to make it suitable for production (from shaping to assembly) and handling at the factory level. To these a natural addition for dealing with Industrial Design should be a set of principles called Design for Usability that should guide designers in providing such features on and around the product that make it easy to use, once basic functioning is achieved.

Work being done around the world on product usability includes the Design for Usability project carried out by Delft University of Technology, TU/Eindhoven, and the University of Twente in the Netherlands. As part of this project, its investigators have studied many aspects of usability, some specifically of electronics products. For example, in [2], the authors identify issues in product development that influence the usability of electronic consumer products. In [3], J. van Kuijk lists 8 rules, based on observation and experience, for guiding a user to purchase a usable product. These rules seem very generic and common-sense, that everyone follows all the time keeping reliability and value of one's time and money in mind. In [4], the authors talk about how users' perceptions about usability change before and after purchase and the experience of using the product. They try to correlate through a small sample the appearance and anticipated usability of a product. The publication at [5] provides a good overview and motivation for the study of usability, introducing terms such as Usability Engineering, and identifying various actors and issues in the practice of usability. In [6], the authors talk about how to obtain data for usability assessment, suggesting that after-sales feedback could be exploited very strongly for the purpose.

In [7], Page, while describing methods of ergonomics evaluation of products, mentions conducting user trials through various methods and in one of them, refers to "preparation of a detailed task analysis" which "involves preparing a list of all of the activities associated with the product in a logical order from initial unpacking right through to emptying of any waste produced and disposal of the item itself." It is exactly this list preparation that the current paper attempts.

In another essay [8], Baber and Mirza also mention "using checklists or expert appraisal" as methods of conducting ergonomics evaluation of consumer products. Surveys conducted by them established measurement of usability of products to be a big outcome of ergonomic assessment. They, however, do not proceed to formulate such a checklist.

Hedge, in [9], discusses the application of ergonomics principles to design of hand-operated devices and draws conclusions about desired geometric details of surface features like handles and various types of grips, etc. This treatment is at a level that is more form-based than the current paper.

Acosta et al. [10] mention considerations of product life-cycle as one of the important factors in the design of consumer products, covering creation, sale, use, up to disposal.

Lastly, Schulze [11] also writes about maintainability and disposal issues as parallel concerns with design for usability.

Little of the literature published from the group above ties the knowledge of an object's expected handling with features that must be provided on the object which this paper seeks to do.

3 Generic Interaction of Users with a Retail-Purchased Product

What handling and interactions does a product need to be designed for? What *are* the interactions that a human user conducts with any consumer product at a generic level? Can a list be drawn up which describes most interactions that a user might ever have with a product so that it can form a master check-list which designers can use to decide whether they have attended to all interactions with future users of the product?

It is of course recognised that the way a user interacts with a water-bottle is going to be different than the way he would interact with a television set, for example. The functions of the two products are entirely different: one is a liquid-carrying contraption, the other an electronic entertainment device. What can be similar between the two for us to claim any commonality between them for design?

Despite their differences, our interactions with the two devices have many commonalities. When purchasing them, both might come packaged in boxes from which we have to extract them. Upon extraction, the TV and the water bottle would both possibly be inspected. The TV set would then be carried to its intended place in the family room, while the water bottle to, maybe, a place in the kitchen closets. In the family room, the TV might be stood up on its stand if it is to be placed on a counter-top, or it might be mounted on the wall if a thin LCD panel model. The water bottle might be placed upon its flat base on a shelf or it might be hung by its strap from a hook.

Before operating the TV, the required connections and settings on it need to be made so that it becomes ready for operation. The screen might need to be adjusted for its angle. Similarly the water bottle might need to have its strap length adjusted for the user's size.

Operating the TV set involves mostly handling the remote (after turning on the power), which therefore needs to be designed for the easiest cognition and best physical feel. "Operating" the water bottle consists of holding it while pouring from it. Simple as it sounds, there are many water bottles that do not satisfy on even this minimal count.

Daily, the TV set and the space around it needs to be cleaned. The remote becomes grimy through constant handling. Similarly, the water bottle needs to be cleaned frequently for hygiene reasons.

Sometimes, the bottle strap breaks or slips loose. A tailor or an odd-jobs person has to be called to fix it. Or the lid falls down and breaks or is misplaced. Sometimes the rubber gasket springs a leak by tearing or hardening. The TV remote might become unresponsive or the TV set might develop faults. The repairman has to be called.

One thus sees that even for products that differ widely in their functions and outputs, there exists a level from which their handling by their users appears common. It is from the vantage point of this level that the industrial design of products is sought to be examined as a serial process of interactions which can be proactively and anticipatively designed for. Facilitating such handing is what industrial design is all about, after all.

4 Product-User Interactions

In a product's life-cycle, the handling it experiences, the interactions it goes through with its user after purchase are captured in the list below:

1. Taking the product out of its packaging (if packed)
2. Handling the product for inspection
3. Moving the product to its place of use
4. Installing the product for operation

(Repeated often:

5. Making adjustments/settings for operation of the product
6. Operating the product
7. Docking the product for rest)
8. Performing maintenance upon the product
9. Repairing the product
10. Transporting the product between locations

These are described below in detail and with examples. For each of these interactions, products should be designed to enable that interaction in the best manner possible. This might translate to availability of specific features on the products.

4.1 Taking the Product Out of Its Packaging

Task breakdown: (i) Grasping it, (ii) Separating it from the packaging, and (iii) Placing it on the floor or on a raised surface.

The act of taking the product out of its packaging is the first physical meeting of the user with the product and should be facilitated through good design of the packaging and/or the product.

A good design facilitating extraction from packaging is shown in Fig. 2: a semi-circular cut in the packaging allows the customer to reach at the product and pull it out. A bad example is the envelope or aerogramme of old, whose glued edge sticks to itself.

As a product peripheral, the packaging should be designed for the usability of the product it packages.

Design implications: The act of removing the article from its packaging would normally require the user to grasp the object with one hand while grasping the packaging with the other. Thus, conveniently located **grasp features** are required to be designed into the object and the packaging to facilitate separation of object from packaging.

Fig. 2 Packaging for a sweet, with enough foresight to provide a notch for ease of removing the box from the cover

4.2 Handling the Product

Taking the product out of its packaging is usually followed by the user holding it in his hands (if small) and inspecting it for defects, taking a closer look at its surface details, peering closely to read text printed on it, etc. In this first (and subsequent) handlings, one doesn't want to be hurt by sharp edges, pointed tips or corners, get ink on one's hands, etc. Similarly, one doesn't want to damage fragile parts of the product by accidental rough handling, neither does one want the product to slip out of one's hands and fall to the floor.

Such common-sense product-safety concerns about what might happen during mere handling of the product translate to well-known caveats for designers that dictate the form of the product, its surface texture, its materials, and a host of other things which are to do with handling (the domain of the industrial designer) and not performance (the domain of the engineer). Some of the issues could be (depending upon the product):

Design implications: (i) Concealment of product parts that pose danger of physical injury (physical cuts, electric shocks, chemical/thermal burns, etc.), (ii) Protection features for the product's fragile and sensitive parts, (iii) Safeguards against breakage through dropping, etc.

4.3 Moving the Product to Its Place of Use

Task breakdown: Grasping it, lifting-carrying or pushing/pulling it.

From the place at which it was unpacked, the article would need to be moved to its intended place of operation. Often, products are provided no geometric features that allow a grasp around and under them, making it difficult to lift them off the surface.

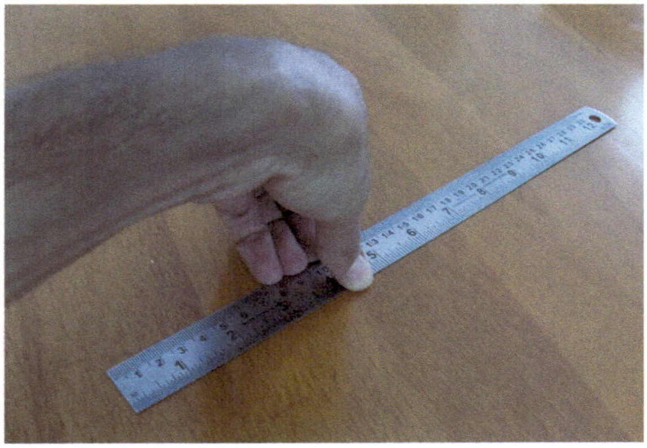

Fig. 3 A ruler with no easy way to lift it off the table

Design implications: (i) Conveniently located **lift features** (protuberances, cavities, notches, projecting flaps, flanges, handles, etc.) should be designed into the object to allow one to reach under or around it and lift it. (ii) For large, heavy products like washing machines and refrigerators, friction-lowering mechanisms such as casters and handles to grasp it to move it should be provided.

Examples of bad design in this regard are: (i) the flat steel foot-ruler (Fig. 3), with no features around or under which one can insert one's fingers to lift it, and (ii) the computer shown in Fig. 1.

4.4 Mounting the Product for Operation

Products that are meant for stationary, immobile use, that users do not have to wield in their hands, such as wall-length mirrors and geysers, need to be placed in a proper position and orientation for operation. Products that are meant for hand-held operation call for a way to fixture or mount the work-piece instead. The "proper position for operation" should ensure a stationary, stable, and ergonomically oriented configuration.

Concerns about the mounting detail of such products are: (1) Orientation of mounting surface: ceiling/wall/floor, (2) Orientation of product elements, (3) Stability in operation: to ensure rigidity of mounting detail, (4) Freedom from vibration effects, (5) Portability.

4.4.1 Orientation of Mounting Surface

Objects need to be designed differently depending upon whether they are going to be used stood upon their base, or mounted upon a wall or hung from a ceiling. For example, ceiling-mounted LCD projectors, desktop, portable ones, and hand-held-sized pico projectors all need to have different mounting features. Similarly with lamps and fans.

Wall-mounting needs to have flanges, ears, mounting screws, holes, etc. A standing arrangement needs flat base, legs, or tripod resting points. Hanging would need hooks and loops. Attention is needed to this aspect of usage at design time.

4.4.2 Proper Orientation of Product Elements

Many products, for example desktop projectors, need to be oriented at varying angles during usage. Rather than stuffing 'packing' under these products as is commonly done, the designer should anticipate this need and provide for it. Good projectors do come with adjusting wheels to set the angle of the projection. But products like TV, wall clocks, ceiling fans, which currently are not made adjustable for orientation, would all be much more useful if they were.

Other than gross orientation of the entire product, product elements like visual displays, printed instructions, connection interfaces, should be placed and oriented to be visible and accessible to the user in as close to his neutral working position as possible. Having to bend, stoop, or peer closely at dials, labels, etc., is a mark of bad design. Desk calculators that have a display inclined towards the user are a good design.

Design implications: (i) Products should be designed with elements properly oriented to the user. (ii) If the orientation is expected to be variable, then provision for adjusting orientation of a product or the relevant parts of a product to its user should be designed into the product.

4.4.3 Stability of Mounting

While nothing can be supported standing on less than 3 points of support, more than 3 can lead to wobble. Excessive support is a great inconvenience in mounting because every extra bolt-hole support requires the exact alignment of the holes of the supports with the holes in the wall.

Scientific determination of support sufficiency and effectiveness of redundancy in support is something that should be consciously done and not be left to guesswork.

Design implications: (i) The number of supports necessary for a product should be properly determined. (ii) Facility for adjustment of length should be provided in cases of redundant support. (iii) Hanging designs should be weighed against standing designs for the different experiences, advantages and disadvantages that the two provide.

4.4.4 Freedom from Vibration Effects

If a stationary product is something with a rotating driver like an electric motor, then it is to be expected that vibrations will result. This vibration might result in a degradation of performance as well as of usability. Washing machines that experience resonant shaking because of vibrations from its motor that couple with an unbalanced load could get damaged, which is a performance hit. Usability-wise, the vibrations could also result in less serious effects of irritations in noise or maybe slippage at the supports.

Design implications: While designing for zero or minimum vibrations is an involved engineering task, it should certainly be the designer's responsibility to anticipate these problems and provide for vibration-, noise-, and slippage-inhibiting mechanisms. These could be as simple as friction pads under the supports of a mixer-grinder or something more complex like brakes on casters for a washing machine trolley.

4.4.5 Portability

The potential need to use a product in more than one place or the need to clean space around it are justifications for making it moveable. 'Portability' is the attribute of designed moveability. The design implications of this need are already covered under 4.3.

4.5 Making Product Adjustments/Settings for Operation

Using a product often entails making settings or adjustments on it, through knobs, levers, screws, wheels provided for the purpose. To iron our clothes, it is necessary to set the fabric setting, to use a toaster, the toast browning control. It is important that the product is designed for the user to be able to ergonomically view the necessary display elements of it and reach out to and handle the control elements dexterously.

Design implications: The user must be given an easily visible, understandable, reachable, and operable interface for making product settings during operation or pause.

4.6 Operating the Product

In the entire life cycle of a product, the time that it is in operation is the period which has—naturally—received the most attention by designers and engineers, since it covers the purpose of the product's creation. Specifics of designing various products are covered under literature pertaining to the class of the product (e.g., power-drills), but any product must provide safety from mechanical, electrical, thermal, chemical danger. Getting cut by a exposed rotating blade, receiving an electric shock, burning oneself are all possibilities against which the designer of the equipment must provide protection to the user.

Design implications: (i) Provide for safety to the user or people around him in all possible eventualities. (ii) Design the operation so that the product enters a safe state if it fails. (iii) Design the operation so the product does not work if in a dangerous configuration.

4.7 Docking the Product for Rest

If the product is a hand-held object, what is one supposed to do with it when not in use, i.e., when the hand is not holding it? Unless it is a piece of machinery that is working around the clock, a product usually spends more time in idle, unused state than in use. Attention should therefore be paid to how it spends its time in the idle state.

In idle condition, a product would need to be placed *somewhere*: either placed down upon or hung from or rested against something. A proper, neat arrangement for 'rest docking' is an absolute must. Products like a hand-held power drill that do not come with any dock pose problems when needing to be put down between uses. It has to be laid down on its side (see Fig. 4), where only one point makes contact, causing it to rock and turn around with the drill-bit in place presenting risk of injury.

A hand-blender is an example of a product that comes with a dock. It is much appreciated by users.

The rest docking arrangement should guard against degradation that can happen because of dust/germs/corrosion, wrong orientation, instability.

4.7.1 Dust

Hand-held products, like hand-blenders, are usually stored between uses inside their original packages or inside cupboards, or outside on their own docks.

Fig. 4 Between uses, a hand-held power drill has to be kept down on its side, resting on any horizontal surface very unstably

Stationary products, like a mixer-grinder, might similarly go without use for days or maybe weeks. In both cases, dust accumulates on the product. A preventive dust-cover is very much a product's required accessory in such cases and should be the designer's responsibility to conceive and the manufacturer's responsibility to provide. This is an avenue for accessory manufacturers to contribute to value addition of the product.

Products that have a health-care association and especially products that are going to make contact with the mouth or eye, should be kept free of contaminants. For example, toothbrush caps should protect not only from dust but also from insects and germs.

Design implications: (i) Products that are expected to stay unused and exposed to the elements for long periods should be provided with dust covers. (ii) Products that are supposed to provide hygiene should be designed to remain completely free of contaminants.

4.7.2 Stability and Balance at Rest

What applies to stationary products in operation applies to hand-held products at rest. Products that are designed for hand-held use are prone to lack of attention paid to their stability upon being stood up between uses. While it is assumed that they are not serving their primary purpose at that time, it is still a matter of some convenience that they should rest stably and not rock about an unstable position.

Consider a soft broom. When not in use, it can be a problem to store. The broom is best hung on a wall, but most brooms do not display the extent of productisation where a loop is provided to hang it. Post-purchase fixes are required.

Design implications: Proper arrangements must be made to store hand-held products firmly and stably in their rest configuration.

4.8 Maintaining the Product

To keep a product running well, there might be need to perform routine procedures like oiling and cleaning to keep it in good condition. Most products are badly designed for performing these, calling for people to hire peripheral services like Annual Maintenance Contracts (AMCs). The author has found products intentionally designed to be impossible to open up, e.g., chair castors, which get jammed with floor litter like hair, have no arrangement to be opened and cleaned, they are made to be junked and replaced by new casters. It is one of the principles of ecologically responsible and sensitive design that products should not be made to be disposed flippantly or to meet their ends of life prematurely when only parts of them are indisposed and can actually be attended to.

Design implications: (i) The product should be designed to be possible to open up. (ii) The product should be designed to be easy to open up without any tools if possible (for example using wing nuts). (iii) Any necessary fasteners should be the basic, commonly available kind which open with standard, easily available tools. (iv) If any non-standard fastener is used, then it should be clubbed with the product. (v) If a part is *required* to be non-user-serviceable for security reasons (only), then its disassembly can be designed to be deliberately difficult using ordinary means. (vi) Every product should be accompanied by a printed maintenance manual.

4.9 Repairing the Product

Like maintenance, the need to repair products arises more frequently than owners would like. Most product owners are not technically competent to conduct any repairs themselves. Most electronics-based products are too complicated to be repaired even by competent users unless they possess special training or tools, let alone lay-persons with little understanding of the working of the product.

Whether it is the owner who repairs it or a trained and equipped technician, the product should be easy to repair and be facilitated by a clear user- and repair-manual. The requirements of easy repairability include all of those of easy maintainability (listed in 4.8 above), and in addition the following:

Design implications: A set of spare parts should be provided to replace those that are known to develop faults.

4.10 Transporting the Product Between Locations

When products need to be carried or sent long distances through the processes of the Post, couriers, or transporters, they need to be packaged well to withstand the roughness of handling and road surface. Some parts might need to be specially protected if fragile or prone to damage. A washing machine with a drum that is loosely mounted and thus prone to vibration or getting knocked around sometimes comes with a restraining bar to immobilise the drum. Thought to this on the part of the designer is very important.

Design implications: The designer must anticipate the special needs of transportation of the product and provide the basic protective mechanism necessary for safe transport.

5 A Check-List for Designers

Based upon the detailed narration presented above of the interactions that users perform with products they buy, it is suggested that designers adopt a check-list approach in which they ask themselves whether the product they are designing fulfils the needs of their future users as they go about purchasing their product, then opening it, installing it, using it, breaking it at length, having to repair it etc. The approach is akin to an expert system guiding the designer through a question-answer process, as follows:

QI. After purchase, how would the user remove the product from its packaging? Have the required grip features been provided to do so?

QII. Having unpacked it, is the product good for tactile handling? Is there any danger in accidentally damaging it in handling? Or getting hurt in doing so? If it is dropped, what will be its fate?

QIII. Is there any issue in moving it from one place to another? Having put it down, is it easy to lift it up again? Have the necessary lift features been provided?

QIV. Is the product stationary or hand-held? (a) If stationary, how is it planned to be installed? On the floor/table-top/wall/ceiling? (b) Are well-designed mounts provided for enabling that? (c) Is there good justification for the number of supports in the mount? (d) Is everything that is required to be visible from the neutral operating position oriented to be just so? (e) Is there a kinetic/thermal energy source on the product? If so, have the required vibration absorption or heat dissipation mechanisms been provided to insulate the user from their effects?

QV. Does the product require any settings to be made before operation? If so, are they easily made and is the state of the settings clearly visible to the user?

QVI. In operation, does the product present any safety hazard to the user? If it fails, does it remain in a dangerous state? Is it possible to start it from the dangerous state?

QVII. When not in operation, how is the product to be placed? Is any dock provided upon which it would rest during pauses in use? Is there any arrangement for it not to collect dust? If your product is a health-care or a hygiene-related one, have you provided for non-contamination during periods of non-use?

QVIII. What is the regular maintenance that the product would require? Is the product dismantlable for maintenance servicing? If it is not obvious how to take it apart and put it together again, has a user's manual been provided for it?

QIX. Are any parts of the product likely to reach end of life or fail sooner than the rest? If so, have you considered including a spare of that part in your product offer?

6 Future Work Potential

The findings and conclusions will later be applied to 30 live cases of design that were conducted here in our design studio of National Institute of Design this year as part of a design project course. Matrices will be created that track the adherence, by the designers, to the usability rules mentioned here.

References

1. Boothroyd, G., Dewhurst, P., Knight, W.: Product Design for Manufacture and Assembly, 2nd edn. Marcel Dekker, New York (2002)
2. van Kuijk, J., Christiaans, H., Kanis, H., van Eijk, D.: Usability in product development: a conceptual framework. In: The British Ergonomics Society Annual Conference, Nottingham, UK, 17–19 Apr 2007
3. van Kuijk, J.: Make it easy on yourself: how to purchase an easy to use product in 8 Steps. Jaargang 33, 2008: nummer 1, februari
4. van Kuijk, J., Preijde, E., Toet, E., Kanis, H.: Expected versus experienced usability: what you see is not always what you get. In: 17th World Congress on Ergonomics, Beijing, Aug 2009
5. van Kuijk, J., Christiaans, H., Kanis, H., van Eijk, D.: Usability in the development of consumer electronics: issues and actors. In: 16th World Cong on Ergonomics, Maastricht
6. van Kuijk, J., Christiaans, H., Kanis, H., van Eijk, D.: Usability in product development practice: after sales information as feedback. In: Conference of International Association of Societies of Design Research: Emerging Trends in Design Research. Hong Kong, Nov 2007
7. Page, M.: Consumer products: more by accident than design? In: Stanton, N. (ed.) Human Factors in Consumer Products. Taylor & Francis, London (1998)
8. Baber, C., Mirza, M.: Surveys of evaluation practices. In: Stanton, N. (ed.) Human Factors in Consumer Products. Taylor and Francis, London (1998)
9. Hedge, A.: Design of hand-operated devices. In: Stanton, N. (ed.) Human Factor in Consumer Products. Taylor and Francis, London (1998)
10. Acosta, G., Morales, K., Lagos, D., Ortiz, M.: Addressing human factors and ergonomics in design process, product life cycle, and innovation: trends in consumer product design. In: Karwowski, W., Soares, M., Stanton, N. (eds.) Human Factors and Ergonomics in Consumer Product Design: Methods and Techniques. Taylor & Francis, London (2011)
11. Schulze, L.: Design, usability, and Maintainability of Consumer Products. In: Karwowski, W., Soares, M., Stanton, N. (eds.) Human Factors and Ergonomics in Consumer Product Design: Methods and Techniques. Taylor & Francis, London (2011)

Assessment of Cause of Difficulty in Assembly Tasks

B. Santhi, B. Gurumoorthy, Amaresh Chakrabarti and Dibakar Sen

Abstract Assemblability is defined as the ease of assembling parts in a product. It is important to address the factors that affect assemblability, so as to generate knowledge with which to take targeted measures to reduce the difficulty of humans involved in assembly tasks and increase productivity. Existing literature addresses few factors that affect assemblability. Our study focuses on identifying a more comprehensive set of causal factors and a more reliable method for assessing difficulties faced in assembly in terms of these factors. In order to understand these in detail, video recording of a case study of a computer panel assembly is used to identify a set of casual factors that affect assemblability. The researchers analysed the videos using RULA method as a basis for identifying factors potentially responsible for the difficulty. Three causal factors were proposed: reach, visibility, and dexterity. Two subjects independently analysed the videos from the point of view of the hypothesized causes of difficulty based on their perception. The results from subject feedback were correlated with the factors identified and were found to have a high level of correlation. The conclusive statement made from the result is that reach difficulty is primarily indicated by torso movement, vision difficulty by head movement, and dexterity difficulty by hand movement.

Keywords Assemblability · Assessment · Difficulty

1 Motivation

An assembly is a collection of independent parts which forms a product. The process of forming an assembly is an integral part of any manufacturing process. Even though assembly accounts for 40–60 % of the total production time and over 50 %

B. Santhi (✉) · B. Gurumoorthy · A. Chakrabarti · D. Sen
Virtual Reality Lab, Centre for Product Design and Manufacturing,
Indian Institute of Science, Bangalore, India
e-mail: santhi@cpdm.iisc.ernet.in

A. Chakrabarti (ed.), *ICoRD'15 – Research into Design Across Boundaries Volume 1*,
Smart Innovation, Systems and Technologies 34, DOI 10.1007/978-81-322-2232-3_49

of the manufacturing cost, it is not a sufficiently well-understood activity [1]. For reducing manufacturing costs and production time, it would be prudent to consider and address assembly related issues at the early design phase of product development. Assembly is considered to be a labour-intensive activity and is carried out manually for complex tasks. In the past, all assembly operations were entirely manual, where assembly operators carried these out using the knowledge and skill they gained through experience. So they often failed to explain how they did an assembly, and why they followed a particular method while carrying out an assembly. Thus, Assembly remained the least understood process in manufacturing [2, 3]. In spite of advances in industrial automation, manual assembly tasks continue to be an important feature of many industrial operations [4]. In a complex manual task, human involvement is critical as it influences feasibility, cycle time, working comfort, cost and safety of an operation. In assembling large scale structures such as in aerospace, ship building and automotive sectors, the structures have to be safely and skillfully handled by an operator. Material handling devices such as Jib cranes or overhead cranes and fixtures are employed to help operators work safer and in some cases, faster. However, during final stage of assembly, involvement of assembly operators cannot be avoided. Some of the manual assembly processes involve situations where parts that are heavy or large can be effectively handled by the assistance of good material handling devices e.g. cranes, jig, etc. [4].

However, in spite of the improvements brought about by the use of tools, human operators continue to play a central role in complex assembly. The work presented in this paper is an attempt to address issues related to ease of assembly in such cases, by identifying and addressing factors that affect assemblability (i.e. ease of assembling a product), so as to generate knowledge with which to take targeted measures to reduce the difficulty of humans involved in assembly tasks and increase productivity.

Existing literature addresses relatively few factors that affect assemblability [5–8]; these are detailed in Sect. 3.2. As can be seen, the factors currently addressed focuses either on parts or processes, and not on human issues. Zha et al. [9] proposed a set of factors that are responsible for difficulty in assembly: (i) part (ii) person (iii) process (iv) tool, and (v) environment and a method to identify portions in the overall assembly process where difficulties are encountered. However, the method proposed in [10] does not consider the causal factors of the difficulty encountered. So, the overall aim of this paper is to identify a comprehensive set of causal factors that take into consideration of human issues, i.e. person related factors.

2 Plan of Work

The research questions to be addressed were:

- What is assemblability?
- What are the factors that affect assemblability?

The first question has been addressed by analyzing existing literature on definitions of assemblability. In order to answer the second question, literature has been used to identify gaps and propose a set of potential factors, and video recording of a case study of a computer panel assembly has been used to corroborate the presence of these factors in this assembly process.

3 Related Work

3.1 *Assemblability*

The term "Assemblability" is defined as probability of successfully assembling product parts [5]. It is also called 'Assembly-Operation difficulty', which can be represented by a fuzzy number between 0 and 1 [7]. Another definition, which focuses on the process aspects is called 'assembly requirement'; it is defined as assembly of individual parts/sub-assembly parts and substances into final assembly of often high complexity and of a given quantity in a given period of time [8]. A definition on assemblability related to part is defined as ease of gripping, positioning and inserting parts in an assembly process [11]. None of these definitions capture human involvement, in case of assembly in which human play a vital role. So for our study, we define assemblability as 'the ease with which a product can be assembled from its parts'. Assemblability has been studied in the literature in the context of design for assembly (DFA) [1, 6, 12, 13]. Different techniques for assessing assemblability have been discussed in the literature. These can be broadly classified into approaches based on: analysis of ergonomic posture [14–20] and analysis of the time required to perform tasks (also referred to as Predetermined motion time analysis) [21, 22].

3.2 *Factors Affecting Assemblability*

From the proposed definition of assemblability, it follows that it is necessary to identify the factors that cause difficulty in assembly. This section reviews related work on such factors. Lee and Chunsik [5] proposed an analytical approach to find clearance and tolerance as the causes of problems for assemblability using Jacobian and sweep operations. Sturges [6] considered boss or groove size, clearance, direction, handling distance, fastening method, insertion path as parameters for the assessment of assembly difficulty. Lu et al. [7] proposed a technique based on stack-up of tolerances to assess assemblability. Sanderson [8] argued that the order of assembly becomes more important for assemblability and he found the order based on Gaussian approximation to actual parts distributions and the use of maximum likelihood as a means of achieving analytical solution.

4 Methodology

As seen above, literature has identified various factors that affect assemblability, such as assembly sequence, quality, tolerances, part/feature size, handling distance, direction, and fastening methods. However, these factors are oriented towards productivity rather than ergonomic difficulty; these do not address assembly difficulty in terms of human difficulties but focuses on improving the process. Our primary goal has been assessment of factors that influence the difficulty of a human operator performing assembly.

Vinayak [23] proposed the decomposition of any task by humans into four phases—*search, locate, reach, execute*. This decomposition is in the context of human cognition related to vision. In the present context of assessing cause of difficulties in assembly, it is believed that the first two phases of *search* and *locate* can be subsumed into one phase of *vision*, implying that the human operator needs to be able to clearly see the objects or parts thereof as the case may be. By the same argument, *execute* is replaced by *dexterity* which is intended to capture '*ability to execute*'. We therefore propose the following factors: (1) Vision, (2) Reach, and (3) Dexterity as the three broad phases in any assembly task to be performed by the human operator. The hypothesis then is that the inability to complete any of these or difficulty with completing any of these would result in the task being labeled as difficult.

Based on the definitions in [23] and the discussion above, we define vision as the ability to clearly see the objects or parts thereof that are relevant to the present task in the assembly process. Reach relates to the ability of the operator to make tactile contact or gather the objects (part, tool or fastener) as relevant to the current task. In keeping with its dictionary meaning [24] the third phase of dexterity relates to the ability to perform a task with hands. Based on a series of experiments in which the first author studied human operators during assembly in order to understand how assembly tasks were carried out in reality, the following hypothesis have been framed:

Hypothesis 1: Reach difficulty is primarily indicated by predominant movement of torso;
Hypothesis 2: Vision difficulty is primarily indicated by predominant movement of head, and
Hypothesis 3: Dexterity difficulty is primarily indicated by predominant movement of hand/wrist.

4.1 Experimental Details

An assembly exercise in a laboratory setting was devised to ascertain the causes of difficulty in assembly tasks. The exercise consisted of fixing a computer panel assembly and was carried out by six subjects. The whole exercise was video recorded with simultaneous verbalization by the subject [25]. Some of the advantages of video recording the exercise are the following: (a) as the speed of the movement of interest increases it becomes progressively less practical to rely on visual observation; (b) even the most careful of observers may miss important aspects of a rapidly executed movement; (c) videos enable repeated viewing and allow even slow motion viewing [26].

The specific features of the experiment were the following:

1. The same experiment was conducted for different subjects.
2. The subjects chosen were of different physical stature varying from a height of 160 cm to 185 cm.
3. The subjects were allowed to freely move to carry out the assembly.
4. The requirements for carrying out the experiment such as the tools, parts, instructions to perform assembly and recording devices were identical for each exercise.
5. During the experiment, the subjects were asked to do simultaneous verbalization. Each experiment was completely recorded.
6. After the assembly exercise task was completed, the video was checked for quality. If the quality was not good, the experiment was repeated and recorded again to obtain better quality.

Recording of assembly operations was done by two independent cameras. One was fixed approximately two meter from the workspace on a tripod on the side of the subject's sagittal plane, facing towards the front of the table. This ensured capturing of at least 60 % of the subjects body proportions. The second camera was handled by the first author and was moved to best capture a closer view. During the assembly operation, it dominantly covered the frontal plane.

The following assumptions were made in this study: (1) Skill level of the subject was sufficient to carry out the assembly operations, i.e. prior experience in carrying out the given assembly was available; (2) All required tools were presumed and supplied to the subjects; (3) It was considered that the causes of difficulty were mutually independent [3, 23]. For example, in the action of opening a lock, it starts with reaching the location i.e. target point which is purely a reach issue, insert the key into the lock is a vision issue and opens the lock by the application of hand and wrist activity which is a dexterity related issue. But in some cases where these factors were combined and become issue. Say for example, in case of reaching out something which is below the table is an issue of both reach and vision. However we were not considering such scenarios for our analysis.

4.2 Processing of Captured Data

Minor editing of the captured videos was done using Windows Movie Maker™ (like removal of idle run time, trial runs) before importing them to files for inviting retrospective feedback from multiple raters. The video was then sliced based on the subdivision of assembly operation into various tasks as proposed by Desai and Mital [27]. These slices of the video were used for further study and analysis. Figure 1 contains images that depict a reach-related issue faced by the subject; Fig. 2 shows a vision-related issue faced; while Fig. 3 depicts a case involving dexterity-related issues. All these cases were obtained from the set of video recordings of the assembly of a computer panel. The raters' received an explanation on how to analyse the video clips. For ascertaining reliability of ratings given, each rater was requested to review the ratings given by him/her after the analysis. A difficulty rating for each task was given by each rater within a scale of 1–3 (1 as very easy; 3 as very difficult) in terms of 3 factors: i.e. Reach, Vision and/or Dexterity, based on the raters' impression from watching the video clips. Further assessment was carried out by the researchers to determine the prominent limb movements for each type of difficulty (Reach, Vision and Dexterity) within the time slots judged by the raters. The details of the analysis were described in the next section.

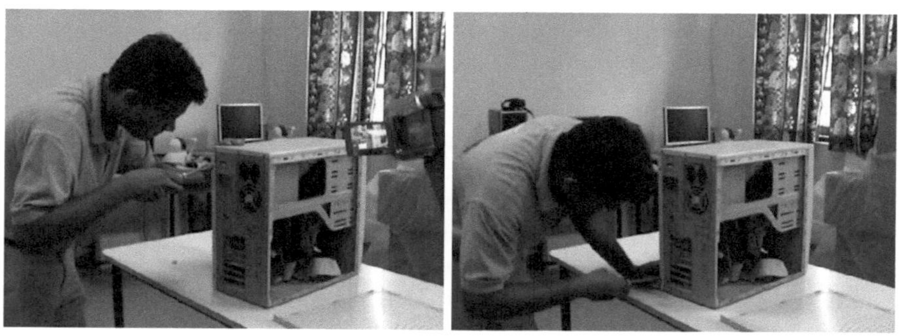

Fig. 1 Issue with reach

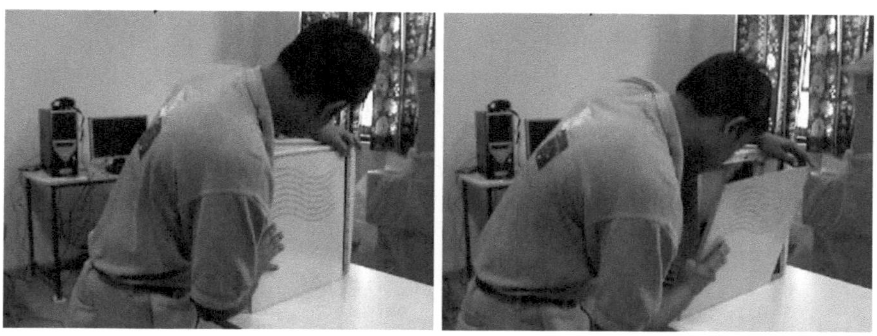

Fig. 2 Issue with vision

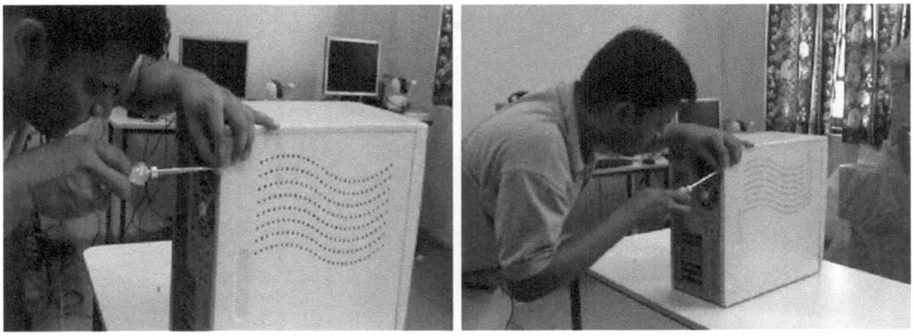

Fig. 3 Issue with dexterity

5 Scheme of Analysis

5.1 Analysis

The analysis of ratings of the difficulty scores for various video clips of recorded assembly obtained from the two raters' and from the first author is discussed in this section. The feedback from the subjects was based on their own perception of Reach, Vision and Dexterity issues. For these issues the raters' will observe the slices of video clips, and rate the task covered in that slice on a 1–3 where 1 implies easy and 3 implies difficult. The subjects were asked to repeat the rating in different days to achieve consistency in the rating. Furthermore, the raters' were requested to rate the video clips with respect to one issue at a time. For example if they rated the video clips with respect to issues related to vision in the morning session then they were asked to rate the clips with respect to reach factors in another session in the evening. This is done to avoid overlap consideration of the different factors, namely reach, vision and dexterity. The authors would like mention that though the rating were done by two raters, the sample data analysed by each rater is much larger as each rater have analysed 84 slices of video on each of the three factors. The first author assessed difficulty in maintaining a particular posture for performing a task in terms of the three hypotheses framed:

Hypothesis 1: reach difficulty is primarily indicated by torso movement;
Hypothesis 2: vision difficulty by head movement, and
Hypothesis 3: dexterity by hand movement.

The researcher used the same set of video clip intervals, that were used by the raters, and analysed these based on the values of change in angles of neck, torso and arm, which were derived on the basis of RULA [14] (Rapid Upper Limb Assessment) scores as shown in Table 1.

Table 1 Interpretation from RULA score used by researcher for analysis

	Low(1)	Medium(2)	High(3)
Head	0°–20°	10°–20°	>20°/<0°
Torso	0°–20°	20°–60°	>60°
Wrist	0°	0°–15°	>15°

Table 2 Sample output of subject and researcher rating

Video clip intervals in time	Subject feedback (S_{1FB})			Video clip intervals in time	Researcher rating (R_{FB})		
	Reach	Vision	Dexterity		Reach	Vision	Dexterity
0.9.100–0.12.600	2	3	3	0.9.100–0.12.600	2	2	2
0.24.700–0.31.900	2	2	2	0.24.700–0.31.900	2	2	3
0.32.700–0.31.900	2	3	3	0.32.700–0.31.900	2	2	3

Table 3 Result of correlation value of subject rating and researcher rating

		S_{1FB}	S_{2FB}
Visibility	R_{FB}	0.413	0.464
Reach	R_{FB}	0.2295	0.444
Dexterity	R_{FB}	0.34	0.337

6 Results

These two rating sets i.e. the rating set obtained from the two raters (S_{1FB}, S_{2FB}) and set based on the researcher's (first author) rating (R_{FB}) were correlated with one another for the three proposed causes of difficulty i.e. reach, vision and dexterity. The raters rated the video clips in terms of their own perception based on three causes of difficulty (Table 2). The researcher rated the video clips based on Table 1 shown above and is shown in Table 2.

Both the ratings are in a 1–3 scale, with 1 as low difficulty score and 3 as high difficulty score. The result of the correlation scores (Table 3) show that at 95 % confidence interval (Acceptance region = 0.1971), the correlations are acceptable in all the three framed hypotheses. (For example, ratings obtained for visibility issue from a rater (S_{1FB}) and first author can be read as 0.413 which is below the acceptance region (0.1971)). This indicates that there is a strong correlation between the rater feedback and researcher ratings, pointing to the possibility that Reach issues are reflected primarily in Torso movements, Vision to head movements, and dexterity to Hand/wrist movements.

7 Discussions and Future Work

The analysis involved answering the questions: Can the cause of difficulty encountered in a particular task be attributed to one of reach, vision and dexterity? Another related question was whether each of these causes have a distinct signature in terms of predominant movement of a particular limb.

The results of the analysis provided an indication of a particular cause of difficulty that arose due to large changes in a particular limb movement. For instance, changes in angle of head was found to be an indicator for vision difficulty, changes in angle of torso was an indicator for Reach difficulty, and changes in angle of wrist/arm was an indicator for dexterity difficulty.

However, even though the three primary movements aligned rather nicely with the three causes of difficulty, other limbs too moved during each of these events, and therefore it may well be that in general, some of these causes might not be unique. In some cases observed, a particular cause of difficulty was found to have more than one kind of indicator i.e. either head or torso in the case of vision. This indicates the need for conducting further experiments before arriving at any general conclusion regarding these indications.

8 Conclusions

This preliminary research work proposes that a set of three causal factors (Reach, Vision and Dexterity) represent ergonomic difficulty in assembly, and suggests a new set of postural indicators (Head, Torso and Limb) respectively for these causes, in the context of manual assembly of computer frames. Experimental results indicate that these causes of difficulty are indicated respectively by movements of one primary limb, viz. changes in angle of head are as indicators of Vision difficulty, changes in angle of torso are an indicator of Reach difficulty, and changes in angle of wrist/arm are an indicator for Dexterity difficulty. As future work, further validation will be carried out to test the assumption of mutual independence of parameters.

Acknowledgments We acknowledge the contribution of the anonymous subjects who performed the assembly exercise and were involved in the analysis part, and the raters who were kind enough to rate the videos.

References

1. Boothroyd, G., Dewhurst, P., Knight, K.A.: Product Design for Manufacturing and Assembly. New York: Marcel Decker Inc. (1994)
2. Andeen, G.B.: Toward a science of assembly. J. Robot. Auton. Syst. **21**, 239–248 (1997)
3. Ding, Z., Hon, B.: Constraints analysis and evaluation of manual assembly. CIRP Ann. Manuf. Technol. **62**, 1–4 (2013)

4. Wongwanich, Y.: Design for assembly method for large and heavy plates: an experimental design. MS thesis, Virginia Polytechnic Institute and State University (2001)
5. Lee, S.S., Yi, C.: An analytic approach to assemblability analysis. In: IEEE International Conference on Robotics and automation, pp. 1484–1489 (1998)
6. Sturges, R.H.: A quantification of manual dexterity: the design for assembly calculator. J. Robot. Comput. Integr. Manuf. **6**(3), 237–252 (1989)
7. Lu, C., Fuh, J.Y.H., Wong. Y.S.: Evaluation of product assemblability in different assembly sequence. In: The 6th IEEE International Symposium on Assembly and Task Planning: From Nano to Macro Assembly and Manufacturing (ISATP 2005) (2005)
8. Sanderson, A.C.: Assemblability based on maximum likelihood configuration of tolerances. IEEE Trans. Robot. Autom. **15**(3), 568–572 (1999)
9. Zha, X.F., Lim, S.Y.E., Fok, S.C.: Integrated intelligent design and assembly survey planning: a survey. Int. J. Adv. Manfac. tech **14**, 664–685 (1998)
10. Santhi, B., Chakrabarti, A., Gurumoorthy.: A new approach for assemblability assessment using time and postural analysis—a case study. International Conference on Research into Design (2009)
11. Fujimoto, H., Ahmed, A.: Entropic evaluation of assemblability in concurrent approach to assembly planning. In: IEEE 4th International Symposium on Assembly and Task Planning, pp. 306–311 (2001)
12. Rempersad, H.K.: The house of DFA. In: IEEE (1995)
13. Whitney, D.E.: Mechanical assemblies. Oxford: Oxford University Press (2004)
14. McAtamney, L., Nigel Corlett, E.: RULA: a survey method for the investigation of work related upper limb disorders. J. Appl. Ergon. **24**(2), 91–99 (1993)
15. Feyen, R., Liu, Y., Chasn, D., Jimmerson, G., Joseph, B.: Computer-aided ergonomics a case study of incorporating ergonomics analysis into workplace design. J. Appl. Ergon. **31**, 291–300 (2000)
16. Kadefors, R., Forsman, M.: Ergonomic evaluation of complex work a participative approach employing video computer interaction exemplified in a study of order picking. Int. J. Ind. Ergon. **25**, 435–445 (2000)
17. Hignett, S., McAtamney, L.: Rapid entire body assessment. J. Appl. Ergon. **31**, 201–205 (2000)
18. Kee, D., Karwowski, W.: LUBA an assessment technique for postural loading on the upper body based on joint motion discomfort and maximum holding time. J. Appl. Ergon. **32**, 357–366 (2001)
19. Massaccesi, M., Pagnotta, A., Soccetti, A., Masalib, M., Masiero, C., Greco, F.: Investigation of work related disorders in truck drivers using RULA method. J. Appl. Ergon. **34**, 303–307 (2003)
20. Santos, J., Sarriegi, J.M., Serrano, N., Torres, J.M.: Using ergonomic software in nonrepetitive manufacturing processes. Int. J. Ind. Ergon. **37**, 267–275 (2007)
21. Laurig, W., Kühn, F.M., Schoo, K.C.: An approach to assessing motor workload in assembly tasks by the use of predetermined motion time systems. J. Appl. Ergon. **16.2**, 119–125 (1985)
22. Laring, J., Forsman, M., Kadeforsa, R., Ortengren, R.: MTM-based ergonomic workload analysis. J. Ind. Ergon. **30**, 135–145 (2002)
23. Vinayak.: A geometric framework for vision modeling in digital human models using 3D tessellated head scans. MSc thesis, IISc, Bangalore (2010)
24. http://www.oxforddictionaries.com/definition/english/dexterity. Referred date 20 July 2014
25. Hoc, J.M., Leplat, J.: Evaluation of different modalities of verbalization in a sorting task. Int. J. Man–Mach. Stud. 18(3), 283–306 (1983)
26. Mares, A., Senderska.: Experience with the application video analysis in the manual assembly (2005)
27. Desai, A., Mital, A.: Incorporating work factors in design for disassembly in product design. J. Manuf. Technol. Manage. **16**(7), 712–732 (2005)

Affect Component and Errors During Numerical Data Entry—A Study

Shrikant Salve, Shanu Shukla and Pradeep Yammiyavar

Abstract The paper focuses on the role of emotions and their influence on errors in computer data entry work. Emotions are important and most pervasive aspect of human behavior including during work. Our focus is to explore the role of 'internal performance shaping factor' like 'emotions' which may affect the work performance during numerical data entry work. It is posited that the user makes less number of errors in a positive emotion during numerical data entry task as compared to negative emotion. A study was conducted using an experimental interface developed to test the frequency of errors in positive and negative emotions induced during the task of numerical data entry. Emotions were induced, before giving the tasks by showing video clips with themes of comedy, anger and sadness. To compare it with the baseline performance, a control group was also designed that had shown a video of abstract images. To assess whether right emotions have been induced, an emotional measure through Self-Assessment Manikin (SAM) scale of participant was taken before and after emotion inducement phase as a validate. Both the valence and arousal dimensions of emotions have been recorded through the scale. Rural Indian participants from middle and lower middle income category familiar with using devices were given a task of numerical data entry in local and English language to perform. The task involved functions of a calculations using a calculator which was designed for this experiment. Data was analyzed for errors committed and speed of data entry by the participants during positive and negative emotional states. Errors and response timing data's were collected as logs during task performance recording of data input. Using this information speed and accuracy were calculated.

S. Salve (✉) · P. Yammiyavar
Department of Design, Indian Institute of Technology Guwahati, Guwahati, India
e-mail: shrikantsalve@gmail.com

S. Shukla
Department of Psychology, Indian Institute of Technology Indore, Indore, India

A. Chakrabarti (ed.), *ICoRD'15 – Research into Design Across Boundaries Volume 1*,
Smart Innovation, Systems and Technologies 34, DOI 10.1007/978-81-322-2232-3_50

Result verify the posits that working under negative emotion induce more errors and require more time than when performing under positive emotion.

Keywords Numerical data entry · Emotion · Human error

1 Introduction

James Reason [1] has identified two types of errors like slips and mistakes. People make slips and mistakes in simple data entry task, which consists of text and numerical entry. In case of numerical entry, humans are prone to make errors. Simple numerical entry errors such as typing numbers many times, typing incorrect numbers can have long term consequences. In many small urban towns of India, computers have become a part of people's livelihood. It has been used in many places like hospitals, banks, railway stations, bus stands, factories, market places/shops, government offices, rural Business Process Outsourcing (BPOs) and Non-Government Organizations (NGOs) (data entry purpose) [2]. In such a context incorrect numerical data entry can leave serious doubts on the efficiency of, for example rural based data entry operators. Errors committed by such operators can be mislabeled as—untrained, poor skills etc. which can affect their sustenance, their confidence level. Not much is known about the affective states of such operators working in remote areas.

Another issue pertaining to errors in data entry, particularly in rural contexts is related to the 'language' it involves. Some researchers [3, 4] have reported that almost 72 % of the population in India reside in villages with 22 regional and 2 national languages i.e. Hindi and English being spoken. Although English is widely spoken, yet it is not an official language [5]. It has been found that about 92.39 % schools in rural part of India impart education in the medium of regional languages (mother tongue) [6]. But, as the language involved in computer interaction and also the language used for data entry in most of above mentioned places is observed to be English (expect in few rural BPOs, NGOs and government offices), this influences the work performance of rural users. Our previous studies on effect of local language on data entry, indicates that rural Indian users are more comfortable with local language while interacting with computers [2]. In that paper more importance was given to external performance shaping factors rather than internal performance shaping factors like emotion [7]. However, language is a sensitive issue that influences affective states of disadvantaged sections.

Emotions guide and direct behavior and influences the perceptual and cognitive aspect of human mind. Researchers [8–11], suggest that an individual's temporary affective state may increase or decrease an individual's cognitive attention which could affect performance. For instance, positive affect increase flexibility in reasoning, greater creativity in problem solving and seeking out variety. Whereas, negative affect seems to lowers the thought action repertoire. Affective state has an influence on interactions with digital devices too.

Data entry (including both numerical and text) mechanism becomes pervasive in the area of Human Computer Interaction (HCI). Youth in non-developed semi urban India, for whom data entry jobs provide employment, get emotionally attached with their jobs and become either complacent or anxious. Designing of effective data entry mechanism are mainly focused on two parameters as speed and accuracy of errors. The present study aims at finding efficiency of numerical data entry by rural users during different emotional states (that is positive, neutral and negative).

Many new forms of devices coming up, many types of interactions are being asked to be designed by designers. A certain change in any one of the several of the parameters involve, May I compound the error or reduce the error? This paper is towards investigating errors in depth such that its results can be used by designers while designing interfaces.

2 State of Art

Numerical data entry is required in many fields including science, healthcare, education, mathematics, finance and government related work. Owing to its ubiquitousness, researchers have address several issues related to numerical data entry performance. For instance, Oladimeji et al. [12] have conducted the study on two main number-entry styles on number entry interface found on medical devices. Their empirical study investigated the effect of interface design on error detection in number entry tasks using two number entry interfaces (serial and incremental interface). The study suggested in giving priority to research number entry styles and their relation to error rate, behavior and performance in the context of safety critical number entry systems. However their study restricted to medical number entry systems. Thimbleby et al. [13] have proposed the user interface to prevent number entry errors in medical devices. Their work have shown that user errors are ignored by many number entry systems in user interfaces for interactive devices to desktop applications in all domains which cause confusion and problems and possibly leads to damage. Barchard et al. [14] have projected the study on impact of human data entry errors on statistical results and calculations. Their results show that in double entry there are significantly fewer errors than both in visual checking and single data entry task. This study shows that the data entry errors can reduce reliability of statistical results and calculations (that is numbers).

Chand [15] has discerned issues pertaining to the designing of the interface for computer driven kiosks used in rural areas of India. His study raises the importance of multilingual text and video contents while developing interface for illiterate and multilingual rural people. In relation to rural users as sample population, there is another study by Patel et al. [16]. They have stressed upon the importance of audio and spoken based modality in designing computer based interaction interface. Through their research they designed and developed voice-based community forum (named *Avaaj Otalo*) interface for Indian rural users.

This application was developed in Gujarati language which allows farmers to receive timely and relevant agriculture information over the phone. They have given importance to the spoken language research in rural areas of the developing world.

Singh et al. [17] have proposed study on the numeric paper forms used by the NGOs for data collection in rural India. They have investigated NGO's form filling requirements which were used to interact with rural people. They proposed the numeric input method for different NGO's form filling requirements which is easy to use for rural people and also machine readable. This paper's context is the data entry jobs in both local and English language, provided for rural people by various NGOs in India.

Several studies have proved the influence of emotions on complex as well as simple tasks in giving rise to human errors. For example, Jeon et al. [18] proved that there are behavior changes when a driver is in different emotional states such as anger and fear, even when those states share the same emotional valence. Similarly, Causse et al. [19] confirmed in their study that negative emotional states can provoke plan continuation errors in pilots. This means that they are more likely to continue acting even when available evidence indicates them to stop. Another study, Cairns et al. [20] have reported the influence of emotions on number data entry on devices like infusion pumps in hospitals. Their hypothesis was that people who are in negative affective state will make more errors than those in a positive affective state. The sample size of the experiment consists of 28 participants. First part involves emotion inducement procedure where participants were shown 24 images for their particular experimental condition and asked to rate each one as they went along. The Microsoft PowerPoint presentation was set up to display 24 images of either positive or negative valence depending on the experimental condition. A standard International Affective Picture System (IAPS)—database of images was used for inducing different level of affect in both the valence and arousal dimensions. In second step the participants moved on to the number entry task. For this the Microsoft Excel was used to randomly generate numbers and display them. Participant had to enter the displayed number into a Google Nexus tablet using number pad touch interface. Finally they concluded that the users in negative affective state are more likely to make number entry errors. This study focuses only on safety-critical environment of number entry in the healthcare domain which involve devices like infusion pump, ventilators having touch screen user interface. Besides, the study did not take into account the arousal dimension of emotions.

In rural India people have minimum access and familiarity with computers because of poverty, low literacy and English language problem [15, 16, 21]. The data entry work done by rural people in local as well as English language at various places like, rural BPOs and NGOs [17] to earn a living. This makes their emotional attachment with work.

The present study focuses on highlighting the effect of different emotional states on rural Indian data entry operators and their local and English language related problems. The performance of participants in three treatment groups (positive, neutral and negative) was studied. Further we analyzed the effect of affective state on their performance in terms of error made and time required. We report our finding on relative accuracies of both affective states (positive and negative) on languages.

3 Hypotheses

The literature study indicates that consideration of the numerical data entry errors is crucial for evaluating efficiency of rural users while performing computer data entry. We are attempting to find the extent of influence of emotions on making these errors in local and English language. The hypotheses are stated as below:

H1 Rural users make more errors in local language numerical data entry without time limit during negative emotion rather than positive emotion.
H2 Rural users make more errors in English language numerical data entry without time limit during negative emotion as compared to being in positive emotion.
H3 Rural users require more time in local language numerical data entry during negative emotion as compared to in positive emotion.
H4 Rural users required more time in English language numerical data entry during negative emotion when compared to being in positive emotion.

The experiment designed to test the above hypothesis is given below.

4 Method

4.1 Participants

Forty eight participants (male/female) belonging to the age group of 16–30 years were selected for the experiment. They were students of 11th and 12th standard, people working in the coffee shop, stationary shop, grocery shop, vegetable market and security guards in the campuses of Indian Institute of Technology Guwahati (IITG). All participants from rural background had educational qualification of 10th–12th standard (that is non-graduate) and used computers or laptops at least one hour in a week. The following Fig. 1 shows the participants performing given experiment.

(a) (b) (c)

Fig. 1 **a** Picture depicts one of the author of this paper explaining the process of experiment to participant and **b**, **c** pictures showing participants performing the numerical entry operation task assigned to them

4.2 Instrument and Materials

A software interface involving a calculator was designed specifically for this experiment. It was designed to input numerical data in Assamese and English language using keyboard and mouse. This interface can perform arithmetic operations like addition, subtraction, multiplication and division—with and without decimal point. Figure 2 depicts the screen shot of the software interface in Assamese and English language.

4.2.1 Stimuli

Videos can be used to induce emotions artificially in the participants [22, 23]. We used, three video clips to influence affective states and for inducing positive and negative types of emotions in both the valence and arousal dimensions. We used three types of video clips like violence video for negative emotion and comedy

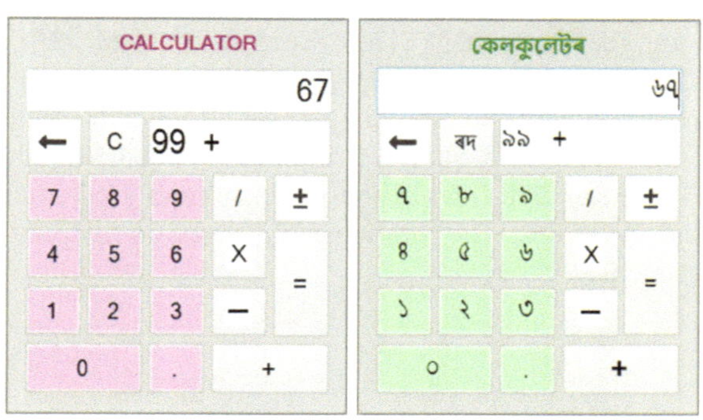

Fig. 2 Screen shot of software interface, which does calculations in Assamese and English language

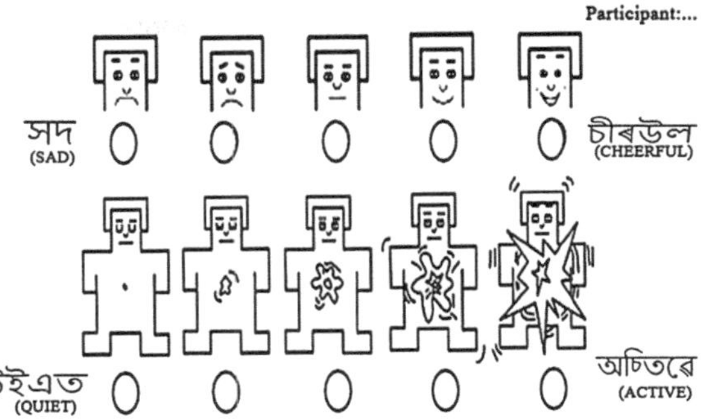

Fig. 3 The valence (*top*) and arousal (*bottom*) scales of self-assessment manikin (SAM) [24]

video for positive emotions each of 5 min duration. The video for control group was an abstract images video for duration of 2 min. To assess whether right emotions have been induced, an emotional measured through Self-Assessment Manikin (SAM) scale, proposed by Bradley et al. [24] was used. Figure 3 depicts the SAM scale having valence (top) and arousal (bottom). The valence was used as typical dimension for checking whether emotional state is positive or negative. At one extreme of the scale one can felt happy, pleased, satisfied, contended and hopeful. The other end it reflects unhappy, annoyed, unsatisfied, melancholic, depressed and/or bored. Similarly the degree of arousal reflects from an excited wide-eyed figure to a relaxed sleepy figure. At one end of this scale one physically felt relaxed, calm, sluggish, dull, sleepy and unaroused. The other end it reflects stimulated, excited, frenzied, Jittery, wide awake and aroused. So a highly aroused negative state corresponds to anger whereas a low aroused positive state would be contentment [24].

4.3 Research Design

4.3.1 Experimental Variables

The experiment was a between subject design. The participants used software interface four times to perform four tasks for calculation in Assamese and English languages. The participants perform calculation in Assamese and English languages using input device—keyboard only. The emotion (positive, negative and

Table 1 Task design

Task no.	Time allotted	Language
Task 1 (T1)	Without time limit	Local
Task 2 (T2)	Without time limit	English
Task 3 (T3)	Within 1 min	Local
Task 4 (T4)	Within 1 min	English

Table 2 Distribution of samples among emotion affective state and task

Sample distribution			Tasks			
Positive	Neutral/control	Negative				
1.2	3.4	5.6	T1	T2	T3	T4
7.8	9.10	11.12	T1	T2	T4	T3
13.14	15.16	17.18	T2	T1	T3	T4
19.20	21.22	23.24	T2	T1	T4	T3
25.26	27.28	29.30	T3	T4	T1	T2
31.32	33.34	35.36	T3	T4	T2	T1
37.38	39.40	41.42	T4	T3	T1	T2
43.44	45.46	47.48	T4	T3	T2	T1

neutral/control group), number entry language (Assamese and English) were the independent variables. The dependent variables were the task completion time, errors made in numerical data entry.

4.3.2 Experimental Design

The experiment was divided in four tasks as given in Table 1 below, Task 1 consists of local language numerical data entry without time limit, Task 2 consists of English language numerical data entry without time limit and so on. Each participant has to perform all tasks, but the sequence/order of the tasks may be different. Table 2 shows how samples distribution was done among each task and sequence of tasks to perform. As shown in Table 2, the samples were equally distributed among three emotional states like positive, negative and control group. The first row of Table 2—positive emotion induced for sample 1, 2, control/neutral emotion induced for sample 3, 4, negative emotion induced for sample 5, 6 and all samples 1–6 perform task 1 (T1) to task 4 (T4) in sequence T1, T2, T3 and T4.

4.4 Procedure

All participants were tested individually. They were briefed about the stages and purpose of the experiment before starting. They were also instructed on the use of SAM scale and the software interface for numerical data entry.

To study empirically, the experiment was divided into two parts. Part one consists of emotional inducement process. When the participants were ready, their emotion was checked by SAM scale before starting actual experiment. Then the participants were shown the 5 min video clip for their particular experimental condition. To assess whether right emotions have been induced, an emotional measured through SAM scale of participants were taken. In second part, participants moved on to the numerical entry task. Prior participants were given orientation session where they could enter 2/3 simple calculations and get familiar with the

Table 3 Results manipulation by statistical analysis

Hypothesis	t-value (independent t-test)	Mean		SD	
		Positive	Negative	Positive	Negative
H_1	−5.90	0.47	8.20	0.74	4.75
H_2	−7.76	2.27	15	3.37	4.74
H_3	−6.12	118.7	176.33	32.01	35.81
H_4	−6.30	124.80	191.80	33.26	33.28

interface. When the participants were comfortable with how the interface worked, they were allowed to proceed to the experiment. The participants were required to enter given mathematical calculations having three different difficulty levels (like very easy, easy and hard) using two interfaces (Assamese and English) in the defined order in Table 2. The participants were provided the experimental sheet including mathematical calculations in English and local language they speak (i.e. Assamese). The participants were instructed to perform the mathematical calculation as quickly and as accurately as possible. The computer based background recording of each participant interaction with designed software interface have taken as a data collection for entry typing speed and errors.

5 Results and Discussion

In this experiment we have observed three types of errors like interface error, interaction error and data entry error. But only the data entry errors (that is numerical data entry) considered for this study. The 48 participants entered around 6,528 numbers in total and made 877 errors, or approximately 13.43 % (overall error rate). Thus participants made a mean number of 4.66 errors and SD = 5.49.

5.1 Emotion Manipulation

As expected with the experiment, a paired t-test showed significant difference in SAM scale ratings for before (pre) and after (post) in case of positive ($t(32) = -7.97$, $p = 0.000$) and negative ($t(32) = 5.59$, $p = 0.000$). Also we found the significant difference in between positive and negative emotions for valence ($t(16) = 18.75$, $p = 0.000$) and arousal ($t(16) = 3.46$, $p = 0.002$).

Table 3 illustrates the statistical analysis done by independent t-test to find the significance. As shown in Table 3, all hypothesis from H_1 to H_4 with their mean, standard deviation (SD) values for both positive and negative affective states and t-values. According to results from Table 3, the hypothesis H_1 and H_2 are significant, that is rural users make more errors in numerical data entry by both Assamese and English language without time limit during negative emotion than in positive emotion. Also rural users require more time in numeric data entry by both Assamese and English language without time limit during negative emotion than positive emotion which proves hypothesis H_3 and H_4.

5.2 Discussion

The participants were able to appropriately attribute the expected valence by watching video clips measured by SAM scale. This suggest that the video clips were influencing the affective state of the participant and the experimental manipulation had worked. As a consequences, there was a significant effect of affective state of the participants on the number of errors made by them. The number of errors made by participants including within and without time limit including two different languages (Assamese and English) are quite high. The use of multiple languages and time pressure component made this is somewhat challenging task to the participants.

The experiment had limitations. There are several issues of environmental validity which were compromised. Participants were required to enter several calculations including addition, subtraction, multiplication and division. This is not the normal job of number entry usually performed in the rural BPOs and NGOs in India. The calculator interface is also not the only style of visual interface seen in rural Indian workplace data entry screens.

6 Conclusion and Future Work

The results show significant difference in error rate and speed of entry for two emotional states—both in Assamese language as well as English language. The rural (middle and lower middle income category) users in negative emotion made more errors and required more time as compared to positive emotional state in both Assamese and English language numerical data entry. The study is helpful from the designer's aspects where it will guide them to incorporate emotional aspects in designing user interfaces to mitigate number data entry error. Could this mean that language itself may not be a factor. This needs another set of experiments to be planned in the future.

References

1. Reason, J.: Human Error. Cambridge Press, Cambridge (1990)
2. Salve, S., Yammiyavar, P.: Towards proposing an intelligent error limiting user interface for rural indian data entry operators. Aust. J. Intell. Inf. Process. Syst. **13**(4) 2014
3. Kam, M., Kumar, A., Jain, S., Mathur, A., Canny J.: Improving literacy in rural India: cellphone games in an after-school program. In: International Conference on Information and Communication Technologies and Development (ICTD), pp. 139–149 (2009)

4. World Bank.: http://go.worldbank.org/8EFXZBL3Y0. Rural India population. Retrieved on 15 May 2014
5. Smith, A., Joshi, A., Liu, Z., Bannon, L., Gulliksen, J., Li, C.: Institutionalizing HCI in Asia. In: Proceedings of 11th IFIP TC 13 International Conference on Human–Computer Interaction, pp. 85–99. Springer, Berlin (2007)
6. Meganathan, R.: English language education in rural schools of India: the situation, the policy and the curriculum. BBC World Service, October, 2009. http://www.teachingenglish.org.uk/blogs/rama-meganathan/english-language-education-rural-schools-india-situation-policy-curriculum
7. Rooney, J.J., Heuvel, L.V., Lorenzo, D.K.: Reduce human error. In: Quality Progress Handbook, pp 27–36 (2002)
8. Ashby, F.G., Isen, A., Turken, A.: A neuropsychological theory of positive affect and its influence on cognition. Psychol. Rev. **106**(3), 529–550 (1999)
9. Aula, A., Surakka, V.: Auditory emotional feedback facilitates human-computer interaction. In: People and Computers, pp. 337–350 (2002)
10. Isen, A.M.: Some ways in which positive affect influences decision making and problem solving. In: Handbook of Emotions, pp. 548–573 (2008)
11. Zajonc, R.B.: Feeling and thinking: preferences need no inferences. Am. Psychol. **35**(2), 151–175 (1980)
12. Oladimeji, P., Thimbleby, H., Cox, A.: Number entry interface and their effects of error detection. In: Human-Computer Interaction (INTERACT 2011), pp. 178–185. Springer, Berlin (2011)
13. Thimbleby, H., Cairns, P.: Reducing number entry errors: Solving a widespread, serious problem. J. R. Soc. Interface **7**, 1429–1439 (2010)
14. Barchard, A., Pace, L.A.: Preventing human error: the impact of data entry methods on data accuracy and statistical results. Comput. Hum. Behav. **27**(5), 1834–1839 (2011)
15. Chand, A.: Designing for the Indian rural population: interaction design challenges. In: Proceedings of Development by Design Conference (2002)
16. Patel, N., Agarwal, S., Rajput, N., Nanavati, A., Dave, P., Parikh T.S.: Experiences designing a voice interaction for rural India. Spoken Language Technology Workshop (SLT 2008), pp. 21–24. IEEE (2008)
17. Singh, G., Findlater, L., Toyama, K., Helmer, S., Gandhi, R., Balakrishnan, R., Numeric paper forms for NGOs. In: International Conference on Information and Communication Technologies and Development (ICTD 2009). IEEE (2009)
18. Jeon, M., Yim, J.-B., Walker, B.: An angry driver is not the same as a fearful driver: effects of specific negative emotions on risk perception, driving performance, and workload. In: Proceedings of Automotive UI '11, ACM, USA, pp. 137–142 (2011)
19. Causse, M., Dehais, F., Péran, P., Sabatini, U., Pastor, J.: The effects of emotion on pilot decision making: a neuroergonomic approach to aviation safety. Transp. Res. Part C: Emerg. Technol. **33**, 272–281 (2013)
20. Cairns, P., Pandab, P., Power, C.: The influence of emotion on number entry errors. In: Proceedings of the 32nd Annual ACM Conference on Human Factors in Computing Systems (CHI '14). ACM, New York, NY, USA, pp. 2293–2296 (2014)
21. Gore, K., Lobo, S., Doke, P.: GappaGoshti^{TM}: digital inclusion for rural mass. In: Fourth International Conference on Communication Systems and Networks (COMSNETS), pp. 1–6 (2012)
22. Neerincx, M., Streefkerk, J.W.: Interaction in desktop and mobile context: emotions, trust and task performance. In: Ambient Intelligence: EUSAI 2003 Lecture Notes in Computer Science, pp. 119–132. Springer, Berlin (2003)
23. Spering, M., Wagener, G., Funke, J.: The role of Emotions in complex problem solving. Cogn. Emot. **19**(8), 1252–1261 (2005)
24. Bradley, M.M., Lang, P.J.: Measuring emotion: the self-assessment manikin and the semantic differential. J. Behav. Ther. Exp. Psychiatry **25**(1), 49–59 (1994)

Part IV
Design Training and Education

The Pedagogy of Self-expression in Animation Film Design

Pooja Pottenkulam

Abstract The ability of an animation filmmaker to express their individual voice through film is crucial to the creation of an animated short film and even an essential constraint. This paper is an overview of the design and delivery methods of an on-going research project that looks at how this essential constraint of self-expression might be taught. It begins with a view of the relevance of self-expression within the field of design and then goes onto consider the role of the animation filmmaker as auteur. The paper then traces the history of animation education and examines the curriculums and teaching methods of the animation schools that have been known to produce animation auteurs. It then goes on to consider a set of requisites outlined by animation professionals that might allow students to gain an awareness of the nature of self-expression, before creating a course to deliver the listed requisites.

Keywords Animation · Auteur · Design pedagogy · Design constraints

1 Introduction

Expressing inner thoughts and feelings in order to realize individuality is considered a requisite for the creation of art while the same in the field of design would be considered inconsequential. Yet, even the strictest of modernist design practices reflect traits unique to the designer. In animation film design, when work does not reflect the sensibilities of the filmmaker, the work tends to be forgotten soon, and the message to be delivered may well be missed. The majority of the animation

P. Pottenkulam (✉)
Animation, Lincoln School of Film and Media, University of Lincoln, Lincoln, UK
e-mail: poojapot@yahoo.com

A. Chakrabarti (ed.), *ICoRD'15 – Research into Design Across Boundaries Volume 1*,
Smart Innovation, Systems and Technologies 34, DOI 10.1007/978-81-322-2232-3_51

films produced by the National Film Board of Canada (NFBC) is about socially relevant subjects such as conflict resolution and children's rights, and can be classified as design films [1–3]. All of the films produced have the indelible marks of their makers and have won important animation awards, including the Oscar (Academy Award).

This paper will look at the place of self-expression within animation film design (as opposed to animation filmmaking as a field of art) and how an ability to express might be inculcated within students of the discipline.

2 Design and Constraints

Since the early 1900s, Design was directed and has evolved in relation to the principles of classical economics. Louis H. Sullivan's quote 'Form ever follows function' would become the basis of modernism in design and design education at the Bauhaus, at Ulm and continues to be used internationally. Modernist design theory and practice developed simultaneously under this tenet in the United States and in Europe [4].

Although design was defined as an activity performed outside of the individual designer's personal tastes and inclinations, it is possible to identify the works of designers; for instance the work produced by Le Corbusier [5], is distinctly different from the work of architect Frank Lloyd Wright [6]. Yet they were both working using the same design principles. In addition to the constraints posed by economics, it is in fact even necessary for the designer to have a distinct personality, in order to create substantial products that offer choice to audiences and consumers.

In animation filmmaking, whether the film is abstract form-driven or narrative-driven, it is the filmmaker's ways of translating a message into the medium of animation that form the content of the work they produce, without which, the film becomes merely a series of manipulated moving images.

3 Animation and Auteurs

Recent revisionist ideas about the Auteur theory have given claim to the director being just one voice amongst a number of contributing voices in a production. However, in the making of a short animation film, the filmmaker has the opportunity to play a number of roles, thereby entitling him/her to more complete ownership of the films than in the case of a live action film [7].

Animation film originated as a technical skill associated with and that aided live-action cinema [8]. In 1960, the first animation film festival was organised in Annecy, France [9]. Following the success of the festival governmental-bodies such as the NFBC started to fund auteur animation [10]. The increase of auteur animation led to animation schools being set up. In the 1990s animation studies became a separate field of study, independent of its association with film studies [11]. With the festivals gaining popularity and with more academic literature about animation, by the 1980s and 1990s, independent or auteurist animation now began to be recognised by mainstream awards such as the Oscars.

3.1 Animation Pedagogy

At the beginning of the nineteenth century, the main principles of animation that were taught through studios have been about the principles of movement in animation [12]. Later, as animation developed in the studios, it moved towards developing sets of formulae for narrative, gags and characters. At the Warner Brother studios, an animator would be able to familiarise himself with the studio's patent stock gags, another set could be learned at the Disney studio that would be associated with Disney shorts and films [13].

In Europe, in the 1920s, Design was being formulated and practiced at the Bauhaus in Germany [14]. Even though film and animation was not part of the curriculum at the Bauhaus, several members of the faculty such as Lazlo Maholy Nagy and Kandinsky, along their students Kurt Krantz, Werner Graeff, Hans Richter and Viking Eggeling experimented with animation as a medium of design [15–20]. In 1933, Len Lye [21] and Norman Mclaren at the General Post Office's Film Unit in the UK began to experiment with form and music to create expressionistic pieces [22]. There was no formal tutoring here and filmmakers were largely self-taught. In Eastern Europe, animation as a medium of self-expression had an early start in the 1940s, but most animation filmmakers graduated from art schools [23].

During the early stages of the development of design pedagogy at the Bauhaus, animation lent itself to explorations made with colour, light, movement and form [24]. Animation later served a more functional purpose; Ray and Charles Eames, began to use animation to bring industrial design work to life and to explain their design philosophy [25]. Self-expression within animation produced at the Bauhaus may have been incidental and on looking at work produced in retrospect, it is evident that the work did reflect the personalities of it's makers, observable especially when the whole body of work of a particular animation filmmaker is taken into consideration.

The films of animators such as Derek Lamb evidence that even works produced within tight design constraints [26] continue to carry a voice that is unique to the filmmaker. Animation filmmaking, even when practiced as design to communicating and deliver a specific message to a specific group of people; the voice of the filmmaker; choice of words, the type of story constructed, the tone of voice, visual treatment used, pacing of the film, use of sound and music would be unique to the filmmaker. It is difficult to extricate the personality and voice of the filmmaker from the films they make.

The first school to offer animation was Sheridan College in Canada, which started an animation department in 1971 [27]. By the 1980s, there were several art, film and design schools that introduced an animation department. Most schools teach animation as a set of skills that could be used either in the creation of independent films or within the industry. It was up to the graduate to decide how the skill-set might be used.

3.2 Festivals as a Way of Assessing Quality of Self-expression

The best way to see which schools might have a method of creating animation auteurs might be to look at those schools that produce films that win the most recognised awards [28] in the field of auteur animation (Tables 1 and 2).

In order to be nominated for the Oscars, a film needs to already have been nominated at the four ASIFA (Association Internationale du Film d'animation) festivals [29], along with other animation festivals that the Oscars list every year. While the ASIFA Festivals tend to remain on the list because they are regarded the most important by the animation community, the rest of the qualifying festivals change without notice [30]. Since the 1980s, the National Film and Television School (NFTS), The Royal College of Art (RCA) in the UK, Cal Arts and the School of Visual Arts (SVA) in the USA have produced the most number of graduates who have been nominated or won awards at the six festivals mentioned above. The RCA and the NFTS are wholly post-graduate colleges that have gained an international reputation for producing quality auteur animators.

An additional method of determining schools that tend to create auteurs would be to study the online curriculums and statements and interviewing alumni and faculty [31] of the main BA animation programmes[1] from which graduates go on

[1] Links to programme overviews of eleven schools:
http://animation.ensad.fr/animation-2000.html. Accessed 28 May 2014;
http://www.rca.ac.uk/schools/school-of-communication/animation/head-of-programme/. Accessed 28 May 2014
http://nfts.co.uk/our-courses/masters/directing-animation. Accessed 28 May 2014
http://www.bezalel.ac.il/en/academics/bachelor_degree/screen/1053.html. Accessed 28 May 2014
http://www.kingston.ac.uk/undergraduate-course/illustration-animation/. Accessed 28 May 2014
http://www.eca.ed.ac.uk/school-of-design/undergraduate/animation-ba-hons. Accessed 28 May 2014
http://www.nid.edu/education/graduate-diploma-programme-in-design/animation-film-design/p-overview. Accessed 28 May 2014
http://www.risd.edu/Academics/FAV/. Accessed 28 May 2014
http://www.filmakademie.de/en/about-us/institute-of-animation-visual-effects-and-digital-post-production/. Accessed 28 May 2014
http://www.animationsinstitut.de/en/curriculum/. Accessed 28 May 2014
http://filmvideo.calarts.edu/programs/experimental-animation. Accessed 28 May 2014
http://www.ucreative.ac.uk/ba-animation/in-depth. Accessed 28 May 2014.

Table 1 Academy awards wins and nominations for animation schools graduates 1984–2013

Animation schools	Wins	Nominations
NFTS, UK	5	8
UCA Farnham, UK	2	1
VGIK, Russia	2	1
Victoria College, Australia	1	0
Sheridan, Canada	1	0
RCA, UK	1	0
Kassel, Germany	1	0
CalArts, USA	0	5
SVA, USA	0	3
Filmakademie Baden-Wurttermberg, Germany UCLA, USA	0	2
Gobelins, France	0	2
RisD, USA	0	1
Total	12/30	24/101

Source en.wikipedia.org/wiki/Academy_Award_for_Best_Animated_Short_Film

Table 2 BAFTA wins and nominations for animation schools graduates 1990–2014

Animation schools	Wins	Nominations
RCA, UK	7	23
NFTS, UK	5	11
UCA Farnham, UK	2	6
Edinburgh College of Art, UK	1	1
Duncan Jordanstone, UK	1	0
Filmakademie Baden-Wurttermberg, Germany	0	1
VGIK, Russia	0	1

Source en.wikipedia.org/wiki/BAFTA_Award_for_Best_Short_Animation

to join the NFTS and the RCA,[2] in addition to interviewing individual auteurs [32], (The animation professionals were interviewed through emails and in person. The main question that they were asked was "What do students need in order to express themselves through animation?"), it was observed that the main requisites for students to excel within the field are/have:

(a) able to, and are interested in understanding themselves
(b) the motivation to make animation films
(c) the ability to translate views and opinions into the medium of film
(d) the time and physical space to develop the craft of making films
(e) able to receive feedback from experienced and practicing animation filmmakers
(f) disciplinary and interdisciplinary exposure (Table 3)

---

[2] http://www.nftsanimation.org/students.htm. Accessed 28 May 2014
www.rca.ac.uk/schools/school-of-communication/animation/graduates/. Accessed 28 May 2014.

Table 3 Responses from 20 animation professionals of six requisites

Animation professional response	Highest			Lowest			Total responses
	6	5	4	3	2	1	
(a) Need to understand themselves	16	3		1			20
(b) Motivation and passion	1	9	5	1	2		17
(c) Translating views into stories		2	5	5	4	2	18
(d) Time and space		3	5	3	2	5	18
(e) Feedback			4	5	5	2	17
(f) Exposure			1	4	5	3	13

Source Results from 20 questionnaires. Copyright PPottenkulam 2014

4 An Animation Course Designed with the Aim of Teaching Self-expression

In response to the six requisites for the education of an animation auteur, I developed a narrative construction course, designed to provide a space to nurture and help develop young animation auteurs—alongside structured lesson plans—that would try to furnish the students with knowledge on the methods of narrative construction. The objective of creating auteurs would be cached within the narrative construction course. An initial one-day version was tried out, followed by an eight-day version taught within the span of 2 weeks. Student feedback from both versions indicated interest in the subject, but also that a longer course would be required to allow the students to understand the idea of the auteur and to develop their ideas further. The course was re-designed to be taught over a period of 12 weeks to include the following methods to teach the six requisites (Tables 4 and 5):

Table 4 12-week narrative Construction course outline

Lessons	Faculty-led Presentations	Student work
Week 1	Narratives, Genres and Auteurs	Identifying areas of interest
Week 2	Idea development	Generating ideas
Week 3	Traditional structures	Researching key themes
Week 4	Visual narratives	Story writing
Week 5	Visual narrative 2	Visual writing
Week 6	Character development	Character development
Week 7	Abstract narratives	Script writing
Week 8	Traditional documentary	Shot construction
Week 9	Cinema verite	Storyboarding
Week 10	Comedy	Animatic
Week 11	Tragedy	Storyboarding
Week 12	Submission	Submission

Source 12-week course outline. Copyright PPottenkulam 2014

Table 5 Structure of each Class

Time	Class
9:30 a.m.–10:30 a.m.	Lecture with film screenings
10:30 a.m.–11:00 a.m.	Class discussion
11:15 a.m.–12:00 p.m.	Short exercises
12:00 p.m.–12:30 p.m.	Class discussion
1:30 p.m.–4:00 p.m.	Tutorials with individual students while remaining students develop film work or watch a subject-related film

Source Structure of each class. Copyright PPottenkulam 2014

1. **Lecture series including short film examples**
   The lectures every week introduced students to the process of animation filmmaking, various genres within it, history with short films that exemplified the subjects being discussed. *Type of learning*: Guided Practice, Guided Exploration
2. **Class discussions**
   The classes included discussions of the student individual work in relation to the subject that was being discussed every week. This allowed students to see different points of view on their ideas as well as different opinions on the subject being studied. *Type of learning*: Guided Exploration
3. **Short class exercises**
   The classes were punctuated with short classroom exercises to help familiarize students with the subject being learnt every week, taking them from being passive to active learners. *Type of learning*: Guided Exploration
4. **Journals**
   The students were required to maintain a journal—they were encouraged to use this journal beyond work on their film and could be used to record ideas, thoughts and plans concerning others aspects of their lives. This was done as an attempt to allow them to immerse themselves and use writing and drawing as a method of personality development. The students were encouraged to see filmmaking, drawing and writing not merely as career choices, but as a way of thinking and living. They could keep the journals confidential, so that they felt the freedom to express their thoughts and feelings without constraint. At the same time it was necessary for the tutor to be able to look at the student's work, so the students were required to highlight and edit the parts of the journals that they felt had potential to be developed into stories. *Type of learning*: Exploratory/Discovery Learning
5. **10–15 min tutorials**
   Students were given 10–15-min slots to present the narrative idea that they had been developing through the length of the course. As the number of students was large, the feedback was recorded through forms, which the students

and the faculty member kept a copy of. The individual tutorials further looked at the ways in which each individual student explored and translated their learning from the modules in their own work. *Type of learning*: Guided Exploration

6. **Narrative development**
   Alongside the faculty-led classroom presentations, discussions, assignments and film screenings, students were required to develop an idea into a fully-finished animatic for a short animated film that would be ready to take into production. During the individual tutorials and the class discussions, students were continually encouraged to make connections between their ideas, their journal writing, the lectures that they had attended and the films that had been screened, so as to be able to write stories through which they could express the ideas and feelings that they considered most important to them. *Type of learning*: Guided Exploration and Exploratory/Discovery Learning

5 Course Implemented

The course was taught in three countries; Egypt, China and in India. With the understanding that all of these countries offered different types of educational establishments, different cultural sets of student groups, different motivation levels of the students, time-frames for the course to be delivered, the student feedback from Egypt and China has been primarily examined with the objective of ascertaining a suitable duration for delivery of the content of the course. Below are the results, based on the documentation of the student work on the course:

5.1 Government-funded Art College in Luxor, Egypt, Number of Students: 8, 2 × 2 h Sessions, February 2012

Research findings (*based on faculty interaction with the students*): The students opted for the course and were eager to learn. They understood basic principles easily. They had a high level of engagement with politics and personal issues. Exposure to animation and other contemporary international art, design and film was low. Practice was somewhat restricted to traditional self-expressive art. The students understood the idea of self-expression through the narrative very well and were able to translate theory into ideas easily. The time to deliver the content of the course was insufficient. The students also did not have enough time to practice, and learning remained largely passive.

5.2 *Animation School Within a Communication University in Beijing, China, Number of Students: 20, 8 × 8 h Sessions, July 2012*

Research findings (based on faculty interaction with the students, translator and other faculty and documentation of work produced): The course was compulsory for all students in the animation dept. The students were highly motivated and were eager to learn. Exposure to animation and other contemporary international art, design and film was low. It was difficult to communicate the role and purpose of self-expression. Students had a high level of skill and exposure to technology and technological trends. Except for one student, the idea development skills of the students tended not to progress much.

5.3 *Private Design School in Bangalore, India, Number of Students: 25, 12 × 6 h Sessions, July–December 2013*

Research findings (based on student feedback forms that were circulated) [33]: Students were not very highly motivated and required high levels of supervision. While the longer duration of the course suited the students and faculty better in delivering the objectives of the course, the number of students was too high for one faculty member to give feedback to. Most students stated that they liked animation because they liked watching animation films. Out of the 25 students, only seven were interested in animation as a profession and these seven were interested in drawing and in producing animation films. It was found that students appreciated the feedback sessions most.

Analysis of student response in relation to the six requisites

1. *Personal motivation of the animation film student*
   It was understood that it would not be possible to select only self-motivated students, as the students to be taught in the three different educational institutions were allowed to choose the course without undergoing an aptitude test.
2. *Time and physical space given to the animation film student to understand themselves and their relation with the world (sense of humour, sensibilities, emotions)*
   The total duration of 12 weeks for the course, as well as the amount of time between the lessons proved to be sufficient for the students to get a grasp of the idea of the auteur and to begin to develop inherent skills and interest in the field of auteur and independent animation.
3. *Ability to translate views and opinions into visual scripts*
   The weekly lessons and screenings alongside short exercises helped the students understand how to develop narratives for animation in a structured manner. While the translation of ideas into visuals began with a slow start, the

majority of the students were able to use drawing as a tool for writing narratives quite easily by the seventh and eighth weeks of the course.

4. *Time and space to develop the craft of making films*
   The 12-module course spread over 12 weeks allowed students to engage with their work better and this was reflected in the animatics that they produced.
5. *Exposure-disciplinary and Interdisciplinary*
   The course included an intense 45-min to an hour-long lecture every week to introduce students to film history and language, alongside the history of literature and literary movements in relation to animation filmmaking. Students found the lectures interesting, but identified that the volume of information was overwhelming and often became a point of distraction in the process of making their own work.
6. *Feedback from experienced and practicing animation filmmakers*
   Every week students had 15–30 min tutorials each with the tutor and occasionally with visiting faculty. They were also encouraged to meet with and discuss their ideas for their films with visiting faculty and begin to identify people amongst fellow students and other faculty members, with whom they could discuss their ideas in order to be able to develop them in the way that they wanted. Students found this a useful way of contextualizing their work within the practice and began to create their own discussion groups outside of the class.

6 In Conclusion

The results of the course implemented, gathered through discussions with students and faculty and student feedback forms showed that

- The students who already had a certain amount of interest and exposure to animation filmmaking were able to progress faster.
- It is possible to direct and help students understand the importance of.
- Personal voice in animation filmmaking by providing them with a platform and structure to explore self-expression.
- If the motivation level of the student group to pursue the profession of animation filmmaking is high, the class presentations could be shortened. And even reduced to reading and film lists.
- It is possible to nurture self-expression in students through tutorials with the tutor and several filmmakers. This would be beneficial to the students, as they would get to hear several points of view from experts, about the quality and standard of their work.
- The tutor's knowledge of the subject areas of narrative construction and understanding of the skills required within auteur filmmaking would largely determine the successful delivery of the course. The tutor would also need to be able to give every individual student a certain amount for feedback that nurtures the inherent skills and abilities of the individual students.

The duration of the course, the number of sessions, the ratio of student-driven and faculty-led class presentations to individual feedback sessions with the faculty would need to be altered according to the context within which the course is delivered. The course is currently being re-designed and offered to a group of 20 animation BA (Hons) students at the University of Lincoln in the UK.

Acknowledgments Simon Gape, University of Lincoln, UK

References

1. Dinner for Two, Janet Perlman, 7:18, col, animation, 1996. https://www.nfb.ca/film/dinner_for_two. Accessed 28 May 2014
2. Elbow Room, Diane Obomsawin, 8:19, col, animation, 2002. https://www.nfb.ca/film/elbow_room. Accessed 28 May 2014
3. Every Child, Derek Lamb, 6:00, col, animation, 1979. http://www.youtube.com/watch?v=6bevAfTcj20&feature=kp. Accessed 28 May 2014
4. Wilk, C. (ed.): Modernist Beliefs, Building Utopia, Modernism: Designing A New World 1914–1939, p. 154. V&A Publishing, London (Reprinted 2008)
5. www.fondationlecorbusier.fr
6. Hauffe, T.: The Road to Modernism, Design, A Concise History, p. 59. Lawrence King, London (1998)
7. Wells, P.: Animation: Genre and Authorship, The Animation Auteur, p. 72 (Chapter 5). Wallflower Press, London (2002)
8. Kanfer, S.: Serious Business: The Art and Commerce of Animation in America from Betty Boop to Toy Story, pp. 18–20. Decapo Press, Boston (2000)
9. www.annecy.org/about/who-are-we-/history
10. onf-nfb.gc.ca/en/about-the-nfb/organization/mandate/
11. Ward, P.: Animation's Status as a Knowledge Area: Is There Such a Thing as "Animation Studies"?, Animation Studies, Disciplinarity, and Discursivity, UK, Reconstruction 3, No. 2, 2003. www.reconstruction.eserver.org/032/ward.htm. Accessed 28 May 2014
12. Bendazzi, G.: Cartoons, One Hundred Years of Cinema Animation. John Libbey and Co. Ltd, London (1994)
13. Kanfer, S.: Serious Business: The Art and Commerce of Animation in America from Betty Boop to Toy Story, Decapo Press, Boston (2000)
14. Wilk, C. (ed.): Film as Modernist Art, Modernism: Designing A New World 1914–1939, pp. 299–302. V&A Publishing, London (Reprinted 2008)
15. Bauhaus, Art as Life, The Radical Screening, The Barbican, London, 2012. www.youtube.com/watch?v=iq1SnKA5kIw. Accessed 28 May 2014
16. Symphonie Diagonale, Viking Eggeling, 7:28 mins, animation, B&W, Germany (1923)
17. Composition I, Werner Graeff, Animation, B&W, Germany, 1922
18. Composition II, Werner Graeff, Animation, B&W, Germany, 1922
19. The Heroic Arrow, Kurt Kranz, Animation, B&W, Germany, 1930
20. Film 1st Rhythm: Rhythmus 21, Hans Richter, B&W Germany, c1921. http://vimeo.com/42339457. Accessed 28 May 2014
21. Dobson, N.: Historical Dictionary of Animation and Cartoons, p. 127. Scarecross Press, UK (2009)
22. Anthony, S., Mansell, J.G.: The Projection of Britain: A History of the GPO Film Unit. BFI, London (2011)
23. Bendazzi, G.: Cartoons, One hundred Years of Cinema Animation. John Libbey & Company Ltd., London (1994)

24. Wilk, C.(ed.): Film as Modernist Art, Modernism: Designing A New World 1914–1939, pp. 299–302. V&A Publishing, London (Reprinted 2008)
25. Powers of Ten, Charles and Ray Eames, col, Animation, 9:00 mins,1977, USA. www.youtube.com/watch?v=0fKBhvDjuy0. Accessed 28 May 2014
26. Speech by Derek Lamb on winning the 1979 Academy awards for Every Child. http://aaspeechesdb.oscars.org/link/052-17/. Accessed 28 May 2014
27. http://www.sheridancollege.ca/academics/programs-and-courses/bachelor-of-animation.aspx. Accessed 28 May 2014
28. Asifa Festivals, Bafta and Oscar and why and how they are considered important
29. http://asifa.net/asifa-wp/asifa/basic-information/asifa-history/. Accessed 28 May 2014
30. http://onlyoscar.com/2013/10/01/road-to-2014-short-film-qualifying-festivals/. Accessed 28 May 2014
31. Faculty and alumni from eleven schools interviewed
32. Nine auteur filmmakers/faculty interviewed
33. Feedback forms collected from twenty-five students

Convergence—Divergence Paradigm in Design Education Curriculum

Ravi Mokashi Punekar

Abstract The Design profession has expanded its scope for intervention in new domains of professional practice in the emergent knowledge economy. There is a generational shift in the aspirations and expectations of design education in India to meet these challenges. This shift may demand a revamping of the undergraduate program in design schools. With the commencement of its design program in 1997, IIT Guwahati is the first amongst IITs' to offer a 4-year undergraduate program in design. Drawing upon data from its alumni and their career patterns, this paper outlines the changes and revisions periodically made in the design educational curricula to meet the evolving challenges of the profession. It highlights the gradual shift in emphasis towards convergence of disciplines required of the new program. These changes in the undergraduate program will lead to changes in the Masters' program in Design presently on offer amongst design schools in India. The paper proposes scope for pedagogic research in design education at this time and juncture.

Keywords Undergraduate design education in india · Curricula planning · Design

1 Introduction

A formal undergraduate program of design was offered in India for the first time only in the early seventies. It is important to trace its origins in order to be able to contextualize the need for change in the design education curricula that meets the expectations and challenges of the present times.

R.M. Punekar (✉)
Department of Design, Indian Institute of Technology Guwahati, Guwahati, India
e-mail: mokashi@iitg.ernet.in

A. Chakrabarti (ed.), *ICoRD'15 – Research into Design Across Boundaries Volume 1*,
Smart Innovation, Systems and Technologies 34, DOI 10.1007/978-81-322-2232-3_52

Based on the famous 'India Report' [1] submitted by the renowned design team of Ray and Charles Eames in 1958, the GoI established the National Institute of Design (NID) at Ahmedabad in 1961. The infrastructural needs required of running an experimental educational program in Design were set up and nurtured under the leadership of Shri. Gautam Sarabhai and Smt. Gira Sarabhai, during its formative years. Led by Prof. Kumar Vyas and Shri. Dasharat Patel, the decade of the 60s was spent in training and nurturing the first generation of design trainers. Some of those trained under them, went on to become the first generation of design teachers at NID. They were instrumental in subsequently establishing the Faculty of Industrial Design, Faculty of Visual Communication and Faculty of Environmental Design in NID. The institute initially offered a Diploma program in Visual Communication. The first formal program offered to school leavers commenced in 1970 and the first batch of 29 design graduates passed out in April 1975.

The UG design education program was based on the educational model program followed in Europe, in particular drawing from the program offered at Ulm in Germany and the Central School of Art and Design in UK. Infrastructure in the form of studios and workshops was set up and the broad based educational inputs drawn from the streams of science and liberal arts balanced those in visual design to train visual sensitivity required of a designer.

The foundation program, which stands as the corner stone of this experimental program was revised for an Indian context by the late 70s. It focussed on laying a strong foundation in basic design by 'sensitising the senses and heighten the students' perception of the physical and social environment by learning about vision, touch, sound and movement'. Most graduates from this school express views that these early inputs of the program added much strength to the fields of specializations they chose to pursue and subsequently in their future professional careers. This is aptly summarized in the words of Prof. Kumar Vyas [2]

> "Starting with 'whys' of design would be difficult because of the subjective nature of enquiry, but it would still be worth tackling. The 'hows' of design, however, can be understood and appreciated in a simple and objective manner"—H. Kumar Vyas

A decade later, the Advanced Entry Program (AEP) commenced in the early 80s. It facilitated the admission to a 2 year post-graduate design program that offered lateral entry of candidates who had acquired an undergraduate degree in Arts, Engineering, Architecture or Science, to select programs in the three faculty streams.

Concurrently in 1971, the Industrial Design Centre commenced its Postgraduate Diploma programs in Industrial Design at IIT Bombay under the leadership of Shri. Adarkar and Prof. Sudhakar Nadkarni.

Both these institutes (NID and IDC) would go on to become the founding institutions for establishing formal design education programs in India. The PG diploma program in design, offered by IDC was subsequently changed in 1979 to the 2 year Masters program in design, leading to the degree of MDes. While there were older and established schools of Art in the country, until the mid '80s', the programs offered at these two institutions would be the hub for formal design education in India that trained candidates for the manufacturing and commercial sector of Indian industry.

The high point of both these programs was a focus on experimentation and exploration in Industrial Design and Visual Communication fields through creativity and innovation. The foundation program at NID was a model worth emulating. 'Go Slow in order to go fast' [3] was the mantra followed in their program. Bringing real life experiences into the classroom was encouraged through projects and pro-active interface with industry and the craft sector. The graduates passing out from these portals of learning today have set up successful design consultancies and are extending design services to leading firms in India. One crop of designers has opted to work in the manufacturing sector; the other has opted to work with NGOs and the craft sector and has contributed significantly to the activities of these organizations. The credentials of these professionals to the overall growth of the design industry in India are commendable.

2 Undergraduate Design Curriculum Framework

The model of curriculum introduced was broad based with a strong focus on visual design considerations in the spirit of 'learning by doing'. The initial two years were divergent with various inputs in basic visual design principles and their attributes that were applied in the following years through project based studio work (Fig. 1).

The structuring of the program broadly captured the spirit of the design process. Courses were offered in the form of block modules. There was emphasis on developing an eye for the visual and enhancement of skill set through studio based

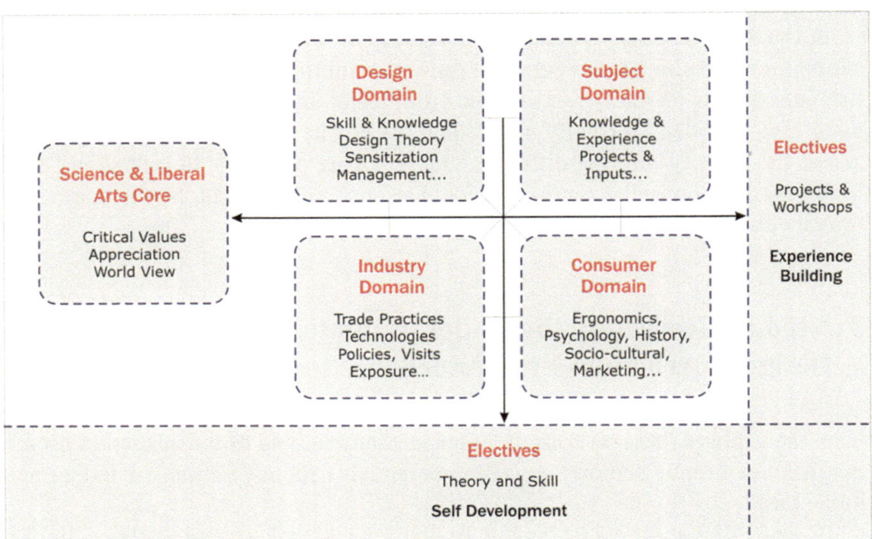

Fig. 1 NID model of undergraduate curriculum structure (*Source* Ranjan, M.P., 'Curriculum Structure of Design Education at NID', Unpublished paper, 1991)

assignments in drawing, colour, form and structure. Exploratory exercises in manipulating materials and understanding their properties and processes gave students an idea of manufacturing processes and technology. These courses helped lay the foundation for strengthening the students' abilities to visualise and conceptualise their design solutions. Inputs of science and technology and their social implications in humanizing technology, were grouped under the head of Science and Liberal Arts (SLA) studies, making the first two years of the program divergent and broad based. These inputs were taken to a higher level of understanding over the following years in the program, where the student chose to specialize in an area under the three faculty viz. Industrial Design, Visual Communication and Textile Design. During this period, the student learnt specifics of the field of specialization and would undergo experience building through application of his learning on the projects one chose to undertake. Design science principles in the form of courses in design process helped the candidate to consolidate his learning to a diverse range of projects successfully. In essence, it rounded off the convergent experience of 'learning by doing'.

As a result, the small intake of students to these two institutes had produced a talented but limited pool of human resource not numbering more than 400 designers by the 90s. For a profession that aims to meet the demands of a large manufacturing sector through its professional training for a developing country like India, these numbers of available resource pool was limiting. During this period, the National Institute of Fashion Technology (NIFT) was established to meet the growing demands of the Textile and Garment Industry. NIFT commenced a design program in Apparel Design, Garment Manufacturing Technology and Accessory Design at 6 locations across the country. Design graduates from NID and other Textile Technology Institutes who joined the faculty would constitute the second generation of design teaching faculty at these institutes. The design studies program adapted broadly followed the NID model.

In the 90s, there was a dramatic liberalization of the Indian economy. The education sector too had been opened to private institutions. This resulted in the establishment of new privately run institutions in which undergraduate design programs were also on offer. Currently, more than 50 private design schools dot the cities across the country offering different undergraduate programs in Industrial Design, Transport Design and Communication Design. The curriculum continues to be broadly outlined on the NID model.

3 Need for Revisiting the Undergraduate Design Curriculum Framework

Why then should there be considerations and a revisiting of the education program that has been replicated and served commendably the aspirations of Indian needs in the past?

If we revisit the general model of the educational system outlined by King and Frick [4] and assess the design program and its content and context in the larger framework of changes in society constituting the Education Negasystem, it becomes evident that the present education system has to reinvent itself from the point

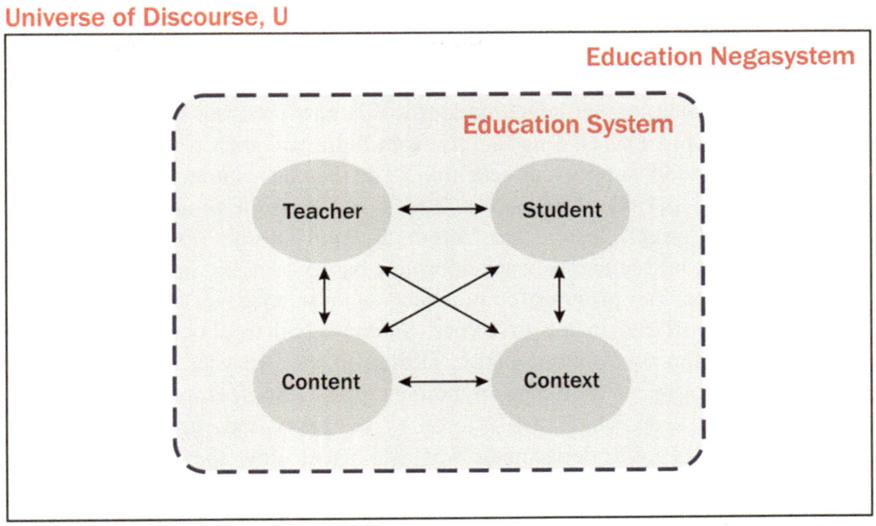

Fig. 2 General Education model proposed by King and Frick (*Source* Frick [4])

of view of the dramatic changes brought about by the advent of the information age where technology is increasingly playing a significant role in our everyday lives. The profiles of students and their aspirations have changed radically. The content of the curriculum has to adapt to the changed context for it to be relevant. New career opportunities have emerged that did not exist in the early 80s. We have a generation of students born in the 90s who have engaged with these technologies from their early schooling days. New tools and methods have evolved that are based on ICT enabled technologies. The choice of media has influenced the methods. Learning experiences are multi-modal today (Fig. 2).

These factors have strongly influenced and resulted in the two revisions to the design curriculum and structure of design education at the Department of Design at IIT Guwahati since it started its design program in 1997.

The need for ushering in concepts in 'design thinking' across all streams of engineering education in institutions of higher technical education is proposed by the MHRD, GoI, in its recent report titled 'Design Manifesto (For a Design Enabled Technical Education)' [5].

4 Experiences in Formulating the Design Curriculum at IIT Guwahati

The first Undergraduate (UG) Design program amongst IITs' began at IIT Guwahati in 1997 headed by Prof. Sudhakar Nadkarni. The first batch comprised of 12 students who had qualified the All India Joint Entrance Examination (JEE) conducted by the IITs. The experimental nature of the design program was seen early with the first batch of students working on a diverse range of projects in Product Design, Transport Design and Graphic Design. A core team of young faculty internally energized the department and a dedicated staff extended full commitment in helping set benchmarks for this new program. During the period 1997–2001 a number of visiting faculty from India and abroad visited the department and extended support in teaching. The batches started small, but have tripled over the period of 10 years. The 4-year Design program offered specialization in Product Design and Communication Design. Taking into consideration the fact that students were admitted based on their qualification of the JEE examination, the program commenced with science, technology and basic design courses offered in the first year. Students had to choose specializations of Product Design or Communication Design in the 3rd year. The program did not offer the option of a change of branch to the engineering stream after the end of their 1st year, which was offered by the other departments. As this was a norm in the IITs, the department was asked to follow the common program with the other streams during the first year. It was therefore time for a review of the program. Being in an IIT, it was only appropriate that we draw strength from the technical expertise that was accessible and needed for the design program. The revisions in the program envisioned that for a generation born in an ICT enabled learning environment, there was need for convergence and flexibility that can balance inputs drawn from technology on one hand and the fields of Art and Aesthetics, Human Factors and Social Sciences on the other. A major revision of the syllabus was outlined and approved by the senate. The program was project based, supported with theoretical inputs and hands on studio work. Technical courses were also drawn from the departments of CSE, Electronics and Mechanical Engineering. Students also had to choose electives from the Department of Humanities and Social Sciences. Core Design courses and Projects helped to consolidate and apply design learning.

This convergent model has found a unique niche in the duration of a decade. The department has established its credentials with the industry as a centre with strength in design with a technology orientation. This is reflected in the excellent placements offered to our design graduates. The fields of ICT enabled design domains emerged as a niche for the department. Design graduates have pursued higher studies in these fields both within the country and abroad. Design students from IITG had a head start in ICT enabled design careers in the rapidly growing software based industries. This sector in particular offers new career opportunities for designers and is in need of specialists in User Experience Design, Interaction Design and Usability Engineering. Most students are drawn and navigate towards this field because of the excellent job prospects it offers, leaving the other domains unattended. For the opportunities in the domain of product design, one is concerned that the balance of having a multi-domain strength of design in the department sometimes gets rather one-sided.

An analysis of the career paths of our alumni highlights the career paths of these new opportunities in the fields of User Experience (UX), Usability and Interaction design (UI) available in the software sector. Of the 164 graduates,

Table 1 Alumni data of department of design, IIT Guwahati

Sector	Legend	BDes	MDes	Total
User experience and interaction design	IN	109	8	117
Automobile/Transport design	AU	1	1	2
Product/Industrial design	PD	6	2	8
Management studies	MA	22	0	22
Graphic design/ advertisement	GA	2	3	5
Entrepreneurship	EN	10	0	10
Education sector	ED	3	5	8
Students currently studying		11	0	11
Total		164	19	183

Source http://www.iitg.in/design

nearly 109 of them are employed in this sector. They have moved away from the conventional fields of Industrial Design that were on offer earlier. The design program seems to also help them to pursue higher studies amongst the highly competitive leading schools of management (Table 1).

4.1 Design Education and the Transformative Phase in the Indian Landscape

With the increasing number of design programs that are on offer in the country and new ones that are expected to commence, it is time to factor in these new opportunities in the design education domain. With the ushering in of the knowledge economy primarily due to an ICT enabled intervention, the changes of social aspirations and economic factors experienced on the Indian landscape have transformed and are different from the earlier opportunities of the 'industrial era'. Classrooms in institutions of higher learning comprise of a generation of kids born in the early 90s—a generation that has been exposed to a canvas of learning drawn from and strongly influenced by the Internet. The medium for communication and the forms of knowledge assimilation is an open book. The expectations and aspirations of this generation from the program lean towards diversity bordering on technological adventurism that is exploratory, innovative and experiential in spirit.

In a recent survey conducted amongst 70 students of the 3rd and 4th year of the undergraduate program in Design at IIT Guwahati, the students were asked to reflect on the program they have undergone. The courses offered were grouped under nine categories: Courses in Sciences and Technology; ICT enabled courses; Visual design courses; Design projects; Design and management courses; Humanities and Social Science courses; Design Science courses; Industrial training and internships; Special projects and workshops. In addition, they were asked to reflect and give opinion on the following modes of assessment: Mid and End semester examinations; In-class studio evaluation; Project presentations, External design competitions. Opinion was sought on the modes of teaching—lecture format for theoretical design courses, studio based learning, workshop assignments, special short duration design workshops from experts, etc.

Responses were sought using a pen and paper format. They were asked to respond stating their opinions on courses they 'must have' and need to be continued in the existing format and those that must be continued but 'with revisions' and the ones they considered irrelevant and that needed to be 'dropped'. In addition, they were asked to suggest 'new courses that need to be introduced'.

A summary of their opinions resulted in the following findings:

Most of the core design course were very essential and should be retained. They expressed that their school education laid more emphasis on the science course with the domains of art and social sciences being neglected. While they struggled with their skill sets, they realised the importance of these courses for design during the later years of their studies.

Early introductions of courses and hands-on studio work in electronics and computing were very essential in design. The inputs from the streams of social sciences strongly influenced design thinking and 'must be' offered in the program. Courses in statistics, psychology, ethnography and economics must be offered during the program as they found them of great relevance for analysis and establishing insights during their project work.

Inter-disciplinary projects must be introduced that encourage team work with students from other branches of technology particularly Electrical and Electronics, Mechanical and Computer Science streams. They asserted that during projects, they most often collaborated with students from these streams informally and they mutually benefitted from these collaborations.

Courses that needed strengthening included cognitive ergonomics and skill courses in visual design. Hands on training in application software for visualization and 3D modelling were desirable and must be introduced through formal assignments during project work.

They expressed the need for the re-ordering of the sequence of some of the courses in design management and design process that they felt must be introduced at an earlier stage of the program.

Short workshops with experts, participation in design competitions and collaborative projects with industry were very enriching and brought dynamism to the learning experience and must be encouraged.

Classroom presentations and studio evaluations were extremely useful as they enhanced soft skills in effective communication and social interactions.

A few felt that there should be more options for students to choose their elective courses and projects.

After a review of the pen and paper feedback session, group discussions that followed also brought out the following aspects.

Classrooms deliberations should encourage discourse in a spirit of 'chalk the talk' rather than 'teacher and taught'.

Conventional boundaries of disciplines need to be necessarily transformed into seamless platforms of inter-disciplinary discourse. Knowledge decimation includes sources of not only 'what' to learn but also to know 'where' to source it from. It is apparent from these views that the expectation in such a milieu of learning has increasingly transformed into the domain of the experiential. The world of objects and their design involves understanding of domains that are as much cognitive as they are about understanding technology and material science. The shift is distinctly away from consideration of the aesthetics of material, form and structure to that of an aesthetic of the understanding of 'understanding' and of human thought processes involved in the engagements of products and product systems to navigate through every day life.

Revamp of the education processes has to factor in these considerations and challenges. This brings to the fore issues that need to be addressed at a level that may definitely focus on the revamp of the Design curriculum and go beyond into the realm of relooking at the structure of the UG program of Technical Education itself. A few questions that one may pose include:

1. Is it an opportune time that the all India examinations conducted for admission to engineering colleges including the Joint Entrance Examination for the IITs be restructured to include awareness and knowledge of subjects beyond Physics, Chemistry and Mathematics that form the primary focus of testing and evaluation for the selection of candidates? Should the admission test be inclusive of the candidate's broader understanding, intelligence and aptitude for science and technology and their influence on society?
2. After the candidates are given admission, should the program of engineering at the UG level train them to be a generalist? Should the program offer flexibility for navigation between disciplines and streams encouraging students to mix and match coursework and projects that encourage innovation, analytical skills and understanding of processes with a focus on an experiential understanding of science and technology in construction and implementation of ideas? A move towards such a possibility is already evident in the option of 'Minors' offered to the students by the different departments within the IIT. They can now 'Major' in a stream of their choice and in addition, opt to undertake a set of courses from another department over the duration of their program and get in addition a 'Minor' degree from that department. Is the introduction of such a concept, a sign of the direction of future programs in undergraduate Technical education programs? There is also a document put up by the Department of Higher Education under the MHRD, GoI titled 'Design Manifesto' that highlights the need to introduce a 'Design Spine' across all departments of Engineering education that will encourage students to undertake learning with a spirit of 'Innovation'.
3. ICT should be taken as a new paradigm and foundation on which to build modes of content delivery. Planning for online courseware should be an enabler for complementing the present modes of teaching. Video conferencing through

Skype or other platforms offers interesting possibilities of collaborative and live projects being introduced into the class room with continuous online connect between experts in the field and the youthful participation of students engaged in such 'Live' class room projects.

4. In the rapid rate of technology driven environments and the anticipated change it brings to the living environment in our times, what remains understated is the coexistence of nearly three generations who are subject to experiences of these changes during one lifetime. This is resulting into the need for evolving an educational system that encourages the demand for being constant and perpetual learners.

These pointers have a bearing on higher education at the Master's and Ph.D. level of design studies.

The Masters' program in Design on offer amongst all leading technical institutions of higher learning run under the MHRD, GoI has traditionally been structured as a multi level entry extended to students coming from cross disciplinary fields of Arts, Engineering and Architecture. In view of diversity of background of students, the programs have visual and technical design inputs that are modeled on fundamentals offered at the undergraduate level. With the UG programs now covering this ground, the Masters' program will have to be restructured and engage in studies leading towards a higher order of academic research and practice /professional based research. As of now, this is in a phase of transformation. It is perhaps an opportune time to revisit and plan for revisions to the areas of design research in which the masters programs must be oriented towards.

Ph.D. program in design too is at a formative stage. During the current phase, the doctoral program in design globally is in a serious search of formulating its own body of theoretical design framework. In a survey on education in design research conducted in 2003, Durling and Sugiyama [6] indicated that there were just 5 Ph.D. in design awarded amongst 6 leading design schools across the globe. Doctoral program in design has been a new pursuit for the profession. Unlike the Science and Engineering streams, it has grown only after the year 2000. In India, it is interesting to note that in the field of design, Industrial Design Centre, IIT Bombay; Department of Design, IIT Guwahati and Centre for Product Design and Manufacturing, Indian Institute of Science, Bangalore have taken the lead in offering doctoral degrees in their design programs while only a few other IITs have only recently commenced their Ph.D. programs in design.

The Department of Design, IITG, commenced and offered its first formal program leading to a Doctorate in Design in the year 2002. One must commend this early initiative taken by the department since globally, the domain of doctoral studies in design was only picking up. Initially the teaching faculty with a Master's degree in Design registered for their Ph.D. and completed their doctoral research as they were teaching. In a period of 12 years, IIT Guwahati has an impressive output of 12 Ph.D.'s from the department. Currently there are 39 registered candidates pursuing their doctoral research at the department. The program started with candidates from different backgrounds registering for their doctoral degree. However, in the last couple of years, it is heartening to see that Master's candidates have also opted to pursue their Ph.D. on completion of their program. This is a welcome development and in times to come, one can foresee that these doctoral research candidates will go on to contribute to the design program amongst various universities in India and abroad.

5 Conclusions

This paper is perhaps reflective of the design programs and the manner in which they have evolved across three leading schools of design, two of which are established as founder schools of design in India and the third, only under two decades old, has attempted to model its design program drawing advantages of learning from the former and going beyond, in experimenting with an integrated convergent model of design education. In times to come, there will increasingly be demands upon institutions of higher learning to evolve programs of learning that are technology driven and encourage convergence of multi-disciplinary streams with flexibility for cross over between streams and disciplines. This paper anticipates that initiating a pedagogic research in design education can help evolve a new program that will encourage experimentation and experiential learning leaning towards innovation and cross disciplinary team work.

Acknowledgments We acknowledge the senior students of the undergraduate and Master's program of the Department of Design, IIT Guwahati, for their constructive feedback given to the author during the academic session of January–May, 2014.

References

1. Charles, E ., Ray, E.: India Report, (2nd edn.). NID Press (2004)
2. Shilpa, D. (ed.): 50 years of the National Institute of Design—1961–2011. NID Press (2013)
3. Balaram, S., Ranjan, M.P., Satwalker, S., Patel, K.: Revised curriculum-national institute of design. Unpublished report of curriculum review and development group NID (1994)
4. Frick, T.W.: Restructuring education through technology. Phi Delta Kappa Educational Foundation, Bloomington, Indiana, USA. www.education.indiana.edu/-frick (1991). ISBN 0-7367-326-3
5. Sharma, A.: Design manifesto (for a design enabled technical education). Report, MHRD, GoI. www.deconstructingdesign.iith.ac.in/documents/DesignManifesto.pdf (2014)
6. David, D., Sugiyama, K.: Results of the survey of education in design research. In: 3rd Doctoral Education In Design Conference, Tsukuba, Japan (2003)

Moving with the Times in India Re-thinking the Foundation Course in Design

Indrani De Parker

Abstract 'Design is a protean, open ended and versatile term, which, over the years, has gathered many dimensions and definitions within the folds of its discourse. The impetus to formulate a manifesto for a pluralistic discourse such as design, comes from the growing critical reflections on modern dominant paradigms of education as embodied in Indian institutions of technical education, vis-à-vis the demands of a larger developmental paradigm of the state. Leading institutions of technical education, while striving towards excellence in developing competencies in specialized fields of engineering and technology, are now increasingly concerned with augmenting opportunities for holistic education……This, in turn, compels an interrogation and re-imagining of academic processes and structures, curriculum and pedagogy for enriching the existing design departments as well as the engineering, sciences, architecture, humanities and management streams.' (The Design Manifesto by Ministry of Human Resource Development, India, 2013).

The social, cultural, economic, technological and ecological challenges facing us today requires a new discourse on how we define design. This includes the changing role of the designer in the future. The world is changing rapidly, and so too are designers who continually adapt to these changes to define new roles for themselves. What implication does this have for the future of design education? This in-progress research into design education, proposes that as designers continually expand the boundaries of the design discipline, adding new dimensions and adapting to the changing circumstances, so too must design education and more specifically, the Foundation Course for Design. The foundation year originated at the Bauhaus in 1919 as the Basic or Preliminary Course and evolved after 1945 at Ulm and Basel. In its emerging period, design was focused on individual products. Today, however, to be relevant to contemporary society, designers need to be able work on complex issues that are interdisciplinary and much broader in scope. 21st century design education needs to be able to apply design and develop strategies to solve real issues and not assume that all solutions should culminate as a 'form'.

I. De Parker (✉)
IDC, IITB, Mumbai, India
e-mail: indranideparker@gmail.com

A. Chakrabarti (ed.), *ICoRD'15 – Research into Design Across Boundaries Volume 1*,
Smart Innovation, Systems and Technologies 34, DOI 10.1007/978-81-322-2232-3_53

Keywords Design education · Foundation course · 21st Century indian context

1 The 'Basic Course' or 'Foundation Course' in Design Eduction

The Bauhaus was a state sponsored initiative of an early model of a design school, which integrated the artist and the craftsman while bridging the gap between art and industry. All entering undergraduate students in design are required to take the Foundation Course in the first year. Basic design as it is offered in the 'Basic' or 'Foundation' Course has evolved from a need that was originally perceived and dealt with at Bauhaus and Ulm as a critical orientation to design thinking and action.

1.1 The 'Vorkurs' or the 'Basic Course' at the Bauhaus

The first Basic Course (Fig. 1) planned for one term in 1919, presented Johannes Itten with three tasks [3]:

1. To liberate the creative forces and thereby the artistic talent of the students. Their own experiences and perceptions were to result in genuine work. Gradually, the students were to rid themselves of all the dead wood of convention and acquire the courage to create their own work.
2. To make the students' choice of career easier. Here exercises with materials and textures were a valuable aid. Each student quickly found the material with which he felt the closest affinity; it might have been wood, metal, glass, stone, clay or textiles that inspire him most to creative work. Unfortunately, at that

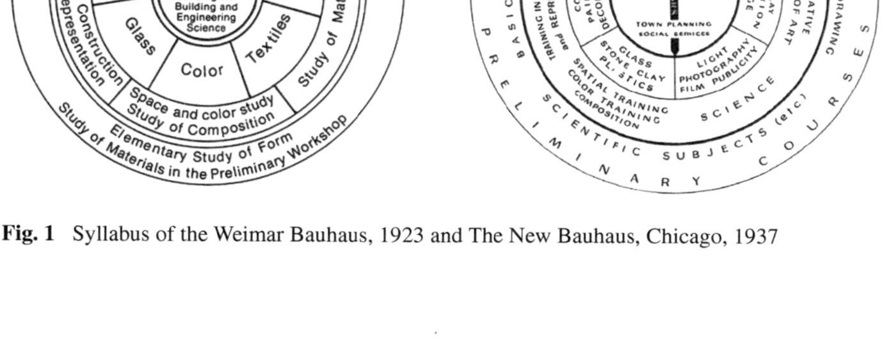

Fig. 1 Syllabus of the Weimar Bauhaus, 1923 and The New Bauhaus, Chicago, 1937

time the Basic Course did not have a workshop, where all the basic skills such as planning, filing, sawing, bending, gluing and soldering could be practiced.
3. To present the principles of creative composition to the students for their future careers as artists. The laws of form and colour opened up to them the world of objectivity. As the work progressed it became possible for the subjective and objective problems of form and colour to interact in many different ways.

The graduates of this Basic Course were expected to [8]:

1. To learn a craft in the workshops of the Bauhaus
2. Prepare for future cooperation with industry

Key design slogans of this era included:

(1) Truth to Materials; (2) The House—A Machine for Living; (3) Form Follows Function

When Bauhaus closed down in 1933, most of the master educators migrated to the United States of America, where the established professional institutes of higher learning absorbed them. While many Bauhaus ideas were integrated into the American culture, the reception of these ideas reduced a complex and multifaceted phenomenon to a simple formula [5]. Lerner claimed that the context, writing and teaching of the master educators of 'Vorkurs' were largely buried by time.

1.2 The 'Grundkurs' or 'Basic Course' at HfG, Ulm

After the World War 2, a new German design school, HfG, an initiative of The Scholl Foundation, was founded in Ulm, based on some of the tenets of the Bauhaus. The course of study at HfG lasted four years—one year of the Basic Course and three years of specialization and would end with a diploma in either Product Design, Visual Communication, Building and Information (Fig. 2). The teaching comprised one half of practical design work and the other half of lectures and seminars.

1. Products were meant to be as long-lived and functional as possible, acceptable with respect to social and ecological criteria and to take into account changing political condition and production technologies.
2. The school's pedagogical concept, the so-called Ulm Model was characterized by a new system oriented design methodology and the encouragement of interdisciplinary teamwork.

Herbert Lindinger in his forward to the book "Ulm Design" tells us that the HfG Ulm had been through six phases of development. They had already developed a critical approach to design education and design theory that was well documented and disseminated by the Ulm magazine 1–21 from 1955 to 1968. He states:

> The third phase, 1956-58, was dominated by the teaching of Otl Aicher, Maldonado, Gugelot and Vordemberge-Gildewart. These instructors tried to build a new and markedly closer relationship between design, science and technology. This was the first manifestation of the Ulmer Modell, the Ulm model

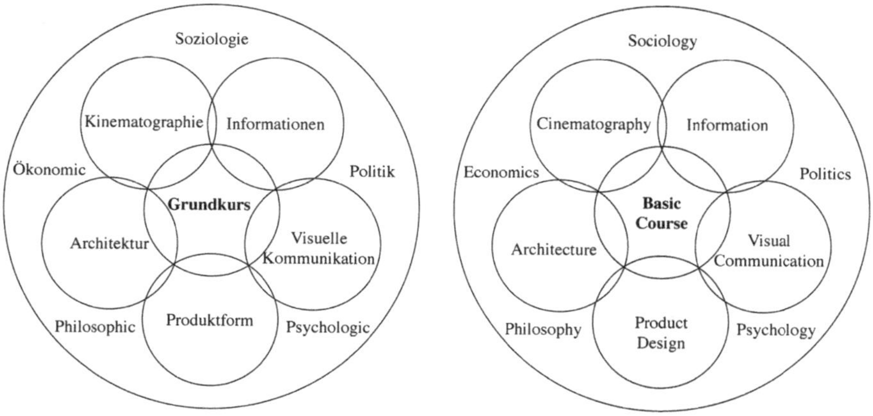

Fig. 2 Hochschule für Gestaltung curriculum (1951)

Under the guidance of Aicher and Maldonado a strong effort was made to create a new balance between theory and practice and between science and design. An attempt was made to reintegrate the departments by introducing interdisciplinary project themes and ideas. The HfG evolved a model of training that aimed to give designers a new, and rather more modest and cautious, understanding of their own role.

> The first ecological themes were starting to appear and the conception of the Basic Course underwent an enormous change. In the realm of theory, the first steps were taken toward design analysis. There was a new profession to be defined and the instructors in the Product Design department exerted considerable influence by formulating job definition and outlining a framework for professional training (as per original, pg 5, Lindinger 1991).

As design was now to concern itself with more complex things than chairs and lamps, the designer could no longer regard himself, within the industrial and aesthetic process in which he operated, as an artist, a superior being. He must now aim to work as part of a team, involving scientists, research departments, sales people, and technicians, in order to realize his own vision of a socially responsible shaping—Gestaltung—of the environment. Under Maldonado, "a new Basic Course came into being, which broke away more and more clearly from Bauhaus concepts and absorbed the lessons of perceptual theory and semiotics [6]."

1.3 The National Institute of Design, Ahmedabad (NID)

It was this Basic Course that Kumar Vyas understood deeply at Ulm and introduced to the new batch of Product Design students when the Postgraduate course was offered to graduate engineers in 1967 at NID. Early examples of the Basic Design assignments as well as the early projects and the methods used in these projects that echo the Ulm paradigm as well as the muted shades of grey and colours that were a hallmark of the HfG Ulm way.

An extensive photo-documentation of students of the Foundation Course from five design schools in India over a few years was conducted as data collection for a rich description of the current Foundation Course. The data demonstrates a universality that exists in design schools today. This partial inert state could well be attributed to the pedagogy borrowed, repeated and replicated over decades.

2 The Story of Design Moving with the Times in Post Independent India

Design as an activity in India is as ancient as its civilization. India has had rich and immense traditional and extremely skilled craftspeople whose skills have been passed on from one generation to another through several centuries. India has such a depth of handcrafted skills, techniques and artifacts. For centuries, handicraft items that combined beauty and utility were crafted in a sustainable way serving the needs of the local markets without any apparent damage to the environment. During the independence movement, the ideals of simple living and right thinking was advocated along with the rejection of British made goods, the promotion of Swadeshi and the spinning and weaving of the Khadi. Post independence, the government proceeded to look for means to make available a source of income to huge population. Acute power shortages, forced an effort to revive the handloom and handicrafts industries [1].

2.1 Craft Sector and Caste Factor

Kamaladevi Chattorpadhya and later, Pupul Jayakar took up the challenge to revive the craft traditions and in making them competitive in local and export markets. Because of their herculean efforts, many traditions of textile and crafts were saved from becoming extinct and many traditions and techniques were revived. A rich heritage in crafts especially with craftspeople with generations of indigenous knowledge and skills, may well have naturally transformed them into modern designers. But this did not happen in India because the traditional craftsman was not given access to the kind of education that would have enabled this transition.

> …..educated castes and communities condemned those who laboured as stupid and unworthy of being treated as human beings. They were treated as people not worthy of education; not worthy of becoming writers, manager, engineer, doctors or accountants. The priestly community supported the attitude of treating hardworking people as bad people. Thus, the practice of humiliating the labouring communities received the sanctity of religion. [2, p. 93].

Caste barriers secluded artisans from moving upwards into professions dominated by superior castes. The situation with regards to art was similar. Hundreds of styles and schools of art flourished in the Indian subcontinent from the third century onwards. British rule had a significant impact on art in India. Under the British rule, art schools were set up which distinguished for the first time, between fine art and craft. With fine art in the British mold, acquired a high status. Thus the students of modern art schools became artists while traditional Indian artists who had learnt their art from family and community were called craftsmen. Postcolonial artists, trained in the British style, continued this appreciation of identity with contemporary artist, mostly coming from the educated elite while craftsmen continued to be from less affluent classes and lower castes. Thus it happened that the first generation of postcolonial designers, were drawn from the educated elite rather than artisanal backgrounds.

2.2 Modern Design Education in Post Independent India

In 1947, after independence from British colonial rule, India's leaders, particularly the first Prime Minister Jawaharlal Nehru, worked towards revitalizing the nation in all spheres. Almost every aspect written on the Nehru's vision of independent India, and nation building, focused on his stress on introducing western science and technology and modernizing agriculture. Naturally, this required the establishment of institutions of higher education in the natural sciences, technology and management. An equally important position of Nehru's nation building ideas was the rebuilding of small, rural and cottage industries as well as handicrafts. As part of the initiatives taken in this direction, was the move to introduce institutions for design education [7].

In India, modern design education began in the late nineteenth century with the opening of schools in architecture and art (fine and commercial art). On the request of the then Prime Minister Jawaharlal Nehru, Charles and Ray Eames' 'India Report' initiated Industrial Design practice and education in the post independence period. Charles Eames who had drafted the guidelines based on which the National Institute of Design (NID) was founded, had spent some time at HfG Ulm. In spite of the focus on Eames' report on Indian design tradition and sensibilities, the design education programs in India, like in many other countries, actually borrowed its pedagogy and thinking from Bauhaus as well as HfG Ulm school tradition especially the Foundation Course.

2.3 Changing Circumstances from 1950s to 1990s in Independent India

During the 1950s and 1960s the economic policy in India was based on Swadeshi or self-reliance. Companies were focused on building a potential domestic market. Foreign direct investments were limited. Exports were discouraged and imports were restricted. During this period, considerable government support was given to the textiles and the craft sectors. The advent of textile mills with power looms, did not affect handlooms and it continued to hold its share of the market because there was already a huge population of people engaged in it and of course it had government support. As India began to rapidly industrialise, Indians dreamt of machines and the idea of modernity. There begun a trend to move away from tradition and embraced the language of modern economy.

From 1966 to 1984, during Indira Gandhi's tenure as the Prime Minister, the state continued to be the chief patron of design. In tune with the socialist agenda, Mrs Gandhi nationalized all commercial banks, airlines, hotels, industries and certain other sectors including oils, mines and minerals, textiles and insurance. All prime sector industries were placed under strict regulatory control. This period was discouraging for design and innovation in India with a very low growth rate and a per capita income growing at a sluggish pace. In the automobile sector, Indians had two cars to choose from—the Ambassador and the Fiat and both had long waiting periods. The need to improve design was never felt by the manufacturers.

Finally, in the early 1980s Indian design, was displayed majestically by the government at the Festivals of India in London, Paris and America. These festivals conceived by Pupul Jayakar showcased India's 2000-year cultural heritage on a massive scale. Most design activities at the time were within the public sector. The fresh design graduate was armed with idealism rearing to contribute to the progress of the young nation.

In 1991, during P V Narasimha's tenure as the Prime Minister, India liberalized its previously protected economy, opening its doors to the outside world. This historic change radically altered the playing field of design. The Indian markets slowly saw the inflow of foreign made material and products, which were of better quality. For the first time in the history of independent India was there an access to better designs than what people had been accustomed to and consumers had a choice and could reasonably expect a superior quality from the products they bought. During the same time, Indian companies too could venture out to the world and find new markets for their products and services. Forced to compete with international products and brands Indian companies started to recognize that design could no longer be ignored. This was a big opportunity for designers to apply their learning and practice their profession [4].

2.4 Turn of the Century and the Emerging Economy in Independent India

At the turn of the century, the quality of Indian products improved remarkably, both in terms of manufacture and design. Trade fairs and the advent of the Internet aided manufactures understand the importance of creating brands out of their products. People's expectation went up. Retailing format staring changing from down market grocery stores to clean brightly lit supermarkets and departmental stores where customers browsed through merchandise at leisure. Besides products, packaging of products and branding took on a new physical appearance. Everything needed to be designed and there were not enough designers to get it done. It soon became obvious that there is a severe shortage of qualified designers in India. For the first time people started seeing the value of design.

In the early 1990s, in a developing country like India, price was the main buying factor. As the economy improved, people turned into consumers and as consumers turned more affluent, they started looking for quality and were willing to pay more for it. Today, in an emerging economy, besides quality and price, customers are willing to pay for design. Their expectation included both function and aesthetics.

Towards the turn of the century, liberalization in India brought about several changes in India. The image of the middle class evolved experiencing the change of aesthetic preference from hand crafted adorned style of products to machine made western style of simple forms. A general consensus of 'whatever originates from the west must be the best' crept into the middleclass mindset. McDonalds, Levis and Coca-Cola were eagerly embraced. In keeping of Nehru's vision for a modern nation based on western influences, design too continued, treading the same path.

The liberalization of the Indian economy coincided with the global technology revolution led to the knowledge economy. The colossal growth of the IT (Information Technology) sector provided high quality design services to Indian companies ready to compete globally. The quantum jump in technology transformed the lives of ordinary people in India. The most amazing transformation happened when the mobile phones became accessible to millions of rural Indian. The mobile phones were their first brush with technology and their first connect with modern economy. India today has one of the largest mobile phone user base in the world. On the other hand access to the Internet has brought enormous amount of information and knowledge into the reach of millions of Indians. Meanwhile, the television surged from one channel to the opening up of networks that hundreds of the world's television channels bringing the western influences to the villages. A highly social able culture, India has taken quickly to social media. Facebook, Twitter etc. has redefined the ways that people communicate. With much greater use of images and items that cut across geographic and linguistic boundaries—ideal in a country like India. It improved massive access to knowledge; access to networks; exposure to quality merchandise and materials etc.

On the product front, access to superior materials, machines and technology led to superior production standards of products. Designers spend minimal time, supervising vendor quality and can now use that time more productively. The methods of production of design virtually are made possible with software like 3D printing. With the surge of economy and technology, which in turn affected the social, cultural and environmental scenario, unfortunately shaded most people's understanding of design, which is generally viewed as luxury or decoration.

2.5 *The Current Context*

Today, well into the 21st century, there seems to be less connects between good designers and real problems that need their intervention. India today needs radical and creative solutions to solve increasingly complex problems. In the next couple of decades more and more people will want to migrate to urban areas from the village. The overpopulated cities are already collapsing under the pressure of inadequacies in housing, water, sanitation, transport, waste disposal etc. Almost seven decades after becoming an independent, democratic nation, almost 300 million Indians are illiterate and almost half of the rural Indians do not have access to basic healthcare and toilets.

In the 1970s and 1980s, design activity was dominated by the public sector. Today it has significantly moved into the private sector. As a result, design in India has become slave to consumerism without designers paying enough attention to the consequences of their interventions. Designers, take decisions that may be successful in fulfilling market needs but unfortunately may have colossal unintended environmental consequences. The government, on the other hand has the most complex problems that need the most innovative solutions and the best design minds to solve them. Unfortunately, the government tendering systems evaluate the creative industries just as any other commodity and this includes the setting up of four new national design institutes [4].

3 Conclusion

Design pedagogy, having operated in the past in the shadow of art, craft and engineering education, has evolved as the field of design and continues growing. Having started as craft based training with rather narrow vocational aims, design education is developing into an interdisciplinary academic field emphasizing research and preparedness for the emerging future. Design has diversified itself into various subfields and different academic levels. The problems of the Indian people, both nationally and locally, within the mesh of cultural diversity with economic disparities, including health, transportation, housing, agricultural support, safe water provision etc., are areas which offer potential for the designer to make a contribution. It is important for designers to understand the complexity of issues at stake as well as being aware of "intangibles" like values, social responsibilities, empathy, humility, and local/global relevance.

In today's context, design should cease to be perceived as a profession for star individuals but the result of collaborations of many minds coming together to arrive at solutions that offer the best possible outcomes. If design is to be useful, it must reach more to the real problems of our times. Its time to take advantage of an open economy and free society no longer look to the government or the structured institutions to provide top down frameworks that once determined initiatives. Improved connectivity, can allow collaborations and networks of small enterprises to start forming ecosystems of innovations. To be able to work across sectors on a variety of issues designers could, to quote Charles Eames's India Report, "help generate that attitude that would appraise and solve problems of our time with service, dignity and love."

The research steadily attempts to establish that from the industrial production, with an emphasis on materiality, design has steadily developed into a wider, interdisciplinary discipline. The research attempts to understand how the Foundation Course in design responds to the current contexts. A chronological description of the social, economic, political, technological and environmental changes, in India since independence establishes that the current context warrants a re-look at the current learning process in the course. This in-progress research into design proposes that as designers continually expand the boundaries of the design discipline, adding new dimensions and adapting to the changing circumstances, so too must design education and more specifically, the Foundation Course for Design.

References

1. Guha, R.: India after Gandhi: The History of the World's Largest Democracy. HarperCollins, New York (2007)
2. Ilaiah, K.: Turning the Pot, Tilling the Land: Dignity of Labour in Our Times, p. 93. Navayana Publishing, New Delhi, India (2009)
3. Itten, J.: Design and Form: The Basic Course at the Bauhaus and Later. Wiley and Thames Ltd, London (1975)
4. Keshavan, G.S.: Towards Global Histories of Design: Postcolonial Perspectives. In: Keynote speech at Design History Society (DHS) Conference, National Institute of Design, Ahmedabad, India (2013)
5. Lerner, F.: Foundations for design education: continuing the Bauhaus Vorkurs vision. J. Issues Res. **46**(3):211–226 (2005) (Spring 2005; Research Library)
6. Lindinger, H.: Ulm: Legend and Living Idea, in Ulm Design: The Morality of Objects, Lindinger, H. (ed.) Translated by David Britt, Mass.: MIT Press, Cambridge (1990). Original German Version Published in 1987
7. Sheth, S.: Genesis Of Design Education In India: The Warp And Weft Of Local—Global Contexts, Faculty Of Doctoral Studies, CEPT University (2012)
8. Wingler, H.M.: Bauhaus Weimar Dessau. The MIT Press Cambridge, Massachusetts and London, England (1978). Originally Published in German in 1962

Quality Education Over Quantitative Education at Primary Level in India

Priyanka Bharti and Bishakh Bhattacharya

Abstract An attempt is made to study the quality of education at primary level in India. Primary education is the first stage of compulsory education which is been considered as the most crucial in a child's education and social development. With an extensive user survey, the significant characteristics of the Indian primary education system have been identified. It is brought out that instead of imparting knowledge, a merit-award based education system is prevalent, due to which the quality of education is falling predominantly across the states. Moreover, the amalgamation of new teaching practices like ABL (Activity Based Learning), MGML (Multi Grade Multi Level) has not created the desired results due to slow and improper implementation. More interesting ways and means of schooling and teaching are required to attract students for completing their basic education successfully. A new technology based interface has been designed and developed which may bridge this gap and enhance knowledge sharing.

Keywords Quality education · Primary education · Education system · School

1 Introduction

India is the seventh largest (area wise) and second most populated country in the world (1.2 billion) after China. China has 94 % literacy rate whereas India has only 74.04 % (as per 2010–2011 census). In India if we assume that 100 kids take birth in a day then, 15 of them will never attend the school in their life, and 50 % out of the left students will dropout before completing class Vth and hardly 10 % out of them will reach till class Xth. Thus, this results into such a condition that 40 % out of all India's kids, are not even able to write their names.

P. Bharti (✉) · B. Bhattacharya
Design Programme, Indian Institute of Technology Kanpur, Kanpur, Uttar Pradesh, India
e-mail: bhartipr@iitk.ac.in

B. Bhattacharya
e-mail: bishakh@iitk.ac.in

A. Chakrabarti (ed.), *ICoRD'15 – Research into Design Across Boundaries Volume 1*, Smart Innovation, Systems and Technologies 34, DOI 10.1007/978-81-322-2232-3_54

2 Primary Education in India

India has almost 0.66 million primary schools (within a distance of 1 km each, 2001–2002). The age span for primary education in India is 6–14 years (World Bank report 2011).

2.1 History

Indian education system is one of the oldest education systems in the world, which believes in spreading the importance of knowledge and religion. In India, education was the way to attain salvation (Moksha). Thus, we can conclude that Indian education system was psychological, which included verbal knowledge, that can be preserved for thousands of years. "A survey of Indian History" says that "A unique degree of verbal authority has been maintained up to in the form the utterance and the mantras of the sacred texts.

After the British invasion, Hindi language was replaced by English as the medium of instruction. This sudden change led to rote learning practice in India and punctured/broke the foundation of Indian learning system. This education system was unplanned and unsuited to the need of our country. It's after post-independence, by the joint effort of Dr. S. Radhakrishnan and Dr D.S. Kothari, in July 1968 that the Government of India declared its National Policy on Education. After implementing new education system the literacy rate from 16.6 % (1951) went up to 29.45 % (1971). During 1975–1976, our education system was further revised and was categorized into elementary stage (primary education; from class I–VIII), secondary stage (class IX–X) and higher secondary stage (class XI–XII).

2.2 Current Scenario

The literacy rate in India has dramatically changed in last few years, from 29.45 % (1971) to 74.04 % in 2010–2011, but it still lacks behind in universalization of primary education. Literacy, in Article 26 of the 'Universal Declaration of Human Rights states' states "Everyone has a right to a basic education and it should be free and available to all." As per census 2001, India produces only 7–8 million literate instead of 15 million per year which means only 46.66 % gets educated, each year. As per the Programme of International Assessment (PISA) test in 2009, India ranked 71st among 73 countries. Another international assessment test Trend in International Mathematics and Science Study (TIMSS) responsible for measuring student's capabilities of class IV (reading) and Class IV–VIII (mathematics and science). In 2003, India ranked 46 among 51 countries because India scored only 392 versus average of 467. Annual Status of Education Report (ASER) is a part of Mumbai based non-government organization Pratam which conducts a nationwide survey of children's ability to read simple text and basic arithmetic. ASER 2012 report reflects the following data (Tables 1 and 2).

Table 1 ASER data 2012: children by class and reading ability (in %age)

Std	Not even letter	Letter	Word	Level 1 (Std I text)	Level 2 (Std II text)
1	43.4	37.6	12	3.8	3.3
2	20.3	35.9	22.8	10.9	10.1
3	11.9	26.2	23.2	17.2	21.4
4	7	17.6	19.9	20.9	34.7
5	4.6	12	15.3	21.4	46.8

Table 2 ASER data 2012: children by class arithmetic level (in %age)

Std	Not even 1–9	Recognize 1–9	Recognize 10–99	Can subtract	Able to divide
1	39.6	39.4	16.8	3.2	1
2	16.3	39.3	31.3	10.3	2.8
3	8.7	30.3	34.7	19.6	6.7
4	4.9	20.8	32	27.1	15.1
5	3.2	14.7	28.6	28.7	24.8

3 Primary Education in India-Problems

To improve the condition of primary education in India, the Constitution of India supported the Right to universal education till the age of 14, under the act of RTE which supports free and compulsory education for all the children. Despite of these efforts, the primary education still lacks in quality education. The major issues related to the quality of primary education in India are:

3.1 Low Enrolment Rate

India's failure for universal primary education becomes more apparent when compared to its neighbor country China, which has almost attained 100 % literacy. The gross enrolment rate in China is 99 % whereas in India its only 94 %.

3.2 High Dropout Rates

8 million children in India have never enrolled in a school and 80 million children have dropped out before completing elementary education (UNICEF 2013). Even after the implementation of RTE still, there is a huge dropout of children. Because a huge part of India's population is poor and prefer one more helping hand for their daily wages rather than sending a child to school.

3.3 Absence of School Staffs

Generally, government schools are not very wisely connected via roads/transportation, due to which teachers need to travel an awful distance [1]. These schools have huge number of students in a single room (usually two grades attend classes together) which creates unfavorable condition for a teacher to teach. 25 % of the teachers were absent from duty and only half of them were involved in teaching. The absence rate of teachers varies from 15 % in Maharashtra to 42 % in Jharkhand [2, 3] (higher rates in poor states).

3.4 Quality Teachers

As per the provision of sub-section (1) of section 23 of the RTE Act, National Council for Teacher Education (NCTE) introduced Teacher Eligibility Test (TET), as a mandatory exam to clear to qualify as a teacher. It was first conducted by CBSE in June 2011 which had following result (Table 3).

As per the above data provided by Ministry of State for Human Resource Development, D. Purandeswari, only 9 % candidates qualified the test. In 2010 there were 91 % teachers who failed to qualify TET, which further grew by 93 % in 2012, declared by "Firstpost.India" in April, 2012.

3.5 Teacher Student Ratio

An ideal teacher student ratio set by RTE is 1:30 up to classes I–V as per Indo Asia News Service (IANS), in 2012. Only 40 % of the total schools in India meet this ratio on 1:30. According to a survey (conducted by Pratham; NGO) Uttar Pradesh (lowest literacy rate) only 15.65 % schools, meets this ratio. Lack of teachers also contributes in, not meeting an ideal teacher pupil ratio in the country. Data related to number of teacher requires meeting the RTE norms in the following states are in Table 4.

Table 3 Teacher eligibility test paper—1 (classes I–V)

Category	Appeared	Qualified	% Qualified
General	255,480	35,768	14
SC	100,837	3,601	4
ST	30,435	515	2
OBC	213,001	14,864	7
Total	599,753	54,748	9

Table 4 Vacancies for primary teacher

State	Primary teachers required	Year
Jharkhand	18,600	2011
Uttar Pradesh	72,637	2013
Karnataka	4,000	2013
Bihar	1.25 lakh	2013
Maharashtra	34,000	2012

3.6 Rote Learning

English has been widely accepted as the medium of instruction in India. Being a foreign language sometimes, it becomes difficult for the students to comprehend. Lack of parental support and guidance; leads to the need of rote learning in order to pass the exams. This problem is also found in top ranked schools too [4]. Students who belongs to a good family background also under goes cramming practice because school management are under pressure to complete the syllabus and make their candidates eligible for exams. Rote learning deceives the thinking skills from a student.

3.7 Lack of Facilities

95 % of the total Indian schools lacks RTE infrastructure [5, 6] (2012, IANS data). One among the RTE norms, all the schools must have the listed basic facilities (specifically from class I to V); building (Office/store/office cum store, playground and boundary walls), drinking water, separate toilets for boys and girls, library and Mid-day meals [1, 7, 8]. ASER 2012 reports on above mentioned categories are [6]:

- *Building*: 73.5 % schools have the facility of Office/store/office cum store, 61.1 % school has playground and only 54.7 % schools have boundary walls.
- *Drinking water*: 16.6 % schools have no facility for drinking water, 10.4 % schools have the facility but no drinking water is available and only 73 % has the availability of drinking water facility.
- *Toilet facilities*: 8.4 % school has no toilet facility, 35.1 % has toilet but not in a working condition and 56.5 % schools have usable toilets among which only 21.3 % schools have separate provisions for girl's toilet.
- *Library*: 23.9 % schools have no library, 32.2 % schools has library but no books being used by children on the survey day, 43.9 % schools have library books being used by children on day of visit.
- *Mid-day meals*: 84.4 % school have kitchen shed for cooking mid-day meal and 87.1 % school were serving mid-day meal on day of visit.

4 Findings from Survey

Hundred eighteen teachers and fifty three parents participated in this survey which included questionnaires, personal interviews and casual talks.

- Many teachers fall under the age group of 25–40 years, with the educational qualification of B.Ed. in addition to the job experience ranging from 5 to 15 years.
- Teachers believe that students achieve fewer grades because of parents negligence at home whereas parents claims that there are lack of skilled teachers, so they need to put their kids for tuitions.
- Almost every student opts for tuitions for math's and science but very few prefer English tuition (approx. 0.1 %).
- Students are not able to read and comprehend due to which it becomes difficult for them to compete any competitive exams. Thus, concluding that all are getting educated but not literate.

5 Proposed Solution

Though, exams have proved to be a check point for the students before they jump into a new grade. But ultimately, text on the sheets decides the promotion level or their level of understanding which is biased. No one can be sure that the answers on sheets are result of student's understanding or rote learning. So, to cater the data's like literacy rate in India is baseless because literacy is meant for lucidity and skill. Several measures are being undertaken in order to solve the issues of education but not literacy. RTE, which talks about compulsory education, successfully met the goal of gathering students (to a limit) at a common place but failed in educating them.

English is the main medium of instruction in India so, it becomes mandatory for every student, that he/she shall be able to comprehend and pronounce it well. Lack of comprehension might lead student's career into an unfavorable condition after elementary level. Lack of skilled and experienced teachers do let the level of quality of education fall sharp. Currently, technology based products and services are making things a lot more efficient and easier for humans. So, after getting in depth insight into the various prevalent problems (Sect. 3), which are acting as hurdles in the path of imparting quality education, the author selected reading and pronunciation of English as the problem to work upon, using support of existing technology. Technology based methods can help in making students literate. One of the major drawbacks of the traditional method is that these are not interactive as per the demand of present situation. Technology based stationaries (e.g. gel pens) are replacing most of the traditional stationaries (e.g. ink pen). Hence, same can be in-cooperated for the methods of education with various interactive tools.

An e-book video, where text can be read aloud by the student/child with the help of an external device like head phones, and the animation gets initiated in order to describe the meaning of the sentence. A web portal consisting of English language books including course books (based on CBSE and ICSE) which can be downloaded on computers or on mobiles; story wise. This e-book video will be a collaboration of animation and voice recognition tool. A person reading a book from the website/app needs to wear a head phone consisting of microphone. The screens of the story will consist of texts, which will start getting highlighted once the user starts reading it correctly followed by the animation. This will help the user to comprehend the meaning of the sentence while reading. Reading more sentences will make the animation go on. The story will stop if the user pronounces the word in a wrong way. The words will stop getting highlighted from the point where the wrong pronunciation started and so the story/animation. There will be an instruction displayed on the screen that the pronunciation made is wrong so, '*listen to the voice carefully and correct it*'. There will be a pre-fed voice, to instruct a child for a correct pronunciation. Once a child pronounces it correct, the story will start unfolding with the text read. Voice and animation are inter-related in the video. Thus, the story will be running along with the voice of the reader. It holds child interest to know about the story. Animation excites a child as it makes them feel that they are building the story as per their wish.

5.1 Targeting Issue

Students reading skills will get better without their knowledge with addition to their interest. Usually, children learn best when they don't know they are learning. The curiosity level is maintained in each step which keeps them involved during the whole process. Nowadays, new technology is always fascinating for kids and mobile is one of them. Major concerns behind the proposal are the instructiveness, involvement of a child, self-assessment, intuitiveness and comprehension.

5.2 Students Perspective

Proposed solution has a feedback feature in it which actually helps in self-assessment. Feedback is bias free. There are many self-reading books available in market but they lack in self-assessment issue. Audio CDs are also available but they too do not fill the gap of comprehension and pronunciation as well as full time user's involvement.

5.3 Scope

Fulfilling basic facilities to a school is a challenge and getting new systems for literacy sounds an impossible scenario that too with continuous electricity in rural areas. There are other modes like Akash tablets which is a good alternative of computers. Low cost mobiles with latest technology in market are a good supplement too.

5.4 Limitations

The following are few limitations in developing the proposed design solution:

- Availability of good sound calibration software.
- Regional based accent of same language.
- Hardware is a must requirement (mobile phone/tablet/computer).
- Huge data conversion from text books to application.

Fig. 1 Logo

Fig. 2 Website screen shot

Fig. 3 App screen shot 01

Fig. 4 App screen shot 02

5.5 Visuals

Youtube link for the video: http://www.youtube.com/watch?v=9EfnUkLd6mw (Figs. 1, 2, 3 and 4).

References

1. The Earth Institute of Columbia, CGSD Working Paper No. 11, Feb 2004, Nirupam Bajpai and Sangeeta Goyal, Primary Education in India: Quality and Coverage Issues
2. Kremer, M., Chaudhury, N., Rogers, F.H., Muralidharan, K., Hammer, J.: Missing in action: Teacher and health worker absence in developing countries. J. Econ. Perspect. **20**(1), 91–116 (2006)
3. Chaudhury, N., Hammer, J.S., Kremer, M., Muralidharan, K., Rogers, H.: Teacher absence in India: a snapshot (2004)
4. The Times of India, Bengaluru, Rote learning prevalent in top schools too, 13 Aug 2012
5. Right to Education, School Infrastructure
6. Annual Status of Education Report (Rural) 2012, ASER, 17 Jan 2013
7. Rte Maharashtra, Rte Schools, Norms for schools and school facilities
8. Chandra, S.: The Times of India, Government Primary Schools Lack Basic Facilities, 3 Apr 2013

The Difference in Design Problems and Its Effect on Divergent Thinking in Middle School Children

Anisha Malhotra and Ravi Poovaiah

Abstract Problems are not equivalent, in content, form, or process. Still very less research exists on the nature of design problems children should be solving to learn the design process or to build a certain kind of creative ability for divergent thinking. In this paper we will concentrate on exploring the differences of solving different design problems and its result on divergent thinking in children. The paper presents an investigation of middle school students' difference in design thinking process and design output for four different kinds of design problems. Twenty children of the age 11–14 participated in this study in four different groups. Each group solved a different kind of design problem. A qualitative analysis of the responses was done and participating groups were judged on the parameters of understanding the design problem and ideational flexibility in their responses. Preliminary analysis of the experiment revealed students' lacked motivation to think beyond one solution. But the comparative study shows difference between design outputs and number of variety of solutions generated for different design problems. The findings reveal that each design problem has its unique traits and it is difficult to choose one over the other. The paper discusses insights based on the outcome of this study and suggests a novel way of designing the instruction and the design problem in such a way that it encourages children to think creatively and to generate more solutions. The present study would provide opportunities to develop design curriculum and try out more and different design problems with students.

Keywords Collaboration · Divergent thinking · Problem solving · Design pedagogy

A. Malhotra (✉) · R. Poovaiah
Industrial Design Centre, Indian Institute of Technology—Bombay, Mumbai, India
e-mail: anisha.malhotra.iitb@gmail.com

R. Poovaiah
e-mail: ravi@iitb.ac.in

A. Chakrabarti (ed.), *ICoRD'15 – Research into Design Across Boundaries Volume 1*,
Smart Innovation, Systems and Technologies 34, DOI 10.1007/978-81-322-2232-3_55

1 Introduction

Problem solving is not a uniform activity. Problems are not equivalent, in content, form, or process. For many years, researchers [1, 2] have characterized design problems as ill-structured due to the complexity inherent in them and the difficulties associated in determining their constraints and requirements. An important aspect of problem solving is problem framing as suggested by Schön [3]. Problem framing involves making sense of the problem by the designer by imposing a frame on the situation and discovering consequences and implications of those chosen frames.

Most of the current research in design education seems to be defining the nature of design problems for high school or undergraduate design students with a focus on entering the professional world of design. Very less research exists on the kind of design problems children should be solving and why?

Many educators and researchers argue that design by students in schools is different from design by professional designers at workplace. Anning et al. [4] contended that the constraints present in schools are different from those in the workplace. With constraints such as curriculum, examination and assessment requirements, Anning et al. mention that although the activities in schools are aimed at individual students, the resources are shared among twenty or more students unlike the designer's workplace. It has also been observed that while designers work on real problems in a highly contextualized situation for which they have considerable knowledge [5, 6] and thereby have predefined goals to resolve, the design problems in schools are artificially constructed and are not relevant to students, thereby becoming meaningless to them.

Children learning design and design thinking have a different goal i.e. to develop their creative, cognitive thinking and social skills to prepare them as creative individuals and not professional designers always. Largely, research on design problems has been in the area of engineering design or software design, which is well-structured than most design problems. It is not a generalizable model for patterning design problems across age groups and disciplines. Archer and Roberts [7] claim that design in schools is different from design in profession because of their different aims. They suggest that the aims of design educational activities are not to produce things and devices but the development of knowledge and understanding about design and the technological world.

2 Design as a Problem Solving Process

The term 'problem solving' in design education is synonymous to teaching design in schools. This approach has led to the development of stag-wise, algorithmic processes of design, from linear design processes [8] to complex loop of interactive processes (e.g. Kimbell, [9]). Proponents of design to be taught as problem solving tasks claim that it involves a general strategy which can be applied in a variety of differing contexts, and even in differing domains [6]. Design is not always problem solving it can be purely aesthetic in nature. McCormick [6] provides several problematic concerns when teaching design through a general problem solving approach. Teaching design with a general problem solving approach has also led to teachers' emphasis on teaching the procedural understanding rather than the conceptual understanding of design and creativity. Design involves a special kind of problem solving as it resolves ill-structured problems by converting ideas into products or systems. This was the main objective of our study to investigate different kinds of design problems to be explored for children in the age of 11–14 for design education (divergent thinking) and collaborative ideation.

3 Aims of the Study

The aim of the experiment was to investigate the following:

1. To investigate the difference in performance as individual versus collectively solving the problem.
2. To investigate the difference in design thinking process and design output for four different kinds of design problems.

4 Sample Description

Twenty children of the age 11–14 from same socio-economic background, voluntarily participated in the experiment. Four groups were formed with almost equal distribution of girls and boys. Each group was given a different set of problem and each individual got a color coded diary to answer. Each group had five members and was asked to solve the problem individually first. Later, each group was provided a new color coded sheet to be used for collective ideation and presentation of their final solution.

5 Experiment Design

The experiment was divided into four phases:

Phase one (20 min): Introduction to 'What is Design?' Design as a tool for problem solving.

Phase two (30 min): Design Activity (Solve individually): Four groups were formed with five children in each team. Each team was assigned a researcher-team member to record and observe the activity. Each child solved the problem on his/her own in their respective diaries.

Phase three (30–35 min): (Solve together): They were asked to solve the problem collectively as a team by listing down all the ideas they thought of and then selecting one final idea for presentation.

Phase four (20 min): Final presentation: All the groups presented their ideas by first introducing their team members and sharing their problem, followed by the final solution.

The groups were colour coded. They were all provided with a pencil and a colour coded paper diary which was folded in such a way that they could either use it like a page wise diary or a single sheet of paper. The method to open and use the diary was demonstrated to children by the researcher. Four different problems were given to children. Each group then was given one problem each. One group solved only one problem. All the problems were different in nature as explained below. The problem were given verbally in Hindi and Marathi first and then written in Hindi on the blackboard for reference.

Described below are the problem definitions and differences in terms of openness of the problem, *structure* (all problems we chose were ill-structured and open-ended with no particular procedure to solve), domain specificity and complexity

Pink: **Group 1**: **Product based problem**

Design Problem: *Design a new school bag or a water bottle for yourself* (Fig. 1).

Definition: Product based problems are the ones where a final product or an artefact is aimed to be achieved. The product to be designed has an assigned and defined function or use which children are fairly familiar with. The product mostly exists, is known and a familiar object of use which is being asked for a (re)design for better performance. Also, there is an assumption that because you are so familiar and use it every day you will be able to think of improving the product as per your own needs.

Green: **Group 2**: **Situation based problem**

Design Problem: *An old grandmother loves to knit sweaters but her wool roll keeps falling down. Suggest (Design) as many ways to help the old grandmother as she has a severe back problem (she cannot bend)* (Fig. 2).

Definition: A situation based problem is more open-ended than a product problem. This provides the opportunity to the designer to think of a variety of solutions and not restrict himself to a product/artefact- oriented thinking. The solutions are possible methods like a service, product, a combination of both etc. The objective

Fig. 1 Samples from product design team members

Fig. 2 Samples from situation based problem team members

is fairly clear as to what the problem is which needs to be solved. The problem is picked from children's everyday life like a classroom, home, problems of people they can easily relate to like a friend, family members or teachers. The main task is to understand the obstacle or the problem which is situated in an existing situation (most of the times) and think of ways to improve the problematic situation.

Red: **Group 3**: **Edward de Bono's design problem** [3]

Design Problem: *Suggest ways how you would stop a Dog and a Cat from fighting.*

This is a basic political problem. How to stop people with differences from fighting each other? The starting situation is very definite- there are cats and dogs which are distinct and which fight each other. The objective is also very definite- how to stop them fighting. What means would children use to try achieving this objective? As there are no traditional, stereotyped ways of stopping a cat and a dog from fighting, the children would have to think and solve the problem on their own. (Fig. 3)

Definition: According to De Bono [3], problem solving is quite easy if you are given a definite objective which has to be achieved. Problem-solving is also quite easy if there is some obvious deficiency in a design and you are asked to get rid

Fig. 3 Samples from Edward De Bono problem team members

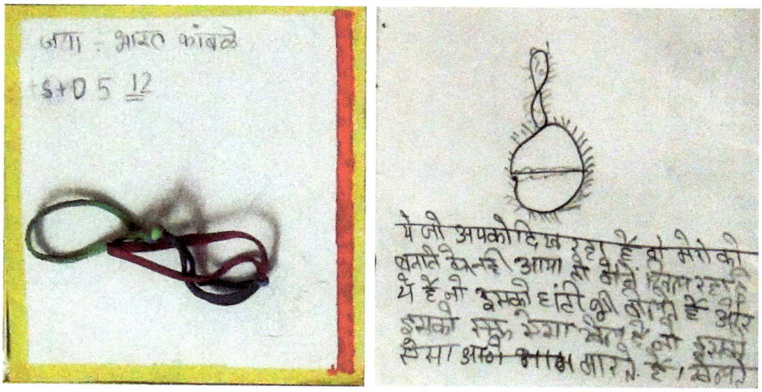

Fig. 4 Samples from game design problem team members

of that deficiency or fill in some gap. Problem-solving is rather difficult when all you are given is a general idea that you should improve an existing design. This becomes even more difficult when the existing design seems to be a very satisfactory one, as in the case of an everyday object, a bicycle.

Yellow: **Group 4: Game design problem** (**Kafai** [10]—**Learning design by making educational games**)

Design Problem: *Design as many games as you can, using a ball and three rubber bands. The game should be played between* 2–3 *children together* (Fig. 4).

Definition: Designing a game is not same as designing a toy (a product). The game will have rules, more participants and social interaction. Games are active experiences which children are involved into almost every day, and they have the capacity to provide intrinsic motivation. In a game design problem children work from beginning till the end, go through the entire process of problem solving, they get to interpret the problem, explore solutions, conceptualize and think of visual design as a package. Hence, as a design problem it offers a diverse variety of thinking strategies and design approaches to enable divergent thinking. Also, a game design problem can be solved in parts and brought together. It also encourages group participation as the task to ideate and produce the game is big and needs to be shared within the team members. Game design also gives more opportunity of combinations of different ideas in a group which may lack in a product design problem.

6 Findings and Discussion

Analysis shows a difference in understanding and performance of each group in different design problem groups. The groups have been analysed on their understanding of the task and the requirement that they had to generate as many solutions for the problem, ideational flexibility within the group and difference between individual and collective sessions.

Table 1 Understanding the problem and the task to think of more than one solution

Problem	Task	More solutions	Observations
Pink: product	✓	×	They did not understand what they have to make. They struggled for a long time to think. Almost all of the solutions are on beautification.
Green: situational	✓	×	One girl thought of a solution and others followed. Only material changes were made. One boy focused on drawing than thinking of alternatives.
Red: De Bono's	✓	✓	Children understood the problem well and came up with multiple solutions.
Yellow: Kafai's game design	✓	✓	Children understood the task and explored with the material. But not all were able to do thetask individually.

6.1 Understanding of Instruction and the Task to Be Performed

Tick mark in Table 1 above means the children followed the instruction. The colour of the tick mark is to indicate the scale of understanding i.e. if the problem was partially understood then a light colour tick mark is used. Whereas, a cross implies they did not follow it completely or understanding was not appropriate.

Children with their limited exposure to formal creative problem solving tasks and open-ended design problems tried to solve the problems. There was a clear difference in the nature of problems which led to generating more solutions like De Bono's cat-dog and game design problem when compared to the product and situation problem where not many solutions were generated. Also, children struggled in understanding the task to be performed and how to solve it.

6.2 Ideational Flexibility: Number of Qualitatively Different Solutions Produced Appropriate to the Problem

Analysis of the sketches, problem solving process and conversations revealed children performed better in problem solving collectively than individually. More discussions and approaches were tried to think of solutions in a group. In the product design problem the score of ideational flexibility was 3 individually and 5 on collective work. The situation team scored a 3 both individually and collective. De Bono's cat and dog fight problem scored an 8 individually and an 11 on the collective session. The game design problem scored 4 in the individual session and a 5 collectively. One of the reasons of better performance in a group could be looking at others solutions in the team and comparing it with your own solution which led to exposure to multiple possibilities. Collectively, they were able to distribute responsibilities, discuss, demonstrate, elaborate and interpret the problem which may be was a limitation when they were solving the problem alone.

6.3 Unique Traits of Different Problems

We found out in the earlier snakes and ladders problem [11] and school bag problem both, the object of redesign was so familiar and probably children were very satisfied with the product that they found it hard to think of alternatives. This may be because a clear gap or a deficiency was not provided as part of the problem and was expected from the children to find their own deficiencies in the product and hence improve. The school bag was sketched as a jhola and no functionality was added to the bag. There were only decoration changes made to the new bag. The reason for this could be the fact that they carry such cotton bags or polythene bags. Later on talking with them we found that was not the case but one of the research team members was carrying a jhola so they discussed and thought of making the same with different designs. Everyone in this group chose to draw as against writing.

In the *situational problem* only one of them (girl) was a thinker for the group and she told everyone to either copy her or make slight changes. The others very comfortably agreed and followed her. All the girls suggested using either a box or a bag (change of material- steel, plastic, cloth) for the old grandmother. In the collective session also, the group did not interact much with each other for ideation or any improvement for their final solution. They did not seem very interested in the task.

The *De Bono problem* group was an active group of five members (two boys and three girls). There was a dislike within boys and girls. It was hard to make them work together. Till problem solving was at the individual level, there was pace at work but collectively they had to compromise and keep their egos aside to work together to win. Main themes were distraction, fear of someone more powerful and greed or by providing something that makes the animal happy (a woolen ball to play with). Another interesting solution was of bonding, love and getting them married. Together in the collective sessions in order to win they worked politely but with differences of opinion which resulted in newer solutions like a see-saw.

In the *game design* group, the girls were stuck with putting rubber bands on the ball and kept drawing the same, whereas, the boys were curious and attempted to explore options and think from a game point of view. But once they all came together, they started with listing their ideas which led to questions, discussions and explorations as one was explaining his idea to the group members. They started out blank but a lot of demonstration happened to explain each other. This led to discussions, arguments and a few but more ideas.

This study led us to think about the role and need for a 'construction set' in design problem solving process? Is this an important factor leading to solutions?

In the game design problem, they may have used the material provided as a lead to design thinking. But on the other hand, there is also a possibility that they were overwhelmed by the materials and kept thinking within the constraints of the material. This may also lead to use of material only once and hence most likely to produce only one solution to an open ended problem. Our aim is to engage children in such a thinking process that they spend more time thinking about the problem and possible ways to solve the problem than to get tangled in the construction sets provided. The design without-make unit was initially proposed by Barlex [12] and is based on Young Foresight. This is a recent design and technology initiative in England. It challenges orthodox approaches to teaching design and technology which rely on design and make assignments where a construction set is required. The design-without-make units are focussed practical tasks and product analysis exercises where pupils design but do not make; pupils design products and services for the future; pupils use new and emerging technologies in their design proposals; pupils write their own design briefs and pupils work in groups.

Barlex's study is a longitudinal study with a formal 6 weeks training module. Whereas, our study with protocols will focus on short term brainstorming kind of sessions and hence generate insights only on such tasks. The same collaborative conditions and proposed structures should be tested separately for longitudinal studies and formal design education.

7 Conclusion

As the first exposure to formal creative problem solving, the children showed a potential for creative work and collaboration. Most of the individuals took time to understand the task and its requirements. The understanding and ideation improved when children formed groups and discussed each other's solutions. Children lacked motivation to think beyond one solution. One solution solves the problem so why think of another one was the attitude. With instruction and exposure to creative problem solving, children should be motivated to think for alternate solutions.

The comparative study clearly shows difference between design outputs and number of variety of solutions generated for different problems. The findings reveal that each design problem has its unique traits and it is difficult to choose one over the other. Each problem type can be designed in a way that it encourages children to think creatively and construct more solutions. Game design as a design problem opens up a range of design elements that can be thought of for creative thinking. In conclusion, we have designed a structure of design problem such that it includes pros of all the problems tested in the study. In our later studies with the same children we are testing combination problems with base as a game design problem which is situation based and familiar to children with a twist of curiosity to solve for the unknown. We also recommend that other problems can also be explored not as combination and tested by improving the instruction and problem.

References

1. Reitman, W.: Cognition and thought: an information processing approach. Wiley, Oxford (1965)
2. Simon, H.A.: The Structure of ill structured problems. In: Artificial Intelligence, vol 4. Springer, Berlin (1973)
3. Schön, D.A.: The reflective practitioner: how professionals think in action. Temple Smith, London, UK (1983)
4. Anning, A., Jenkins, E., Whitelaw, S.: Bodies of knowledge and design-based activities. In: Smith, J.S. (ed.) IDATER 1996. Loughborough University, Loughborough (1996)
5. Hennessy, S., McCormick, R.: The general problem-solving process in technology education: Myth or reality? In: Banks, F. (ed.) Teaching Technology. Routledge, London (1994)
6. Hennessy, S., McCormick, R., Murphy, P.: The myth of general problem-solving capability: Design and Technology as an example. Curriculum J. **4**(1), 73–89 (1993)
7. Archer, B., Roberts, P.H.: Design and technological awareness in education. Stud. Des. Edu. Craft Technol. **12**(1), 55–66 (1979)
8. Banks, F., Owen-Jackson, G.: The role of making in design and technology. In: Barlex, D. (ed.) Design and Technology for the Next Generation. Cliffeco Communications, Whitechurch (2007)
9. De Bono, E.: Children Solve Problems. Harper and Row, New York (1972)
10. Kafai, Y.B.: Playing and making games for learning. In: Games and Culture, vol. 1, pp. 36–40. SAGE Publications, New York (2006)
11. Malhotra, A., Poovaiah, R.: Developing young thinkers: an exploratory experimental study aimed to investigate design thinking and performance in children. In; ICoRD'13 4th International conference on Research into Design, 2013 at IIT Madras (2013)
12. Barlex, D.: Young Foresight. Young Foresight, London (1999)

Design in the School Classrooms—Applying Design Tools to Improve Quality of Education

Kshitiz Anand and Jean Haag

Abstract For Design to change from its elitist image, there is a need to implant the value of design from a young age, because design is not just elite. The impression of Design to be a tool for Problem Solving (in its analytical, rational and intuitive thinking) is of most importance. There can be no better place to inculcate the values of Design than in the schools. Fundamentals of Design, when integrated with the pedagogy, can result in better forms of learning and a shift from the traditional forms of rote learning. Tools like Visual minutes, understanding typography, design thinking patterns, critical analysis, gamification, etc. can result in a better way of problem solving and hence learning. We applied these through different activities held in schools across India and across different demography and take inspiration from the findings of these to present in this paper. In the end the paper proposes tools from the world of design that can and should be adapted to the activities done in the schools, in order to present a better learning environment and quality.

Keywords Design education · Schools · Pedagogy

1 Introduction

Design has traditionally been synonymous with the thing of the elite. However of late there has been enough impetus towards design having a larger say in the improvement of the quality of life and solving the world's problems. There has been extensive emphasis on education itself, with the United Nations also defining the Right to Primary Education being one of the Millennium Development Goals [1]. Moreover, there has been a constant clamor around designers needing to work

K. Anand (✉) · J. Haag
L'Ecole de design Nantes Altantique, Bangalore, India
e-mail: k.anand@lecoledesign.com

A. Chakrabarti (ed.), *ICoRD'15 – Research into Design Across Boundaries Volume 1*,
Smart Innovation, Systems and Technologies 34, DOI 10.1007/978-81-322-2232-3_56

for the other 90 %, that involves creating design solutions (products, systems and services), that provides value to a larger population [2]. Whiteley [3] in his book 'Design for Society' calls for design to come out of its consumerist values, consumer-led design approach too.

The success of Massive Open Online Course (MOOC) platforms like Coursera, Udacity and edX, have made premium educational content available to a larger audience around the world for free. Meng Sang Chew and Kelly Grim explore the effect of MOOCs on six key stakeholders—*course instructors, course providers, students, parents, colleges and universities, and society, including the government agencies that subsidize higher education*. The rise of MOOCs poses a challenge to the educational and business models of traditional universities [4].

The success of Khan academy that uses online videos as a platform tool towards learning by reaching out to millions of students for education is commendable too [5]. IDEO's Toolkit for educators, helps one design meaningful solutions in the classroom, at school, and in community, is primarily aimed at teachers and encourages them to explore it as a key offering in the classrooms [6].

Mitra [7], researcher of Education Technology and best known for the '*Hole in the Wall*' project, explores learning through '*Minimally Invasive Education*', in which children operate in unsupervised environments provided with tools of learning. In the absence of supervision or formal teaching, children can teach themselves and each other, if they're motivated by curiosity and peer interest. Taking his research further, in his TED prize acceptance talk Mitra says "*My wish is to help design the future of learning by supporting children all over the world to tap into their innate sense of wonder and work together. Help me build the School in the Cloud, a learning lab in India, where children can embark on intellectual adventures by engaging and connecting with information and mentoring online. I also invite you, wherever you are, to create your own miniature child-driven learning environments and share your discoveries*." [8].

The common approach in bringing about innovation in education is through usage of technology. By the end of 2014, there will be almost 3 billion Internet users, two-thirds of them coming from the developing world, and the number of mobile-broadband subscriptions will reach 2.3 billion globally. Mobile-cellular subscriptions will reach almost 7 billions by end-2014, and 3.6 billion of these will be in the Asia-Pacific region [9].

One has to be careful about the usage of technology and its adoption. The failure of the 'One Laptop Per Child' project is something that cannot be forgotten easily. With increased cost of operations, and the cost of manufacturing itself going up, unique proposition that the project possessed was lost [10].

Our paper proposes that a creative and '*designerly*' approach to learning is needed in our centers of learning. The emphasis is to enable a shift from association-based to causation-based thinking, which facilitates the fine-tuning and manifestation of the creative work [11]. The paper as aimed not just at designers but also at educators, other stakeholders (government, policymakers, parents, NGO, Company CSRs etc.).

2 Education

2.1 The Present State of Education

With changing lifestyles and pressure of work, parents are lesser involved in their kids' education as before, and leave the responsibility to the academic institutions. With the quality of education being questionable [12], government bodies or private institutions are struggling to find new ways to rectify this [13]. Research also shows that the quality of graduates from colleges is not up to mark, and one need to question the origin of the problem too [14].

What is perceived to be the 'right' method of education is often enforced upon the children, who do it unquestionably, due to lack of self-expression opportunities [15]. Another grave situation is in the shortage of qualified teachers. Even from the ones that are currently imparting education, too many remain under-qualified [16]. Having this problem at the very formative years of education is hugely detrimental to the improvement of the education system [17].

Organizations like *Teach For India* and the *Azim Premzi Foundation*, attempt to solve this through their teacher fellowship programs, but the numbers are still very low. Education should aim at helping kids find their identity, through a more holistic approach towards education, and that is often missing in the current generation of teachers [18].

2.2 Wicked Problem and Why Education Is a Wicked Problem

The state of education is a much more complex problem than it sounds. One appears to be looking at a very linear approach towards problem solving, but in reality it is not, for it is a wicked problem. Other challenges designers face includes the difficulty to extract one particular problem and define it properly, and propose a non-interfering solution regarding co-related problems [19].

3 Education as a System

A system, prior to any design intervention may be considered as composed of two elements: context and behaviors, where the context includes the sociological, political, traditional and psychological background of the said. However a design-enhanced system is a proposal of a new set of behaviors involving the entire community and aiming towards a common goal facilitated by the implementation of a vector of change.

We draw a system based on three core elements. First, the stakeholders (S) are the concerned community where the work is happening. Then the vector (V), or artifact, represents the design intervention, be it a space, a product, or a service. Lastly, actions (A) are the result of the system's influence over stakeholders.

Actions (A) are produced either by (i) stakeholder - stakeholder interaction, (ii) stakeholder—vector interaction or, (iii) vector—vector interaction. Each is different ways to solve the given problem and choosing the right one remains the designer's responsibility, keeping in mind the final goal.

4 Design Thinking

Design Thinking acts as a guideline for various kinds of project, and enables the practitioner to define his/her own angle of attack. In the context of education, it is important to ensure that the various stakeholders understand aspects of Design Thinking.

1. System-based thinking approach
2. Keeping in mind each stakeholder
3. Working with constraints
4. Working with a methodology

5 Design Intervention Points

We analyze the state of education by looking at the different components within the learning environments, and understand where design could intervene.

5.1 Existing Classrooms and Learning Environment

As of today, majority of classrooms can be summed up in three elements. The first one is a large communication surface owned by the teacher, the second is the furniture on which the students sitting face the above mentioned surface. The third is often the neglected part of the learning environment, namely the walls and other surfaces within the school premises [20]. Design intervention is required in all these elements.

5.2 Interactions in the Classrooms

Design plays a role in the '*interactions*' that happen within the educative environment. We look at the possible human-human interactions and the human-object interactions for design intervention. Today the '*knowledge flow*' appears to be very unidirectional: from the teacher, through the board, and finally to the children and their notebook. The children are made to 'remember' things. These form the very basis on rote learning that we seek to reduce.

6 The Tools in Design—*Designerly* Ways of Knowing/Learning

The following sub-sections are presented in a format to provide the importance of the tool in education (the why, what) and to show insights into how the designers have been using it. The tools need to be introduced not just to the students, but also to the other stakeholders.

6.1 Brainstorming

Designers love to brainstorm and it should be encouraged as a core activity of all learning process. Here are some commonly agreed points about it:

- No idea is a bad idea and it encourages everyone to participate
- Generate as many ideas as possible
- Understand doing things with time constraints by having a deadline
- Learn to group together ideas and build onto others' ideas
- Start with the topic (generally a noun, or a sentence explaining the context) at the centre, and write all around it ideas while organizing them in sub-themes

Introducing brainstorming to children at a very early age is important. It promotes the *'divergent mode of thinking'* that is popular amongst design professionals. Introducing a totally *'constraint-free brainstorming'* session to generate a lot of broad, raw ideas shows a more participatory approach to learning. Doing this permits one to generate out-of-the-box ideas that may not be consistent or realistic, but it encourages innovative thinking. Adding constraints progressively can help sort these ideas out, or bend them to fit into the desired solutions and frameworks to facilitate better learning.

6.2 Visual Note Taking

Young children are already used to the aspect of Visual Note Taking, albeit in a different capacity. However, today the act of drawing or doodling is considered bad after a certain age, and children are often discouraged from doing so. Artistic communication such as Visual Note Taking, and acquisition of other skills like reading and writing is key towards overall child development [21].

Visual Note Taking is to take notes by writing keywords, making doodles, illustrations (collectively called information bits) and finding connections between them using nouns, verbs or adjectives. These are represented by arrow marks to give a sequential processing of information. It also allows the information to be remembered for a longer time. Concepts are structured around connections that they make with the other parts of the chapter.

The teacher (often the one who discourages any form of artwork) needs to avoid being self-judgmental about the quality or aesthetic appeal of the visual notes [22]. Embracing children's creative potential and structuring it to serve the class can result in a fruitful primary outcome: children 'play' while taking notes, and for this particular reason the said lesson might stick in their brains more easily [23].

6.3 Understanding Visual Order and Information Processing

When children are introduced to a large amount of information, they feel intimidated. Textbooks and study materials use this classification of information to make the information easy to grasp.

The human brain analyses this knowledge as bits of information either sequentially or in parallel. When information is perceived to be of an established hierarchy (visual order) the mind processes it in a sequence and when it perceives it to be of the same order, the mind analyses it in parallel.

Humans have known to get more interested in consuming information through images rather than text. Even within texts, the ones that are bold (highlighted) get the higher attention and a hierarchy is established.

Understanding Visual Order and fundamentals of typography facilitate the learning process by identifying what information is important and needs to be consumed first. This tool would also encourage the children to know precisely what they need to 'revise' when the time for assessments comes.

6.4 Socratic Questioning

The essence of Socratic Questioning lies in the ability to ask the right questions, rather than telling the answers. In a classroom setup, the teacher assumes ignorance for a certain topic, and let the children answer the questions. This would naturally help the audience build the answer by themselves and extract knowledge out of it [24].

Designers have used this practice to refine and develop concepts with a strong Design Rationale. Often presented as the '*Why Game*', it follows a simple rule, which is to question everything. This is a natural behavior for kids and every answer they are given leads to a new question asking 'Why' the answer is what it is?

6.5 Storytelling

Children are fascinated by stories and it influences their language development and story comprehension at a young age [25]. Meaningful experiences, during these early years, can provide language opportunities to enhance and sustain its growth as well [26, 27]. Early childhood teachers can provide appropriate occasions for young children to play with language, while gaining an appreciation of the sounds and meaning of words [28].

The aspect of storytelling is deeply imbibed in the profession of design, for every idea that needs to be shared, is conveyed through a story and has to be presented before a wider audience.

6.6 Gamification

In a gamified lesson kids actively generate and capitalize knowledge, thanks to a sense of competition pushing them forward. Also, they are more likely to develop strategy elaboration, leadership and teamwork skills. In their activity, designers use principles of gamification to generate engaging and experience-driven solutions.

6.7 Rapid Prototyping

Switching to a hands-on activity (prototyping) is an alternative way to the talking mode of learning, and can be much more appealing for the kids. The fact that they are able to see the lesson being built before their eyes, the strong visual component is of most importance for kids development and confidence building.

We highly recommend classrooms to be equipped with a broad set of stationery and other materials that would help children prototype. This includes paper and cardboard (with numerous variations of size, color, thickness), as many different pens and pencils as possible, raw materials (clay, wood, fabric, plastic), paint, scrap items (water bottles broken household items).

6.8 Critical Thinking

Across countries, a common trend can be found regarding the aim of education: it focuses on accumulating raw knowledge. In such a system rote learning becomes the key skill to master. The curriculum becomes the main motivator for the teacher, whose concern lies in completing it by the end of the school year.

A widespread trouble, consequently to this, is the very short life span of the acquired knowledge: the endeavor is placed on the next test, not on remembering it for one's own sake and a less emphasis on the application of the accrued knowledge.

By introducing Critical Thinking to schools, we believe that children would take ownership over their instruction [29].

Proposal of a new class template It is important that education be looked upon with a System-based approach. This is also very different than just trying to improve the quality of the study materials or add more human resources (teaching staff), as has been the case often and repeatedly attempted in the past.

In order to improve the quality of education we look at three different things. We question the process of learning and what is learning itself (a), the second is to look at the stakeholders involved in the learning process (b) and the third is the space in which the learning happens (c). Learning for us therefore is defined as a function of the three above-mentioned components.

We are confident that the overall development of the child will be higher as compared to the outcomes of the existing education system. What we propose is not a new, but stronger application of the tools of design into education that could be used both by the different stakeholders. The tools when implemented would have both a short term and long term impact. We map out the tools discussed with the impact it would have on the child's learning.

Student's cognitive ability: *Brainstorming, Critical Thinking, Socratic Questioning*
Student's skills: *Rapid prototyping, Gamification, Understanding Visual order and Information Processing, Visual Note Taking*
Student's personality development: *Storytelling*

7 Case Studies

We present here two case studies of different workshops, where we tried to apply our proposed design tools to existing education systems. It should be noted that we do not aim to conduct only short-term workshops. The workshops give us first hand experience of the acceptance or rejection of our ideas being proposed. The long-term impact will be felt when schools at the system level adopt the proposed tools and we evaluate it after a few months. The workshops are the first step in a long journey but with a clear direction.

7.1 Design Thinking and Innovation Workshop with 9–10 Year Old Underprivileged Children in Pune

Design was introduced on a broad level, as a process that can help find better solutions to daily life problems. A participative talk was launched at the beginning of the workshop, with questions like: *How to solve a problem? What is the very first thing to do when facing a problem? How to know if a solution is good? What to do when a solution doesn't work?* The idea behind this introductory exercise was to understand the traditional approaches to problem solving that was being taught in the school. This concluded and summarized with a simplified three steps Design Thinking process, which was explained to them: *Inspiration, Ideation, Implementation*. During the workshop, one also discussed on involving local stakeholders, did rapid prototyping, looked at embracing constraints and emphasized on out-of-the-box thinking. Innovation was also introduced with few game-based activities.

To test the newly transmitted knowledge, we did a design problem solving session, with the topic "*How to make their schoolbag less heavy/less hard to carry*?" Pupils were given problems requiring them not to focus on the most obvious solution, but rather spend few minutes beforehand to think about the best way to access it. In groups of 5–7 pupils, they brainstormed, proposed several solutions, chose one, and developed it. At the end, they were asked to make an elevator pitch for the others (along with a drawing exposing their project) while teachers turned into examiners, asking questions and giving marks.

7.2 'Fun with Type' Typography and Visual Order Workshop with 10–15 Year Old Children in Delhi

A workshop on understanding different types and fonts was organized for students from the class 3–10. The workshop intended to sensitize the students with the nuisances of Typography, as an essential tool to understanding Visual order and the representing of information. The students had to relate the understanding of Visual Order to their own academic books.

To test the newly imparted knowledge, we did a poster-making workshop with the theme being India. The students were to take decisions on what fonts to use, where to use and had to come up with answers on why they use it. Immense workshops materials were provided in the form of old newspapers, magazines, glue, A3 paper, colored paper etc. The students had to cut the relevant fonts, text, images from the newspapers to create a poster that depicts the theme.

Our observations from both these workshops were that the children were immensely engaged in the activities and looked very excited when introduced to these new tools of design. Almost every student participated in the activities and it confirmed our belief on using the tools of design in education in these short workshops.

8 Conclusion

The rise of initiatives by organizations and individuals attempting to bring about this change, by virtue of proposing and trying out new things, with or without the help of technology is highly encouraging. It is with this hope that we propose 'Design' to have a larger say in the schema of things.

The tools of design, as outlined in this paper, are tried and tested by designers worldwide and have been in use since long. When we look at adopting and leveraging these tools in the domain of education, and primary education in particular, emphasis has to be towards more teachers and students believing in them and using them on a regular basis, to the point that it is a core element of the education system. The results will not be felt immediately and it could also be faced with resistance. This is to be truly worth it, for its time is right. In that perspective design becomes a great enabler towards ways to create innovative educative solutions.

9 Future Work

The work we have started is not finished, for have barely scratched the surface. What started out as experiments is now a proper research agenda. There remains a lot to be done, for the scale that it exhibits. We are just looking at the primary school education system for now, but with proper contextualization; these proposed tools could be then adapted to the education at the secondary, college levels of education and also at the skills-based learning. Measuring the implementation's impact of these frameworks would need to be done in order to make a stronger case for design and its importance in the education and pedagogy. Necessary tie-ups would be required with a lot more non-government organizations (NGOs) and social organizations interested in and working in the field of education. We would be reaching out to the main stakeholders in the government and their run schools, in order to outreach to a larger number of schools and children, that can adopt the proposed frameworks to bring about a different way of teaching in the schools on a larger scale. It is only by doing so; we would be able to objectify our results and findings.

References

1. United Nations Millennium Development Goals. (www.un.org/millenniumgoals)
2. Juul-Sorensen, N.: Designers design for the 1 %—It's time to start designing for the 99 %. The Guardian (http://www.theguardian.com/sustainable-business/designers-design-one-percent-sustainable-future)
3. Whiteley, N.: Consumer-led design. In: Design for Society. Reakton Books Ltd, Clerkenwell (1997) ISBN 0948462655
4. Gaskell, A.: New Paper Explores the Success of MOOCS. (www.adigaskell.org/2014/02/04/new-paper-explores-the-success-of-moocs)
5. Noer, M.: One Man, One Computer, 10 Million Students: How Khan Academy Is Reinventing Education. Forbes (www.forbes.com/sites/michaelnoer/2012/11/02/one-man-one-computer-10-million-students-how-khan-academy-is-reinventing-education)
6. IDEO Design Thinking For Educators. (www.designthinkingforeducators.com)
7. Mitra, S., Dangwal, R., Chatterjee, S., Jha, S., Bisht, R. S., Kapur, P.: Acquisition of Computing Literacy on Shared Pubic Computers: Children and the 'Hole in the Wall'. Aust. J. Edu. Lit. **21**(3), 407–426 (2005)

8. Mitra, S.: School in the Cloud—Sugata Mitra. (www.ted.com/participate/ted-prize/prize-winning-wishes/school-in-the-cloud-sugata-mitra)
9. ITU: Mobile-broadband penetration approaching 32%, three billion Internet users by end of this year, 2014 ICT Figures
10. Shah, N.; A Blurry Vision: Reconsidering the Failure of the One Laptop Per Child Initiative. WR 3 (2011). (www.bu.edu/writingprogram/journal/past-issues/issue-3/shah/)
11. Gabora, L.: Cognitive mechanisms underlying the creative process. In: Fourth International Conference on Creativity and Cognition, Loughborough University, UK, pp 126–133 (2011)
12. Lockheed, M. E., Verspoor, A. M.: Improving Primary Education in Developing Countries (1991)
13. Kingdon, G.: The progress of school education in India. Oxford University Press, Oxford (2007)
14. PTI: 47% of Graduates of 2013 Unemployable for Any Job: Study. The New Indian Express (www.newindianexpress.com/business/news/47-of-Graduates-of-2013-Unemployable-for-Any-Job-Study/2013/12/26/article1966550.ece)
15. Twain, M.: I Have Never Let Schooling Interfere with My Education
16. Ramachandran, V., Pal, M., Jain, S., Sekar, S., Sharma, J.: Teacher motivation in India. In: Knowledge and Skills for Development (2005)
17. DC: India faces shortage of 1.4 million trained teachers. Deccan Chronicle (archives.deccan chronicle.com/130707/news-current-affairs/article/india-faces-shortage-nearly-14-million-trained-teachers)
18. Ergas, O.: Reclaiming the missing links of Western education. In: Religion and the Body, Donner Institute, vol. 23 (2011)
19. Conklin, J.: Wicked Problems and Social Complexity. CogNexus Institute
20. Jonasen, D.H., Rohrer-Murphy, L.: Activity theory as a framework for constructivist learning environments. Edu. Tech. Res. Dev. **47**(1), 61–79 (1999)
21. Wright, S.: Young children's meaning-making through drawing and 'telling': Analogies to filmic texutal features. National Institute of Education, Singapore (2007)
22. Cohn, N.: Framing 'I can't draw': The Influence of Cultural Frames on the Development of Drawing. Centre for Research in Language, University of California, San Diego (2014)
23. Keengwe, J., Onchwari, G., Oigara, J. N.: Promoting Active Learning Through the Flipped Classroom Model. IGI Global, Hershey (2014)
24. Paul, R.W., Martin, D., Anderson, K.: On Socratic Questioning. In: Critical Thinking Handbook ISBN 0-944583-03-2
25. Isbel, R., Sobol, J., Lindauer, L., Lawrence, A.: The effects of storytelling and story reading on the oral language complexity and story comprehension of young children. Early Childhood Educ. J. **32**(3), 157–163 (2004)
26. Fillmore L. W., Snow, K. E.: What teachers need to know about language. In: Centre for Applied Linguistics, Washington DC (2000) (Report N.FL026371)
27. Genishi, C.: Young children oral language development (1988)
28. Rubin, P. C., Wilson, L.: Enhancing Language Skills in Four and Five Years Old (1995)
29. Parsons, S., Nuland, L.R., Parsons, A.W.: The ABCs of student engagement. George Mason University, Fairfax (2014)

Investigation of Key Principles of Game-Play and Their Abstraction to Enhance Learning in Rural Children

Abdul Sameer Ashraf, Aditi Padhi and Ravi Mokashi Punekar

Abstract The paper outlines the outcome of an undergraduate design project that focused on user-centric design approach in designing an activity-based learning solution for upper primary students in rural Assam. The advantages of introducing games to classroom learning were studied through literature review. Ethnographic research methods were followed. Interviews were conducted amongst students, parents and teachers in rural schools in Guwahati, and insights drawn to design a set of activities that were implemented amongst identified schools. Feedback received from the initial pilot study was further incorporated into the solution and the improvised version was tested for its usability. Results obtained from the usability test were analyzed for their significance and effectiveness. In conclusion, the paper highlights the experience in design methods for the design of educational aids for learning that incorporate the principles of game-play and act as a supplement to the current system of school education.

Keywords User-centered design · Design for education · Games · Learning

1 Background

The design project undertaken as part of undergraduate design program by the authors was initiated with the intent to incorporate activity-based learning in municipal schools based on the principles of gaming. Visits to local government

---

All interviews were conducted in Hindi or Assamese, with the help of a local translator, in familiar surroundings of the participants.
In figures where applicable, P1, P2 etc. represent the participants.

A.S. Ashraf · A. Padhi (✉) · R.M. Punekar
Department of Design, Indian Institute of Technology, Guwahati, India
e-mail: padhiaditi@gmail.com

A. Chakrabarti (ed.), *ICoRD'15 – Research into Design Across Boundaries Volume 1*,
Smart Innovation, Systems and Technologies 34, DOI 10.1007/978-81-322-2232-3_57

schools in Guwahati confirmed the poor state of primary education in rural areas as highlighted in ASER Reports by Pratham [1]. Although the syllabus and school curriculum seem well crafted with textbooks being illustrative and activity oriented, there were problems with the method of teaching. This led us to examine the principles of gaming as a possible direction for enhancing the learning experience. Gaming principles are widely used in web design, education technology etc. Research in the field of gaming indicates that the effective use of game mechanics helps to effectively enhance the experience of learning, resulting in an overall increase in the motivation levels amongst students [2]. Based on principles of gaming, collaborative activity-based learning methods were developed for classroom learning. The paper outlines the approach undertaken in the development of a set of card games developed as an aid to English language learning for school children in municipal schools of rural Guwahati. The conclusions and inferences are presented based on the Kirkpatrick method of usability tests conducted at two rural schools. This paper outlines the methodology followed in detail.

2 Literature Review

Literature research focused on the study of games of different genres and media across various platforms, and their features. It was noted that the use of games in education led to enhanced student motivation, and consequently improved learning [3]. More specifically, children needed to find learning meaningful, and enjoy learning only when they have something to look forward to (goal) [2]. Progressive game-based learning systems and environments were preferred, and attracted and motivated them towards learning. There was also a direct co-relation between desirable learning outcomes and the gaming principles. The following set of principles were identified—Fantasy, Rules/Goals, Sensory Stimuli/Dynamic Visuals, Challenge, Feedback, Control, Interactivity, Learner's Role, Multiple Paths to the Goal, Complexity and Strategy [4]. In addition to the literature research done on gaming and education, various theories and concepts in psychology were studied [5].

Following this, some of the existing games in the market were studied and their learnability features compared based on age group and how they assisted in the overall development of a child's personality at different phases of the child's growth. A multitude of videos and articles in favor of a game-based system of learning were reviewed as well [6–8]. Concepts such as peer learning and mutual sharing of knowledge and ideas, avoiding humiliation resulting from failures while learning in a game, and having a reward to look forward to at the end of every game, were inferred to be the key motivators towards learning from a game-based model. Additionally, gaps pertaining to the lack of motivation on the part of students and teachers were analyzed. It was realized that teachers needed incentives too. Monetary, or even simply respect and an elevated status for teaching could result in the teachers working harder in their profession. Further, the current systems of education were reviewed amongst the municipal schools in Guwahati.

There was a lack of connection between the school curriculum and the real world. The student-to-teacher ratio was also generally high resulting in lesser personal attention to the student and a decrease in motivation in the students.

The Annual Status of Education Report released by Pratham [1] highlights the glaring inadequacy of reading and comprehension abilities of upper primary students across various government and private schools in India. The report indicates that only 67.4 % of students were able to read a Class II level text. Also, just 16.8 % students were able to read only up to Class I level text but not Class II level text. Of the students who were able to read up to Class II level text, only 31 % belonged to Government schools. The rest were from private schools.

These facts were kept in mind during field trips undertaken at the Senior Basic School at Amingaon, Guwahati that was selected for conducting the pilot study for this project.

3 Field Research

The Senior Basic School in Amingaon, offers classes up to standard Eight. It is located at a distance of 6 km from IIT Guwahati. Most of the students belong to the lower and lower-middle income category. The aim of the field research was to collect data pertaining to the poor reading and comprehension abilities of the users, and reasons for the same. Lack of motivation, if any, was also analyzed. The primary subjects for the study were students in the age group of 10–13. Ethnographic interviews were conducted among teachers and parents to gain relevant insights. The prime objectives of the field research were:

- To understand the attitude of the students towards education and the roles played by teachers, parents and peers in shaping their attitude.
- To identify weak points of students in the different subjects taught, thereby helping to narrow down the domain for a design intervention.
- To collect data regarding the demographical characteristics of the users and their socio-economic and cultural backgrounds.

The data collected from interviews was analyzed and the following insights and inferences were drawn from these interactions.

3.1 Results and Observations

Students were in general, talented, carefree, energetic and inquisitive. Since most parents had a poor educational background and belonged to the lower economic classes, the school was the only place of learning. However, despite the lack of resources, the motivation towards learning was high and attendance in classes was regular. The concept of no-failure, instituted by the Education Act [9], rendered exams as simply a formality before progressing to the next class. Therefore, there was no check on the completion of the prescribed syllabus. Teachers were not aware about changes in curriculum and lacked motivation and proper training. The sole intention of 'educating the young' was lost in the education system. Teachers were generally well respected and shared an emotional connection with the students.

Most parents being illiterate were largely unaware of the importance of education. Poor financial conditions often forced children to drop out of school and take up casual labor instead of school. In a few instances, parents sent their children to stay with their relatives for the sole purpose of going to school, while they themselves stayed in far-off places to make a living. It was realized that parents had to be motivated to come forward and actively get involved in the education of their child, as it would have a huge impact on the growth of the child.

Textbooks and study materials were quite illustrative and activity-orientated. However, the content was taught in a relatively monotonous manner leading to a lack of interest among the students. Teaching methods were often 'one-way interaction', where the students only 'listened' and did not 'participate'. The lack of practical education further made it difficult for the students to connect the subject matter with real life, thereby deepening the lack of interest. It was realized that focus has to be shifted to better methods of teaching that involve greater participation.

4 Conceptualization

The field trips confirmed the findings of the ASER Report 2012 [1] for the subject of English reading and comprehension. Students faced difficulties in learning the language and any design intervention should aim at improving this shortcoming.

4.1 Activity Design

Three learning activities were designed that included experiences drawn from the everyday lives of the students—Kirana Guide, Word Pool and Word Hunt.

4.1.1 Kirana Guide

It was observed that apart from school, the students used English words related to grocery items quite frequently (without actually knowing their meanings in English) in the immediate environments such as the local markets and *kirana* stores (grocery stores). Hence, the design activity was planned using familiar objects from their immediate surroundings. The concept of 'metarepresentations'

[5], in combination with the principles of gaming (interactivity, goal, rules, challenge, feedback and visual appeal) was used in designing the activity.

In this activity, students are divided into groups of 3 and each group is given a grocery product. A time frame of 5 min is allowed to analyze the text/images on the product and words associated with it. 20 flashcards, 10 related to the given product and 10 unrelated, are given to each group and they are asked to sort the related cards in the minimum possible time. Each flashcard has the equivalent names written in English and the local language Assamese, on one side and has the corresponding image on the other side.

4.1.2 Word Pool

The objective of Word Pool was to teach sentence construction. It was observed that, in spite of lacking communication skills in English, students were creative and were found to translate sentences from Assamese to English in a word-by-word format. This tendency was used to our advantage for introducing common nouns, adjectives and verbs of Assamese and English into their sentences. The Peekaboo Effect [5], along with the principles of gaming (interactivity, strategy, visual appeal and multiple paths to the goal) was used as a basis for this activity.

Students—individually or in groups of 3, are asked to form small sentences combining words after picking one card each from packs of nouns, verbs and adjectives. The activity is carried out in 3 versions:

- Using Assamese words to form Assamese sentences
- Using English words and translating them to Assamese to form Assamese sentences
- Using English words to form English sentences

4.1.3 Word Hunt

Students were capable of reading the English alphabet. Given a word, they could identify the alphabets present, but could not always form the entire word. Such a hurdle can only be tackled through practice. The Word Hunt was therefore designed to aid the students in memorizing words while at the same time, making it fun to play with. The concept of Law of Grouping and the Peekaboo Effect [5] were combined with the principles of gaming (goal, rules, challenge, feedback, visual appeal) to design this activity.

Students were given a chapter from their Science or Social Science textbook as homework. The following day, they were given a sheet containing a word maze and a set of 8 images corresponding to the same chapter. They were asked to identify the maximum number of words in the least possible time. This was followed by showing them just the images and they were asked to recollect the English words from memory (Fig. 1).

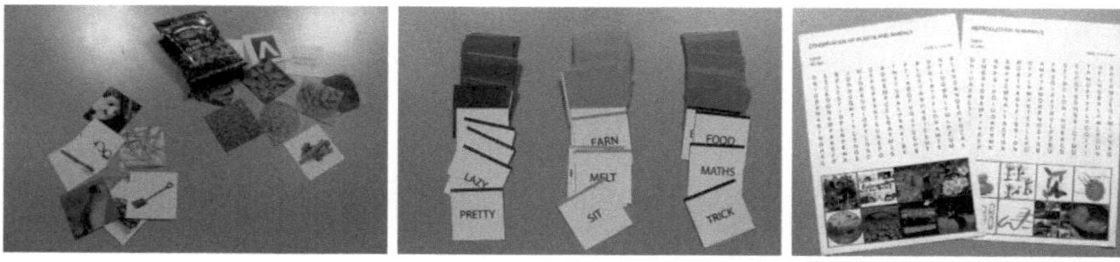

Fig. 1 (From *left* to *right*) kirana guide, word pool, word hunt

5 Pilot Study

Samples of the flash cards were printed for the different design activities and a pilot study was carried out over a two-week period to gain preliminary insights regarding the proposed learning aids. The design activities were performed every day for 2 h with the help of teachers and were carried out in the following sequence—1 round of Kirana Guide, 4–5 rounds of Word Pool, 1 round of Word Hunt.

5.1 Results and Observations

For the *Kirana* Guide exercise, the students were excited to be shown familiar products. Asking them to read the products triggered inquisitiveness. The act of looking closely at a familiar product from a new point of enquiry seemed comfortable and acceptable. The children first attempted to identify the items pictorially and then read the text on the other side of the card, if necessary. They succeeded in identifying most of the related items. However, they were not familiar with symbols or logos of the manufacturer/brand, the vegetarian symbol etc.

In case of the Word Pool, after picking 3 cards, the students randomly formed small paragraphs of 2–3 sentences that used all the 3 words but as they became familiar with the different parts of speech, they started connecting the words together and forming single sentences. They succeeded in thinking out of the box and came up with extremely creative and thoughtful statements. Initially, the children preferred to form sentences in Assamese. Further, Hindi was preferred to English. However, after a week, they started showing interest in learning English and forming sentences as well. They were eager to add more words to their vocabulary and were quick at rectifying their mistakes.

For the Word Hunt, the kids were familiar with the terminology used in a chapter in Assamese only, even though NCERT provides a list of selected words translated in English as well. In fact, they learnt English words in Assamese as well and were unaware of the language of which it is originally a part. For e.g. they learnt AMOEBA in Assamese but did not know that it was an English word and was spelled as A-M-O-E-B-A. The activity therefore, had to be modified to suit the needs of the children. Words were written in English above the corresponding images and the students were then asked to find them out from the maze.

6 Usability Testing

The objective of conducting a usability test was to analyze how effective the proposed solution was, in fulfilling the project goals. The aim of the usability test therefore, was to assess the effectiveness of the design solution in increasing the motivation and knowledge levels of students.

Changes in knowledge levels can be measured objectively from mean differences between pre-test and post-test measures, associated with anticipated learning outcomes [10, 11]. Changes in motivation, on the other hand, have to be derived subjectively through self-assessment. Relevant models were considered and the Kirkpatrick's model was finalized [12]. The reaction of the student is a self-assessment tool that records the feelings and attitude of the student towards the activities, and can be used to measure changes in motivation. The second level, or the learning of the student, measures the increase in the knowledge of the student as a result of the activities. The third and fourth levels were not considered because of their lack of relevance to the context at hand.

6.1 Sample Specifics

Tests were conducted at two separate schools—the Girls' High School and the Boys' High School—in Amingaon, Guwahati. Demographic profile of students in both schools was found to be similar. Students from the 8th standard were randomly selected to form two groups—the gaming group and the control group. Each group consisted of 12–15 students in the age group of 10–13 years with approximately equal male-female ratio.

6.1.1 Measures

Based on the Kirkpatrick's model 2 attributes—motivation and knowledge—were measured. After conducting the design activities, the gaming group was given a reaction questionnaire to measure changes in motivation based on a 5 point Likert scale. The questionnaire also included a few subjective questions.

Knowledge being an objective measure was computed through pre-assessment tests and post-assessment tests in English. The tests were administered to both the groups to assess the language skills of the students viz. vocabulary building, sentence construction, creating general awareness about one's surroundings and better memorizing capabilities. Scores obtained for the two groups in the two tests were computed and compared for their significance.

Assamese being the native language for these students, English was taught as the second language in school. Changes in anxiety levels in learning a second language were therefore an important measure in this preliminary test. The Foreign Language Classroom Anxiety Scale (FLCAS) is a 5-point Likert survey consisting of 33 questions, aimed at measuring communication apprehensions w.r.t the foreign language learning in the classroom. For the purpose of this study, only 15 questions from the FLCAS [13] were chosen as relevant to this context and administered to both the groups in the pre-test and post-test sessions.

6.2 Procedure

The usability test was carried out for four weeks, beginning with a pre-test session and ending with a post-test session. Activities were conducted between the sessions for the gaming group whereas the control group wasn't subjected to any such intervention. Each pre-test session consisted of an assessment test and the FLCAS, whereas the post-test session included these in addition to a reaction questionnaire for the gaming group only. Instructions and/or questions in all the assessment tests and questionnaires were written in Assamese with the help of a local translator. The assessment tests were given out to both the groups for a time limit of 30 min each, whereas the FLCAS and the responses of the student participants were recorded in the questionnaire by interviewing each participant individually.

6.3 Results

The results obtained from the different tools have been listed below and are mostly quantitative in nature.

6.3.1 FLCAS

Figure 2 displays the feedback from the FLCAS. Each cell is divided into 4 sub-cells. The upper row for each cell represents the feedback for the gaming group while the bottom row gives the results for the control group. Additionally, the left column shows the pre-test results and the right column depicts the post-test results for each group. It can be seen that the variance is decently low, indicating that all the participating students were at the same levels of anxiety w.r.t English.

The mean values for the gaming group have shown a slight change between the two sessions indicating lesser anxiety, though the range remains pretty much the same. A similar comparison for the control group shows that the mean values have shown negligible change, indicating no change in the anxiety levels of the students.

QUESTION	MEAN		VARIANCE		S.D.		RANGE	
I never feel quite sure of myself when I am speaking in my English class	3.75	4.10	0.56	0.51	0.75	0.71	3.0 - 4.5	3.4 - 4.9
	3.60	3.50	0.24	0.26	0.50	0.51	3.1 - 4.1	3.0 - 4.0
I don't worry about making mistakes in English class	4.16	4.20	0.33	0.38	0.57	0.62	3.6 - 4.7	3.6 - 4.9
	4.50	4.60	0.42	0.26	0.65	0.51	3.8 – 5.0	4.0 - 5.0
I tremble when I know that I'm going to be called on in English class	3.33	3.80	0.60	0.51	0.77	0.71	2.5 - 4.1	3.1 - 4.5
	2.60	2.20	0.40	0.33	0.63	0.57	2.0 - 3.3	1.6 - 2.8
It frightens me when I don't understand what the teacher is saying in English	4.10	4.10	0.44	0.33	0.66	0.57	3.4 - 4.7	3.6 - 4.7
	4.10	4.30	0.44	0.33	0.66	0.61	3.5 - 4.8	3.7 - 4.9
During English class, I find myself thinking about things that have nothing to do with the course	1.90	1.60	0.44	0.44	0.66	0.66	1.2 - 2.6	0.9 - 2.2
	1.60	1.60	0.24	0.26	0.50	0.51	1.1 - 2.1	1.0 - 2.1
I keep thinking that the other students are better at English than I am	4.00	4.50	0.54	0.45	0.73	0.67	3.2 - 4.7	3.8 - 5.0
	4.00	4.30	0.60	0.37	0.78	0.61	3.2 - 4.8	3.7 - 4.9
I am usually at ease during tests in my English class	2.10	1.70	0.44	0.24	0.66	0.49	1.4 - 2.7	1.1 - 2.1
	2.30	2.10	0.37	0.43	0.61	0.66	1.7 - 2.9	1.5 - 2.8
In English class, I can get so nervous I forget things I know	2.25	2.10	0.56	0.62	0.75	0.79	1.5 – 3.0	1.3 - 2.9
	1.90	1.60	0.44	0.40	0.66	0.63	1.2 - 2.5	1.0 - 2.3
I would not be nervous speaking English with native speakers	2.25	2.20	0.56	0.69	0.75	0.83	1.5 – 3.0	1.3 - 3.0
	1.80	1.80	0.33	0.74	0.57	0.86	1.2 - 2.3	1.0 - 2.7
Even if I am well prepared for English class, I feel anxious about it	2.25	1.90	0.93	0.81	0.96	0.90	1.3 - 3.2	1.0 - 2.8
	2.10	1.70	0.84	0.37	0.91	0.61	1.1 – 3.0	1.1 - 2.3
I am afraid that my English teacher is ready to correct every mistake I make	2.16	1.70	0.87	0.42	0.93	0.65	1.2 - 3.1	1.0 - 2.3
	2.00	1.70	0.46	0.37	0.67	0.61	1.3 – 2.7	1.1 - 2.3
The more I study for an English test, the more confused I get	3.66	3.90	0.42	0.62	0.65	0.79	3.0 - 4.3	3.1 - 4.7
	3.50	3.70	0.57	0.68	0.75	0.82	2.7 - 4.2	2.9 - 4.5
I don't feel pressure to prepare very well for English class	2.10	1.70	0.44	0.60	0.66	0.77	1.4 - 2.7	0.9 - 2.4
	2.20	2.00	0.49	0.61	0.69	0.78	1.5 - 2.9	1.2 - 2.8
English class moves so quickly I worry about getting left behind	2.00	1.70	0.36	0.56	0.60	0.75	1.4 - 2.6	1.0 - 2.5
	2.30	2.10	0.24	0.53	0.49	0.73	1.8 - 2.8	1.3 - 2.8
I feel overwhelmed by the number of rules you have to learn to speak English	4.00	4.10	0.36	0.62	0.60	0.79	3.4 - 4.6	3.3 - 4.9
	4.30	4.30	0.40	0.40	0.63	0.63	3.7 – 5.0	3.7 - 5.0

Fig. 2 Feedback from FLCAS

Comparing the anxiety feedback from the post-test sessions between the 2 groups shows that the mean values are significantly different for certain anxiety questions, clearly indicating the reduced levels of anxiety among the gaming group.

6.3.2 Assessment Tests

Scores obtained by both the groups in the pre-assessment tests and the post-assessment tests have been displayed in Fig. 3. The blue line represents the pre-assessment test scores while the red line depicts scores from the post-assessment tests.

For both the groups in the pre-assessment tests, it can be noted that almost all the students fared well in the 'Match Word to Picture' question, and mediocre results were observed in the 'Picture Study' part. However, the entire batch performed very poorly in the 'Jumbled Sentences' question indicating that the vocabulary of the kids was decently good, but they lacked in sentence formation skills.

Between the 2 sessions for each group, negligible improvement was noticed in the 'Jumbled Sentences' question though the scores improved in the 'Match Word to Picture' and the 'Picture Study' questions. It can be further observed from Fig. 4 that for both pre-assessment and post-assessment tests for both the groups, the mean scores lie around the same value. However, there is a slight increase in the mean scores in the post-assessment test, but not a drastic change.

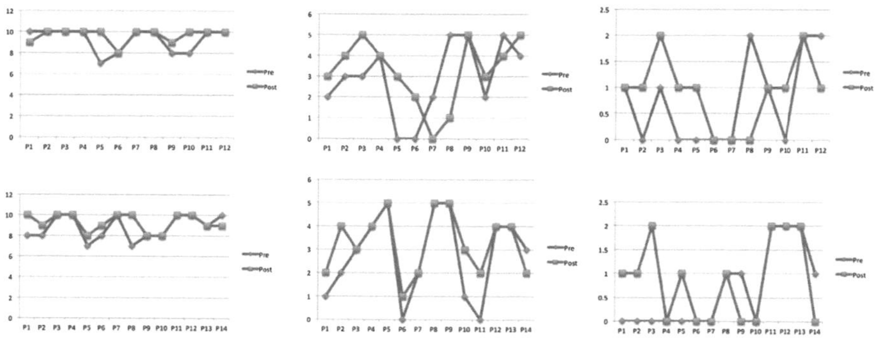

Fig. 3 Question-wise comparison of pre-assessment and post-assessment tests for gaming group (*top*) and control group (*bottom*); questions from *left* to *right*: match word to picture, picture study, jumbled sentences

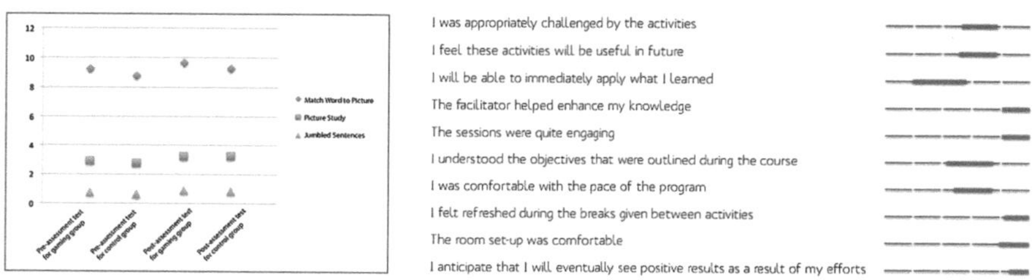

Fig. 4 (*Left*) comparison of scores from pre-assessment and post-assessment tests from both groups; (*Right*) feedback from reaction questionnaire

6.3.3 Reaction Questionnaire

The reaction feedback turned out to give positive results. From Fig. 4, it can be observed there is very little variance for almost all the questions, indicating that the impact of the activities was uniform across all the students. The activities were well conceived by the students, and they did succeed in intriguing and capturing the interest of the students.

7 Conclusion

The overall response to our proposed solution was positive, with both students and teachers getting actively involved in the design activities. However, the improvement in English levels was not very significant owing to the short time span for the test. Language learning is a time-taking process as evident from the results that showed that while there was a remarkable improvement in the vocabulary, the students continued to fail in sentence formation. Another major limitation was the language barrier. The students, not being very fluent in Hindi or English, the design activities were of a basic level, with the use of minimum words for communication. Further, despite the effort to involve teachers, there is still much to be achieved. Efforts are required to encourage teachers to incorporate these methods or to come up with similar newer methods for teaching.

As a project involving the design of novel methods in learning in the domain of school education, one could appreciate the intellectual challenges faced in understanding systemic complexities in designing, involving young minds. Diversity of parameters including cultural, multi-lingual and learnability can be very demanding but intellectually very stimulating. Use of games and understanding of principles of gaming in education was something new and novel for the team. Engagement with young students proved very unpredictable but was very stimulating. The project overall added a new dimension to design education and their processes from a new perspective, that is different from that of dealing with the material world.

Acknowledgments We would like to acknowledge the children, their parents, teachers and officials who were either subjects in our experiment, or respondents in our field research. We thank Apurva, Ashish, Tigmanshu, Satyan, Neha, Abhishek and other volunteers who helped us in the fieldwork. We also acknowledge the help of Ms. Umme Hani, Mr. Rhitupon Bora, Mr. Jyotirmay Nayak and Mr. Ratul Deka, for assisting with the translation into Assamese language. We thank our guide and mentor, Professor Ravi Mokashi Punekar, who has been a source of inspiration for us throughout this study.

References

1. Annual status of education report—Rural, Pratham, India (2013)
2. Sedighian, K., Sedighian, A.: Can educational games help educators learn about the psychology of learning Mathematics in Children. In: 18th Annual Meeting of the International Group for the Psychology of Mathematics Education—the North American Chapter, Florida, USA (1996)
3. Whitehall, B., McDonald, B.: Improving learning persistence of military personnel by enhancing motivation in a technical training program. Simul. Gaming **24**, 294–313 (1993)
4. Grarris, R., Ahlers, R., James, E.: Games, motivation and learning: a research and practice model. Simul. Gaming **33**, 441 (2002)
5. Ramachandran, V.: The Telltale Brain: A Neuroscientist's Quest for What Makes us Human. W.W. Norton & Company, New York (2010)
6. Robert, P.: Comparing education systems around the world [Online]. Available: http://mbctimes.com/en/comparing-education-systems-around-world. 01 June 2013
7. TED Talks, Re-imagining school [Online]. Available: http://www.ted.com/playlists/24/re_imagining_school.html
8. Huffington Post. Best education in the world: Finland, South Korea top country rankings, U.S. rated average [Online]. Available: http://www.huffingtonpost.com/2012/11/27/best-education-in-the-wor_n_2199795.html. 27 Nov 2012
9. Government of India, Right to education [Online]. Available: http://mhrd.gov.in/rte. April 2010
10. Castellar, E.N., Looy, J.V., Szmalec, A., Marez, L.D.: Improving arithmetic skills through gameplay: assessment of the effectiveness of an educational game in terms of cognitive and affective learning outcomes. Inf. Sci. **264**, 19–31 (2014)

11. Kraiger, K., Ford, J.K., Salas, E.: Application of cognitive, skill-based, and affective theories of learning outcomes to new methods of training evaluation. J. Appl. Psychol. **78**, 311–328 (1993)
12. Business balls [Online]. Available: http://www.businessballs.com/kirkpatricklearningevaluationmodel.htm
13. Horwitz, E.K., Horwitz, M.B., Cope, J.: Foreign language classroom anxiety. Mod. Lang. J. **70**(2), 125–132 (1986)

Effectiveness of Precedent Based Design Pedagogy—A Case for Sustainability Precedents

Rupa Agarwal and Ravi Poovaiah

Abstract Precedents are used by instructors as well as students to illustrate a point; or as justification for the design decisions made. In spite of its effectiveness, reliable methods to retrieve and access precedents, has not been laid down. This paper demonstrates how the precedents can be retrieved, categorized and mapped into a usable form. Once precedents appear in usable form, it becomes apparent how precedent based technique can lead to making effective design decisions. The argument therefore is that, the effectiveness of this technique can be leveraged for sustainability precedents as well. A discussion amongst a group of design students has been recorded and transcribed to extract precedents. These precedents have been analyzed and categorized to show the linkages with design decisions and problem interpretation. Precedents thus extracted are retrospective in nature. This study will restrict itself to extraction of retrospective precedents only. However, a model has been proposed for turning retrospective precedents into usable prospective sustainability precedents. This is part of an ongoing research, the purpose of which is to map a structure of prospective sustainability precedents, to a student project.

Keywords Precedent based pedagogy · Sustainability precedent · Design decisions · Design process

R. Agarwal (✉)
Department of Fashion Communication, National Institute of Fashion Technology, Mumbai, India
e-mail: rupa.n.agarwal@gmail.com

R. Poovaiah
Industrial Design Center, Indian Institute of Technology Bombay, Mumbai, India

A. Chakrabarti (ed.), *ICoRD'15 – Research into Design Across Boundaries Volume 1*, Smart Innovation, Systems and Technologies 34, DOI 10.1007/978-81-322-2232-3_58

1 Introduction

Precedent is the previous knowledge and experience, which a designer draws upon in order to take appropriate design decisions. Precedents are in the form of formal, structural, syntactic or systematic features [1] in a pedagogic situation. Precedents have been used historically and are in use in contemporary design pedagogy also. Previous research shows that precedent based pedagogy is effective in guiding design process towards desired goals. Therefore it is hypothesized that sustainability precedents will guide design process towards sustainability. The hypothesis is based on two main theoretical concepts as discussed in the following paragraphs.

The first is the term 'exemplar' which is possibly the earliest attempt to capture this concept. It has roots in science but extends its understanding to design; in a designerly way of knowing [2]. Kuhn [3] moved away from the rigid term 'paradigm' to the more flexible 'exemplar' to explain the 'symbolic generalization' and 'previous knowledge' within a discipline. Kuhn laid down specifications for a discipline. Even though design is a contingent discipline, it earns its place by fulfilling Kuhn's requirements [4] and exemplars are a vital part of it being so. The description of exemplar as 'previous knowledge' works well for design. Previous knowledge has been termed as precedents, and thereby the knowledge generated through this technique is Precedent based Knowledge (PBK).

Another major theoretical basis for Precedent based knowledge and its use in sustainability pedagogy is the Conjecture/Analysis (C/A) model. This model is derived from Sir Karl Popper's account of the scientific method. The C/A model challenges the Analysis/Synthesis (A/S) model followed traditionally in design process. The C/A theory emphasizes the role of testing conjectures, which means pre-structuring problems in a process of reflective design [5]. These pre-structuring elements are akin to the precedents suggested in this paper. Further, the C/A nature of design process is reinforced when environmental sustainability issues are integrated in it [5].

1.1 Background

Precedent based reasoning is used very effectively in study and practice of law, medicine and management. Several databases exist in these domains, from which to tap precedents. There is an interesting account of how precedent based pedagogy is used in engineering education to teach concepts of structures. The students were asked to describe the "relationship of new experiences to prior experiences and beliefs" [6]. One of the successful methods used by Hong in student projects is the "study of rationale behind structures that are found from nature and explicitly used in precedent architecture" [6]. Tzonis [7] made frames to understand transfer of precedent to a new project by studying Le Corbusier's conception of Unite d'Habitation. Historically, precedents do play a small role in art and design schools, where pattern books were developed for reference. But these references were mostly visual in nature where a visual vocabulary was being developed. This focused mainly on the end product. In contrast sustainability may require precedents, which are process based. Databases for sustainability precedents, which are organized and retrievable are yet to be developed. A further objective is to inculcate an attitude of respect for gathering precedents and to develop the skills to do so. This is an important area for further research and practice.

The larger purpose of this inquiry is to align the ontological nature of design with sustainability discourse. This has been operationalized through the use of Precedent based knowledge (PBK) in design pedagogy.

Precedents are used by instructors as well as students to illustrate a point; or as justification of design decisions made. PBK has been effectively used to improve design and instruction skills. It has also been known to increase innovation potential of design process. Design precedents are also valuable as it acquaints us with design ontology; the nature and intent of design thinking. Design is replete with exemplars at every stage. Lawson [8] describes precedents as "either whole or partial pieces of designs that the designer is aware of", but concedes that, the way in which designers use precedents while working on a new design project is still unclear.

In an attempt to answer this question, the authors conducted a protocol study with two groups of students of National Institute of Fashion Technology, Mumbai. The problem was posed to the students by the researchers as instructor. The students were asked to deliberate on the given problem. This paper will limit itself to identifying the precedents in a discussion of one group of students.

1.2 Design Pedagogy and Precedents

Design instruction is different from other instruction in as much as that it does not follow any immutable principles and theories. What it does, is borrow from a corpus of precedents, which are generalized from previous experiences [9].

There are terms, which are close to the meaning of precedent and precedent based approaches.

1. Case based studies, which are based on distinctive cases.
2. Analogy and Metaphor—the domain of drawing parallels between entities or events. This is used predominantly in idea generation.
3. Patterns—a discourse initiated by Alexander [10], who articulated patterns as a set of relationships influencing design. Deriving emergent design products from precedents as a design strategy has its roots in various fields of knowledge, which include theory of dynamic memory, design methods, architectural typography, prescriptive compositional principles and finally studies in decision making [1].

In the study presented, there are several precedents, which have been retrieved from the protocol. But only some have been identified as useful. The useful precedents are those, which match the design decisions taken by students. By doing this, the study aligns itself to the antecedent in the decision-making field of knowledge.

The nature of precedents in design pedagogy is of various kinds. (1) Notes taken in the field by student, notes taken in class during lectures and interactions, and that collected from secondary sources. (2) Visual references collected from secondary sources or documented by the student/instructor in the field or presented to the students by instructor. (3) Actual physical examples used by the instructor as teaching aid or collected by student in the field.

1.3 Barriers to Integrating Sustainability in Design Pedagogy

There are barriers in integrating sustainability in design education because of a threefold reason—(1) lack of sustainability evaluation parameters in design. (2) Limited focus of sustainability as a domain. (3) Lack of trained instructors in the area of sustainability.

Ramirez [11] found that 'although academics consider sustainability as a topic that design students need to be exposed to, the classic design assessment criteria—aesthetics, functionality, ergonomics, manufacturability, etc.—are still deemed to be the overriding determinants of what constitutes good design, and impacts to society or to the environment do not hold the same significance. Ramirez, in an earlier paper [12] lists various design education surveys and studies done in Australia for various design disciplines. They have generally shown that sustainability issues are hardly penetrating into core design programs. Industrial design Society of America in 2001 reported that among American industrial design educators, only 12 % reported ecodesign to be integrated in some fashion in their curricula [12]. Though such empirical studies have not been conducted for the Indian design schools the teaching experience of the author has experienced the barriers to integrating sustainability approaches within design academia. But there is a great deal of enthusiasm shown by instructors as well as students in integrating sustainability approaches into projects.

Clune's [13] insights in his PhD thesis, points towards a sound understanding of unsustainability, in order to design sustainably. Education's focus on technical design skills, without adequately defining the problem of unsustainability, negates effective improvements required to move towards sustainability.

At a curriculum and pedagogic level Boyle [14] identified six key problems of incorporating sustainability into Engineering education: lack of textbook, lack of examples, lack of time, knowledge of sustainability by staff and maturity of students. This holds true for design education too.

From the line of thinking in this and the preceding sections, the emergent question is; to what extent is it possible to guide student projects towards sustainability through precedent based pedagogy?

1.4 A Case for Sustainability Precedents

As a corollary to the fact that Precedent based knowledge is effective, one can use it to overcome barriers of integrating sustainability into design pedagogy. Therein lies the gap in knowledge, where precedent based approaches have not been applied to sustainability issues in design and design pedagogy.

This study is one part of the series of inductive studies conducted to answer the research question mentioned above—To what extent is it possible to guide student projects towards sustainability through precedent based pedagogy? This protocol study was initially conducted with the intention of finding precedents/sustainability precedents in the problem-solving paradigm. But it was found that the scope for sustainability precedents actually occurred at the time of decision-making. Action research therefore guided the formulation of the model by situating sustainability precedents in the decision-making process.

Below is the model (Fig. 2) proposed by the first author in the doctoral thesis for converting retrospective precedents into sustainability prospective precedents. The validation of the model, is work in progress. It is presented here with the purpose of strengthening the case of sustainability precedents. This paper however concentrates on the first stage of the model. It comprises of extracting precedents from the protocol and selecting useful and potential prospective precedents.

In order to make this a robust technique, the design pedagogic community needs to rigourously document sustainability cases. To build such a corpus from which to draw from, is in itself is a worthwhile exercise towards strengthening the discourse.

2 Method

The research design consists of a combination of theoretical, action, content and protocol researches.

The approach was to gather insights inductively. A protocol of a group of students has been conducted where they were given a multilayered problem. The discussion of the problem had to conclude in formulation of a design brief. A non-participant observation by the researcher revealed several insights, which are discussed in the sections below. The problem statement; was preset by the instructor. It incorporated broad sustainability issues, such as production-consumption dichotomy and social well-being.

The problem is as below.

> *"A parent bought a Barbie for her daughter. In a week's time, the child wanted Barbie's bedroom, then a bathroom. Another week later, the child wanted new clothes for Barbie. Barbie has also become the child's identity. A group of concerned parents get together, to demand a solution to reduce Barbie's needs and therefore the child's desires. You have been invited as a team member to find a solution for the above scenario."*

The group of three students were chosen from senior years of different design disciplines of Fashion design, Communication design and Master's program in design from National Institute of Fashion Technology (NIFT), Mumbai. They had no exposure to sustainability issues in a formal course. Their understanding was from informal secondary sources only. During the entire length of the protocol the researcher was a silent observer except when the students needed assistance. This happened once in the entire protocol, when they wanted the brief repeated. The students were also provided with paper and pencils, incase they needed to note down or sketch something.

The discussion of the group lasted around 35 min. This was video recorded and transcribed. The entire transcription consisted of one hundred and seventy (n = 170) segments out of which, there were forty six (n = 46) instances where precedents occurred.

3 Results

3.1 Breaking Down the Protocol into 'Useful' Precedents

The retrieved precedents along with its linkages to problem statement and design decisions form an interesting pattern. This pattern can be mapped to provide insights for further analysis. Out of the forty six (n = 46) precedents, only the useful precedents have been chosen for analysis. The most important indicator of the usefulness of a precedent is its link with design decisions and problem interpretation. By this argument, twenty eight (n = 28) instances were found to be useful. Out of these, seventeen (n = 17) instances are positively linked to design decisions and twelve (n = 12) to problem interpretation. This has been represented in the Fig. 1. Figure 1 also shows a design decision, which is not connected with precedents. It is still important due to the fact that it summarizes the decisions taken in the entire protocol—"those are our two solutions do you agree?". The 'solutions' are the design decisions for formulating the brief. In this case they appear as below.

(a) the image of Barbie should be changed
(b) the toy should be more interactive with the child

As seen in Fig. 2, the design decisions are of two levels. One is at a more abstract level such as:

> *"So our solutions is, what I think is first is the personality lift giving a more specific, more realistic personality".*

This is then broken into more descriptive expressions such as:

> *"Say if Barbie comes with a helmet instead of a tiara....in stead of a gown she comes with a jacket" or "I think its time we should go ahead from the princess thing and probably give a little more realistic personality ".*

Some of the descriptive decisions are repeated. The repetition makes the precedents important. The repeated utterances did not appear in the same order in the protocol as they have been shown in Fig. 2. The students meandered into other directions of discussion before reiterating an already mentioned decision. Thus the

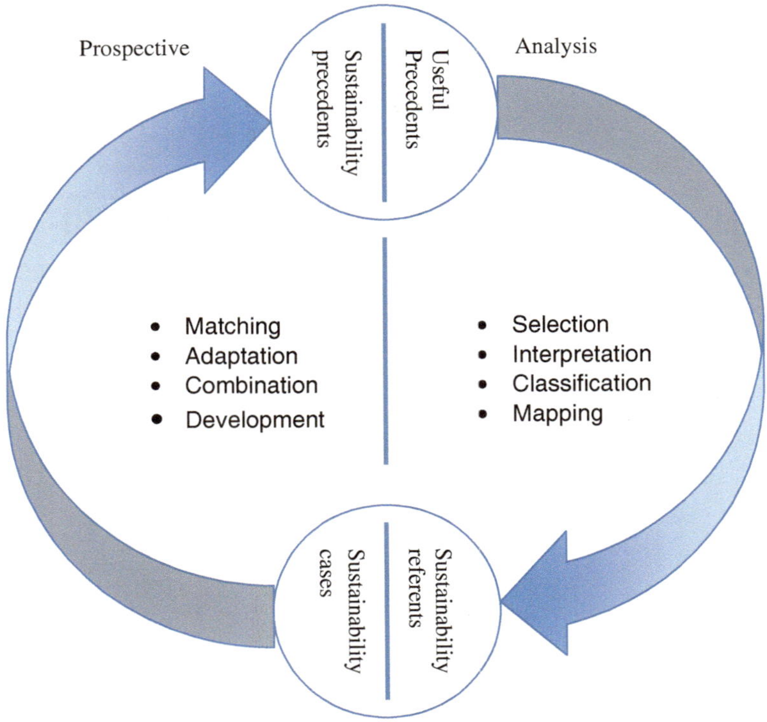

Fig. 1 Model for converting a precedent into a sustainability precedent

repeated decision and the associated precedent make it a stronger candidate for turning into a prospective precedent (explained in Sect. 3.2.2).

3.2 Categories of Precedents

3.2.1 Useful Precedents

Useful precedents are the operational precedents, which are the relevant candidates to be used in the design process. The listing of all the useful precedents is given in Table 1.

Precedents as 'previous knowledge'

(a) *Example of another product similar to the one being discussed*
Myscene doll, Hannah Montana, GI Joe and Robin, Hotwheels, Uno have been mentioned in the protocol. Instances such as the following were mentioned, "I fell for the whole thing, you know I want a tank, a chopper, I want… If you want superman, I buy a batman I don't want Robin maybe,

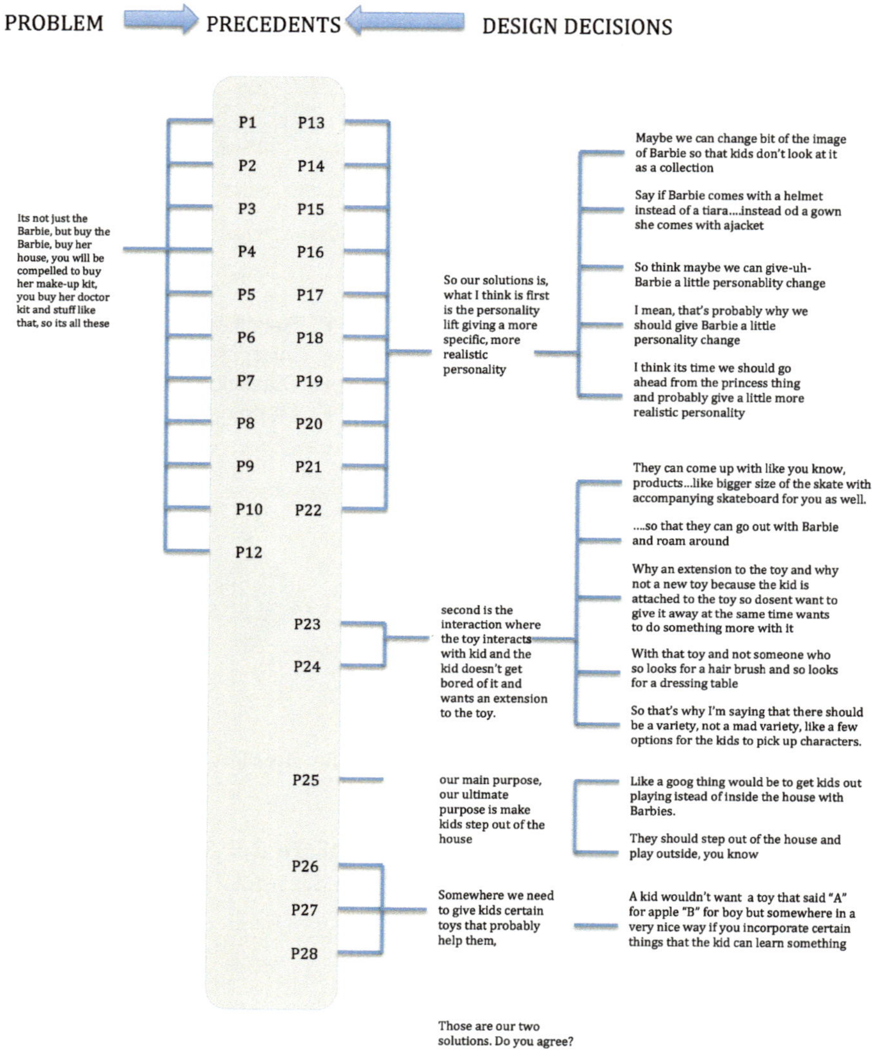

Fig. 2 Mapping of useful precedents from student protocol

I can do without the Robin. And, I've seen this recent phenomena of Hannah Montanna Like that was an idol, now its on every bag, every pencil, every T-shirt, every roller skates. Its just the same thing…"

(b) A similar situation in the student's own personal experience. *The students spontaneously related personal situations from their childhood, there was a time when I was obsessed with GI Joes, they didn't solve any purpose, you know, I would buy them, they were expensive. And this is from my personal this thing that my parents were never too happy seeing me playing with GI Joes but they were happier to see me playing with Uno.*

Table 1 List of precedents

Precedent	Retrospective
P1	You know probably the reason why Barbies sale has gone down is that is a new set of Barbie dolls kind, they call Myscene dolls
P2	I fell for the whole thing, you know I want a tank, a chopper, I want
P3	Ya, you want more and more
P4	Want everything around
P5	You build your own world and everything around
P6	If you want superman, I buy a batman I don't want Robin maybe, I can do without the Robin
P7	But if I buy a GI Joe torpedo, I want the …
P8	But you would not want to have three different Hannahs
P9	I've seen this recent phenomena of Hannah Montanna
P10	Like that was an idol, now its on every bag, every pencil, every T-shirt, every roller skates. Its just the same thing
P11	Thing, is that you want a Hannah bag and Hannah bottle and a Hannah … but that is something you cannot control, you know
P12	You build your own world and everything around
P13	That they have shifted the whole idea of girls is shifting from that very pretty looking doll to a very heap sort of a thing
P14	Those Barbie dolls, they come with bikes and not like brushes and make-up kits they come with bikes and dogs
P15	Like with this generation we see that most of the girls, they're more tomboyish
P16	Like they want to wear jackets and they want their hair to be cut.
P17	Like how a Batman is a Batman, he doesn't need his car around, he has a very strong identity
P18	So and if you see that kid is not just, she's not a princess
P19	She's a … A normal girl
P20	No, she's a rock star, it definitely shows that there is a way to build up such a personality for a cartoon that people are happy with it
P21	But you would not want to have three different Hannahs
P22	So there is an option to have two different Hannahs dolls, because Hannah is such a strong personality that they can have only one Hannah doll
P23	There are talking Barbies
P24	Had GI Joes been more fun guys and not serious with guns and all, had it been more you know
P25	Mostly you see, they're all sitting inside
P26	There was a time when I was obsessed with GI Joes, they didn't solve any purpose, you know, I would buy them, they were expensive
P27	Same with hot wheels
P28	This is from my personal this thing that my parents were never too happy seeing me playing with GI Joes but they were happier to see me playing with Uno

3.2.2 Retrospective and Potential Prospective Precedents

All the precedents from 1 to 28 (ref. Table 1 and Fig. 2) retrieved from the protocol are retrospective in nature, but precedents 13–28 have the potential to convert into prospective precedents. Therefore we call it potential prospective precedents. When they get converted into sustainability precedents will they be prospective precedents in the true sense. They can be used in the design process of the new project.

4 Discussion

The analysis until now explains in detail how useful precedents are retrieved, categorized and mapped. This is only the retrospective part of the analysis. The other stages of the proposed model, is work in progress. In order to align the retrieved precedents with sustainability approaches the potential prospective precedents need to be matched with sustainability referents. Sustainability referents are terminologies such as concepts, approaches and tools, which the students may be inclined towards. These matched pairs of useful precedent and sustainability referent need to find a suitable case. The derived precedents from thus chosen case are the sustainability precedents, which can be used in a project.

For example the precedent, "So there is an option to have two different Hannahs dolls, because Hannah is such a strong personality that they can have only one Hannahdoll" can be matched with emotional design and applied in the new project. Or the precedent," there was a time when I was obsessed with GI Joes, they didn't solve any purpose, you know, I would buy them, they were expensive" can be matched with "Reduce" to embed in the new project brief. This becomes the prospective part of the analysis, which is not within the purview of this paper. The proposed model however, is shown in Fig. 2.

This merged entity of the useful precedent and sustainability referent needs to find an appropriate sustainability case. Scanning such an appropriate sustainability case yields sustainability precedents, which can be used readily in new projects of students. This study is restricted to finding, selection, classification and mapping of the precedents. This falls under the analytical part of the design process. The innovativeness of the model is that this way sustainability can be integrated into a plethora of design problems.

Acknowledgments We would like to thank the students of National Institute of Fashion Technology for their support.

References

1. Eilouti, B.H.: Design knowledge recycling using precedent based analysis and synthesis models. Des. Stud. **30**, 340–368 (2009)
2. Cross, N.: Designerly Ways of Knowing. Springer, Berlin (2006)

3. Kuhn, T.: The Structure of Scientific Revolutions. The University of Chicago, Chicago (1996)
4. Wang, D.: Holding creativity together—a sociological theory of Design professional. Des. Issues **25**, 5–21 (2009)
5. Trebilock, M.: Integrated design process: from analysis/synthesis to conjecture/analysis. PLEA2009–26th Conference on Passive and Low Energy Architecture (2009)
6. Hong, P.: Quantitative and qualitative understanding in structural engineering. In: ASEE Southeast section Conference (2014)
7. Tzonis, A.: Huts, ships and bottleracks: design by analogy for architects and/or machines. In: Proceedings of workshop 'Research in Design thinking', pp. 139–164 (1991)
8. Lawson, B.: What Designer's Know. Elsevier, Amsterdam (2004)
9. Akin, Omer: Case-based instruction strategies in architecture. Des. Stud. **23**, 407–431 (2002)
10. Alexander, C.: Notes on the Synthesis of Form. Harvard University Press, Cambridge (1964)
11. Ramirez Jr. M.: Sustainability integration in industrial design education: a worldwide survey. In: International Conference of Design Education, pp. 9–12 (2007)
12. Ramirez Jr. M.: Ecological sustainability in Australian industrial design education. In: Future Ground International Conference of the Design Research Society, pp. 17–21 (2004)
13. Clune, S.: Developing sustainable literacy in industrial design education. PhD Thesis, University of Western Sydney (2009)
14. Boyle, C.: Considerations on educating engineers in sustainability. Int. J. Sustain. High. Educ. **5**, 147–155 (2004)

Design Training via Industry Sponsored Projects in Undergraduate Engineering Program

Swaminathan Balachandran

Abstract The industrial engineering (IE) program at University of Wisconsin-Platteville (UW-Platteville) provides training to students via a wide variety of activities that are incorporated into the undergraduate curriculum. These activities are outlined and discussed. Activities in courses include case studies, class participation problems, and industry sponsored projects. The process for defining industry sponsored projects is presented. The above details will be useful to the audience in implementing these ideas in other curricula. The paper also presents reflections from students and these provide a summary skills and knowledge gained by students while designing and/or improving integrated systems of people, equipment, and all other essential resources. This paper will stress that design training based upon hands-on training is essential for all engineers.

Keywords Design training · Industrial projects · Case study · Design projects

1 Introduction

In the U.S.A. the Engineering Accreditation Commission (EAC) of the ABET Inc. (EAC/ABET) specifies the criteria for accrediting engineering programs. Accreditation by EAC/ABET is a non-governmental, peer-review process that assures the quality of the postsecondary education that students receive. Educational institutions or programs volunteer to undergo this review periodically to determine if the above criteria are being met. These criteria emphasize continuous improvement of programs and provide general guidelines for various aspects of a program [1].

S. Balachandran (✉)
Department of Mechanical and Industrial Engineering, UW-Platteville, Platteville, WI, USA
e-mail: balachas@uwplatt.edu

A. Chakrabarti (ed.), *ICoRD'15 – Research into Design Across Boundaries Volume 1*,
Smart Innovation, Systems and Technologies 34, DOI 10.1007/978-81-322-2232-3_59

The EAC/ABET defines engineering design as the process of devising a system, component, or process to meet desired needs. It is an iterative decision-making process in which the basic sciences, mathematics, and the engineering sciences are applied to convert resources optimally to meet these stated needs. Students must be prepared for engineering practice through a curriculum culminating in a major design experience based on the knowledge and skills acquired in earlier course work and incorporating appropriate engineering standards and multiple realistic constraints such as economic, environmental, social, political, ethical, health and safety, manufacturability, and sustainability [1]. Most engineering programs use industrial projects to train students [2].

2 Industrial Engineering Program at UW-Platteville

The IE program at the UW-Platteville achieved its first accreditation by EAC/ABET in 1988, even though the first Bachelor of Science degree in industrial engineering had been awarded in 1970. The program serves currently about 144 students and comprises of four full time faculty members. The program requires students to complete about 133 semester credits for graduation. The program provides hands-on experience to students in three IE laboratories: IE Systems Design Laboratory, Human Factors Engineering and Work Measurement Laboratory, and Computer Integrated Manufacturing Laboratory. In addition, industrial projects are integrated into some IE courses to train students in IE design principles.

The author has incorporated industrial projects and case studies into the following courses he has taught since 1985 in order to provide IE design training to students: INDSTENG 3430—Human Factors Engineering, INDSTENG 3530—Operations Research I, INDSTENG 4030—Production and Operations Management, INDSTENG 3630—Work Measurement and Design, INDSTENG 4230—Facilities Design, INDSTENG 4330—Materials Handling and Warehousing, INDSTENG 4430—Quality Engineering, INDSTENG 4730—Engineering Management, and INDSTENG 4930—Industrial Systems Design. The primary focus of this paper is INDSTENG 4930 which is a three-credit capstone design course that must be taken by each student in the last year before graduation. Each student group of size three or four is assigned a large industrial project to be completed that semester. In other lower level courses, industrial projects and case studies of smaller size are undertaken during the last month of the semester. This paper focuses on the training of students in IE design concepts via industrial projects in the capstone design course and case studies in lower-level IE courses.

3 Industrial Design Projects and Case Studies in Lower-Level Courses

The author created design case studies using his work at Pella Windows Inc. and used these in the course INDSTENG 3430—Human Factors Engineering course. The following design problems were incorporated into the course on the basis of his consulting experience:

1. Design a human-machine interface for a specific target group
2. Design a human-machine interface for a the general public
3. Design and analysis of manual material handling tasks
4. Design of workstations

These problems taught students the use of anthropometric database when designing systems for the public and development of anthropometric data for a target group when designing systems for that group.

The course INDSTENG 3530—Operations Research I deals with mathematical models for analyzing and optimizing systems. This course uses case studies to teach design concepts in modeling and analyzing systems. Both Linear, Interactive, Discrete Optimizer (LINDO) and Linear, Integer, Nonlinear, General Optimizer (LINGO) software packages are used in this course to solve linear programming, transportation, assignment, network, and queuing models to design, analyze, and improve systems.

The author developed laboratory projects in the course INDSTENG 3630—Work Measurement and design to train students in designing and analyzing work processes and methods. In addition he also incorporated projects sponsored by industries into this course to provide realistic hands-on training to students in this area of study. The projects required students to visit a company that made die castings to study, analyze, and improve work processes. Students had numerous opportunities to apply qualitative and quantitative methods that were taught in this course. The qualitative methods consisted of charts, tables, graphs, and diagrams to document current processes. The analytic methods consisted of mathematical and statistical procedures to analyze current operations and identify alternative ways to improve the processes.

4 Industrial Design Projects and Case Studies in Upper-Level Courses

The courses INDSTENG 4230—Facilities Design and INDSTENG 4330—Material Handling and Warehousing deal with design topics that are of interest to many industries that strive to take advantage of free technical assistance. Industrial projects sponsored by industries in these courses provided hands-on design experience to students in the following areas. During the past four decades, more than fifty of the following projects have been completed by students under the author's guidance and supervision.

1. Evaluate and redesign raw material storage in food processing, die casting, furniture manufacturing, and other industries
2. Evaluate and design layout of manufacturing and service facilities
3. Design automatic data collection system to track inventory in finished goods storage

The author has relied on case studies in the course INDSTENG 4430—Quality Engineering to teach six-sigma process design concepts in manufacturing and service organizations. These case studies were used as laboratory projects and students use Minitab software for statistical process control and design of experiments. In addition, students review papers and present them in class to learn about application of design for six-sigma in various disciplines and organizations.

5 Senior Capstone Design Course

EAC/ABET criteria for accrediting engineering programs requires engineering programs to prepare students for engineering practice through a curriculum culminating in a major design experience based on the knowledge and skills acquired in earlier course work and incorporating appropriate engineering standards and multiple realistic constraints such as economic, environmental, social, political, ethical, health and safety, manufacturability, and sustainability [1]. In the IE program at UW-Platteville, this culminating design experience is provided in the course INDSTENG 4930—Industrial Systems Design. Graduating senior must take this course during the last year of their study. Table 1 provides an overview of the organization of this course.

During the past three decades, the author has maintained very close and continual contact with local industries and alumni. In addition, industries frequently call the author and request technical assistance via senior design projects. Therefore the author keeps updating the list of potential industrial project sponsors and respective projects every semester and the current list of projects will serve the IE program for the next few years. The author usually visits industries during the summer of each year to finalize the problem statements and set up the project timeline. In addition, the author prepares and submits proposal for each project in the summer to obtain funds from a campus program for reimbursing students' travel and other project expenses. When university funds are not available, the industrial sponsor usually agrees to reimburse the student team for all their travel, printing, copying, and other expenses related to the project. Students submit their expense report to the sponsor directly and get reimbursed for these expenses. Usually, the college of engineering will allow students to use university vehicles after they are approved for driving university vehicles. Some industries contribute additional funds for college scholarships.

Table 1 Capstone design course organization

Lec/lab	Activity	Work due
1-Lec.	Assign project, discuss schedule	Submit student and group information
1-Lab.	Project, ethics, lifelong learning (LL)	IA1 (1 %): resume
2-Lec.	Industry visit	
2-Lab.	Lifelong learning and career planning	TA1 (1 %): letterhead and planning
3-Lec.	Industry visit	Initial portfolio review (1 %)
3-Lab.	Team presentation-textbook	TA2
4-Lec.	Industry visit	Problem statement and contract (5 %)
4-Lab.	Graduate school information	
5-Lec.	Industry visit/project consultation	Draft #1 of project report (2 %)
5-Lab.	Personal budget planning and investments	IA2 (1 %): interview questions
6-Lec.	Industry visit/project consultation	IA3 (1 %): interview book report
6-Lab.	Team presentation—textbook, project work	IA4 (1 %): lifelong learning
7-Lec.	Intermediate progress report	IA5 (1 %): career and life planning
7-Lab.	Intermediate progress report	Class/sponsor (5 % each); report (10 %)
8-Lec.	Intermediate progress report	TA3 (1 %): team photo CD
8-Lab.	Current topic presentation	Mid-term evaluation (1 %)
9-Lec.	Speaker from industry	IA6 (1 %): speaker and assessment
9-Lab.	Social, environmental, and other issues	Draft #2 of project report (3 %)
10-Lec.	Industry visit/project consultation	IA7(1 %): graduate school
10-Lab.	Current topic presentation	IA8 (1 %): budget, investment, etc.
11-Lec.	Current topic presentation	TA4 (1 %): current topics
11-Lab.	Industry visit/project consultation	IA9 (1 %): ethics case study
12-Lec.	Current topic presentation	TA5 (1 %): current topics
12-Lab.	Project team meetings	
13-Lec.	Current topic presentation	IA10 (1 %): soc., env., and other issues
13-Lab.	Current topic presentation	Draft #3 of project report (3 %)
14-Lec.	Final project presentation	Final portfolio review (1 %)
14-Lab.	Final project presentation	Class (5 %); sponsor (5 %)
15-Lec.	Final project presentation	Project report final draft (10 %)
15-Lab.	Final project presentation	Sponsor evaluation (2 %)
16-Lec.	Final evaluation (2 %); report (10 %)	Letter of transmittal (8 %)
16-Lab.	Project binder (5 %); exit survey	Career center survey

In all senior design projects sponsored by industries during the past three decades, initial project formulation by the contact person in the industry was very brief. In one particular project the objective of the industrial project was stated by the industry as the reduction of changeover time in assembly area. However, the instructor views the course as an opportunity to provide an integrated design experience to graduating seniors. The above project was redefined as listed below to include specific analyses that had been taught in the curriculum. The instructor's embellishment of this project ensured that the scope of the project was large enough for a three-credit course and a group of four graduating seniors had to work together for about 400 h during the semester to complete it. It is important for the faculty member to balance the needs of the industry with the educational objectives of the program.

- Tour the facility, and work with assembly supervisor to understand problems associated with wet changeovers from one product line to another product line. Wet changeover occurs when water is introduced to the assembly line in cleaning operations during changeovers.
- Study changeover operations of the line and classify them into at most ten major categories.
- Break each changeover operation category into basic time study elements and conduct a time study or a work sampling analysis to determine the percentage of total changeover time devoted to each changeover operation category.

Table 2 Industrial projects from 2011 to 2014

#	Year	Company and project details
1	2011	Frito-Lay, Beloit, WI: track waste, estimate cost, and improve process
2	2011	Bodine Electric Co, Peosta, IA: reduce design variations and waste
3	2011	Sauer Danfoss, Freeport, IL: design Kanban system for assembly area
4	2012	SSI Technologies, Janesville, WI: analyze, redesign, and improve cells
5	2012	SSI Technologies, Janesville, WI: analyze, redesign, and improve cells
6	2012	URT, Janesville, WI: analyze, redesign, and improve inbound area
7	2012	URT, Janesville, WI: analyze, redesign, and improve process line
8	2012	URT, Janesville, WI: safety engineering assessment and improvement
9	2012	UW-Platteville Campus Planning: design layout and relocate human performance lab
10	2012	UW-Platteville Campus Planning: design layout and relocate Robotics Lab
11	2012	UW-Platteville Campus Planning: design layout for phase I expansion of Eng. Hall
12	2012	UW-Platteville Campus Planning: design layout for phase II expansion of Eng. Hall
13	2012	Sulzer Machine and Manufacturing, Mosine, WI: redesign layout of current facility
14	2012	Sulzer Machine and Manufacturing, Mosine, WI: redesign layout of new building
15	2012	John Deere Dubuque Works, Dubuque, IA: analyze, redesign, and improve warehouse
16	2013	NewPage Corp., Stevens Point, WI: analyze, redesign, and improve trimmed core area
17	2013	Nord Gear, Waunakee, WI: analyze, redesign, and improve storage/retrieval of parts
18	2013	Eagle Windows, Dubuque, IA: analyze, redesign, and improve sliding door line
19	2013	Frito-Lay, Beloit, WI: analyze, and improve line changeover operations
20	2013	Southwest Health Center, Platteville, WI: analyze, and improve surgical department
21	2014	Flexsteel Industries, Dubuque, IA: design system to automatically track inventory

Draw a Pareto Chart of percentage of total changeover time devoted to each changeover operation category. Draw a spaghetti chart to portray the movement of personnel and or equipment. Draw C-E diagram to enumerate causes for delay in changeover.
- Learn about single minute exchange of dies and identify internal and external elements of changeover. Discuss, analyze, develop, and implement methods to increase the external elements and decrease the internal elements.
- Identify vendors to select new equipment that may be necessary to reduce the changeover time. Provide cost details. Work with the contact person in industry to order and install these.
- Improve movement of personnel, tool, and parts. Allocate responsibilities to the changeover team personnel in an optimal manner. Develop detailed changeover preparations and schedule. Compute the savings in changeover time and the total cost.
- Identify and eliminate muda (wastes).
- Analyze, recommend, and implement Kaizen 5S and visual controls.
- Implement all recommendations by working closely with the contact person in industry.

The author has supervised more than two hundred senior capstone design projects during the past three decades. The Table 2 provides the list of industrial projects that were completed in the IE senior capstone design course from 2011 to spring 2014. These projects reveal that IE students dealt with a wide variety of projects from diverse industries. Class presentations provided opportunities for students to learn IE design principles and their applications from each other.

6 Team Work

Each team is provided an introduction to team and project management. The team is required to keep a record of its communications with industry, meetings, and task assignments to members along with due dates, and send weekly progress report along with the Gantt chart to the faculty member. In addition the course is structured to include five different team activities to give opportunities for the team to develop team dynamics and function as one unit. The Table 1 provides an overall view of the project organization and planning. This table highlights the following aspects of the course:

- at the beginning of each week a team submits a detailed progress report
- by the end of the second week, a team designs the team letterhead and completes team planning
- by the end of fourth week, a team submits the problem statement, contract, work breakdown structure and a Gantt chart of project plan
- a team submits at least 4 drafts of project report during the semester before the final report is submitted

- a team reports progress at the middle of the semester to the class and sponsor
- a team reads and gives a presentation on a current topic related to the project
- a team presents the final report to the class and sponsor

7 Results and Conclusions

The industrial sponsor survey forms completed by sponsoring industries during the past two decades show an overall score of 4.7 out of 5 for satisfaction of sponsors with the work completed by industrial engineering seniors. The graduate exit surveys and alumni surveys also reveal that graduates gain experience and benefit from the capstone design course. The demand from businesses and industries for projects to be completed in the course has been increasing. In addition, as new graduates transitioning into industries tend to sponsor more industrial design projects. Therefore the demand for such service from the program is not expected to decrease in the near future. In addition to student oral and poster presentations to the project sponsor and university community, students are required to submit their reflections on their work. A few of student reflections are presented below:

> It was a good feeling to know that Frito Lay trusted UW-Platteville students enough to let them develop a change plan that will hopefully allow them to gain control of their spare parts room inventory and save money by not ordering more parts than needed.—Ryan Steuck
>
> I also got the chance to work with a team on a real-world project. This is important because I had to combine my engineering skills along with good communication skills to complete this project on time while still doing professional quality work… The Frito Lay project that my team and I did this semester was a great experience that will helped me develop professional engineering skills and knowledge. I was able to learn about inventory control, layouts, and spare parts rooms in general.—Joe Lange
>
> It showed me that I need to be ready for unexpected events. I had no knowledge of power plant operations coming into this project, so that is something that I needed to research at the beginning of the project. The refection of group members was also something to learn from. That was something that was unexpected, but I couldn't let the group be affected too much by it.—Ryan Dow
>
> This will help me work with people in industry and work with team members. It will also help me understand different data points and how to interpret them. We also learned a lot about the machines in this project and being able to learn how to ask the right questions to really understand how a machine that I am unfamiliar with works. …. Finally, being able to do a hands-on project like this with a company outside of our academic setting was useful. We talk about having to have people skills and the ability to work with other members of the workforce in class, but with this project, we practiced this in the field.—Danny Tisdale
>
> This project experience allowed me to work directly with a professional client and analyze a problem within a company. I was able to apply concepts that I have learned in classes to develop a solution to the issue. This experience will influence my future career by helping me understand how to approach a problem and talk with individuals to solve it.—Acacia Myers
>
> I have improved many skills over the course of this project. I have improved my team communication skills as well as my professional communication skills. I also have improved my problem solving skills. This experience has also furthered my confidence in my work which will influence my future career.—Pauline McCarty

I have learned that not all companies have up to date software, and I would have to deal with this challenge and make the best of what is available. I also learned different levels of acceptance when it comes to communicating ideas to a group of people. Some ideas the group generated for solutions were accepted well, while others were turned down without even the chance for a trial run… During this project I enhanced my knowledge and also gained new skills. I never worked in a food packing industry before, so this was a great new experience to learn about a common industry that I may one day work in. I had previous experience creating work instructions, but this was only in the classroom setting. Having this experience in Boscobel allowed me to take the classroom experience and apply it to a real world application. I expanded on my communication skills by speaking with workers at the factory and working with my team on the project. I feel I also grew somewhat as a leader.—Colin Beay

The project showed me how communication between everyone in the facility, from floor workers to upper management, is important to meet demand. Throughout the project we got to deal with different levels of personnel throughout the facility. This was very important to gain this experience, because as an industrial engineer I will be dealing with multiple levels throughout a facility. Also, I got to see firsthand how teamwork can ease the difficulty in a large project. Teamwork is a very important aspect of industrial engineering… Working on this project really helped to relate to material that I have learned in the classrooms over the past few years. It was finally nice to get the chance to see what engineers actually get the chance to do in a facility.—Peter Jecklin

The project was very rewarding due to the fact that we got to actually develop something that we would if were working in the real world. It is hard to recreate this setting in the classroom. Not having participated in an internship or co-op yet, any outside experience dealing with Industrial Engineering that I can get is very beneficial. I look forward to sharing this experience with future potential employers during my quest to find a job after graduation, and I feel like it was a very important experience to have.—Ryan Dow

This was the first project that I have worked on without the close supervision of a teacher or experienced engineer. So this project really pushed me to be more professional in everything that I did for this project. Without the experienced guidance of a teacher or supervisor to help me through difficulties and answer my questions this project pushed me to work through difficulties and to figure out the answer to my questions by myself. This project helped me to become more of an independent worker and it taught me to be precise with the question that I do ask.—Travis Glaser

I am a foreign student, hence I came here to learn from zero the way the Industrial Engineers communicate with each other, all the concept they use, and the point of view they use toward to work in a company and make improvements. In conclusion, I learned a lot about IE concepts, teamwork, how a real company works, giving presentations, and mostly, how the American culture works.—Luis Peralta-Cervantes

In order to provide such design training to students, the author had to establish a strong network of alumni and industries within about 120 miles from the campus. The network building involves periodic communications with alumni and industries, attending on campus events for alumni, and continuous use of social media to maintain the network. The course planning had to be completed at least a month before the semester began so that industrial project sponsors could be visited and problem statements may be finalized. From 1985 to 2008, the author had to depend on industries to support student travel expenses. After 2008, the author has been able to secure funds by submitting project proposals to a program on campus. Therefore, the integration of industrial projects into a course takes a lot of effort first to get industry sponsors and then to secure funding for students' expenses.

As projects vary from semester to semester, the instructor has to revise the lecture slides each semester so that the current industrial projects could be used to illustrate lecture topics. In addition course assignments are also restructured so that students may add these coursework as appendices to their final project report.

The most time consuming aspect of using industry sponsored design projects in a course is the need for meetings with the team, monitoring contribution of each student to a project via weekly progress reports or meetings, travel to industry once a month, and resolving conflicts within a team. As the author had worked as a manufacturing engineering consultant in industry for more than 6 years and contributed more than 10,000 h of work in ergonomics, quality improvement, manufacturing cell design, simulation, etc. he did not have any problem in obtaining and incorporating industry sponsored projects in his courses. He has a backlog of industry sponsored projects that can be used in his courses during the next five years.

The design experiences provided to students via industrial projects are recognized by alumni and employers. This feedback is provided by them through alumni and employer surveys which are used in the assessment and evaluation process to continuously improve the curriculum. The design training of students via industrial projects in many courses strengthens the curriculum and distinguishes the program from other similar programs, in addition to allowing it to maintain its accreditation status from 1988 to the present.

References

1. Criteria for accrediting engineering programs, effective for evaluations during the 2014–2015 accreditation cycle, October 26, 2013, Engineering Accreditation Commission, ABET, Inc., 111 Market Place, Suite 1050, Baltimore, MD 21202
2. Howe, S.: Lessons learned in the inaugural capstone design course at Smith College. In: International Conference on Engineering Education, Gainsville, 16–21 Oct 2004

Author Index

A. Chakrabarti (ed.), *ICoRD'15 – Research into Design Across Boundaries Volume 1*, Smart Innovation, Systems and Technologies 34, DOI 10.1007/978-81-322-2232-3